Society of Economic Geologists Annual Conference (SEG 2025)

Brisbane, Australia
26-29 September 2025

ISBN: 979-8-3313-2917-4

Printed from e-media with permission by:

Curran Associates, Inc.
57 Morehouse Lane
Red Hook, NY 12571

Some format issues inherent in the e-media version may also appear in this print version.

Society of Economic Geologists (SEG)
7811 Shaffer Parkway
Littleton, CO 80127-3732
USA

Phone: +1(720) 981-7882
Fax: +1(720) 981-7874

www.segweb.org

Additional copies of this publication are available from:

Curran Associates, Inc.
57 Morehouse Lane
Red Hook, NY 12571 USA
Phone: 845-758-0400
Fax: 845-758-2633
Email: curran@proceedings.com
Web: www.proceedings.com

Technical Program - September 26, Friday | 5:00 PM - 9:00 PM

= Great Hall Q1 & Q2 = Mezzanine M3 = Mezzanine M4

5:00 to 9:00 PM	Opening Event
5:00 PM to 5:15 PM	Welcome to Country...N/A *Steven Coghill (Invited Speaker)*
5:15 PM to 5:45 PM	Mineral Security in a Changing World...N/A *Daniel Franks (Invited Speaker)*
6:00 PM to 6:30 PM	Value of Economic Geology to Society (Panel Discussion)...N/A *Rick Valenta (Chair) Brad Welsh, Caoilin Chestnutt, Peta MacRae, Marina Costelloe, Daniel Franks*
6:30 PM to 7:00 PM	SEG Presidential Address...N/A *Anne Thompson, SEG President 2025*
7:00 PM to 9:00 PM	Welcome Reception in Exhibit Hall Sponsored by BHP Xplor

Technical Program - September 27, Saturday | 8:00 AM - 10:30 AM

8:00 AM to 9:30 AM	Welcome and Plenary Sessions
08:00 AM to 08:10 AM	Opening Comments...N/A *Jennifer Craig, SEG Executive Director and Rick Valenta, SEG 2025 Co-chair*
08:10 AM to 08:15 AM	Geology and Mining: Introduction to the New SEG Publication...N/A *Dan Wood, Co-editor*
08:15 AM to 08:30 AM	Opening Address...N/A *Tim O'Connor, Group Exploration Officer, BHP*
08:30 AM to 09:00 AM	Mineral Systems Models for Clastic-Dominated Zn Deposits: From the Geodynamic to the Microscales...1 *Sarah Gleeson (Plenary Speaker)*
09:00 AM to 09:30 AM	Mineral Systems in a Plate Tectonic Context...2 *Dietmar Müller (Plenary Speaker)*
9:30 AM to 10:30 AM	**Oral Session 1: Lithosphere and Mineralization**
9:30 AM to 9:45 AM	Probabilistic Assessment of Australia's Lithospheric Architecture and Its Relationship to Mineral Systems...3 *Mark Hoggard*
9:45 AM to 10:00 AM	Targeting the Biggest Discoveries along the Biggest Faults: How Hard Can it Be?...4 *Nick Hayward*
10:00 AM to 10:15 AM	Linking Orogenic Gold and Orthomagmatic PGE-Ni-Cu Deposits in the Archean Yilgarn Craton...5 *Matthew Demmer*
10:15 AM to 10:30 AM	Magmatic Activities Directly Contributed to the Formation of Giant Jiaodong Gold Prov-ince: Evidence from the Xiejia Diorite...6 *Zhan-Ke Li*
9:30 AM to 10:30 AM	**Oral Session 2: Tectonics and Porphyries**
9:30 AM to 9:45 AM	Linking Porphyry Cu Formation to Tectonic Change in Postsubduction Settings: A Case Study from the Giant Yulong Belt, Eastern Tibet...7 *Ming-Liang Huang*
9:45 AM to 10:00 AM	Slab Tearing as a Driver of Porphyry Ore Deposit Formation in Indonesia and the Philippines...8 *Jack Ward*
10:00 AM to 10:15 AM	Magmatism and Post-Emplacement Uplift at the Red Chris Cu-Au Deposit, BC Canada: Implications for Porphyry Copper Formation and Preservation...9 *Brian McNulty*
10:15 AM to 10:30 AM	Magmatic Fertility and Timing of the Dalli Gold-Rich Porphyry Deposit, Central Iran: Geological Insights and Exploration Implications...10 *Hooshang Asadi Haroni*

Technical Program - September 27, Saturday | 9:30 AM - 2:00 PM

9:30 AM to 10:30 AM	**Oral Session 3: 3D Exploration and Prospectivity**
9:30 AM to 9:45 AM	Loop: An Open-Source Framework for Probabilistic Geological and Resource Modelling...11 *Lachlan Grose (Invited Speaker)*
9:45 AM to 10:00 AM	Using the Loop platform to Build Structurally Complex 3D Geological Models: Towards Better Resource Models?...12 *Laurent Ailleres*
10:00 AM to 10:15 AM	Unlocking Deep-Time Prospectivity: Integration of Multiple Paleogeographic Models with Geochemical, Geochronological, and Thermochronological Big Data...13 *Fabian Kohlmann*
10:15 AM to 10:30 AM	Unlocking the Potential of Core Photography Leveraging SOTA Open-Source Computer Vision Models...14 *Ankita Singh*
10:30 AM to 11:00 AM	Break, Poster and Speed Talks in Exhibit Hall | Coffee Sponsored by Geologic AI
11:00 AM to 12:30 PM	**BHP Xplor**
	Idaho's Untapped Mineral Potential (Cu Au)...N/A | *Curtis Johnson*
	Royalty Creation Through Prospect Generation and Acquisition...N/A | *Marc Tran*
	Project Generation and Exploration of Critical Mineral Resources, Argentina...N/A | *Jason Ward*
	Discovering Labrador's Critical Metal Deposits, Primarily Copper...N/A | *Tyrell Sutherland*
	Multi-Commodity Exploration Prospectivity of Saudi Arabia...N/A | *Oliver Jones*
	Cu Au Exploration in the Western Tethyan Belt...N/A | *Elena Clarici*
	Cu Exploration in Canada, Peru and Australia...N/A | *Glenn Poole*
12:30 PM to 02:00 PM	BHP Xplor Showcase and Sponsored Lunch in Exhibit Hall

Technical Program - September 27, Saturday | 2:00 PM - 8:00 PM

2:00 PM to 3:30 PM	**Oral Session 4: Orogenic Gold**
2:00 PM to 2:15 PM	Structural Controls on a High-Grade BIF-Hosted Gold System: Umwelt Deposit Case Study, Back River Gold District, Nunavut, Canada...15 *Cam Bartsch (Invited Speaker)*
2:15 PM to 2:30 PM	Linking Geoscience Research to Orebody Knowledge at the Epizonal Fosterville Gold Deposits...16 *Wessley Edgar*
2:30 PM to 2:45 PM	"Fuchsite" and Chlorite in Carbonated Komatiites at the Kerr-Addison Deposit, Ontario, Canada: Genesis of Listvenites in Orogenic Gold Settings...17 *Derek Leung*
2:45 PM to 3:00 PM	Mafic Magmatic Origin of Orogenic Gold Systems...18 *Daniel Wiemer*
3:00 PM to 3:15 PM	Red Hill: Unveiling the Geochemical and Hydrothermal Evolution of a Granitoid-Hosted Orogenic Gold System in the Yilgarn Craton, Western Australia...19 *Renan De Souza*
3:15 PM to 3:30 PM	Sunday Creek: High-Grade Gold-Antimony Discovery in Clonbinane, Victoria...20 *Kenneth Bush*
3:30 PM to 4:00 PM	Break and Posters in Exhibit Hall | Coffee Sponsored by Ma'aden
4:00 PM to 5:45 PM	**Oral Session 7: Orogenic and Other Au and Cu**
4:00 PM to 4:15 PM	Unearth Tomorrow: Exploration and Technology Delivering the New Vision for the Kingdom of Saudi Arabia...N/A *Giuseppe LoGrasso (Sponsor Talk)*
4:15 PM to 4:30 PM	Unearth Tomorrow: Exploration and Technology Delivering the New Vision for the Kingdom of Saudi Arabia (continued)
4:30 PM to 4:45 PM	Retrograde Overprint of the MEM2 Gold Deposit: Evidence for Polyphase Mineralization Along the Antimony Line, Murchison Green-stone Belt, South Africa...33 *Luke Carlton*
4:45 PM to 5:00 PM	IOCG, ISCG and Genetically Affiliated Deposits in the Cloncurry Dis-trict, Queensland, Australia: Diverse Products of One Mineral System...34 *Vladimir Lisitsin*
5:00 PM to 5:15 PM	From the Outside Looking In: An Exploration Perspective of the Mount Isa Cu and Zn-Pb-Ag System...N/A *Alex Brown*
5:15 PM to 5:30 PM	Sulfur Isotope Fingerprints of Sediment-Hosted Copper Deposits in Mt Isa Western Succession, Australia...35 *Jemi Ahnaf*
5:30 PM to 5:45 PM	Orogenic Copper Systems: A Global Perspective and Exploration Framework...36 *Matthew Hales*
5:45 PM to7:30 PM	Exhibit and Poster Reception, NGEA Panel Discussion in Exhibit Hall *Marc Gasparotto, Tony Knight, Britney Russell, Alison Morley*

= Great Hall Q1 & Q2   = Mezzanine M3   = Mezzanine M4

September 27, Saturday | 2:00 PM to 8:00 PM

Oral Session 5: Paterson to Global Cu (Au-Pb-Zn)	Oral Session 6: Machine Learning and AI
Expanding the Chronological Evolution of Winu Cu-Au Deposit (Western Australia): Insights from Modern In Situ Titanite-Apatite-Calcite Petrochronology...21 *Bruno Vieira Ribeiro*	Integrating Machine Learning and Conditional Simulation for Geometallurgical Block Modelling: A Case Study from the Resolution Copper Project...27 *Cecilia Artica*
The Geology and Breccia Evolution of the Havieron Breccia-Hosted Au-Cu Deposit...22 *Isaac Brown*	A Machine-Learning Workflow for the Automated Stratigraphic Interpretation of Pilbara Iron Ore Stratigraphy...28 *Elisabeth Scibiorski*
What Sulfur Isotopes Reveal About Ore Formation in the Southwest Yilgarn Craton...23 *Paul Duuring*	Challenges in Validating High-Resolution In-Hole Geoscience Data in Metalliferous Open Pit Mining...29 *John Jackson*
Quantifying Rifting History of the Proterozoic Mt Isa Superbasin and Its Relationship to Base Metal Mineralisation...24 *Edgar Leong*	AI-Driven Core Logging and Drill Optimization Using High-Resolution Core Scanning: Enhancing Geological Models and Reducing Exploration Risk...30 *Grant Sanden (Sponsor Talk)*
Magma and Salt: The Discovery of the Oktyabrsky Supergiant and the Role of Salt-Magma Location on Genesis and Value...25 *Steve Beresford*	How Confident Are We in Our Models? Unidentified Dataset Dependencies May Inflate Machine Learning Performance Metrics...31 *Taryn Scharf*
A Remote-Predictive Approach to the Geodynamic Evolution of the Coriolis Troughs...26 *Elizabeth Xu*	A New Paradigm for Geoscientific Insights in the Digital Era: Large Scale Modelling of Orogenic Systems...32 *Jean-Phillipe Paiement (Sponsor Talk)*

Oral Session 8: Orthomagmatic	Oral Session 9: Exploration Technology
Geodynamics Control the Variability of Magmatic Ni-Cu-PGE Sulfide Mineral Systems Through Time...37 *David Holwell*	Go Small or Go Home... The Evolution of Using the Ultrafine Soil Fraction for Mineral Exploration...43 *Ryan Noble (Invited Speaker)*
Mafic-Ultramafic Ore Systems: From Genetic Insights to Exploration Tools...38 *Margaux Le Vaillant*	Using Automated pXRF Data Collection to Model Gold Deportment and Litho-Geochemistry through Regolith in the Yilgarn Craton, Western Australia...44 *Aidan Kitchener*
Controls on PGE Mineralization in the Lac des Iles Intrusive Suite: Insights from Trace Element Geochemistry and δ34S in Sulfides...39 *James Tolley*	Field-Based Mineral Chemistry Using Portable Laser-Induced Breakdown Spectroscopy (pLIBS) for Exploration and Mineral Processing...45 *Cassady Harraden*
Dual Horizon Ni-Cu-PGE Mineralisation in the Deep Platreef, Mogalakwena Mine Complex, South Africa: Insights into Formation of High PGE Tenors...40 *Kate Canham*	Utilising Indicator Minerals to Narrow the Search Space for Greenfields Exploration...46 *Walid Salama*
The Ni-PGE Mineralisation in the Paleoarchean Bacuri Mafic-Ultramafic Layered Intrusion, Guyana Shield-Amazonian Craton, Brazil: Insights from Bulk-Rock Geochemistry and Mineralogy...N/A *Carlos Spier*	The Application of Radiogenic (Sr and Pb) and Stable Metal (Cu) Isotopes in Undercover Exploration, Northwest Queensland...47 *Joseph Tang*
The Thomson River Cu-Ni-PGE Deposit of the Southeast Lachlan Fold Belt: Implications for Cu-Ni-PGE Prospectivity in Silicic Dominated Magmatic Systems...41 *Steven Boger*	LandScape+® - A new Software Providing Machine-Learned Landscape Context and Target Definition from your Exploration Surveys...48 *Anicia Henne*
Cumulates and the Orthomagmatic to Porphyry Transition: New Insights from the Lorraine Cu-Au-Ag Deposit, Quesnel Terrane, Central British Columbia...42 *Matthew Manor*	Data Before Algorithms: An Open-Source Approach to Legacy Geochemical Method Categorization...49 *Sam Scher*

Technical Program - September 28, Sunday | 8:00 AM to 12:00 PM

8:00 AM to 8:30 AM	**Plenary Session**
8:00 AM to 8:30 AM	Olympic Dam IOCG-U deposit: Significant New Additions to the Mineral Endowment...50 *Kathy Ehrig (Sponsor Talk)*
8:30 AM to 10:30 AM	**Oral Session 10: IOCG**
8:30 AM to 8:45 AM	The Intersection of the Geoscientist with Business...51 *Jim McCluskey (Invited Speaker)*
8:45 AM to 9:00 AM	Structural Insights of the Olympic Dam Cu-U-Au-Ag Deposit, South Australia Provided by Hardrock 3D Seismic Data...52 *David Haddow*
9:00 AM to 9:15 AM	A Temporal Framework for the Genesis of the Carrapateena Iron Oxide Copper-Gold (IOCG) Deposit, Northern Gawler Craton, South Australia...53 *Richard Lilly*
9:15 AM to 9:30 AM	New Tectono-Stratigraphic Model with Revitalised Discovery Potential for Olympic Dam-type Haematitic Iron Oxide Copper Gold (HIOCG) Deposits...54 *John Anderson*
9:30 AM to 9:45 AM	Critical Ingredients of IOCG Mineral Systems: Insights from a New National Mineral Potential Model of Australia Using a Hybrid Approach...55 *Jonathan Cloutier*
9:45 AM to 10:00 AM	Giant IOCG Deposits Formed in Late Paleoproterozoic to Early Mesoproterozoic Rift Basins After the Assembly of the Columbia Supercontinent...56 *Xin-Fu Zhao*
10:00 AM to 10:15 AM	Deciphering the formation of Andean Iron Oxide-Copper-Gold Deposits: New Advances in their Geology and Exploration...57 *Irene Del Real (Keynote Speaker)*
10:15 AM to 10:30 AM	Deciphering the formation of Andean Iron Oxide-Copper-Gold Deposits: New Advances in their Geology and Exploration (continued)
10:30 AM to 11:00 AM	Break, Posters, and Speed Talks in Exhibit Hall | Coffee Sponsored by Equinox Gold
11:00 AM to 12:30 PM	**Oral Session 13: IOCGs, Cu/Co and Replacement**
11:00 AM to 11:15 AM	The Role of Iron-Rich Hydrosaline Liquids in the Formation of Kiruna-Type Iron Oxide–Apatite Deposits: Evidence from Fluid Inclusions...72 *Li-Ping Zeng*
11:15 AM to 11:30 AM	Experiments Supporting IOA Deposit Formation by Interaction of Evaporites and Silicate Magmas...73 *Michael Anenburg*
11:30 AM to 11:45 AM	Alteration-Derived Volume Changes Control the Location of Iron Oxide-Cu-Au Mineralization...74 *Tobias Schlegel*
11:45 AM to 12:00 PM	What Controls the Geometry of Ernest Henry Ore Shoot?...75 *Adrian Corvino*

☐ = Great Hall Q1 & Q2 ☐ = Mezzanine M3 ☐ = Mezzanine M4

Oral Session 11: Porphyry Exploration and Technology	Oral Session 12: Pegmatites
The Tujuh Bukit District: An Emerging Cluster of Cu-Au Porphyry Systems in East Java, Indonesia...58 *Alexsei Robert Taube*	Application of Satellite and Proximal Hyperspectral Sensing to Target Li Mineralization in Volcano-Sedimentary Deposits: The McDermitt Caldera (USA) Case Study...66 *Francesca Corrado*
Geology and Geochemistry of an Emerging Copper – Gold Discovery in the Autonomous Region of Bougainville, Papua New Guinea...59 *Patrick Highsmith*	Geology and Geochronology of the Mount Marion Lithium Pegmatite Deposit, Yilgarn Craton, Australia...67 *James Porter*
The Significance of the Mineralising Magma Composition on the Diagnostic Exploration Criteria for Porphyry Copper Deposits...60 *Ross Large*	The Crust-Mantle Structure of West Kunlun-Songpan-Ganzi Giant Pegmatitic Li Belt and Its Implications for Metallogenic Processes...N/A *Yongbao Gao*
Synchronous Carbonate-Base Metal Gold and Skarn Copper-Gold Mineralization New York-Wellington Area, Ok Tedi District, PNG: Implications for Genetic Models...61 *Peter Pollard*	Isotopic Constraints on Tungsten-Beryllium Mineralisation at Mahuaping, SW China...68 *Bin Fu*
An Updated Appraisal of the Mineral Prospectivity of the Arabian Shield...62 *Jamie Price*	
Sulphur Flushing as a Driver of Copper Accumulation in Magmatic Arcs...63 *Teresa Ubide*	Melt-Melt Immiscibility as a Key Mechanism for Spodumene-Quartz Mineralisation in Pegmatites, Insights on a Novel Concept from Andover, Western Australia...69 *Richard Cubitt*
Advanced Mineralogical, Spectral, and Geochemical Tools for Locating Upflow Zones and Mineralization in the Lithocap Environment...64 *Lejun Zhang*	Innovating Discovery: Field Sensors and Breakthrough Technology at the Sinclair Cesium Mine, Western Australia...70 *Nigel Brand*
Close-Range Hyperspectral Sensing of Mineralized Outcrops in Open-Pit Mines: Analysis of the Lithocaps of the Allumiere-Tolfa Epithermal System (Central Italy)...65 *Anna Sorrentino*	Multiple Mineralisation Events Demonstrated in Complex Li-Cs-Ta Pegmatites at Buchanan Creek, Queensland...71 *Ben Hines*

Oral Session 14: Collaborative Geoscience 1	Oral Session 15: Machine Learning Workflows to Prospectivity
Building Bridges in Geoscience: EuroGeoSurveys in Fostering International Resource Collaboration...77 *Julie Hollis (Invited Speaker)*	Identification and Measurement of Minerals on Thin Sections of Intrusive Igneous Rocks Using Deep Learning...83 *Stalyn Paucar*
The Critical Minerals Geoscience and Data (CMGD) Initiative of the Geological Survey of Canada...78 *Geneviève Marquis*	Exploring the Digital Terrain: The Power of Geospatial Technology and GeoAI in Modern Mining...84 *Aune Kamosho*
Environmentally Sustainable Production of Critical Metals in Tasmania: A Collaborative Multidisciplinary Initiative...79 *Owen Missen*	The Rock Whisperer: Integrating Automated EDA Work-flows with LLM...85 *Putra Sadikin*
The Global Critical Minerals in Ores (CMiO) Database – A Review and Update...80 *Evgeniy Bastrakov*	Characterisation of Ore Deposits: Overcoming Cross- and Multi-Scale Challenges...86 *Elena Belousova*

12:00 PM to 12:15 PM	Exploration Success and Refinement of Mineralisation Models in the Western Forelands of the Democratic Republic of Congo...76 *Amanda Stoltze (Keynote Speaker)*
12:15 PM to 12:30 PM	Exploration Success and Refinement of Mineralisation Models in the Western Forelands of the Democratic Republic of Congo (Continued)
12:30 PM to 02:00 PM	Lunch in Exhibit Hall
02:00 PM to 03:30 PM	**Oral Session 16: IOCGs, Cu/Co and Replacement**
02:00 PM to 02:15 PM	Structural Architecture of the Lufilian Arc: Arc-Scale Structural Controls on Mineralization Based on a New GIS Compilation...89 *Ian Basson*
02:15 PM to 02:30 PM	Skarn-Porphyry Transition...90 *Zhaoshan Chang*
02:30 PM to 02:45 PM	Carbon and Oxygen Isotope Analyses with Paired Geochemical Data in Exploration for Cu-Zn Skarn Mineralization at Antamayo, Central Peru...91 *Lucas Marshall*
02:45 PM to 03:00 PM	The Timing and Geochemistry of Dolphin W Skarn, King Island, Tasmania: A Micro-Scale Study of Garnet and Scheelite...92 *Angela Isaura Santos Costa*
03:00 PM to 03:15 PM	Subtle Potassic Alteration as a Vector for Mineralization in the Silvertip CRD, Northern British Columbia...93 *Max Hohl*
03:15 PM to 03:30 PM	Unlocking the Timing of Au-Sb-Hg Mineralization in the Kyrgyz South Tien Shan: Constraints from In-Situ U–Pb Dating of Hydrothermal Calcite...94 *Aizat Zhaanbaeva*
3:30 PM to 4:00 PM	Break and Posters in Exhibit Hall | Coffee Sponsored by VRIFY
4:00 PM to 5:30 PM	SEG Distinguished Lecture Presentation and Awards Ceremony...N/A *Mei Fu Zhou (Distinguished Lecturer)* **See p. 8 for 2025 SEG Awardees**
05:30 PM to 07:30 PM	Exhibit and Poster Reception, Exhibitor Spotlight Talks in Exhibit Hall
8:00 PM to 11:00 PM	Conference Celebration (ticketed)

September 28, Monday | 8:00 AM - 9:30 AM

8:00 AM to 8:45 AM	**Plenary Session**
8:00 AM to 8:45 AM	How Do We Fix Metals Exploration? (Panel Discussion)...N/A *David Lawie (Chair), Richard Schodde, Ned Howard, Sarah MacKenzie*
8:45 AM to 10:30 AM	**Oral Session 19: Applied Orebody Knowledge**
8:45 AM to 9:00 AM	What Is the Value of Ore Body Knowledge for the Mining Industry?...N/A *Shaun Barker (Invited Speaker)*
9:00 AM to 9:15 AM	Teaching AI the Language of Rocks: Transforming Orebody Knowledge...107 *Dave Lawie*
9:15 AM to 9:30 AM	The Flow of Geological Information in the Mining Value Chain and Its Use for Operational Optimization...108 *Carolina Marin Suarez*

☐ = Great Hall Q1 & Q2 ☐ = Mezzanine M3 ☐ = Mezzanine M4

September 27-28, Sunday 12:00 PM - Monday 9:30 AM

Disruptive Thinking and Transformational Research to Shape the Future Resource Workforce...81 *Marco Fiorentini*	Multi-Scale Characterisation of Intrusion-Related Gold Systems in NE Queensland: Insights from the Ravenswood Deposit...87 *Al-tamini Tapu*
Collaboration in Action: Advancing Geoscience Through Government, Research and Industry Partnerships...82 *Janelle Kerr (Invited Speaker)*	Mapping Alteration Footprints of IOCG Systems: A Case Study of the Ernest Henry Camp, Cloncurry District, Queensland...88 *Alkis Kontonikas-Charos*

Oral Session 17: Collaborative Geoscience 2	**Oral Session 18: Fluid Flow and Metal Deposition**
Government Geoscience Accelerating Resource Discovery and Development...95 *Marina Costelloe (Invited Speaker)*	Mineral Systems Modeling: New Technologies for Exploration Vectoring...100 *Daniel Palmowski*
Academia-Industry-Government Collaborative Projects to Explore the Proterozoic Basins in Australia and their Sediment-Hosted Resources...96 *Juraj Farkas*	The Architecture and Dynamics of Coupled Failure and Fluid Flow in Magmatic-Hydrothermal Systems: Insights from Swarm Seismicity, Hakone Caldera, Japan...101 *Stephen Cox*
Assessing Proximity to Cu and Au Deposits Using Resistate Mineral Chemistry...97 *Caroline Tiddy*	Identifying Mineralisation Within the Sunrise Dam Gold Mine Using Geomechanical Targeting Technology...102 *John Mclellan*
A Continental-Scale Heavy Mineral Reference Framework for Mineral System Detection...98 *Brent McInnes*	Suspension-Driven Transport of Noble Metals in Hydrothermal Fluids...103 *Nestor Alfredo Cano Hernandez*
MinEx CRC's Model for Collaboration in Exploration Geoscience...99 *David Giles (Invited Speaker)*	Electrical Processes in Charge of Ore Formation...104 *Chris Voisey*
Collaborative Geoscience Discussion	Colloidal Flocculation and the Mechanical Transport of Gold: Causes of the 'Nugget Effect' in Hydrothermal Ore Deposits...105 *Duncan McLeish*

September 28, Monday | 8:00 AM - 9:30 AM

Oral Session 20: Porphyry Tracers 1	**Oral Session 21: From Geomet to Remediation 1**
Delamerian Porphyry Mineral Systems in South Australia: What Can Magmatic Mineral Chemistry Deliver on Fertility and Exploration?...113 *Wei Hong*	Geometallurgy of the Olympic Cu-Au Province Deposits, South Australia...119 *Vanessa Liebezeit*
Alteration Mineralogy Controls on Using Geochemical Data for Rock Classification: Insights from the Camp Creek Cu-Mo-(Au-Ag) Porphyry Deposit, Canada...N/A *Maxwell Porter*	Improving Remediation Strategies for Legacy Mines: Case Study of Mt. Morgan Gold and Copper Mine, Queensland, Australia...120 *Adithyanti Febriana*
Zircon and Whole-Rock Trace Element Indicators of Magmatic Hydration State and Oxidation State Discriminate Copper Ore-Forming from Barren Arc Magmas...114 *Gonzalo J. Henriquez*	Bacteriologically Driven REE Recovery from Mary Kathleen Uranium Mine Tailings: Potential Formation of Biogenic REE-Minerals...121 *Samadhi Gunathunga*

9:30 AM to 9:45 AM	Magnetite Ores and the Green Steel Revolution: A Global Outlook on Resources and Future Prospects...109 *Yoram Teitler*
9:45 AM to 10:00 AM	The Application of Data Fusion Algorithms and Machine Learning to Increase Orebody Knowledge...110 *Penny Stewart*
10:00 AM to 10:15 AM	The Critical Role of Ore Body Knowledge in Mining Success: A Systems Leadership Perspective...111 *Andrea Rutley*
10:15 AM to 10:30 AM	Modelling the Future of Copper Production in Queensland's NW Minerals Province: A New Foundation for Policy Development and Infrastructure Planning...112 *Tom Evans*
10:30 AM to 11:00 AM	Break, Poster and Speed Talks in Exhibit Hall | Coffee Sponsored by Equinox Gold
11:00 AM to 12:30 PM	**Oral Session 22: Porphyry Tracers 2**
11:00 AM to 11:15 AM	Sericite and Chlorite Type and Composition within and Around Porphyry Deposits: Tools to Vector Towards K-Silicate Alteration at Depth...126 *Farhad Bouzari*
11:15 AM to 11:30 AM	Trace Element Compositions of Micro Inclusions Extracted from Mixed LA-ICP-MS Analyses and Their Use in Exploration...127 *Ivan Belousov*
11:30 AM to 11:45 AM	Assessing the Precision and Accuracy of In Situ Re-Os Dating of Molybdenite and Considerations for Ore Systems Research...128 *Duane Petts*
11:45 AM to 12:00 PM	New Developments in Geochronology to Place Mineralisation Processes into Better Constrained Tectonic and Geodynamic Context...129 *Dr. Sarah Gilbert*
12:00 PM to 12:15 PM	The Geological Footprint of Alteration and Mineralization at the High-Grade Resolution Porphyry Cu-Mo Deposit, Arizona...130 *Takeshy Coaquira*
12:15 PM to 12:30 PM	Adakitic Signatures from Permian Granitoids Around the Georgetown Region – Potential for Porphyry Cu Mineralisation and Implication for Tectonic Setting...131 *Jonghyun Lee*
12:30 PM to 02:00 PM	Lunch, Posters and Speed Talks in Exhibit Hall
02:00 PM to 04:00 PM	**Oral Session 25: Porphyry Americas**
02:00 PM to 02:15 PM	New Insights for the Porphyry Model from Complex Overprinted Porphyry Systems...143 *Angela Escolme (Invited Speaker)*
02:15 PM to 02:30 PM	Magmatic Plagioclase as a Porphyry Copper Fertility Indicator in Arc Systems...144 *Daniela Parra-Encalada*
02:30 PM to 02:45 PM	Insights Into the Petrogenesis of the Taca Taca Bajo Cu-Mo Porphyry Deposit (Argentina) and Its Significance Within the Andean Retro-Arc ...145 *Madeleine Ince*
02:45 PM to 03:00 PM	Anatomy of a Moderately-Telescoped Porphyry–High Sulfidation System: Insights of a 2-km Vertical Profile of Alteration at Valeriano Cu-Au Deposit, Chile...146 *Jaime Osorio*

☐ = Great Hall Q1 & Q2 ☐ = Mezzanine M3 ☐ = Mezzanine M4

Redox-Sensitive Trace Elements in Apatite: Insights into Porphyry Copper Mineralization Environments...115 *Maria Alejandra Rodriguez Mustafa*	From Penalty to Opportunity: Geometallurgical Characterisation of Gold Concentrates and Tailings for Critical Metal Recovery Using Novel Solvents...122 *Lara Tritton*
Zircon Trace-Element Indicators of Magmatic Gold Fertility...116 *Robert Loucks*	Responsible Mining – the Pyrite Challenge...123 *John Thompson*
Zircon and Melt Inclusion Geochemistry from the 1.1 Ga Jogran Porphyry, Canada: Implications for Porphyry Formation in a Non-Arc Setting...117 *Pete Hollings*	Selective Bioleaching of Critical and Precious Metals from Northern Queensland Mine Waste Using Indigenous Bacteria...124 *Eric O. Ansah*
Exploration and Genesis Information Revealed from Comparative Copper Isotope Analysis of Sulfides from PCDs, IOCGs, and Mantos...118 *Ryan Mathur*	Interrogating LA-ICP-TOF-MS Mineral Maps: Methods and Applications to Critical Metal Research...125 *Jeffrey Oalmann*

Oral Session 23: ESG	Oral Session 24: From Geomet to Remediation 2
Global Perspectives on Exploration and ESG...132 *Sarah Mackenzie (Invited Speaker)*	Cobalt Potential from IOCG Tailings and Copper-Cobalt Deposits of the Atacama Region, Chile: Bioleach Alternatives for Sustainable Recovery...138 *Brian Townley (Keynnote Speaker)*
Practitioner Perspectives on Current Approaches to Community Relations at Exploration: Highlights from Northern Europe, Latin America, North America, and Central Africa...133 *Kathryn Kochan (Invited Speaker)*	Cobalt Potential from IOCG Tailings and Copper-Cobalt Deposits of the Atacama Region, Chile: Bioleach Alternatives for Sustainable Recovery (continued)
Geospatial Modelling of Environmental, Social and Governance (ESG) Risks for Mineral Exploration and Mining Feasibility Assessments in Finland...134 *Bijal Chudasama*	Environmental Impact of Abandoned Mining Activities in the Monte Amiata District (Italy) and Implications for a Sustainable Environmental Management...139 *Guia Morelli*
Biomass Carbon Emissions from Nickel Mining Have Significant Implications for Climate Action...135 *Evelyn Mervine*	Software Tool Development for Evaluating the Beneficiation Potential of Mining Waste...140 *Dandara Ataide Salvador*
Mineral Landscapes: Connecting Geoscience to Stakeholders...136 *Karin Olson Hoal*	Rare Earth Elements' Secondary Prospectivity of Mine Waste in South Australia...141 *Francesco Colombi*
Evidence-Based Solutions for Improving Equity, Diversity and Accessibility: Informing Policy and Governance in Economic Geology...137 *Fernanda Alvarado-Neves*	Carbon Dioxide Removal Using Serpentine-Rich Mine Residues...142 *Greg Dipple (Invited Speaker)*

Oral Session 26: REE and Carbonatites	Oral Session 27: Big Workflows
Occurrence of Invisible Au in Carlin-Type Au Deposits...151 *Zhuojun Xie*	Automating Lithology Prediction Using Python and ioGAS: A Streamlined Workflow for Geochemical Analysis...158 *Ingrid Flemons*
Inwards-Younging of the Annular Twihinate Carbonatite-Aillikite Complex (Southern Morocco) and Its Implication for REE Distribution...152 *Lucas Tatnell*	From Data to Discovery: Targeting Sedimentary Copper in South Australia Using a Mineral Systems Approach...159 *Genevieve Luketina*
Data Integration of Hyperspectral and Whole-Rock Geochemistry: New Insights into Mt Weld Carbonatite-Hosted REE Deposit...153 *Siyu Hu*	Technological Innovations in Exploration and Mining – Success Stories from a Copper Miner...160 *Michaela Young-Mitchell*
Geology and Genesis of the REE-Bearing Ironstones of the Yin Complex, Gascoyne Province, Western Australia...154 *Ross Chandler*	Prospectivity Mapping of Gold Deposits in the Qinling Orogen Belt: A Data-Driven Framework...161 *Haicheng Wang*

Technical Program - September 29, Monday | 3:00 PM - 6:30 PM

03:00 PM to 03:15 PM	Geology, Mineral Alteration and Mineralization of the Santa Cecilia Porphyry Au-Cu Deposit, Maricunga Belt, Northern Chile...147 *Jose Franco Moraga*
03:15 PM to 03:30 PM	Discovery of the Ferrobamba Deeps Deposit, Las Bambas, Southern Peru...148 *James Cannell*
03:30 PM to 03:45 PM	New Discoveries in a Mature Terrain Through Holistic Exploration Practices: A Story of Safford, Arizona...149 *Matthew Wetzel*
03:45 PM to 04:00 PM	Reevaluating the Schultze Granite: A New Temporal Model for the Globe-Miami Magmatic-Hydrothermal System and Its Implications for Porphyry Exploration...150 *Lawrence Carter*
4:00 PM to 4:30 PM	Break in Great Hall Foyer
4:30 PM to 7:30 PM	**Closing Events**
4:30 PM to 5:30 PM	Closing Ceremony and Concluding Remarks *Rick Valenta, SEG 2025 Conference Chair* *Alan Wilson, SEG President 2026* *Stephanie Mills, SEG 2026 Conference Chair*
5:30 PM to 6:30 PM	Closing Social, Mezzanine Level of Great Hall (Ticket Required)

September 29, Monday | 3:00 PM to 6:30 PM

Preservation of Na-REE Mineralisation at Kangankunde Carbonatite Complex, Malawi, with Implications for Magma Evolution...155 *Nicholas Mappin*	Using Open-Access Remotely Sensed Data to Support Greenfield Mineral Exploration: Methods and Lessons from Case Studies...162 *Mana Rahimi*
Rare Earth Element Ion Adsorption Clay Deposits: A Mineral Systems Approach for Exploration...156 *Samantha Russo*	SMX - Sustainable Mining Explorer Platform: A Geoscience-Based Approach for ESG Risk Management in Mining...163 *Vinicius Louro*
The Cradle and Future of Rare Earth Elements...157 *Ignacio Gonzalez-Alvarez (Keynote Speaker)*	From Ore to PV: A Sustainable Tellurium Value Chain for Net-Zero Goals...164 *Homay Fath*
The Cradle and Future of Rare Earth Elements (continued)	Critical Minerals Potential in the Delamerian Orogen Margin: Exploring the Sustainable Prosperity...165 *Yanbo Cheng*

☐ = Great Hall Q1 & Q2 ☐ = Mezzanine M3 ☐ = Mezzanine M4

Posters

Posters will be on display during conference hours. Titles and presenting authors are listed below. **Posters listed in blue are Speed Talks.**

Collaborations

P1.001 Evaluating the Tectonic and Climatic Controls on Sediment-Hosted Ore Deposits in Deep Time...166
Sheree Armistead

P1.002 Insights into the Delamerian Orogen Mineral Systems Through Pyrite Chemistry...167
Adrienne Brotodewo

P1.003 Solubility of Alkali Elements and Chlorine in Sulfide Melts Under Magmatic Conditions: Implications for Ore Genesis...168
Maria Cherdantseva

P1.004 The Queensland Mineral Deposit Atlas Series...169
Tom Evans

P1.005 Alkaline Magmatism in the Adang Volcanic Complex, Mamuju District (Western Sulawesi, Indonesia): A Preliminary Study on Rare Earth Elements Enrichment...170
Fadlin Fadlin

P1.006 National-Scale Critical Mineral and Strategic Material Potential Assessments to Support Decision-Making in the Net Zero Transition...171
Arianne Ford

P1.007 Phreatomagmatic Ash as a Natural Drill Core: Unraveling Subvolcanic Hydrothermal Systems in Active Arc Volcanoes...172
Muhammad Andriansyah Gurusinga

P1.008 Empowering Exploration with National Scale Maps...173
Nichole Knepprath

P1.009 Chromite and Sulphide Mineralization of the Uitloop Ultramafic Bodies in the Northern Limb of the Bushveld Complex, South Africa...174
Mabatho Mapiloko (Student)

P1.010 Archaean BIFs from Karnataka, India: Genetic Insights from the Dharwar craton, India, and Implications for Post-Depositional Hematitic Iron Deposits...175
Perumala Venkata Sunder Raju

Deposits

P2.011 Deciphering the Role of Remobilisation in the Formation of High-Grade Gold Deposits...176
Sumail

P2.012 Geochemical Signature of Au-Cu Mineralization in Soil Profiles of the Wadi Al Jaww Prospect and Applications for Exploration, Saudi Arabia...177
Hassan Abbas

P2.013 Applications of Mineral Chemistry to Petrogenesis and Exploration in the Borralan Carbonatite Complex and Associated Alkaline Silicate Rocks...178
Abdulkadir Mohamed Abdulkadir (Student)

P2.014 Exploring for Cu-Au Alkalic Porphyry Deposits from the Fringes...179
Fabrizzio Abello (Student)

P2.015 Mineralization and Ore Genesis of the Neoproterozoic Pan-African Belt, Eastern Ghana: Constraints on Lithology and Mineralogy...180
Rauda Addae (Student)

P2.016 Sulfide Dendrites in Silica Scales at the Salton Sea Geothermal Field, California...181
Aaron Adsit (Student)

Posters, continued

P2.017 What Do the Pyrites Tell Us About the Young-Davidson Orogenic Gold Deposit, Matachewan, Ontario, Canada?...182
Oya Ak (Student)

P2.018 Geological Features and Diagnostic Characteristics of Gunung Akmil HSE Prospect in Central Java, Indonesia: Vectoring to Concealed Porphyry System...183
Andreas Rama Alfario (Student)

P2.019 Initial Investigation of Geology and Litho-Geochemistry in North Halmahera, Indonesia: Implications for Porphyry Copper and Epithermal Deposit Potential...184
Achmad Zulfan Almahdy (Student)

P2.020 Nature of the Parental Source Material(s) for the Placer Gold-Bearing Banket Conglomerate, Ghana: Constraints from δ18O Water in the Quartz Pebbles...185
Edwin Animah Obeng (Student)

P2.021 Geology and Geochemistry of Nikolauz Mafic-Ultramafic Intrusions, Peru: A Gondwanide Ni-Cu Sulfide Mineralized Conduit System in an Andean Orogenic Environment...186
Abraham Arana (Student)

P2.022 REE and HFSE Fractionation and Mobility in the Tapira and Salitre Carbonatite Complexes Southeast, Brazil...187
Jo Hannah Asetre (Student)

P2.023 Unlocking the Hidden Potential of the Cargo Deposit: New Geochronological and Geochemical Insights from the Macquarie Arc...188
Jessica Askew (Student)

P2.024 The Geology and Paragenesis of the Akyem Orogenic Gold Deposit, Ghana...189
Fuseini Atanga (Student)

P2.025 μ-XRF Atlas of Epithermal and Porphyry Deposit Textures...190
Aaron Atkins (Student)

P2.026 Mineralogical and Geochemical Characteristics of the Epithermal Au-Ag Mineralization at the Felicia Orebody, Lepanto, Mankayan District, Philippines...191
Yasuhiro Awano (Student)

P2.027 Textures and Geochemistry of Pyrite and Arsenopyrite from the Leon Sediment-Hosted Gold Deposit, Ratatotok District, Indonesia: Implications for Ore-Forming Conditions...192
Muhammad Arba Azzaman (Student)

P2.028 Iron-Oxide-Copper-Gold (IOCG) Type Hydrothermal-Alteration Recorded in the Lawa Gold Deposit: North Singhbhum Mobile Belt, Eastern India...193
Paromita Banerjee (Student)

P2.029 Late Au-Bearing Quartz - Carbonate Veins System Overprinted the Base Metal Skarn Mineralization in the E44 Deposit, Northparkes District, NSW...194
Billy Beas Caceres (Student)

P2.030 Geochemical and Mineralogical Signatures of the Rocklands Cu-Au-(Co) Deposit, Northwest Queensland, Australia...195
Elena Belousova

P2.031 Tectonic Controls on Long-Lived Sn Mineralization in the Martinamor Extensional Dome (Salamanca, Spain)...196
Daniel Bermejo López (Student)

P2.032 Clay-Hosted REE Deposits Associated to Eucla Basin Paleovalleys in Southern Australia...197
Gabriel Valentim Berni

P2.033 A Prolonged Magmatic History to the Dismembered Mid Miocene Arc, Manus Island, Papua New Guinea...198
Joel Blake

P2.034 Exploring a Sleeping Giant: Chlorite Chemistry as a Vectoring Tool in the World-Class Tennant Creek Au-Bi-Cu Province...199
Damian Braize (Student)

P2.035 Orebody Knowledge from the Sinclair Deposit: A New Template for Cesium Exploration...200
Nigel Brand

P2.036 The Prominent Hill Deposit: A Modified, High-Grade IOCG System Hosted in Hematite-Altered Sedimentary Breccias...201
Riccardo Brigante

P2.037 Statistical Analysis of Geochemical Data from In-Bearing Mineralization in the Marta Centro Vein, Pingüino Deposit, Patagonia, Argentina...202
Juan Matías Buratti (Student)

P2.038 Regolith-Hosted REE Systems in South America: Mineralogical and Geochemical Insights Across Contrasting Geological and Climatic Settings...N/A
Nicolás Bustos (Student)

P2.039 From Regional Insight to Discovery: Integrated Targeting for Sapphire Mineralization in Southeastern Madagascar...203
Evelyn Caiza (Student)

P2.040 Exploration Vectoring Tools in Complex Magmatic-Hydrothermal Systems: Application to the Potrerillos District in Northern Chile...204
Yamila Cajal

P2.041 Assessing the Timing of the Magmatic to Hydrothermal Transition and Onset of Porphyry Mineralization Across the Yerington Porphyry System, USA...205
Lawrence Carter

P2.042 High-Grade Gold Controls and Spatial Chemical Zonation at The Snip North Prospect of The Iskut Project, Canada...206
Daniel Castano Madrigal (Student)

P2.043 Cobalt Enrichment in Sulfides from Iron Oxide Copper-Gold Mount Colin Deposit, NW Queensland...207
Gabriel Cellier (Student)

P2.044 Structural Controls on Epithermal Ag-Pb-Zn Vein Systems in the Eastern Erzgebirge, Germany...208
Jan Cerny

P2.045 Contrasting Styles of REE Mineralisation Within the Southern and Central Gifford Creek Carbonatite Complex...N/A
Ross Chandler

P2.046 Link Between Deformation and Pegmatite Emplacement in the Mataki Greenstone Belt, North East Zimbabwe Craton...209
Pharisie Chibaya (Student)

P2.047 Multi-Stage Fluid Alteration of the Agua Santa Pegmatite (Brazil): Implications for Mineral Zonation and Exploration...210
So ia Chinaglia (Student)

P2.048 A New Process-Based Mineral System Classification Linked to Commodities...211
Jonathan Cloutier

P2.049 Oxidized Mantle Sources in Cretaceous Komatiites from Gorgona Island: Insights into Mantle Redox Structure and Impacts on Ore-Forming Systems...212
Camilo Andrés Conde Carvajal (Student)

P2.050 Chemical and Isotopic (Sr, Nd) Analysis of the Nurra Basin Bentonites (NW Sardinia, Italy) ...213
Sebastiano Coticelli (Student)

P2.051 Structural Characterisation of the Ribeirão Grande Shear Zone, Southern Brazil...214
Graciany da Silva

P2.052 Geochemical and Structural Controls of the Gold-Bearing Carbonaceous Veins Hosted Malenggang Project, West Kalimantan, Indonesia...215
Arif Zardi Dahlius

P2.053 Resolving the Syn-Genetic Age of the King Metamorphosed VHMS Deposit, Yilgarn Craton, Western Australia...216
Cendi Dana

P2.054 Garnet Geochemistry as a Tracer of Timing and Fluid Pathways in Pb-Zn Systems...217
Joëlle D'Andres (Student)

P2.055 Genesis of IOA-IOCG deposits, Coastal Cordillera, Peru: Insights from Iron ($\delta^{56}Fe$) and Triple Oxygen ($\delta^{18}O$, $\Delta^{17}O$) Isotopes in Magnetite...218
Jose Carlos Davila

P2.056 Relative Timing of Au Mineralization at the Epithermal Calm Before the Storm Zone, Treaty Creek Area, Golden Triangle, British Columbia...219
Nikola Denisova

P2.057 Contribution of Auger and Reverse Circulation to the Definition of Waraba Mineralization (Loulo Permit)...220
Zoumana Diallo (Student)

P2.058 Polyphase Cobalt Mineralisation in the Rajapalot Au-Co Deposit, Finland...221
Lily Dickson (Student)

P2.059 Morphochemical Analysis of Gold in the Pinalejo Placer: The First Quantitative Analysis of Gold Wear Rate During Fluvial Transport...222
Kelvin Dos Santos Alves (Student)

P2.060 Is the Zabrnjica Prospect on Ober License in Southwestern Serbia a Previously Unrecognized Reduced Intrusion Related Gold Systems (RIRGS)?...223
Dragan Dragic

P2.061 Geological Controls on Critical Metal Endowment in Sedimentary-Hosted Zn-Pb Deposits: Insights from the Canadian Cordillera...224
Foteini Drakou

P2.062 Oak Dam West IOCG deposit, South Australia – Evolving Geological Knowledge as Resource Definition, Advanced Exploration and Mining Studies Continue...225
Kathy Ehrig

P2.063 Quantitative Analysis of Mineral Orientation in Sulphide Ore Deposits...226
Tom England (Student)

P2.064 Geochemical and Mineral Inclusion Analysis of Alluvial Gold as Exploration Tools: Experiences from the Central Iberian Zone, Spain...227
Vanessa Leonor Escobar Duche (Student)

P2.065 Constraining Magma Evolution Associated with Alkalic Porphyry Mineralization in British Columbia, Canada: A Melt Inclusion Study...228
Maria Paula Estrella (Student)

P2.066 Global Controls on the Formation of Porphyry Systems: Insights from Machine Learning and Plate Tectonic Reconstructions...229
Ehsan Farahbakhsh

P2.067 Decoding Structural Influences on Vein Formation in the Japudali Prospect, Indonesia...230
Dian Yesy Fatimah

P2.068 Pyrite Chemistry as a Tracer for Metallogenic Evolution in the Au-Ag-Te Kochbulak Deposit (Chatkal-Kurama Region, Uzbekistan)...231
Danis Ionut Filimon (Student)

P2.069 Integrated Exploration Approach Unlocks a Deep-Seated High-Grade Copper Mineralization in the Jervois Range, Northern Territory, Australia...232
Jalu Bias Firdausi

P2.070 Geometallurgical Behavior at Increasing Depths in the Cuiabá Underground Gold Mine, Brazil...N/A
Beatriz Fontes Andrade (Student)

P2.071 Integration of Petrographic, Structural, and Geochemical Data in Weathered and Challenging Prospective Areas: Case Study of the Descoberto Prospect, Brazil...233
Beatriz Fontes Andrade (Student)

P2.072 Structural Controls, 3D Geometry and the Evolution of Vein-Hosted Copper Mineralization at Frontier Mine, Democratic Republic of Congo...234
Heike Fourie

P2.073 The Epithermal to Skarn-Porphyry Transition at La Colorada Deposit (Zacatecas, Mexico)...235
Pedro Francisco (Student)

P2.074 Geochemistry Modeling and Mineralogy of the Banded Iron Formation, West Moeda Syncline, Quadrilátero Ferrifero Region...236
Carolina Frésca Alves (Student)

P2.075 Chile's First Defined Intrusion-Related Gold System: Ternera's Hydrothermal Footprint in the El Zorro Gold District...237
Eduardo Fritis (Student)

P2.076 Unveiling Diamond Potential in the Meya Kimberlite Dyke: Petrographic and Thermobarometric Insights...238
Tenoch Galeana Cornejo (Student)

P2.077 Long-Lived High-Temperature Metamorphism in the Late-Archean Granulite-Hosted Katanning Gold Deposit, Southwest Yilgarn Craton...239
Jonathan Garcia (Student)

P2.078 Geochemical and Structural Comparison of the Golden Pike and Dunlops Crossing Stockwork Orebodies, Golden Mile, Kalgoorlie, WA...240
Ana Sofia Garcia Saltzmann

P2.079 Carboniferous Epithermal Gold Deposits in the Khundii Metallogenic Province, Southwest Mongolia...241
Bat-erdene Gendenjamts

P2.080 Critical Metal Distribution in the Renison Bell Tin Deposit, Western Tasmania...242
Javier Gil-Rodriguez (Student)

P2.081 Geology and Age of the Granite Related Metasomatic Overprinting of the Rosebery VHMS Deposit...243
Vinicius Godoi Pereira Da Cruz (Student)

P2.082 Trace Elements of Pyrite, Chalcopyrite and Sphalerite Across Distinct Ore Types and Lenses at the Rosebery VHMS Deposit, Western Tasmania...244
Vinicius Godoi Pereira Da Cruz (Student)

P2.083 Advanced Argillic Alteration and Geochemistry of Alunite in the Iriki Gold Exploration Area, Hokusatsu, Japan...245
Yuji Gonoi

P2.084 Rodinia Breakup and a New Undercover Neoproterozoic Magmatic Cu-Ni-PGE Province at Cape York, Australia...246
Peter Gregory (Student)

P2.085 Quantifying the Financial Implications of Orebody Knowledge Deficiencies in Mining Projects...247
Haruna M. Grema (Student)

P2.086 The Archean Troilus Au-Cu-Ag Deposit: An Atypical Syn-Volcanic World-Class Deposit Reworked During Amphibolite-Grade Metamorphism and Deformation (Quebec, Canada)...248
Pierre-Arthur Groulier

P2.087 Alteration Zoning and Exploration Implications in the Freiberg Pb-Zn-Ag Epithermal Vein System, Germany...249
Marie Guilcher (Student)

P2.088 New Calibration of the Al-in-Hornblende Geobarometer: Evidence for Lower-Crustal Fluid Saturation in Porphyry Copper Magmas...250
Gonzalo Henríquez

Posters, continued

P2.089 Clumped Isotope Thermometry in MVT Carbonates: Insights into Ore-Forming Temperatures and Fluid Evolution...251
Hashindra Herath (Student)

P2.090 Mineralogical, Geochemical, and Sulfur Isotopes Studies of the Partolang Volcanogenic Massive Sulfide Deposit, Wetar Island, Indonesia...252
Tomy Herawan (Student)

P2.091 Developing Soil Sampling Workflows for Lithium Exploration in SW England...253
Edward Hill (Student)

P2.092 Geochemical Characterisation of the Orion Massive Sulfide Deposit and Surrounding Prospects, Wunaamin Miliwundi Orogen, Western Australia...254
Sam Hill

P2.093 REE Mineralization at Koppamurra, South Australia: Insights into the Transport, Vertical and Lateral Distribution of REEs in a Low-Temperature Environment...255
Jasmin Hiller (Student)

P2.094 Multiple Mineralisation Events Demonstrated in Complex Li-Cs-Ta Pegmatites at Buchanan Creek, Queensland...256
Benjamin Hines

P2.095 The Comparison of Sulfuric Isotope Results of Gold Deposits in Mongolia: Emphasising the Western Part...N/A
Khaliunaa Iderbat

P2.096 Supporting Geometallurgical Prediction: 3D Modeling of Mineralization and Alteration in Low-Grade Gold Ore Using Multi-Element Data...N/A
Khaliunaa Iderbat

P2.097 Zircon and Apatite Geochemical Insights into Porphyry Cu Fertility in the Artvin District During Subduction–Collision Transition...258
Ali Imer

P2.098 Characterisation of Clay-Hosted Rare-Earth Element Deposits in Australia...259
Madeleine Ince (Student)

P2.099 Geochemical Characteristics and Zircon U-Pb Geochronology of the Late Triassic Granite in the Pilok Tin-Tungsten Deposit, Kanchanaburi, Thailand...260
Mallika Intachai (Student)

P2.100 Origin of the Aït Abdellah Copper Deposit (Bou Azzer, Anti-Atlas, Morocco): Perspectives from Mineralogical, Stable Isotope, and Fluid Inclusion Analysis...261
Marieme Jabbour (Student)

P2.101 Classification of Au Systems in the Mazaruni Greenstone Belt, Guyana, Using Pyrite Trace Element Chemistry...262
Tramaine James (Student)

P2.102 Bantug Lithocap, Negros Island, Philippines: Mineralogy, Textures, and Chemistry...263
Carlos Jimenez

P2.103 Patchy-Wormy Textures in Lithocaps: Examples from Bantug, Philippines and La Zanja, Peru...264
Carlos Jimenez

P2.104 Gold-Sulphide Mineralization within Sheared, Subvertical Banded Iron Formation, Jaspilite and Dolomite, Pickstone-Peerless Mine, Chegutu Greenstone Belt, Zimbabwe...265
Caitlin Jones

P2.105 Eocene Intrusion-Related Gold in the Golden Triangle of British Columbia...266
Randall Karcher (Student)

P2.106 Sulfide Sponges — Sulfide Associations with Spinels and Other Minerals...267
Kathryn Keane (Student)

P2.107 The State and Fate of Noble Metals in Bornite and Digenite...268
Samuel A King (Student)

P2.108 New Insights into Ore Genesis and Exploration Models from Multi-Element Geochemistry...269
Aidan Kitchener

P2.109 Structural Evolution and 3D Modelling of the Giant Vasilkovskoye Gold Deposit, Northern Kazakhstan...270
Corne Koegelenberg

P2.110 Structural Evolution and Constraints on Gold Mineralisation of the North-East Section of the Kibi-Winneba Gold Belt...271
Samuel Asare Konadu (Student)

P2.111 The Formation of Granite-Type Uranium Deposits in South China: Insights from Hydrodynamic and Chemical Modelling...272
Zenghua Li

P2.112 Carrapateena IOCG Deposit: Discovery to Operating Sub Level Cave Mine and Beyond...273
Shaun Light

P2.113 Intrusion-Related and Epithermal Gold Systems in North-East Queensland, Australia: Multiple Metallogenic Districts and Epochs, Diverse Geochemical and Mineralogical Signatures...274
Vladimir Lisitsin

P2.114 Evolution of Eocene Magmatism at Swales Mountain, North-Central Nevada, and the Relationship to Carlin-Type Gold Deposits...275
Shasta Longo Florez

P2.115 Weathering at Depth - Key Process Forming REE-Rich Regolith at the Mt Weld Carbonatite (Western Australia)...276
Anderson M. Santos (Student)

P2.116 Mechanisms of Nb and Ta Enrichment in Ferricrete at Mt Weld Carbonatite, Western Australia...277
Anderson M. Santos (Student)

P2.117 Trace Element Variations of Pyrite and Chlorite Related to Cu Mineralization in the Osarizawa Epithermal Deposit, Akita, Japan...278
Jonathan Macuroy

P2.118 In-situ U-Pb Dating of Monazite and Rutile: Implications to Depositional Age of QPC and U-Au Mineralization Potential, Singhbhum Craton, India...279
Smruti Prakash Mallick (Student)

P2.119 Genesis of the Hybridized Awakmas Gold System, Indonesia...280
Ikrar Mandiri

P2.120 Prolonged Mineralization for the Formation of the Southeast Asian Tin Province...281
Wei Mao

P2.121 The Distribution of Co and Ni in Sulfide and Arsenide Minerals at the Eloise ISCG Deposit, Queensland...282
Elizabeth de las Mercedes Marino Morejon (Student)

P2.122 Fingerprinting Intrusion-Related Versus Orogenic Gold, Western Lachlan Orogen, Tasmanides, Australia...283
Helen McFarlane

P2.123 Critical Minerals in Laramide Porphyry Copper and Paleogene Molybdenum-Tungsten Deposits in Southwestern New Mexico, USA...284
Virginia McLemore

P2.124 Alteration, Mineralization and Geochemical Characteristics of the Arakompa Intermediate Sulfidation Epithermal Gold Deposit, Kainantu District, Papua New Guinea...285
Owen Missen (Student)

P2.125 Geochemical Characteristics of Copper Mineralization and Exploration History at the Otavi Mountainland, Namibia...286
Takao Miura (Student)

Posters, continued

P2.126 Magnetite Trace Elements as Petrogenetic Indicators in Exploring Savage River Deposit, Northwest Tasmania...287
Pratichee Mondal (Student)

P2.127 Miocene Low-Sulfidation Epithermal Deposits of the Colorado Extensional Corridor, USA...288
Thomas Monecke

P2.128 Lithium Brine Exploration in La Unión, Ahumada, Chihuahua, México...289
Héctor Montes Carreón (Student)

P2.129 The Role of Oligocene Volcanism to Develop Ag-Au-Pb-Zn-Cu Mineralization at San Marcial Area, Western Sierra Madre Occidental, Sinaloa, Mexico...290
Paula Montoya Lopera

P2.130 The State of Lithium Resources in 2024: A Deposit-Type Analysis for Supply Potential...291
Brock Moody (Student)

P2.131 Determination of Migration Events of Present Hidrocarbons in the Eastern Llanos Basin, Through the Characterization of Fluid Inclusions...292
Gabriel David Moreno Aguilar (Student)

P2.132 Understanding the Importance of Heavy Mineral Sands as a Source of Rare Earth Elements, Using a Mineral Systems Approach...293
Sidy Morin-Ka (Student)

P2.133 When Dates Lie: The Impact of Mineral Porosity on Isotopic Dating of Base Metal Deposits...294
Michelle Moxey (Student)

P2.134 Lithium Oxide in a Weathered Pegmatite in Nyanza District, Rwanda: A Preliminary Data...295
Theophile Mugerwa (Student)

P2.135 Structural Architecture of the Kirkland Lake Mining District...296
Taylor Mugford (Student)

P2.136 On the Nickel-Sulfide Potential of Late-Archean Komatiites from the Dharwar Craton: A Case Study from the Shankaraghatta Complex (Southern India)...297
Ria Mukherjee

P2.137 Geology and Polymetallic Mineralization of the Tsagaan Tolgoi Skarn Deposit, South Mongolia...298
Otgonbayar Munkhtsol (Student)

P2.138 Exploring Two-Liquid Separation of Fluorosilicate Melts: Advancing Phase Relationships and Mineralization Insights for Granite Hosted REE Deposits...299
William Munro (Student)

P2.139 Genesis of a 3.53 Ga Volcanogenic Massive Sulfide Deposit in the Nondweni Greenstone Belt, Kaapvaal Craton...300
Thendo Netshidzivhe

P2.140 Two-Stage Ore-Forming Process for Bonanza-Grade Gold Veins in the Brucejack Au-Ag Deposit, British Columbia, Canada...301
Kevin Man Hoi Ng (Student)

P2.141 A Remote-Predictive Geologic Map of the Mid-Atlantic Ridge Between 22.6°N - 19.6°N: Implications for Tectonic Evolution and Massive Sulphide Mineralization...302
Sam Nichols (Student)

P2.142 Characteristics of the Epithermal Veins in the South Teresa Deposit, Lepanto, Mankayan District, Philippines...303
Shion Nohara (Student)

P2.143 Rhenium Deportment in Samples from the Spremberg-Graustein-Schleife Kupferschiefer Deposit...304
Zahra Nourizenouz (Student)

P2.144 Tracing Cu-Au Ore-Forming Processes in the Proterozoic Elaine Dorothy Skarn Deposit, Mount Isa (Queensland, Australia), Using In-Situ Sulphide Mineral Compositions...305
Catherine Nyakecho (Student)

P2.145 Multi-Method Comparison of Mineral Abundance Estimates for the Sandsloot PGE-Ni-Cu Deposit, Platreef, Northern Bushveld Complex...306
Giovana Oliveira Pimentel (Student)

P2.146 Geochemical Signatures of Apatite in IOCG Systems in the Cloncurry District, Queensland...307
Ernest Boamah Opoku

P2.147 Basin Reconstruction and Dissecting Vein Population: Implications for Exploration of Sedimentary-Hosted Copper Deposits...308
Chirantan Parui

P2.148 Using Apatite Chemistry to Identify Proximity to Copper at the Nifty Sediment-Hosted Cu Deposit, Western Australia...309
Jennifer Porter

P2.149 Magmatic-Hydrothermal History at the Antakori Cu-Au-Ag Project, Northern Peru: New Ages from U-Pb and Re-Os Geochronology...310
David Portocarrero

P2.150 Multi-Scale Structural and Geochemical Analysis of Gold Mineralization at the REN Deposit, Carlin Trend, Nevada...311
Amber Prevallet (Student)

P2.151 Influence of Serpentinization and Mass Balance Constraints on Ni-Co Enrichment in Laterite Profiles in Pomalaa, Sulawesi, Indonesia...312
Muhammad Fatih Qodri

P2.152 Structural Controls and Tectonic Setting of Gold Mineralisation in the Broken River Orocline, Northeastern Australia...313
Raiza Rodrigues (Student)

P2.153 Structural Analysis of Polyphase Deformation in the South Peruvian Andes and Its Basement: Implications for Cu-Au Mineral Deposit Localization...314
Maryi Rodriguez Cuevas (Student)

P2.154 Newly Discovered Cu-Ni-Sulfide Mineralisation in the 1.8 Ga Stendalen Mafic Intrusion, Ketilidian Orogen, South Greenland...315
Matthew Roper (Student)

P2.155 Understanding Hazelton Group Metallogeny Using Detrital Zircon Geochronology Across Multiple Mineral Districts, Golden Triangle, NW British Columbia...316
Kate Rubingh

P2.156 Epidote as a Discriminator of Skarn Deposits: Insights from Dolphin W Skarn, King Island, Tasmania...317
Angela Isaura Santos Costa (Student)

P2.157 Paragenesis and Trace Element Signatures of Pyrite from Paramanahalli Gold Deposit, Chitradurga Greenstone Belt, Karnataka, India: Implications for Ore Genesis...318
Manju Sati

P2.158 Ore-Forming Fluid Characteristics and P–T Condition of Gold Mineralization at Paramanahalli, Chitradurga Greenstone Belt, Karnataka, India...319
Manju Sati (Student)

P2.159 Porphyry Cu-Au-Ag Mineralization of the Haman Alkaline-Carbonatite Intrusive Complex, South Korea...320
Chris Sennitt

P2.160 Intermediate Sulphidation Polymetallic Epithermal Mineralization at Dongil, South Korea...321
Chris Sennitt

P2.161 The Role of Structural Inheritance in Gold Mineralisation Along the Gondwana Supercontinent Margin: Insights from Ravenswood, Queensland...322
Hugo Serra (Student)

P2.162 Spatial Variation of Flake Graphite Grade and Crystallinity in the Namapwia Prospect, Ruangwa District, Lindi, Tanzania...323
Mikidadi Shaha (Student)

P2.163 Apatite Geochemistry and Geochronology to Distinguish Between the Fluids Responsible for the Formation of Eloise ISCG Deposit, Australia...324
Swastik Shinde (Student)

P2.164 Lithostructural Study of Mineralized Formations in the Kéniéba District...325
Nouhoum Noumouké Sidibé (Student)

P2.165 Mineralogy and Geochemistry of the Yellowknife Lithium Pegmatite Field...326
Philipp Siebold (Student)

P2.166 Using In Situ Mineral Chemistry to Extend the Detectable Lithogeochemical Footprint of Magmatic Sulfide Orebodies...327
William Smith (Student)

P2.167 Nb-Ta Mineralisation in Pegmatite from the Amapá State, Brazil: First Constraints from U-Pb LA-ICP-MS Geochronology and Mineral Composition on Columbite-Tantalite...328
Carlos Spier

P2.168 Fourier Transform Infrared Spectroscopy for Geometallurgy Modelling - GRE46 Gold Deposit (NSW, Australia)...329
Markus Staubmann (Student)

P2.169 Andradite U-Pb Dating at the Ernest Henry IOCG deposit, NW Queensland: Constraints on the Timing of Cu-Au Mineralization...330
Je rey Steadman

P2.170 Development of an Empirical Nickel Targeting Model for Exploration in a Structurally Complex, Poly-Deformed Terrane...331
Benjamin Stoch

P2.171 Structural Analysis of the Hyde-Macraes Shear Zone Hosted Gold Deposit, Aotearoa-New Zealand...332
Madi Styles (Student)

P2.173 Geological Characterization of an Archean Oxidized Intrusion-Related Au-Cu(-Mo)...333 Deposit: The Upper Beaver Deposit, Kirkland Lake, Canada
Jonathan Sutton (Student)

P2.174 Trace Element and Sulfur Isotope Characteristics of the Star of Mangaroon Gold Deposit, Western Australia...334
Halik Rifki Daeng Taha (Student)

P2.175 Geochemical Characteristics and Behavior of Hydrothermal Fluids Related to Porphyry Cu-Au Mineralization of the Tujuh Bukit Deposit, Banyuwangi, Indonesia...335
Ran Takeda (Student)

P2.176 A Structural Ore-Control Model for Carlin-Type Gold Deposits, Youjiang Basin, China...336
Qinping Tan

P2.177 What Controls High-Grade Carlin-Type Gold Deposits?: Structural and Geochemical Factors from the Leeville Underground Complex, North Carlin Trend, Nevada, USA...337
Koray Tasbicen (Student)

P2.178 Sulfur Mineralizations of the Tichibanine Deposit (Bou-Azzer Inlier, Central Anti-Atlas, Morocco): Geology, Petrographic Investigations, and Sulfide Ore Chemistry...338
Lucas Tatnell

P2.179 Reinterpretation of Field Data from Ag-Pb-Zn (F-Ba) Mineralization in the Sierra de Santa Rosa, Melchor Muzquiz, Coahuila, Mexico...339
Jose Tavizon (Student)

P2.180 Unravelling Rubidium Enrichment in Highly Fractionated Pegmatites: Insights from the High-Grade Mt Edon Deposit, Western Australia...340
Yoram Teitler

P2.181 Molybdenum Isotope Compositions and Formation Mechanisms in Metallic Sulfides in Porphyry Deposits...341
Yamei Tian (Student)
P2.182 Geochemical Signatures from Porphyry Cu Systems in Sonora, Mexico...342
Oscar Daniel Torres-Alarcon (Student)
P2.184 Quantitative Mineralogical Characterisation of Complex Ores to Interpret Behavior in Processing...343
Djoan Kate Tungpalan
P2.185 Mineralogy and Geochemistry of REE-Enriched Cryolite-Molybdenite Melt in a Porphyry Molybdenum Deposit, Trans-Pecos Texas, USA...344
Mert Ugurhan
P2.186 Interrogating the Plutonic-Volcanic Connections for Economic Copper Deposit Potential Across the Sunda-Banda Arc...345
Sri Budhi Utami
P2.187 Defining the Syn-Kinematic, Amphibolite-Facies Sequeirinho-Style IOCGs, Carajás Mineral Province...346
Charlys V. S. Neves (Student)
P2.188 Nb and Zr Mineralisation Driven by Carbonatite Metasomatism at the Mt Weld Carbonatite, Western Australia...347
Arthur Vicentini de Oliveira (Student)
P2.189 Beneath the Surface: The Geology of the Magdalena VHMS Deposit, Iberian Pyrite Belt, Spain...348
Poliana Vidal Salgado (Student)
P2.190 Paragenesis and Hydrothermal Alteration Characterization Using Short-Wave Infrared (SWIR) in the Intermediate Sulfidation Epithermal Deposit of Ajaruni, Huacullani, Puno, Peru...349
Brayan Paul Vilcanqui (Student)
P2.191 Exploration of Lithium Minerals in Lead-Zinc-Barites Mineralization at Ubaru Area, Ebonyi State, Southeastern Nigeria...350
Paul Visaa (Student)
P2.192 Apatite Petrochronology and Geochemistry – Tracking Mineral System Evolution at the Ernest Henry Iron Oxide-Copper-Gold (IOCG) Deposit, Northwest Queensland, Australia...351
Xin Wang (Student)
P2.193 Fertile or Futile? Using Plagioclase Composition and Texture to Assess Copper Fertility in Magmas...352
Catharine Wang (Student)
P2.194 In-Situ Geochemical Signatures of Apatite in Carbonatite-Related REE Deposits: Insights into Magmatic Sources and Fluid Overprints...353
Chunwan Wei
P2.195 Episodic Magmatism Contributes to Sub-Seafloor Copper Mineralization: Insights from Textures and Geochemistry of Zoned Pyrite in the Ashele VMS Deposit...354
Bing Xiao
P2.196 Structural and Hydrothermal Controls on Mineralization of the Bronson Slope Porphyry Cu-Au-Mo Deposit, British Columbia: Implications for Exploration Targeting...355
Luis Yagual (Student)
P2.197 Structural Model of the La Plata VHMS, in the Western Cordillera of Ecuador...356
Luis Yagual (Student)
P2.198 Magma Fertility Assessment in the Atacama Region: Implications for Cu and Au Prospectivity and Geodynamics During the Cretaceous...357
Colomba Trinidad Yañez Labarca (Student)
P2.199 Late Jurassic Epithermal Ag-Au Mineralization in Southeast Coastal China: Evidence from U-Pb Dating of Hydrothermal Monazite at the Dongji Deposit...358
Lina Yi

P2.200 Syn- and Late-Orogenic Hydrothermal Si-Al-Ca Alteration of Ultramafic Rocks: Evidence for Ni-Co Downgrading in Outokumpu, Eastern Finland...359
Christoforos Zamparas (Student)

P2.201 Devolatilization of Carbonate-Rich and Carbonaceous Rocks as Fluid Source for the Outokumpu Cu-Zn-Co-Ni-Ag-Au Deposit, Eastern Finland...360
Christoforos Zamparas (Student)

P2.202 Fluid-Rock Interaction Recorded in Apatite: Implications for Source and Evolution of Ore-Forming Fluids in the World-Class Jiaodong Gold Deposits, China...361
Liang Zhang

P2.203 Genesis of Pegmatite-Type High-Purity Quartz Deposits in Tysfjord, N Norway...362
Haoyang Zhou

Investment and Policy

P3.204 Zr-Nb-REE Enriched Alkaline Silicate Magmatism in Victoria...363
Tom Andrews

P3.205 Building Recognition, Trust and Confidence through Good Practice: An Example from the Research Ireland Centre for Applied Geosciences (iCRAG)...364
Shonny Lehane

P3.206 Quantifying Economic Value of Air Quality Improvement Schemes in Quarry Mining: A Social and Environmental Perspective...365
Tochukwu Ngwu (Student)

P3.207 The Immortality of Hydrocarbons in an Era of Green Industrialisation: The Case of 'Drill Baby Drill'...N/A
Shepherd Nyambe (Student)

P3.208 Systems-Thinking to Tackle the Impending Economic Geologist Crisis...366
Elizabeth Sharman

P3.209 Resourcing Western Australia - Mineral, Geochemical and Isotopic Datasets to Accelerate the Discovery of Critical Mineral Resources in Western Australia...367
Rose Turnbull

Secondary Metals and Remediation

P4.211 Secondary prospectivity of mine wastes from the Thalanga, Highway-Reward and Balcooma volcanic-hosted massive sulphide deposits...368
Rosie Blannin

P4.212 Cobaltite-Pyrite-Pyrrhotite Association: Triassic Metasedimentary Rocks in the South Kitakami Belt, Japan...369
June Born Born (Student)

P4.213 Authigenic Iron Oxyhydroxide Rims Attenuate Deleterious Element Fluxes During Sulphide Oxidation in Historical Gold Mine Tailings...370
Steve Chingwaru (Student)

P4.214 Process Mineralogy of Historical Sn Mine Waste: Recovery of W, In, and Sn from the Herberton Mineral Field, Australia...371
Steve Chingwaru (Student)

P4.215 Geoenvironmental Characterisation of the Abandoned Elverdton Mine Tailings, Western Australia...N/A
Pamela Curi Rivera (Student)

P4.216 Understanding Bismuth as Critical Mineral: From Source to Sink...372
Cendi Dana (Student)

P4.217 Mineralogical Controls on Critical Elements (Bi, In, Te) and Ag Deportment in the Ruwai Zn-Pb-Ag Skarn Deposit, Central Kalimantan, Indonesia...373
Shelly Mardhia Faizy (Student)

P4.218 Natural Leachability of Critical Metals - Investigating AMD and Metal Recovery at Normanby Mine...374
Afreen Iqbal (Student)

P4.219 From Source to Sink, Geochemical and Mineralogical Dynamics of Sediment-Associated Metal(loid)s in an Acid Mine Drainage-Affected River...375
Elin Jennings

P4.220 Extracting Critical Metals from Mine Waste Using Molten Alkali Salts...376
Kathryn Keane (Student)

P4.221 DERT (Diagnostic Environmental Reporting Tool): A Unified Geospatial Platform Transforming Mine Remediation and Environmental Compliance...377
Fabian Kohlmann

P4.222 Using Vegetation as a Tool for Environmental Monitoring in Legacy Mines...378
Maria Mavroudi (Student)

P4.223 Mine Waste as a Resource: Critical Minerals in Mine Wastes in New Mexico U.S.A....379
Virginia McLemore

P4.224 Cobalt Trace Element Mineralogy and Geometallurgy in the Mount Lyell Copper Mine Tailings, Tasmania, Australia...380
Owen Missen (Student)

P4.225 Recycling Lithium-Rich Wastes: Perspectives from Mine Waste Rock and Li-Rich Enamels and Lithium Aluminosilicate Glass (LAS) Industrial Non-Compliant Materials...381
Guia Morelli

P4.226 Process Mineralogical Constraints on the Recovery of Platinum Group Metals via Direct Leaching of UG2 Ore in the Bushveld Complex...382
Nonhlanhla Ndlovu (Student)

P4.227 Antimony Fate During Schwertmannite Transformation: Implications for Passive Remediation in Acid Mine Drainage–Impacted Environments...383
Mohammad Rastegari (Student)

P4.228 Evaluation of the Geochemistry and Mineralogy of Mine Waste in Nevada, USA...384
Joseph Rosal (Student)

P4.229 Geoenvironmental indicators using line scan hyperspectral and XRF core analyses for early AMD prediction in Northern Queensland ore deposits, Australia...385
Enrique Saez Salgado (Student)

P4.230 Predictive Geometallurgy and Mine Waste Valorisation: Circular and Eco-friendly Recovery of Gold in the Witwatersrand Goldfields (South Africa)...386
Xolile Carol Simelane (Student)

P4.231 Metal Isotopic Monitoring of Environmental, Extraction, Distribution of Metal Associated with Mine Practice...387
Adam Simon

P4.232 Indian Iron Ore Waste Mine tailings for Critical (Li) and REE Minerals -Case Studies...388
Perumala Venkata Sunder Raju

P4.233 Unlocking Australia's Re-Mining Potential...389
Jane Thorne

P4.234 Drone-Based VNIR–SWIR Hyperspectral Imaging for Mapping and Monitoring Environmental Conditions in Legacy Mine Sites...390
Victor Tolentino

P4.292 Repurposing Historical Iron Slag as a Sustainable Source of Rare Earth Elements: A Case Study from Centre Furnace in Pennsylvania...391
Jacqueline Silva (Student)

Technology

P5.235 A New Discovery Frontier for Copper/BHT Deposits in the Thomson Orogen Revealed by Merging Public Research and Precompetitive Data...392
John Anderson

P5.236 REE Pattern Shape Analysis Using Lambda Coefficients...393
Michael Anenburg

P5.237 Spectral and Structural Analysis of Satellite Imagery for Lithium Pegmatite Exploration in Northwestern Somalia...394
Hooshang Asadi Haroni

P5.238 Rapid Fingerprinting of Cassiterite Paragenesis via µXRF for Exploration of Li Pegmatites...395
Jason Bennett

P5.239 Fast-Tracking Brownfields Exploration at Geita: Unleashing the Power of 3D Seismic and Cutting-Edge Seismic Innovations...396
Sian Bright

P5.240 Machine Learning Approach to Automated Mineral Identification from RGB Reflected Light Microscopy...397
Kevin Brown (Student)

P5.241 Exploring Regolith-Hosted Rare Earth Element Deposits: A Practical Methodology for Mineral Exploration...398
Nicolás Bustos (Student)

P5.242 Hydrothermal Alteration Zones Wstimated by ASTER Satellite Image and ASD TerraSpec on the Eastern Zone of Cerro de Pasco, Peru...399
Evelyn Caiza (Student)

P5.243 LoopConverter: A Tool to Standardise Geological Data for 3D Geological Modelling...400
Rabii Chaarani

P5.244 New Insights of Alteration Mineral Mapping in the Skarn-Porphyry Systems...401
Huayong Chen

P5.245 Microinclusion Geochemistry in White Micas from Porphyry Deposits and Transitional Environments: A Case Study from Globe-Miami Copper District, USA...402
Axel Cima

P5.246 Breaking Through the Basalt Barrier: Unveiling the Mt Read Volcanics with Ambient Noise Tomography...403
Tjaart De Wit

P5.247 Regolith Landform Mapping Using Support Vector Machine and Artificial Neural Network: Study Case of Sissingue Gold Project in Côte d'Ivoire...404
Hamed Aly Diaby (Student)

P5.248 Planning, Calibrating and Running a Drone Magnetic Survey on a Gold Project Côte d'Ivoire Central Part...405
Hamed Aly Diaby (Student)

P5.249 Mineral Mapping Through a Ni-Cu Ore-Bearing Intrusion...406
Lily Dickson (Student)

P5.250 *Why a Foundational Data Layer is Critical to Unleash AI for Mineral Exploration and Mining...407*
Mason Dykstra

P5.251 Rapid Cu-Porphyry Indicator Mineral Characterization by µXRF: A Case Study Investigating Benchtop-µXRFs as a Prospective Automated Indicator Mineral Analytical Tool...408
Ben Eaton (Student)

P5.252 Fluorite Deposits in Mongolia: Geochemical Characterization and Exploration Implications...409
Jamsran Erdenebayar

P5.254 Selective Calibrating Phosphatic Ore Resource Models to Mill Feedstock Head Grade Accurately Using Machine Learning...410
David M. First

P5.255 Thermodynamic Modelling and Trace Element Partitioning as Predictive Tools for REE-Enrichment in Alkaline-Silicate Igneous Systems...411
Corinne Frigo

P5.256 Smarter Use of Geochemistry and Machine Learning to Support Geological Domaining and Geometallurgical Modelling of Porphyry Cu Deposits...412
Melissa Gregory

P5.257 Ore Vectoring Using Machine Learning: Case Study of Copper Occurrences in Northern Ireland...413
Felix Hasudungan (Student)

P5.258 Anomaly Detection Using Variogram and Machine Learning Can Reveal Potential Orefields: Case Studies from the United Kingdom and Australia...414
Felix Hasudungan (Student)

P5.259 Spectral and Compositional Characterization of Lithium-Bearing Zinnwaldite Mineralization in Eastern Tasmania...415
Wei Hong

P5.260 Utilising In-situ Lu-Hf Dating to Constrain Timing of Mineralisation in the Mount Isa Inlier: A Case Study from Maronan...416
Helena Hordern (Student)

P5.261 Revealing High Geological Spatial Variability at Bench Scale – Implications Upstream and Downstream in the Mining Value Chain...417
John Jackson

P5.262 Innovative Remote Sensing and GIS-Based Approach for the Identification and Mineralogical Characterization of Mine Tailings in South Africa...418
Gaone Joseph (Student)

P5.263 Capabilities of Advanced Mineral Identification and Characterisation System (AMICS)...419
Andrew Kostryzhev

P5.264 Novel Applications of Time of Flight (TOF) Mass Spectrometry to Compositional Mapping...420
Alice Macdonald

P5.265 Getting The Most out of Drill Hole Data: Meaningful and Robust Boundary Detection Tools...421
Ryan Manton

P5.266 Systematics and Mineral Systems: How Controlled Vocabularies Enable System Mapping for Sedimentary-Hosted Copper Systems...422
Helen McFarlane

P5.267 Identifying Mineralisation Within the Sunrise Dam Gold Mine Using Geomechanical Targeting Technology...423
John Mclellan

P5.268 Meso-Scale Imaging and Whole-Rock Quantification of Indium in Mine Waste from a Granite-Related Deposit Using Laser-Induced Breakdown Spectroscopy (LIBS)...424
Olivia Mejías (Student)

P5.269 Enhancing Usability of Public Data: 3D for Free...425
Gem Midgley

P5.270 Automated Quantitative Mineralogy Applied to the Hydrothermal Footprint of the Paulo Afonso Cu Deposit, Carajás Mineral Province...426
Victor Nogueira (Student)

P5.271 Refining the understanding of the hydrothermal systems footprint using automated quantitative mineralogy in the Carajás Mineral Province...427
Victor Nogueira (Student)

P5.272 Artificial Intelligence in Mining: Evaluating Efficiency, Safety, and Productivity through Technological Integration for Global Sustainability...N/A
Shepherd Nyambe (Student)

P5.273 Uncovering Subsurface Polymetallic Mineralization with Machine Learning: A Geochemical Approach in Rudabánya, Hungary...N/A
Mohamed Abdelnaby Oraby (Student)

P5.274 Mineralogical and Textural Characterization of the Ag-Pb-Cu-Au Maronan Deposit (NW-QLD) Using Laser-Induced Breakdown Spectroscopy (LIBS)...428
Daniel Patias

P5.275 Advanced Hydrogeological Modeling and Sensitivity Analysis to Optimize Dewatering and Characterize Water Flow at the Mingomba Cu Deposit...429
Gloria Quispe Oruro (Student)
P5.276 Data-Driven Geometallurgy at Tujuh Bukit: Integrating Hyperspectral Core Imaging, SEM/MLA, Dynamic Rebound Hardness, and Machine Learning for Processing Design Optimisation...430
Mochammad Hafid Rahadi
P5.277 Using PHAST Reaction-Transport Modeling to Study the Formation of Zebra Limestone...431
Andrea Salas Rivera (Student)
P5.278 From Basement Rocks to Mineral Systems: Demonstrating the Value of In-Situ Zircon (U-Th)/He Thermochronology...432
Paola Salas Rivera (Student)
P5.279 UAV-borne Hyperspectral Data Across Tailings and Waste at Memi Mine in the Republic of Cyprus...433
Katerina Savinova
P5.280 The Application of Unsupervised Machine Learning to Identify Geochemical Associations with Geological Domains of Nickel Laterite Deposit in Sorowako, Indonesia...434
Alya Sevy Setiawan (Student)
P5.281 Determination of Iron Oxidation State in Biotite Using Electron Probe Microanalysis...435
Shahed Shahrestani (Student)
P5.282 Automated Coal Picking in MWD Logs...436
Katie Silversides
P5.283 Breakthrough Hydrogeochemical Techniques for Exploration and Heap Leaching...437
Adam Simon
P5.284 High Velocity = High Prospectivity? A Seismic Window into Southeast Tasmania's Natural Hydrogen Potential...438
George Taylor
P5.285 Multispectral Survey Based on Remote Sensing for Chromite Prospecting in Oman Ophiolite...439
Cezar Augusto Teixeira Falcão Filho
P5.286 Mapping Hydrothermal Alteration for Gold Prospecting in British Columbia, Canada: Integrating Multispectral Remote Sensing and Geophysical Data...440
Cezar Augusto Teixeira Falcão Filho
P5.287 Machine Learning-Driven Geometallurgical Evaluation of the Lowermost Manganese Bed in the Kalahari Manganese Field, South Africa...441
Louis van Eeden (Student)
P5.288 A Multi-Scale Integrated Geological Modeling Approach of the Northern Norrbotten Ore Province, with a Focus on Iron Oxide-Apatite Mineral Systems...442
Ervin Veress (Student)
P5.289 Application and Optimization of Blasting Parameters in Selected Rock Quarries of Southwest and Northcentral Nigeria...443
Paul Visaa (Student)
P5.290 Pyrite Trace Element Behavior in IOCG Deposits: An Unsupervised Learning Approach for Vectoring and Ore Paragenesis...444
Shouchun Yu (Student)
P5.291 An Apatite to Unravel Magma Chamber Dynamics, Mantle Sources and Metallogenesis in the Bushveld Complex...445
Peace Zowa (Student)

Additional Posters

The Genesis of Podiform Chromitites From the Baryulgil Serpentinite, NSW, Australia...257
Tim Huxley

Connection and Collaboration - Building a Resilient Geoscience Community and Future Workforce...446
Ned Howard

Speed Talks (1-Minute Breaks Between Talks)

Saturday, September 27

Morning Talks 10:35 AM - 11:00 AM	
10:35 AM - 10:40 AM	Andradite U-Pb Dating at the Ernest Henry IOCG deposit, NW Queensland: Constraints on the Timing of Cu-Au Mineralization (see page 330) *J. Steadman*
10:41 AM - 10:46 AM	Geology and Age of the Granite Related Metasomatic Overprinting of the Rosebery VHMS Deposit (see page 243) *Vinicius Godoi Pereira Da Cruz*
10:47 AM - 10:52 AM	Tracing Cu-Au Ore-Forming Processes in the Proterozoic Elaine Dorothy Skarn Deposit, Mount Isa (Queensland, Australia), Using In-Situ Sulphide Mineral Compositions (see page 305) *Catherine Nyakecho*
10:52 AM - 10:53 AM	Morning Wrap-up
Late Afternoon Talks 3:35 PM - 4:00 PM	
3:35 PM - 3:40 PM	The Application of Unsupervised Machine Learning to Identify Geochemical Associations with Geological Domains of Nickel Laterite Deposit in Sorowako, Indonesia (see page 434) *Alya Sevy Setiawan*
3:41 PM - 3:46 PM	Ore Vectoring Using Machine Learning: Case Study of Copper Occurrences in Northern Ireland (see page 413) *Felix Sihombing*
3:47 PM - 3:52 PM	Breakthrough Hydrogeochemical Techniques for Exploration and Heap Leaching (see page 437) *Adam Simon*
3:53 PM - 3:58 PM	Thermodynamic Modelling and Trace Element Partitioning as Predictive Tools for REE-Enrichment in Alkaline-Silicate Igneous Systems (see page 411) *Corinne Frigo*
3:58 PM - 3:59 PM	Late Afternoon Wrap-up

Sunday, September 28

Morning Talks 10:35 AM - 11:00 AM	
10:35 AM - 10:40 AM	Deciphering the Role of Remobilisation in the Formation of High-Grade Gold Deposits (see page 176) *Sumail*
10:41 AM - 10:46 AM	Applications of Mineral Chemistry to Petrogenesis and Exploration in the Borralan Carbonatite Complex and Associated Alkaline Silicate Rocks (see page 178) *Abdulkadir Mohamed Abdulkadir*
10:47 AM - 10:52 AM	Chile's First Defined Intrusion-Related Gold System: Ternera's Hydrothermal Footprint in the El Zorro Gold District (see page 237) *Eduardo Fritis*
10:52 AM - 10:53 AM	Morning Wrap-up
Afternoon Talks 12:45 PM - 1:45 PM	
12:45 PM - 12:50 PM	National-Scale Critical Mineral and Strategic Material Potential Assessments to Support Decision-Making in the Net Zero Transition (see page 171) *Arianne Ford*
12:51 PM - 12:56 PM	Empowering Exploration with National Scale Maps (see page 173) *Nichole Knepprath*
12:57 PM - 1:02 PM	Archaean BIFs from Karnataka, India: genetic insights from the Dharwar craton, India, and implications for post-depositional hematitic iron deposits (see page 175) *Perumala Venkata Sunder Raju*
1:03 PM - 1:08 PM	Chromite and Sulphide Mineralization of the Uitloop Ultramafic Bodies in the Northern Limb of the Bushveld Complex, South Africa (see page 174) *Mabatho Mapiloko*

1:09 PM - 1:14 PM	Evaluating the Tectonic and Climatic Controls on Sediment-Hosted Ore Deposits in Deep Time (see page 166) *Sheree Armistead*
1:15 PM - 1:20 PM	Building Recognition, Trust and Confidence through Good Practice: An Example from the Research Ireland Centre for Applied Geosciences (iCRAG) (see page 364) *Shonny Lehane*
1:21 PM - 1:26 PM	Zr-Nb-REE Enriched Alkaline Silicate Magmatism in Victoria (see page 363) *Tom Andrews*
1:27 PM - 1:32 PM	Resourcing Western Australia - Mineral, Geochemical and Isotopic Datasets to Accelerate the Discovery of Critical Mineral Resources in Western Australia (see page 367) *Rose Turnbull*
1:33 PM - 1:38 PM	Systems-Thinking to Tackle the Impending Economic Geologist Crisis (see page 366) *Elizabeth Sharman*
1:39 PM - 1:44 PM	The Immortality of Hydrocarbons in an Era of Green Industrialisation: The Case of 'Drill Baby Drill'...N/A *Shepherd Nyambe*
1:44 PM - 1:45 PM	Afternoon Wrap-up

Monday, September 29

Morning Talks 10:35 AM - 11:00 AM	
10:35 AM - 10:40 AM	Secondary Prospectivity of Mine Wastes from the Thalanga, Highway-Reward and Balcooma Volcanic-Hosted Massive Sulphide Deposits (see page 368) *Rosie Blannin*
10:41 AM - 10:46 AM	Cobalt Trace Element Mineralogy and Geometallurgy in the Mount Lyell Copper Mine Tailings, Tasmania, Australia (see page 380) *Owen Missen*
10:47 AM - 10:52 AM	Geoenvironmental indicators using line scan hyperspectral and XRF core analyses for early AMD prediction in Northern Queensland ore deposits, Australia (see page 385) *Enrique Saez Salgado*
10:53 AM - 10:58 AM	Geoenvironmental Characterisation of the Abandoned Elverdton Mine Tailings, Western Australia...N/A *Pamela Curi Rivera*
10:58 AM - 10:59 AM	Morning Wrap-up
Afternoon Talks 12:45 PM - 1:45 PM	
12:45 PM - 12:50 PM	From Source to Sink, Geochemical and Mineralogical Dynamics of Sediment-Associated Metal(loid)s in an Acid Mine Drainage-Affected River (see page 375) *Elin Jennings*
12:51 PM - 12:56 PM	Process Mineralogical Constraints on the Recovery of Platinum Group Metals via Direct Leaching of UG2 Ore in the Bushveld Complex (see page 382) *Nonhlanhla Ndlovu*
12:57 PM - 1:02 PM	Machine Learning Approach to Automated Mineral Identification from RGB Reflected Light Microscopy (see page 397) *Kevin Brown*
1:03 PM - 1:08 PM	Getting the most out of drill hole data: meaningful and robust boundary detection tools (see page 421) *Ryan Manton*
1:09 PM - 1:14 PM	Smarter Use of Geochemistry and Machine Learning to Support Geological Domaining and Geometallurgical Modelling of Porphyry Cu Deposits (see page 412) *Melissa Gregory*
1:14 PM - 1:15 PM	Afternoon Wrap Up

SEG 2025 CONFERENCE PROGRAM

SEG
2025

BRISBANE, AUSTRALIA

September 26-29, 2025

seg2025.org

SOCIETY OF ECONOMIC GEOLOGISTS
FOUNDED 1920

Conference Venue

Brisbane Convention & Exhibition Centre
Merivale St
South Brisbane, QLD 4101
Australia

Wi-Fi:
Sponsored by AngloGold Ashanti Australia

ANGLOGOLD ASHANTI
AUSTRALIA

Wi-Fi details at registration

Parent's room: See registration desk for locations

Speaker Ready Room (M10 Mezz level)

Thursday, September 25:	3:00pm-6:00pm
Friday, September 26:	7:00am-5:00pm
Saturday, September 27:	7:00am-5:00pm
Sunday, September 28:	7:00am-5:00pm
Monday, September 29:	7:00am-5:00pm

Registration (Main Foyer)

Friday, September 26:	7:00am-6:00pm
Saturday, September 27:	7:00am-6:00pm
Sunday, September 28:	7:00am-6:00pm
Monday, September 29:	7:00am-12:00pm

EXHIBITS AND POSTERS (EXHIBIT HALL, Q1 & Q2)		
Thursday, September 25	3:00pm–6:00pm	Exhibit Move In (Poster and Exhibitor Only Access)
Friday, September 26	8:00am-3:00pm 6:00pm-6:30pm 7:00pm-9:00pm	Exhibit Move In (Poster and Exhibitor Only Access) Exhibitor Only Access Posters and Exhibits Open-Welcome Reception
Saturday, September 27	9:00am-10:00am 10:00am-7:30pm 5:30pm-7:30pm	Exhibitor Only Access Exhibits Open Exhibit and Poster Reception
Sunday, September 28	9:00am-10:00am 10:00am-7:30pm 5:30pm-7:30pm	Exhibitor Only Access Posters and Exhibits Open Exhibit and Poster Reception
Monday, September 30	9:00am-10:00am 10:00am-2:00pm 2:00pm-4:00pm	Exhibitor Only Access Posters and Exhibits Open Exhibits Dismantle (Exhibitor Only Access)

Acknowledgment of Country

We respectfully acknowledge the Traditional Custodians of the land on which we will gather, the Jagera and Turrbal of Meanjin (Brisbane). We honor their enduring connection to the land, waters, and skies and pay our respects to their Elders past, present, and emerging. As we convene for this conference, we recognize the importance of Indigenous knowledge in fostering sustainable and respectful resource practices. May this conference be an opportunity to foster understanding, collaboration, and shared stewardship of our natural resources.

The information in this program is current as of September 15, 2025.
Please see the SEG 2025 Attendee Hub for updated information.

seg2025.org/hub

SEG 2025 Schedule at a Glance

THURSDAY, SEPTEMBER 25

8:30am-5:30pm
Pre-Conference Workshops

6:00pm - 10:00pm
Student and Early Career Social

FRIDAY, SEPTEMBER 26

8:00am-5:30pm
Pre-Conference Workshops

8:30am-4:30pm
Student and Early Career Events

5:00pm-5:15pm
Welcome to Country

5:15pm-6:30pm
Presentation and Panel Discussion

6:30pm-7:00pm
SEG Presidential Address

7:00pm-9:00pm
Exhibits and Posters Open

7:00pm-9:00pm
Welcome Reception - Sponsored by BHP Xplor

SATURDAY, SEPTEMBER 27

8:00am-9:30am
Opening and Plenary Sessions

9:30am-10:30am
- Session 1: Lithosphere and Mineralization
- Session 2: Tectonics and Porphyries
- Session 3: 3D Exploration and Prospectivity

10:30am-11:00am
Break, Poster and Speed Talks in Exhibit Hall

11:00am-12:30pm
BHP Xplor

12:30pm-2:00pm
BHP Xplor Showcase and Sponsored Lunch

2:00pm-3:30pm
- Session 4: Orogenic Gold
- Session 5: Paterson to Global Cu (Au-Pb-Zn)
- Session 6: Machine Learning and AI

3:30pm-4:00pm
Break, Posters and Speed Talks in Exhibit Hall

3:30 pm-6:00pm
- Session 7: Orogenic and Other Au and Cu
- Session 8: Orthomagmatic
- Session 9: Exploration Technology

6:00pm-8:00pm
Exhibit and Poster Reception, NGEA Panel Discussion

SUNDAY, SEPTEMBER 28

8:00am-8:30am
Plenary Session

8:30am-10:30am
- Session 10: IOCG
- Session 11: Porphyry Exploration and Technology
- Session 12: Pegmatites

10:30am-11:00am
Break, Posters and Speed Talks in Exhibit Hall

11:00am-12:30pm
- Session 13: IOCGs, Cu/Co and Replacement
- Session 14: Collaborative Geoscience 1
- Session 15: Machine Learning Workflows to Prospectivity

12:30pm-2:00pm
Lunch, Posters and Speed Talks in Exhibit Hall

2:00pm-3:30pm
- Session 16: IOCGs, Cu/Co and Replacement
- 17: Collaborative Geoscience 2
- Session 18: Fluid Flow and Metal Deposition

3:30pm-4:00pm
Break and Posters in Exhibit Hall

4:00pm-4:30pm
SEG Distinguished Lecturer 2025

4:30pm-5:30pm
Awards Ceremony

5:30pm-7:30pm
Exhibit and Poster Reception

8:00pm - 11:00pm
Conference Celebration

MONDAY, SEPTEMBER 29

8:00am-8:45am
Plenary Session

8:45am-10:30am
- Session 19: Applied Orebody Knowledge
- Session 20: Porphyry Tracers 1
- Session 21: From Geomet to Remediation 1

10:30am-11:00am
Break, Posters and Speed Talks in Exhibit Hall

11:00am-12:30pm
- Session 22: Porphyry Tracers 2
- Session 23: ESG
- Session 24: From Geomet to Remediation 2

12:30pm-2:00pm
Lunch

2:00pm-4:00pm
- Session 25: Porphyry Americas
- Session 26: REE and Carbonatites
- Session 27: Big Workflows

4:00pm-4:30pm
Break

4:30pm-5:30pm
Closing Ceremony and Closing Remarks

5:30pm-6:30pm
Closing Social

TUESDAY, SEPTEMBER 30

8:30am-5:00pm
Post-Conference Workshops

WEDNESDAY, OCTOBER 1

8:30am-4:15pm
Post-Conference Workshops

Conference Information

Conference Organizing Committee

Conference Chairs:	Rick Valenta and Mark Noppé
SEG Executive Director:	Jennifer D. Craig
Technical Program:	Nick Oliver
Poster Program:	Corey Jago
Early Career Program:	Nick Dyriw, Rocio Vargas, and Kam Bhowany
Workshops:	Kaylene Camuti and Andrea Rutley
Field Trips:	Vladimir Lisitsin and Ioan Sanislav
Social Program:	Loren Nicholls
Sponsorships and Exhibitors:	Helen Degeling and Kylie Prendergast
Marketing and Media:	Evelyn Mervine
Publications:	Pat Hayman
Advisory Panel:	Stuart Simmons, Anne Thompson, and Steve Piercey

SEG Contact Information

7811 Shaffer Parkway
Littleton, CO 80127-3732
Tel.1.720.981.7882
E-mail: seg@segweb.org

Conference Organizer

Sea to Sky Meeting and Association Management, Inc.
conferences@segweb.org

The information in this program is current as of September 15, 2025.
Please see the SEG 2025 Attendee Hub for updated information.
seg2025.org/hub

Welcome to the SEG 2025 Conference!

Thank you for joining us for SEG 2025 in Brisbane. The world in 2025 is a complex and challenging place, and the need to secure critical and strategic resources for the future is greater than ever before. This conference is a great opportunity to get together and learn from each other, to swap lessons and approaches, and to meet both old friends and people from our region who have never before been to a conference like this!

We have put together a fantastic technical program that combines new insights on the major deposits from the Asia Pacific region and worldwide, the new technologies and workflows to improve discovery, the latest pre-competitive data sets, and the social, environmental, and market aspects of our fascinating business.

We also have a great lineup of field trips to key areas in Australia and more broadly in our region; workshops addressing some of the key development needs of our members; and an exciting mix of social activities. We will further have an ambitious program to bring early-career professionals from around the world together with each other and with more experienced professionals, as well as activities to ensure a combination of personal development and fun.

The conference will take place at the Brisbane Convention and Exhibition Centre at Brisbane's Southbank. The venue is ideal for a conference of this type and is located in an area full of activities, restaurants, and attractions.

Thank you again for joining us for the conference. We hope you have a great experience and come away energized and informed—with new ideas and approaches that you will find useful in the future. And of course, we hope you have the chance to make new connections and to reconnect with friends and colleagues.

Rick Valenta,
SEG 2025 Co-Chair
Director, Sustainable Minerals Institute, The University of Queensland

Mark Noppé,
SEG 2025 Co-Chair
Director, WH Bryan Mining Geology Research Centre, The University of Queensland

Jennifer Craig
SEG Executive Director
Society of Economic Geologists

SEG Awardees for 2025

Awards will be presented at 4:30 p.m. on September 28, 2025,
at the Brisbane Convention and Exhibition Centre

R.A.F. PENROSE GOLD MEDAL

Sarah-Jane Barnes
(Université du Québec Chicoutimi)

SEG SILVER MEDAL

Hilke Dalstra
(University of Utrecht)

SEG WALDEMAR LINDGREN AWARD

Xuyang Meng
(China University of Geosciences, Beijing)

BRIAN J. SKINNER AWARD

Robert R. Loucks
(University of Western Australia)

SEG DISTINGUISHED LECTURER

Mei-Fu Zhou
(Chinese Academy of Sciences)

SEG THAYER LINDSLEY LECTURER

Stephen Roberts
(University of Southampton)

REGIONAL VICE PRESIDENT LECTURER

Juan Carlos Castelli
(Geovectra)

SEG AWARD OF SPECIAL RECOGNITION

Lawrence D. Meinert

Pre-Conference Field Trips

Cu-Au Deposits of Java Island, Sunda Arc, Indonesia
Dates: September 14-23, 2025
Leaders: Lucas Donny Setijadji, Adam C. Simon, Andreas Rama Alfario, and supported by members of the Gadjah Mada University SEG Student Chapter

IOCG, ISCG, and Sediment-Hosted Cu (±Zn-Pb-Ag) Deposits of the Mount Isa Province - North West Queensland
Dates: September 19-22, 2025
Leaders: Vladimir Lisitsin, Elena Belousova, Alkis Kontonikas-Charos, and Ioan Sanislav

Porphyry and Epithermal Deposits of the Macquarie Arc, New South Wales
Dates: September 20-25, 2025
Leaders: David Cooke, Alan Wilson, Jonathon Hoye, Sebastien Meffre, and Jessica Askew

World-Class Archean Gold Deposits of the Eastern Yilgarn, Western Australia, and Insights into Au, Ni, Li, and REE Deposits
Dates: September 22-24, 2025
Leaders: Gerard Tripp and Alicia Verbeeten

Sn-W Deposits and Rare-Metal Mineral Systems of Northeast Queensland, Australia
Dates: September 23-25, 2025
Leaders: Avish Kumar, Benjamin Hines, and Jaime Poblete

World-Class Intrusion-Related and Epithermal Gold Deposits of Northeast Queensland
Dates: September 23-25, 2025
Leaders: Ioan Sanislav, Hugo Sera, Al-Tamini Tapu, and Vladimir Lisitsin

Post-Conference Field Trips

Tasmania's Diverse Array of Base, Precious, Critical, and Strategic Metal Deposits
Dates: September 30-October 6, 2025
Leaders: David Cooke, Owen Missen, and Emrecan Yurdakul

Iron-Oxide Cu-Au Deposits of the Olympic Cu-Au Province, South Australia
Dates: October 1-2, 2025
Leader: Adrian Fabris

Pre-Conference Workshops

Micro to Macroscale XRF Core Scanning Technologies in Mineral Exploration and Mining
Date: September 25, 2025
Presenters: Shaun Barker, Cassady Harraden, Gustav Nortje, Ayesha Ahmed, Nigel Kelly, Carolina Marin Suarez, Jason Bennett, Fernando Fontana, and Brian McNulty

Structural Analysis, Deformation Processes, and Fluid Flow in Fracture-controlled Hydrothermal Ore Systems
Dates: September 25-26, 2025
Presenter: Stephen F. Cox

Deposit Styles and Ore Textures of the Mount Isa and Cloncurry Districts
Date: September 25, 2025
Presenter: Richard Lilly

How to Extract Value from Ore - An Introduction to Mineral Processing
Dates: September 25-26, 2025
Presenter: Mohsen Yahyaei

From Orebody Knowledge to Geometallurgy: Discovery to Closure
Date: September 26, 2025
Presenters: Cassady Harraden, Pia Lois-Morales, Nathan Fox, and Lucas Pereira

Post-Conference Workshops

Introduction to Python for Geoscientists
Dates: September 30 - October 1, 2025
Presenters: Taryn Scharf, Luc Doucet, and Ben Knight

Diversity in High-K Calc-Alkaline to Alkaline Porphyry Copper-Gold Deposits: Practical Observations for Effective Exploration
Date: September 30, 2025
Presenter: Alan J. Wilson

Interpreting Ore Textures, Breccias, and Alteration: From Fundamental Observations to Advanced Technology
Dates: September 30 - October 1, 2025
Presenters: Gavin Clarke, Vladimir Lisitsin, Ioan Sanislav, Al-Tamini Tapu, Alkis Kontonikas-Charos, Benjamin Hines, and Kaylene Camuti

All workshops are held off-site.

Student and Early Career Events

Thursday, September 25, 2025
Student and Early Career Social
(Ticketed event)
6:00pm - 10:00pm at The Plough Inn
Sponsored by iCRAG, University of Queensland, Resource Technology and Critical Minerals Trailblazer

Friday, September 26, 2025
Student and Early Career Day

8:30am-9:00am
Icebreaker/coffee

9:00am-9:30am
Inside Xplor: How Majors Spot the Next Generation of Leaders
Presenter: *John Miller, Technical Lead for BHP Xplor*

9:30am-10:45am
Panel Discussion
From Aspiration to Reality: Navigating Our Early Career Days as Geoscientists
Moderator: *Chirantan Parui (CSIRO)*
Panelists: *Sheree Armistead (CODES), Jessica Askew (University of Tasmania), Richard Lilly (NExUS), and Dipanshu Sharma (Consultant)*

10:45am-11:00am
Morning Break

10:00am-12:00pm
How to Stand Out in a Competitive Pool Workshop
Presenter: *Jacob Sipos (Boston Resourcing)*

12:00pm-1:00pm
Lunch

1:00pm-2:00pm
SEG Opportunities for Students and Early Careers

2:00pm-2:30pm
Speed dating - Networking
With an introduction by Iain Dalrymple (AngloAmerican)

Social Events

Student and Early Career Social
Thursday, September 25, 2025
6:00pm - 10:00pm
The Plough Inn
Sponsored by iCRAG, University of Queensland, Resource Technology and Critical Minerals Trailblazer
[Ticketed Event]

Conference Celebration
Sunday, September 28, 2025
8:00pm - 11:00pm
Lina Rooftop
[Ticketed Event]

Closing Social
Monday, September 29, 2025
6:00pm - 8:00pm
Brisbane Convention & Exhibition Centre
[Ticketed Event]

Invited Speakers

Shaun Barker is the director of the Mineral Deposit Research Unit, within the Department of Earth, Ocean and Atmospheric Sciences at the University of British Columbia in Vancouver, Canada. Shaun has a background of geochemistry, mineralogy, and structural geology. Shaun is involved in a diverse range of research projects and has worked on a variety of gold and copper deposits in collaboration with the minerals industry, and in particular works on orebody knowledge and innovative methods for mineral exploration.

Cameron "Cam" Bartsch is a structural geologist with B2Gold in their corporate office in Vancouver. He has over 20 years of combined experience in mineral exploration and production, with a strong emphasis on orogenic gold systems throughout North America and Africa. Prior to joining B2 Gold in 2023, Cam worked as a consulting geologist specializing in 3D deposit modeling focused on mine-site ore controls and geotechnical analysis. Cam obtained his bachelor of science degree from the University of Saskatchewan in 2001, and a master's degree studying the tectonic history of the Appalachian orogen in eastern Canada from Acadia University in 2005. He is a registered professional geologist with Engineers and Geoscientists British Columbia.

Caoilin Chestnutt is head of technical services at Thiess, leading global teams in engineering, geotechnical, geologic, and geospatial disciplines. With nearly 30 years of postgraduate experience, she has worked across gold, copper, nickel, coal, and coal seam gas projects worldwide. Caoilin is nonexecutive chair of Australian Securities Exchange-listed Pacgold, a member of RFC Ambrian's critical minerals investment committee, and deputy chair of the Queensland Exploration Council. Her previous roles span senior positions at BHP, Xstrata, Anglo American, and Arrow Energy, as well as resources investment commissioner for Queensland. She is a recognized industry leader and advocate for women in exploration and mining.

Marina Costelloe is head of the mineral systems branch at Geoscience Australia, leading work to support a responsible transition of Australia's resource sector toward a low-emissions future. With a career spanning mineral exploration, groundwater, critical infrastructure, earthquakes, nuclear test monitoring, and space weather, Marina brings deep technical expertise and strategic insight. She holds a BSc and a graduate diploma in science in geology and geophysics from the University of Sydney and an MSc from James Cook University. A former president of the Australian Society of Exploration Geophysicists, Marina is a committed STEM advocate, mentor, and champion for gender equity and inclusion.

Irene del Real has a geology undergraduate degree from the University of Chile, Chile, an MSc degree from the University of British Columbia, and a PhD from Cornell University, Ithaca, New York, USA. She worked as a postdoctoral researcher at the University of Chile, and currently she is an assistant professor at the mining engineering department of the Catholic University of Chile. She has worked in Chile and in North America, and her research focuses on understanding the genesis of Andean iron oxide copper-gold deposits, specifically in the Candelaria-Punta del Cobre district porphyry Cu and iron oxide-apatite deposits. Her research is characterized by understanding the processes involved in the formation of mineral deposits—integrating structural geology, field geology, and geochemistry—and using this knowledge to develop new approaches for their exploration.

Greg Dipple is the head of science and co-founder of Arca Climate Technologies. He is also emeritus professor of geological sciences at the University of British Columbia (UBC), where he studied mineral-fluid reactions, including those that modulate Earth's long-term climate through chemical weathering. The UBC CarbMin Lab pioneered research on carbon mineralization in mine tailings, establishing it as a key pathway for mitigating greenhouse gas emissions. Today, Arca partners with mining companies to remove tonnes of carbon dioxide from the atmosphere and store it permanently in minerals.

Kathy Ehrig completed her PhD in geology at the University of California-Berkeley in 1991. She left San Francisco in 1992 to join the former Western Mining Corporation as a research geologist to work on the genesis of the Olympic Dam deposit in Australia and to provide mineralogical support for the Olympic Dam processing plant. In 2006, she moved to Adelaide to lead the development of the Olympic Dam geometallurgy program. She has remained focused on using mineralogy to solve processing issues, predicting metallurgical performance, unravelling the complex geological history of the Olympic Dam deposit, and using deposit-scale geological/mineralogical insights as inputs into discovering new IOCG deposits. She has co-supervised 16 PhD students and 10 postgraduate researchers working on Olympic Dam-based projects. She has shared the geological/geometallurgical knowledge gained from Olympic Dam and surrounding prospects by authoring or co-authoring >125 published papers and delivering >80 national and international presentations.

Angela Escolme is currently a senior lecturer in geology and geometallurgy at the Centre for Ore Deposits and Earth Sciences (CODES), University of Tasmania, and an independent geometallurgy consultant. She has recently returned from a three-year career break with her young family. Prior to this, her research activities at CODES were centered around developing new approaches to ore deposit characterization for improved geometallurgical domaining and predictions of processing performance, as well as advancing our understanding of ore deposits. She enjoys working closely with the minerals industry to find innovative solutions to real-world problems. Prior to completing her PhD at the University of Tasmania in 2016, Angela spent four years working in regional, near-mine, and underground gold exploration in Western Australia. She also holds a master of earth sciences from the University of Manchester, UK.

Daniel Franks is a professor and the director of the Global Centre for Mineral Security at The University of Queensland's Sustainable Minerals Institute and a leading authority on the role of minerals in sustainable development. His work has shaped global understanding of how minerals contribute to poverty reduction, inclusive growth, and environmental responsibility. He introduced influential concepts such as mineral security and development minerals and has advised the United Nations, UNESCO, and the World Bank on mineral governance. Daniel has led global initiatives to reduce mine waste, improve social outcomes, and support artisanal miners, positioning him as a key voice in responsible resource development worldwide.

David Giles is the John Ralston Chair of Minerals and Resources Engineering at the Future Industries Institute, University of South Australia, and chief scientific officer of MinEx Cooperative Research Centre (CRC). He has worked at the interface between academia and the minerals industry for his entire career; as an exploration and project geologist (1993–1996); as a PhD candidate and industry-linked postdoctoral researcher (1996–2005), and as the inaugural State of South Australia Chair of Mineral Exploration (2006–2016). Between 2010 and 2018 David was Program 3 (Targeting) Leader in the Deep Exploration Technologies CRC. As chief scientific officer of MinEx CRC (2019-2028) David's role is to manage, support, and champion the research efforts of a multidisciplinary team distributed across industry, government, and academia.

Sarah A. Gleeson is director of iCRAG, Research Ireland Professor at University College Dublin (UCD), and full professor in sustainability geoscience at the UCD School of Earth Sciences. She received a BA (mod.) in geology from Trinity College, Dublin, and a PhD in geochemistry from the Royal School of Mines, Imperial College, London. Subsequently, she held postdoctoral positions at the Natural History Museum, London, and the University of Leeds before taking up a professorship at the University of Alberta, Canada, and more recently at the Freie Universität Berlin. She led the inorganic and isotope geochemistry section at the GFZ, German Research Centre for Geosciences from 2015 to 2024. She serves on several advisory boards to scientific institutions and on international grant funding panels. Sarah has broad research interests in mineral deposit genesis, hydrothermal fluid flow, and water-rock interaction. Throughout her career, her research has focused on the genesis and detection of base metal mineral deposits in sedimentary basins with an emphasis on mineral system models, sedimentary geochemistry, diagenesis, and ore-forming processes.

Ignacio (Nacho) González-Álvarez is a Spanish-born geoscientist and the principal geochemist at CSIRO in Australia. He has extensive experience in rare earth elements (REEs) and high field strength elements (HFSEs). Ignacio has led mineral exploration and geoscience research across six continents and currently oversees multidisciplinary, nationally coordinated projects focused on critical minerals. His research focuses on the mobility of REEs and HFSEs in the Earth's crust, analysis of mineral systems, and the geochemistry of surface and sedimentary processes. Additionally, Ignacio contributes to the scientific community as an associate editor for international journals, and he serves as an adjunct research fellow at the University of Western Australia.

Lachlan Grose is a research fellow at Monash University and CTO of the Loop3D Foundation. He has a PhD in 3D structural geologic modeling and is the principal developer of Loop-Structural—an open-source library for 3D structural geological modeling—and has led the development and research for the Loop project. The Loop project is an open-source, interoperable, integrative, probabilistic 3D geologic modeling platform. Lachlan's work focuses on integrating structural geology concepts into the implicit modeling algorithms that form the basis of most 3D modeling workflows. His work prioritizes building workflows where geologic uncertainties can be propagated through the model-building process and by ensuring interoperability with modern data-science tools.

Julie Hollis is secretary general of EuroGeoSurveys, a not-for-profit association representing 37 European geologic survey organizations. She has a background in regional geologic mapping and more than 20 years of experience in leadership roles in government geoscience for geologic survey organizations in Australia, Denmark, Greenland, and Belgium. Julie has a PhD in geology and a master of science in communication and public engagement from the University of Edinburgh. Her passion for geoscience communication has a human-centric and storytelling approach. Julie's primary focus is delivering EuroGeoSurvey's vision of a geological service for Europe—a permanent, central, and sustainable service that provides foundational geoscientific data, information, and knowledge at a continental scale.

Ned Howard is a geologist with 20 years' experience in gold and copper exploration and mining across Australia and internationally. He is currently manager geoscience at Evolution Mining Ltd., a leading mid-tier gold-copper producer. Ned is presenting on behalf of the Australian Institute of Geoscientists (AIG) education committee, whose members include chair Kaylene Camuti, Doug Young, Martin Robinson, and Jenna McGovern. The committee recently completed a new strategic plan designed to strengthen AIG's role in supporting and advancing geoscience education across Australia.

Janelle Kerr is the director of minerals geoscience with the Geological Survey of Queensland (GSQ) in the Department of Natural Resources and Mines, Manufacturing and Regional and Rural Development. A geophysicist by training, Janelle has worked with the GSQ for 14 years, managing acquisition and modeling of geophysics, particularly magnetotelluric surveys, across Queensland. In her current role, she oversees the delivery of large-scale geoscience programs for the benefit of Queensland, particularly the mineral exploration sector. Janelle is also the past-president of the Australian Society of Exploration Geophysics, which supports geophysicists across Australia and the world, and promotes the role of geophysics in society.

Kathryn Kochan has a background in economics and international development. She has been with the Centre for Social Responsibility in Mining for more than five years. Kathryn recently completed her doctoral research in the field of community relations in the mining sector, focusing on the processes mining companies use to address and manage community level grievances, conflicts, and incidents. Kathryn's research focused on a global set of public inquiries and case studies to understand the triggers, precursors, and moderating factors that influence the visibility and escalation of mining-related community issues. She is determined to apply her doctoral findings to understand how company and community approaches to the grievance landscape are shifting in response to geopolitical, social, and regulatory challenges.

Sarah Mackenzie is an international development professional with deep field experience in Africa and Latin America. She has worked extensively with civil society and nongovernment organizations, particularly in the field of water, sanitation, and hygiene. While based in Burkina Faso, Sarah began collaborating with a mineral exploration company in 2011 as the country's gold mining sector boomed. She has since worked with other mining and exploration companies in Botswana, Ivory Coast, Liberia, and Niger. Her current areas of disciplinary expertise include community engagement, community development, social investment, and gender as a cross-cutting theme. Sarah holds a bachelor of arts in Spanish and history from Auckland University and a master of science in development management from the Open University in the UK. She is fluent in French and Spanish.

Peta MacRae is the mayor of Mount Isa and a dedicated community leader with a strong background in education, business, and local government. She holds a bachelor of education degree from James Cook University, a diploma of education from the University of Southern Queensland, and is a graduate of the Australian Institute of Company Directors. She has also completed MBA units through the Australian Institute of Management. With over 13 years in business and eight years as a councilor, Peta is passionate about youth mentoring, sport and recreation, tourism, and economic development. She is committed to building a vibrant, resilient future for Mount Isa.

Jim McCluskey has worked for more than 27 years in many different parts of the mining industry, from underground and open-pit mining, brown and greenfields exploration, prefeasibility and feasibility phase capital projects, business development activity in mergers, acquisitions, and divestments, as well as long-term strategic planning and short-term delivery planning. He has also done due diligence across diverse commodities and managed to get to every continent through work (barring Antarctica!). He is motivated by being part of something larger than himself and the energy that being part of, or leading, a strong team brings. His focus is on maximizing value through focus on key value drivers and on working at the right level for the opportunity at hand. He also believes that how we do things is just as important as what we do and accomplish, and that our values and performance are intimately linked.

Dietmar Müller received his undergraduate degree from the University of Kiel, Germany, and his PhD in earth science from the University of California, San Diego, in 1993. After joining the University of Sydney, he built the EarthByte Research Group, which pursues geodata synthesis through space and time. He is leading the construction of a virtual Earth laboratory, assimilating the wealth of disparate geological and geophysical data into an experimental planet. His group's virtual Earth models have numerous applications, ranging from resource exploration to linking deep Earth to surface processes. He has held an Australian Laureate Fellowship and led an ARC Industry Transformation Research Hub. Currently, his EarthByte team and industry partners are creating prospectivity maps for a range of critical minerals including copper, nickel, cobalt, and chromium, as well as rare earth elements, using spatial and spatio-temporal machine learning methods. He is a Fellow of the American Geophysical Union and the Australian Academy of Science.

Ryan Noble is a senior principal research scientist and the group leader of predictive mineral systems science at the Commonwealth Scientific and Industrial Research Organisation (CSIRO). For the past 20 years, Ryan has worked on numerous regolith and groundwater geochemistry projects related to gold, base metal, Ni, U, and rare earth element (REE) mineral exploration. His current project interests are focused on getting the maximum amount of information from soil samples with the least amount of effort, machine learning-driven landscape classification, and developing the UltraFine+ technique. Ryan is a past president and a Fellow of the Association of Applied Geochemists, and Fellow of the Australian Institute of Geoscientists. He is a board member of the Australian Geoscience Council and the past chair of Australian Earth Science Education.

Tim O'Connor is the group exploration officer at BHP. In his current role, Tim has accountability for BHP's global Exploration efforts – including the innovative Xplor incubator. Prior to stepping into the exploration, Tim led the Resource Centre of Excellence and Legacy Assets—with functional accountability for Geoscience, Resource Engineering, Tailings and Processing across the group, as well as operational accountability for 24 mine sites in various states of closure. A strong believer in the solving complex challenges across the value chain, Tim led the integration of the OZ Minerals business into BHP following its acquisition in 2023. Tim has been at BHP since 2006 and has worked in both the petroleum and minerals businesses, holding senior roles in strategy, exploration and production. Tim has degrees in both geology and geophysics from Boston College.

Jean-Philippe "JP" Paiement is a geostatistician with more than 20 years of experience in mineral exploration and mining, and he has applied his expertise across diverse geological environments, gaining insight—through hands-on experience—into how technology can, and will, shape the mineral exploration industry. JP is recognized as a pioneer in applying machine learning and deep learning to geoscience, and his groundbreaking innovations have transformed mineral exploration by leveraging AI to reduce bias and risk while enhancing targeting accuracy. In 2016, JP made history by applying machine learning to mineral exploration, winning the prestigious Integra GoldRush Challenge. Since then, he has led the development of cutting-edge AI applications, including supervised learning for targeting, automated geologic mapping, and deep learning for geophysical data enhancement. As CTO at VRIFY, he is further redefining mineral exploration with DORA, helping exploration teams unlock critical mineral discoveries faster and more efficiently than ever before.

Grant Sanden is the CEO of GeologicAI. Prior to founding GeologicAI, Grant had over 15 years of experience in various engineering and geoscientific roles working on resource development projects across the globe. He founded GeologicAI in 2013, with the mission of exploiting emerging technologies to improve the science and practice of resource geology. GeologicAI has grown into a global leader in mining innovation, applying robotics, machine vision, artificial intelligence, cloud computing, and big data to revolutionize the discovery and recovery of subsurface resources. In 2023 GeologicAI announced an investment from Bill Gates–founded Breakthrough Energy Ventures, and Export Development Canada. In 2025 GeologicAI recently announced the successful closing of a $44 million US Dollar Series B funding round. The round was led by Blue Earth Capital, with participation from BHP Ventures and Rio Tinto, representing two of the world's largest mining companies, and existing investors such as Breakthrough Energy Ventures.

Amanda Stoltze is exploration manager for Ivanhoe Mines in the Democratic Republic of the Congo (DRC). She completed an undergraduate degree in applied geology from Curtin University in 1997 and a PhD from The Australian National University in 2006. Since that time, she has worked extensively in Africa exploring for gold, nickel, and copper. She also worked as senior geochemist with ALS from 2019 to 2022. The current exploration program in the DRC is focused on the Western Forelands region with more than 2,000 km^2 of licences and a 10-diamond-rig drilling program.

Brian K. Townley is an associate professor in the Department of Geology at the University of Chile, an associate researcher at AMTC, and academic coordinator for mining engineering diploma programs. With more than 30 years of applied research experience, his work spans ore deposits, Andean metallogenesis, applied geochemistry, geo-mineral metallurgy, and the geology of vine cultivation. He leads major research and development projects on mineral recovery from tailings, sustainable mining, and geo-metallurgical modeling. Over the past decade, he has authored more than 35 papers on mineral exploration and vineyard geology, and he has been a consultant for major mining companies such as BHP-Billiton, Codelco, and Anglo American. Brian has also collaborated with the wine consortium of Chile, including the VitisGeoClima platform. His research has been funded by state, private, and international sources like AMIRA. He actively promotes knowledge transfer through diploma programs, workshops, and short courses across Chile and the Americas, including Argentina, Brazil, Mexico, and Canada.

Brad Welsh is chief executive and managing director of Energy Resources of Australia. Since being appointed in 2022, Brad has focused on leading the organization through highly complex transactions, stakeholder engagement, and operational activities while delivering sustainable outcomes. He was the first Indigenous Australian to become a chief executive and managing director of an Australian Securities Exchange-listed company in the mining sector. He also serves as a non-executive director for nib Holdings Limited, and he is the director of Mawal Pty. Ltd., a private company specializing in uniting investors and host communities to deliver mutual value and accelerate projects. Brad holds a master's in mining engineering (mine management) and a bachelor of law from the University of New South Wales, as well as a bachelor of welfare (Aboriginal community studies) from the University of Western Sydney. He is also a graduate of the Australian Institute of Company Directors. Brad is a proud Muruwari man from northwestern New South Wales who grew up in the Aboriginal community of Redfern, Sydney. He believes in integrating indigeneity into the balance sheet, recognizing it as a vital contributor to organizational success. In 2020, Brad was recognized as the Exceptional Indigenous Person in Queensland Resources.

Dan Wood AO spent 24 years with BHP and 18 years with Newcrest Mining Limited, during which time he was associated with a number of significant discoveries. After joining Newcrest at its formation in 1990, he was executive general manager of exploration from the mid-1990s, leading the company's exploration team, which was judged by the Metals Economics Group of Canada to have been the world's most successful gold explorer from 1992 to 2005. For a period in 2010 to 2011, he was director of the WH Bryan Mining and Geology Research Centre at the University of Queensland, where he is presently an adjunct professor. In 2015, Dan was appointed an Officer of the Order of Australia by the Australian government for his services to the mining industry and academia.

Mei-Fu Zhou is a research professor at the Institute of Geochemistry, Chinese Academy of Sciences, China. Professor Zhou obtained a BSc degree from Nanjing University (China) and a PhD degree from Dalhousie University (Canada). He was a postdoctoral researcher at Laurentian University (Canada) before joining the University of Hong Kong, where he taught for more than 25 years. His research includes ophiolites, large igneous provinces, mineral deposits, and regional geology. He has published more than 450 papers in scientific journals with more than 45,000 citations and an H-index of 118 (data from Google Scholar).

Student Presenter Support

Support for 91 student presenters was made possible through the generous contributions of our conference sponsors. Thank you!

- Abdulkadir Mohamed Abdulkadir, University College London
- Fabrizzio Abello, University of British Columbia
- Aaron Adsit, Colorado School of Mines
- Elizabeth Agyekum, University of Mines and Technology (UMaT)
- Oya Ak, University of Toronto
- Achmad Zulfan Al-Mahdy, Jenderal Soedirman University
- Muhammad Arba Azzaman, Akita University
- Jo Hannah Asetre, Macquarie University
- Fuseini Atanga, University of Tasmania
- Aaron Atkins, University of Texas at El Paso
- Paromita Banerjee, Indian Institute of Technology (IIT) Roorkee
- Billy Beas Caceres, University of Tasmania
- Damian Braize, University of Tasmania
- Kevin Brown, The University of British Columbia
- Juan Matías Buratti, Universidad Nacional de La Plata
- Nicolás Bustos, Pontificia Universidad Católica de Chile
- Néstor Alfredo Cano Hernandez, National Autonomous University of Mexico
- Axel Cima, University of Tasmania
- Camilo Andres Conde Carvajal, University of Geneva
- Francesca Corrado, University of Naples "Federico II"
- Sebastiano Coticelli, University of Naples Federico II
- Lily Dickson, University of Leeds
- Kelvin dos Santos Alves, Universidad de Salamanca
- Ben Eaton, University of British Columbia
- Vanessa Leonor Escobar Duche, Universidad de Salamanca
- Maria Paula Estrella, University of Alberta
- Shelly Mardhia Faizy, TU Bergakademie Freiberg
- Danis Ionut Filimon, University of Pisa
- Pedro Francisco, Colorado School of Mines
- Jose Franco Moraga, The University of Texas at El Paso
- Carolina Fresca Alves, Federal University of Minas Gerais
- Eduardo Fritis, The University of Auckland
- Javier Gil-Rodriguez, University of Tasmania
- Vinicius Godoi Pereira Da Cruz, University of Tasmania
- Felix Hasudungan, University of Oxford
- Hashindra Herath, University of Arkansas
- Edward Hill University, of Exeter
- Jasmin Hiller, University of South Australia
- Madeleine Ince, University of Western Australia
- Mallika Intachai, Akita University
- Tramaine James, Laurentian University
- Gaone Joseph, University of Witwatersrand
- Aune Kamosho, Namibia University of Science and Technology
- Randall Karcher, Colorado School of Mines
- Jonghyun Lee, James Cook University
- Edgar Leong, Australian National University
- Derek Leung, Laurentian University
- Anderson M. Santos, Murdoch University
- Smruti Prakash Mallick, Indian Institute Of Technology, Kharagpur
- Mabatho Mapiloko, University of the Witwatersrand
- Nicholas Mappin, University of Exeter
- Carolina Marin Suarez, The University of British Columnbia
- Maria Mavroudi, Montanuniversitaet Leoben
- Pratichee Mondal, University of Tasmania
- Brock Moody, University of Nevada, Reno
- William Munro, University of Toronto
- Luis Navarrete, Curtin University
- Nonhlanhla Ndlovu, University of the Witwatersrand
- Sam Nichols, University of Toronto
- Catherine Nyakecho, James Cook University
- Giovana Oliveira, Pimentel University of Tasmania
- Mohamed Abdelnaby Oraby, King Fahd University of Petroleum and Minerals (KFUPM)
- Jaime Osorio, University of Tasmania
- Stalyn Paucar, Central University of Ecuador
- Maxwell Porter, The University of British Columbia
- Amber Prevallet, University of Nevada, Reno

- Gloria Quispe Oruro, Stanford University
- Maryi Rodriguez Cuevas, James Cook University
- Samantha Russo, James Cook University
- Paola Salas Rivera, University of Texas at El Paso
- Andrea Salas Rivera, University of Texas at El Paso
- Angela Isaura Santos Costa, University of Tasmania
- Manju Sati, Indian Institute of Technology, Roorkee, Uttarakhand, India
- Mikidadi Shaha, Akita University
- Swastik Shinde, James Cook University
- Philipp Siebold, Colorado School of Mines
- Anna Sorrentino, University of Naples "Federico II" - DiSTAR
- Halik Rifki Daeng Taha, The Australian National University
- Koray Tasbicen, Colorado School of Mines
- Yamei Tian, China University of Geosciences (Beijing)
- Oscar Daniel Torres-Alarcon, University of Illinois - Chicago
- Lara Tritton, University of Leicester
- Louis van Eeden, Stellenbosch University
- Arthur Vicentini de Oliveira, Curtin Univeristy
- Brayan Paul Vilcanqui, Universidad Nacional Del Altiplano Puno (UNAP)
- Catharine Wang, The University of Queensland
- Elizabeth Xu, University of Toronto
- Luis Yagual, University of Texas at El Paso
- Colomba Trinidad Yañez Labarca, Universidad del Desarrollo
- Shouchun Yu, University of Tasmania
- Christoforos Zamparas, Karlsruhe Institute of Technology

Conference Accessibility Program (CAP)

SEG's Conference Accessibility Program (CAP) offered funding to 104 local students, postdoctoral researchers, and those early in their careers in order to participate in the SEG 2025 Conference. Generous support from our CAP sponsors (+CAP) assisted in their travel, conference registration, and SEG membership costs.

- Moses Ninkama Ain, University of Papua New Guinea
- Syaukan Akhdhar, University of Indonesia
- Andreas Rama Alfario, Gadjah Mada University
- Achmad Zulfan Almahdy, Jenderal Soedirman University
- Maya Josephine Almirol, Adamson University
- Rifa Bintang Anggareksa, University of Indonesia
- Jake Aris, University of Papua New Guinea
- Nickyta Aur, University of Papua New Guinea
- Annisa Azzahra Ariyanto, University of Indonesia
- Affada Badzar, University of Indonesia
- Adniwan Banuzaki, University of New South Wales
- Doreen Bibila, University of Papua New Guinea
- Bioantika Bioantika, University of Queensland
- Silveria Bobogare, University of South Pacific
- Sekonaia Buadromo, University of South Pacific
- Matereti Buren, University of South Pacific
- Adivuna Rukalisi Cagituevei, University of South Pacific
- Chaandvi Chand, University of South Pacific
- Kritik Chand, University of South Pacific
- Rufina Chong, University of South Pacific
- Cendi Dana, University of Edinburgh
- Dauba Dauba, University of Papua New Guinea
- Joseph Espi, University of Papua New Guinea
- Annalise Fa'Abasua, University of South Pacific
- Fadlin Fadlin, Jenderal Soedirman University
- Dian Fatimah, Pertamina University
- Frenz Aivereen Flores, Curtin University
- Extivonus Fransiskus, Freiberg University
- Roan Joshua Frias, University of Southeastern Philippines
- Lydia Garae, University of South Pacific
- Henry Gaugau, University of South Pacific
- Hawking Gonapa, University of Papua New Guinea
- Jethro Guinto, Adamson University
- Andriansyah Gurusinga, Akita University
- Tomy Herawan, Akita University
- Muhammad Ibnu Hibban, University of Indonesia
- Jackson Hopulaiwhi, University of Papua New Guinea
- Wa Ode Jasmine Syifa, University of Indonesia
- Wenni Jesica, University of Indonesia
- Toniella Kafolea, University of South Pacific
- Nathaniel Kame, Charles Sturt University
- Nevio Kamel, Imperial College London
- Gideon Kamori, University of Papua New Guinea
- Stephany Vanessa Kapissa, Cenderawasih University
- Jonathan Kensie, Pertamina University
- Railyne Kink, University of Papua New Guinea
- Ariza Khan, University of South Pacific
- Joseph Kui, University of Papua New Guinea
- Rishil Kumar, University of South Pacific
- Shohel Kumar, University of South Pacific
- Joeli Kurualeba, University of South Pacific
- Cleodette Lagata, Ateneo De Manila University
- Newman Leo, University of Papua New Guinea
- Jonathan Macuroy, Akita University
- Ardian Baroto Murti, University College London
- Mispa Gaius Mwatapa, Newmont Cooperation - Lihir Gold Mine
- Taraivini Nailigamasei, University of South Pacific
- Daniel Nelson, University of Papua New Guinea
- Lyn Helen Nick, University of Papua New Guinea
- Rosmalia Nugraheni, Australian National University
- Sekar Aurellia Oksatiawan, University of Indonesia
- Ruth Paika, University of Papua New Guinea
- Shekinah Pilamp, University of Papua New Guinea
- Natasha Poli.Sahin, University of Papua New Guinea
- Daniel Prang, University of Indonesia
- Divisha Prasad, University of South Pacific
- Morien Prasad, University of South Pacific
- Muhammad Fatih Qodri, Kyushu University
- Mochammad Hafid Rahadi, Bandung Institute of Technology
- Meresine Raikoro, University of South Pacific
- Arishma Ram, University of South Pacific
- Ghina Suci Ramadhanti, University of Science And Technology Beijing
- Kalesi Ravela, University of South Pacific
- Marama Rokara, University of South Pacific
- Gabriella Roring, Pertamina University
- Immanuela Martha Irianti Rumbiak, Cenderawasih University
- Utreck Frans Rumbiak, Padjadjaran University
- Elizah Irai Seiweni, University of Papua New Guinea
- Kireimercy Mema Palagi Seru, University of South Pacific
- Alya Sevy Setiawan, Jenderal Soedirman University
- Shristi Sheetal, University of South Pacific
- Kelevi Stark, University of South Pacific
- Muhammad Rahaldi Taher, Aalto University
- Longfield Tai, University of Papua New Guinea
- Lolo Tambai, University of Papua New Guinea
- Ginalyn Tanilon, University of Southeastern Philippines
- Diandra Tauauvea, University of South Pacific
- Shennill Tina, University of Papua New Guinea
- Intan Tsabita, Alzahra University of Indonesia
- Pauliasi Tubui, University of South Pacific
- Augustine Uga, University of South Pacific
- Raynon Ungere, University of Papua New Guinea
- Keresi Vasikali, University of South Pacific
- Sekove Vodo, University of Fiji
- Michael Collockson Waisime, University of Papua New Guinea
- Hernah-Maree Wally, University of Papua New Guinea
- James Wapyer, University of Papua New Guinea
- Ratu Tevita Waqaiseurobo, University of South Pacific
- Ivan Wareke, University of Papua New Guinea
- Keziah Wianjembi, University of Papua New Guinea
- Zariah Renee Divine Wong Ling Warren, University of South Pacific

SEG 2025 Affiliates

Thank you to our SEG 2025 Affiliates—professional societies and organizations that share our commitment and vision for the future of geology. Affiliate organizations assist SEG with promoting the conference to their members.

Amira

Australian Institute of Geoscientists

Economic Geology Research Centre (EGRU) at James Cook University

Queensland Government Geological Survey of Queensland (GSQ)

International Association on the Genesis of Ore Deposits (IAGOD)

Society of Geology Applied to Mineral Deposits (SGA)

Government of South Australia Geological Survey of South Australia (GSSA)

Exhibitor Booths

Booth	Exhibitor
1, 2	Ma'aden
3, 4	Society of Economic Geologists
5, 6	Rio Tinto
7	Geo Spectra Pty Ltd
8	VRIFY
9	ALS Geochemistry
10	Gnomic Exploration Services
11, 12	Geological Survey of Queensland
13	Resource Technology and Critical Minerals Trailblazer
14	Gap Geophysics Exploration Services
15	CSIRO
16	iCRAG
17	Portable PPB
18, 19	GeologicAI
20	The University of Queensland
25, 26	TECT Geological Consulting
27, 28	BHP Xplor
29	MDF Global Limited

Booth	Exhibitor
30	SpecTIR
31	DDH1 Drilling
32	SRK
35	Brunton
36	Association of Applied Geochemists
37	Australian Institute of Geosciences (AIG)
38	Geological Survey of Western Australia
39	Activation Laboratories Ltd
40	CODES, University of Tasmania
41	Datarock
42	PGN Geoscience
43	Intertek
44	Automated Mineralogy Incubator
45	Lithodat
46	Bruker
47	Veracio
48	SciAps

Exhibitor Spotlights (1-Minute Breaks Between Talks)

Sunday, September 28

5:45pm - 5:50pm	The Brunton Axis Transit: Measuring Any Angle in Any Direction *Brunton*
5:51pm - 5:56pm	From Data to Discovery: Unlocking Orebody Insights with Machine Learning *Datarock*
5:57pm - 6:02pm	Application of AI for Porphyry Copper-Gold Targeting in the Woodjam District, British Columbia, Canada *VRIFY*
6:03pm - 6:08pm	Geoscientists and the Critical Mineral Exploration Crisis *GeologicAI*
6:09pm - 6:14pm	The Pyramid Porphyry Cu Mo system, Alaska: Fresh Insights on the Deposit That Shouldn't Exist *MDF Global*
6:15pm - 6:20pm	Advancing Gold and Copper Target Generation Through Multispectral and Hyperspectral Satellite Data: Insights from Easter Iran and Western Australia *Geo Spectra Pty Ltd*
6:21pm - 6:26pm	The Value of Imagery in Core Analysis *TerraCore*
6:27pm - 6:32pm	Automated Mineralogy Incubator

Thank You, Exhibitors

Actlabs

Geo Spectra

PORTABLE PPB

ALS

GeologicAI

TRAILBLAZER RESOURCES TECHNOLOGY & CRITICAL MINERALS

Australian Institute of Geoscientists

Queensland Government

RioTinto

The Association of Applied Geochemists

Geological Survey of Western Australia

SciAps

automated mineralogy incubator

BHP Xplor

GNOMIC EXPLORATION SERVICES

Society of Economic Geologists Founded 1920

BRUKER

iCRAG SFI RESEARCH CENTRE IN APPLIED GEOSCIENCES

SpecTIR® ADVANCED HYPERSPECTRAL SOLUTIONS

BRUNTON

CODES

intertek minerals

srk

CSIRO

LITHODAT

TECT

Datarock

MAADEN

The University of Queensland Australia

DDH1 DRILLING

MDFGLOBAL

VERACIO™

Gap GeoPhysics A DIAS COMPANY

PGN GEOSCIENCE

VRIFY

Thank You to Our SEG 2025 Conference Sponsors

DIAMOND PARTNER

BHP Xplor

PLATINUM PARTNER

MAADEN

PREMIER PARTNERS

GeologicAI

VRIFY

GOLD SPONSOR

EQUINOX GOLD

SILVER SPONSORS

B2GOLD

Teck

Canada

BRONZE SPONSORS

Dawn Zhou

iCRAG
SFI RESEARCH CENTRE IN APPLIED GEOSCIENCES

SOUTHERN CROSS GOLD CONSOLIDATED

THE UNIVERSITY OF QUEENSLAND AUSTRALIA

TRAILBLAZER
RESOURCES TECHNOLOGY & CRITICAL MINERALS

VERACIO

SUPPORTER

AGNICO EAGLE

Australian Government
Geoscience Australia

MinEx CRC

MUNDORO

Thank you to Our 2025 Conference Accessibility Program (CAP) Sponsors

CAP DIAMOND PARTNER

RioTinto

CAP GOLD SPONSOR

AngloAmerican

CAP SILVER SPONSOR

FREEPORT
FOREMOST IN COPPER

CAP COPPER SPONSOR

Newmont™

CAP BRONZE SPONSOR

BRUNTON

ISLAND PASSAGE EXPLORATION

CAP SUPPORTER

ANGLOGOLD ASHANTI AUSTRALIA

DDH 1 DRILLING

MINERSAI

mining3 TRANSFORMINGMINING

Quinton Hennigh

SEG 2026 CONFERENCE

SEG 2026

SALT LAKE CITY, UNITED STATES

September 30-October 3, 2026

The Society of Economic Geologists (SEG) invites you to the 2026 conference in Salt Lake City, Utah, USA, from September 30 to October 3, 2026. This event will deliver the high-quality technical content that is a hallmark of SEG conferences, including innovations, advancements, and discoveries throughout the mining chain. The program will feature a balance of applied case studies, global advances in framework geology, technical innovation, and archetypal ore deposits and geologic processes globally, while highlighting the North American Cordillera. These themes will be present throughout a vibrant slate of field trips, workshops, technical presentations, and unique conference events.

The SEG 2026 conference is focused on facilitating knowledge transfer within the economic geology community, showcasing geologic and career insight from experienced members, providing technical growth and development for mid-career geologists, and introducing a new generation to the fascinating world of economic geology. With a mix of participants from industry, government, and academia worldwide, SEG 2026 offers a unique opportunity to connect, learn, and help shape the future of economic geology.

Stephanie E. Mills
Utah Geological Survey
SEG 2026 Chair

Rachelle Boulanger
Talon Metals
SEG 2026 Chair

Jennifer Craig
Society of Economic Geologists
SEG Executive Director

SESSION THEMES

- The future of the mining workforce
- Orebody knowledge and deposit characterization
- Framework geology for ore deposits
- Keeping up with the commodities: Expanding terranes and targets
- New technology and old deposits

Watch www.seg2026.org for more information

Mineral Systems Models for Clastic Dominated Zn Deposits: From the Geodynamic- to the Micro-Scales

Sarah Gleeson[1,2], Joseph Magnall[2], Anne Glerum[2], Peter Berger[2], Philipp Weis[2], Sascha Brune[2]
[1]iCRAG, University College Dublin, Dublin, Ireland, [2]GFZ Helmholtz Centre for Geosciences, Potsdam, Germany

Clastic dominated (CD) Zn deposits are the product of time integrated processes operating from the geodynamic to the micro-scales. At rifted margins, the interplay between tectonic and surface processes may generate large volumes of sediments which may be the source, or the trap, for metals. The coupling of ASPECT rift models with Fastscape models of surface erosion and sedimentation has allowed us to assess the most prospective part of rifted margins, where faults connect source rocks (rift clastics >250°) with host rocks (organic-rich marine sediments <150°). In Paleozoic Selwyn-type deposits, the most prospective areas are sub-basins closest to the continental margin, formed in narrow asymmetric rifts. In these sub-basins, redox and temperature gradients between ore fluids and host rocks on the macro- and micro-scale control reaction permeability, metal deposition and the nature of alteration. In Selwyn-type deposits, reduced hydrothermal fluids react with, and replace, diagenetic barite (oxidized sulphur). In McArthur type deposits, an oxidized fluid reacts with diagenetic pyrite (reduced sulphur) resulting in the dissolution of carbonate and precipitation of sphalerite. Reactive transport models of replacive mineralization in both deposit types show that ore grade and alteration styles are influenced by local pore-scale mineralogy and geochemistry in prospective horizons. The formation a large CD deposit, therefore, requires the alignment of key processes from the lithospheric to micro-scales. At the district to camp scale, the nature of the redox trap is key for understanding the alteration assemblages produced by the hydrothermal system and should be considered in vectoring strategies.

Mineral Systems in a Plate Tectonic Context

R. Dietmar Müller[1]
[1]The University Of Sydney, Sydney, Australia

Understanding how different ore deposits depend on plate tectonic processes is critical for better understanding a range of mineral systems. The recent development of global plate models with continuously evolving plate boundaries now allows us to implement spatio-temporal “deep time” mineral exploration, leading to the generation of new ideas and improved predictive exploration models. These models include continental paleo-crustal thickness estimates, oceanic crustal age-area distribution, seafloor spreading rates, sediment thickness, and water and carbon contents of oceanic sediments, altered ocean crust and mantle lithosphere. Custom-tailored python software libraries including pyGPlates and GPlately are used to visualise and analyse mineral deposits and other geological data in the context of reconstructed continents, ocean basins and plate boundary evolution. Machine learning is applied to train mineral prospectivity models to learn how particular types of deposits depend on plate tectonic evolution and tectonic plate properties, particularly those of subducting slabs that partially devolatilise while descending in the mantle, fertilising the upper mantle. For Cretaceous to Cenozoic copper porphyry deposits, our analysis shows that the top five important features for deposit formation are Cordilleran crustal thickness, the total carbon density of the subducting slab, the seafloor spreading rate, and the distance to the trench edge. However, we also find distinct regional differences. Where passive margins have experienced subduction initiation and a conversion to active margins, ocean crust with unusually thick sediments is subducted, making this the most important feature. Where subducting ocean crust is relatively sediment starved, as along North America, other parameters like crustal thickness, convergence rate and seafloor age are more important. Our models also reveal a clear influence of subduction proximity and processes on a range of sediment-hosted craton-edge deposits since the Paleo-Proterozoic. These new approaches are powered by AuScope NCRIS open-source software and open-access data infrastructures such as EarthBank.

Probabilistic Assessment of Australia's Lithospheric Architecture and Its Relationship to Mineral Systems

Mark Hoggard[1], Rhys Hawkins[1], Thomas Duvernay[1], James Hazzard[2], Fred Richards[2], Sinan Özyadin[3], Zach Sudholz[1,4], Lynton Jaques[1], Greg Yaxley[1], Marcus Haynes[5], Karol Czarnota[5]

[1]Australian National University, Canberra, Australia, [2]Imperial College, London, United Kingdom, [3]University of Sydney, Australia, [4]University of Cambridge, United Kingdom, [5]Geoscience Australia, Canberra, Australia

It is increasingly recognised that the subcontinental lithosphere plays a first-order control on the distribution and dynamics of geological processes at Earth's surface, with prominent examples including active tectonics, magmatism, sedimentary basins, groundwater systems, and mineral deposits. Due to its importance, academic, governmental, and industry efforts to image lithospheric architecture have recently coalesced. Here, we present a new probabilistic approach to this problem that combines mantle xenolith geochemistry, seismic tomography, laboratory constraints on anelastic deformation at seismic frequencies, and magnetotelluric imaging to build self-consistent thermochemical models of the Australian lithosphere. The models allow us to query properties such as the seismogenic thickness, Curie depth, lithosphere-asthenosphere boundary, and presence of metasomatisation. These properties are correlated with several surface observables, including the distribution of sediment-hosted and IOCG mineralisation, and the associated maps and their uncertainties will underpin future exploration efforts including in regions undercover.

Targeting the Biggest Discoveries along the Biggest Faults: How Hard Can it Be?

Graham Begg[2], **Nick Hayward**[1], Quentin Masurel[2], Nico Thébaud[2]
[1]PredictOre Pty Ltd, East Perth, Australia, [2]University of Western Australia, Perth, Australia

Large ore deposits have a more ordered distribution than smaller ones, which implies that their locations should be more predictable. Examples of ordered spatial patterns include the proximity of giant deposits to translithospheric fault zones (TLFs) and the spatial and temporal periodicity of many camp-scale large deposit clusters. Using more than 120 global case studies of district-scale structural controls on large magmatic Ni-Cu and porphyry Cu-(Au) deposits, we show that TLF intersections provide more predictable controls on giant ore deposit locations than other commonly proposed structural targets (e.g., fault jogs). Large magmatic Ni camps preferentially form proximal (less than 25 km) and aligned parallel to primary TLFs marking paleocraton margins. Large porphyry Cu-(Au) deposits/camps also form near primary arc-parallel TLFs at lithospheric domain boundaries, in contrast to Ni, which frequently form closer (within 5 km) and aligned parallel to cryptic secondary transverse TLFs. We hypothesize that differences in camp-scale structural controls are related to the dynamic permeability of TLF intersections in predominantly transtensional (Ni) versus predominantly transpressional (Cu) settings. Many large porphyry Cu-(Au) deposits in well-explored metallogenic belts show along-strike semi-regular spatial periodicity, which appears to be inherited from the spacing of intersections with a (commonly older) transverse TLF set. In contrast, evidence for spatial periodicity among large magmatic Ni-Cu deposits appears lacking due to either exploration immaturity and/or non-regular spacing of cross faults. The challenge of applying the TLF framework to exploration lies in the fact that, while primary strike-parallel TLF corridors are usually well-defined, intersecting secondary transverse TLFs tend to be cryptic. Moreover, distinguishing prospective translithospheric faults from more common transcrustal faults is challenging, given that seismic reflection profiling does not directly detect steep TLFs. Here we propose leading practices, highlighting a multidisciplinary approach to optimize lithospheric structural targeting at corridor to camp scales for undercover giant discoveries.

Linking Orogenic Gold and Orthomagmatic PGE-Ni-Cu Deposits in the Archean Yilgarn Craton

Matthew Demmer[1,2], Marco Fiorentini[1], Isra Ezad[1]
[1]ARC Training Centre in Critical Resources for the Future, The University Of Western Australia, Australia, [2]Chalice Mining Limited, Australia

The Archean Yilgarn Craton hosts both the world class orogenic gold deposits of the Kalgoorlie–Kurnalpi Rift and the recently discovered orthomagmatic Gonneville PGE-Ni-Cu deposit, Australia's largest PGE resource. Traditionally these separate deposit types are viewed as products of distinct metallogenic processes. However, new observations suggest a deeper connection. This study highlights the remarkable coincidence of gold and orthomagmatic PGE-Ni-Cu mineralisation forming synchronously at ~2.65 Ga on opposite margins of the craton. Despite their differing deposit styles, both systems share four key features: enrichment in semi-metals (Bi–Te), consistent positive mass-independent sulfur isotope ($\Delta^{33}S$) signatures, and geochemical indicators of hydrous, metasomatised mantle sources.

These shared traits challenge prevailing models that treat orthomagmatic and orogenic systems as isolated. Instead, we propose they reflect a common lithospheric control: the presence of a fertile mantle reservoir enriched in volatiles and sulfur, generated through crustal recycling. In this framework, local geology acts as a secondary control, while primary fertility is inherited from mantle-scale processes.

We further suggest that $\Delta^{33}S$ anomalies provide a powerful tool to map these enriched domains across both time and space. Recognising deep-seated controls reframes how Archean mineral systems are explored, advocating a unified view of metallogenesis and integrating models that span different deposit styles. By recognising the importance of metasomatised mantle and its sulfur signatures, we offer a new prospective for targeting Archean mineral systems globally.

Magmatic Activities Directly Contributed to the Formation of Giant Jiaodong Gold Province: Evidence from the Xiejia Diorite

Zhan-Ke Li[1]
[1]China University of Geosciences (Wuhan), Wuhan, China

Gold deposits in Precambrian cratons were mostly generated during the formation and stabilization of the cratons, but the North China Craton (NCC) is unusual in that its gold deposits were mainly formed ca. 1.7 billion years after its stabilization. A magmatic-hydrothermal origin or mantle-derived fluid source has been proposed for the giant gold deposits of the Jiaodong district in the eastern NCC, but direct evidence is sparse and the mineralization processes remain controversial. Here, we present the results of a comprehensive study of the Xiejia diorite beneath the Linglong ore field at Jiaodong to link the gold mineralization to underlying magmatism. Magmatic zircon and titanite grains from the Xiejia diorite have U-Pb ages of 121.3 ± 0.9 Ma to 120.8 ± 1.1 Ma and 121.7 ± 3.9 Ma, respectively. The geochemical characteristics (REEs, Sr-Nd-Hf-Pb isotopes) of the Xiejia diorite are akin to those of contemporaneous mafic dikes in most gold mines at Jiaodong, indicating that it was most likely derived from an enriched lithospheric mantle source. The upper part of the diorite intrusion is pervasively altered and mineralized. Gold-bearing pyrite has a Re-Os isochron age of 122.5 ± 6.7 Ma, which is consistent with the ore-related sericite 40Ar/39Ar plateau age of 122.6 ± 1.3 Ma. Pyrite from the mineralized diorite yielded δ34SCDT values of 2.1 to 9.7 ‰, which are comparable with those of pyrite from gold ores of Dongfeng. Pyrite grains from both groups also have similar Pb isotope compositions. The sulfur and lead isotope data are consistent with values of mafic dikes which are spatially and temporally associated with gold veins. The xiejia diorite is thus supposed to link the shallow gold mineralization to deep-seated mantle-derived magmatism generated during the extensive destruction of the NCC, which was induced by the rollback of the subducted Paleo-Pacific plate.

Linking Porphyry Cu Formation to Tectonic Change in Postsubduction Settings: A Case Study from the Giant Yulong Belt, Eastern Tibet

Ming-Liang Huang[1], Xian-Wu Bi[1], Rui-Zhong Hu[1,2], Massimo Chiaradia[3], Jing-Jing Zhu[1], Lei-Luo Xu[1], Zong-Yong Yang[1]
[1]State Key Laboratory of Critical Mineral Research and Exploration, Institute of Geochemistry, Chinese Academy of Sciences, Guiyang, China, [2]University of Chinese Academy of Sciences, Beijing, China, [3]University of Chinese Academy of Sciences, Geneva, Switzerland

Porphyry deposits in magmatic arcs form coincident with changes to steady-state oceanic subduction conditions, such as changes in plate convergence rate and vector or angle of subduction. However, it remains unclear whether such processes also operated during formation of postsubduction porphyry deposits. The Yulong magmatic belt in the eastern Tibetan Plateau consists of middle to late Eocene igneous rocks (~51–35 Ma) that formed during the India-Asia collision, whereas all known porphyry deposits are associated with late Eocene rocks (43–35 Ma). A synthesis of new and published geochemical data shows marked variations from the middle to late Eocene, including increasing whole-rock La/Yb, Sr/Y, and EuN/EuN* values, as well as zircon EuN/EuN* values. These geochemical variations, together with petrographic observations, indicate a transition from plagioclase-dominated to amphibole-dominated fractionation from the middle to late Eocene.
Coupled changes of magma compositions and porphyry Cu metallogeny from the middle to late Eocene coincided with, or were slightly preceded by, the onset of regional uplift and crustal thickening, triggered by the India-Asia hard collision and rapid deceleration of India-Asia convergence rate at ca. 50–44 Ma. Crustal thickening may have caused prolonged magma differentiation at greater depth and accumulation of dissolved H2O, which both contributed to amphibole-dominated fractionation and generation of hydrous melt that are prospective for porphyry Cu mineralization. Our study highlights the importance of tectonic changes in the formation of the Yulong and other postsubduction porphyry Cu belts, a scenario similar to that operated in subduction-related settings such as the Andes.

Slab Tearing as a Driver of Porphyry Ore Deposit Formation in Indonesia and the Philippines

Jack Ward[1], Gideon Rosenbaum[2], Teresa Ubide[2], Jeffrey Oalmann[1], Alice MacDonald[1], David Cooke[1], Lejun Zhang[1], Jonny Wu[3]
[1]Centre for Ore Deposit and Earth Sciences (CODES), University of Tasmania, Hobart, Australia, [2]School of the Environment, The University of Queensland, Brisbane, Australia, [3]Department of Geosciences, University of Arizona, Tucson, USA

Discovering new porphyry ore deposits is vital to address the rapidly rising demand for strategic and critical metals. On a global scale, porphyry ore deposits are largest and most abundant in mature continental arcs, such as the Andes, where lower-crustal garnet fractionation may produce oxidised magmas suitable for metal enrichment during fluid saturation in the upper crust. Nevertheless, porphyry ore deposits (including giant porphyry deposits, characterised by >2 Mt contained Cu) also occur in island arcs, where the crust is too thin to fractionate garnet. In this study, we test the hypothesis that slab tearing (associated with the subduction of buoyant features such as aseismic ridges) facilitates deep melting within the garnet stability field, emulating the petrogenetic effects of thick continental crust and enabling the formation of Pliocene–Holocene porphyry ore deposits in Indonesia and the Philippines. To do so, we first modelled slab geometry using high-resolution seismic tomography models to establish spatial relationships between deposit distribution and slab tears. To identify possible geochemical contributions of slab tearing to porphyry metallogenesis, we collected new, high-resolution geochemical data (collected via LA-ICP-TOF-MS) from the groundmass of the most primitive and least-altered samples from Batu Hijau (Indonesia), Far Southeast (Philippines), Black Mountain (Philippines), and Santo Thomas II (Philippines), which we integrated with a compilation of published whole-rock data. Our multi-disciplinary approach demonstrates that porphyry ore deposits in Indonesia and the Philippines are concentrated at the margins of slab tears. Despite appreciable inter- and intra-deposit geochemical diversity, geochemical signatures putatively imparted by garnet, either as a residual or fractionating phase, are consistently detected, suggesting that deep, tear-driven melting is a plausible means of generating porphyry ore deposits in island arcs.

Magmatism and Post-Emplacement Uplift at the Red Chris Cu-Au Deposit, BC Canada: Implications for Porphyry Copper Formation and Preservation

Brian McNulty[1], Nichole Moerhuis[1], Shaun Barker[1], Will Clarke[2]
[1]University of British Columbia - MDRU, Vancouver, Canada, [2]Newmont, Vancouver, Canada

The Late Triassic Red Chris porphyry Cu-Au deposit in British Columbia, Canada, hosts proven and probable reserves totalling 5.29 Moz of gold and 1.4 Mt of copper, with average grades of 0.62 g/t Au and 0.52% Cu (Newmont, 2025). Mineralization is concentrated in several zones within the Red Stock, a 6.5 × 1.5 km east-northeast-trending, multiphase dyke swarm (Rees et al., 2015). Chalcopyrite ± bornite ± molybdenite mineralization occurs in quartz veins and as disseminations within plagioclase-hornblende monzonite porphyry intrusions and, to a lesser extent, in mafic volcanic and volcanosedimentary rocks of the Triassic Stuhini Group. Molybdenite Re-Os ages date mineralization to ca. 206 ± 1 Ma (Zhu et al., 2018).

Previously, magmatism at Red Chris was thought to span nearly 10 million years, based on LA-ICP-MS zircon U-Pb ages ranging from 211.6 ± 1.3 Ma to 201.7 ± 1.2 Ma (Zhu et al., 2018). However, preliminary CA-ID-TIMS U-Pb zircon ages suggest rapid emplacement of the Red Stock at ca. 206.5 Ma, over a ~0.5-million-year period.

Stratigraphic constraints further define the pre- and post-emplacement history. The youngest fossils in the underlying Stuhini Group are Late Norian (212–205.6 Ma), while the overlying Hazelton Group, which forms an erosional unconformity with the Red Stock, contains Late Pliensbachian fossils (189.6–186.7 Ma; Evenchick & Thorkelson, 2005). This framework, combined with our new geochronology, reveals a dynamic ~16-million-year interval encompassing emplacement, uplift and erosion of the Red Chris Cu-Au porphyry deposit.

These results highlight the role of short-lived magmatism and tectonic exhumation in the formation and preservation of porphyry systems. When integrated with regional biostratigraphy, high-precision geochronology provides critical insights into the geological processes that shape economically significant mineral systems.

Magmatic Fertility and Timing of the Dalli Gold-Rich Porphyry Deposit, Central Iran: Geological Insights and Exploration Implications

Hooshang Asadi Haroni[1], Yongjun Lu[2], Dariush Esmaeily[3]
[1]Tethyan Gold Exploration, Brisbane, Australia, [2]RSC Global, Perth, Australia, [3]Geological Survey and Mineral Exploration of Iran, Tehran, Iran

Dalli is a recently discovered, gold-rich porphyry copper deposit with a proven reserve of 17 million tonnes grading 0.7 g/t Au and 0.5% Cu. It is situated within Iran's main volcanic arc, part of the central segment of the Tethyan orogenic belt. The deposit was identified through spectral analysis of ASTER satellite imagery. Dalli comprises two principal porphyry centres composed of quartz diorite intrusions emplaced into porphyritic andesites within a northeast-trending tectonic corridor. These rocks exhibit intense potassic, quartz-sericite, and silica alteration associated with Cu-Au mineralization. Mineralization occurs primarily as disseminated sulfides and within quartz-hematite-magnetite stockworks and veins. Gold is present as free grains and as inclusions within pyrite and chalcopyrite. A strong magnetic signature associated with potassic alteration, and a distinct Cu-Au geochemical association serve as a key exploration vector at the Dalli deposit. This study presents a comprehensive LA-ICP-MS U–Pb geochronology and trace element analysis from the magmatic suites at the Dalli deposit to characterize zircon geochemical signatures indicative of magmatic fertility. The results exhibit elevated Sr/Y (49–61) and Eu/Eu (0.89–0.92) ratios, consistent with signatures typical of porphyry systems. Zircon trace element compositions are also characterized by high Eu/Eu (0.3–0.8), (Ce/Nd)/Y (0.01–0.3), and 10,000 × (Eu/Eu*)/Y (2–15), comparable to those of world class fertile porphyry suites such as the Sar-Cheshmeh and Qulong Cu deposits along the Tethyan metallogenic belt. The zircon geochronology of mineralized quartz diorite and andesite porphyry units yields magmatic crystallization ages of 15.4–16.0 Ma. These findings demonstrate that the Dalli Oligo-Miocene intrusions are magmatically fertile and justify further deep drilling to evaluate their full mineral potential. Zircon trace element ratios effectively fingerprint fertile magmatic suites and provide new constraints on magmatic events' timing and metallogenic fertility, enhancing exploration models for gold-rich porphyry-style mineralization.

Loop: An Open-Source Framework for Probabilistic Geological and Resource Modelling

Lachlan Grose[1], Laurent Ailleres[1], Ayla Edwards[1], Axel Menge-Nicole[1,2], Rabii Chaarani[1,2]
[1]Monash University, Clayton, Australia, [2]Universite de Lorraine, CNRS, GeoRessources, Nancy, France, [3]Northern Territory Geological Survey, Darwin, Australia

Efficient and reproducible construction of geologically realistic 3D models of the subsurface is critical for the sustainable management of natural resources — including mineral exploration, groundwater, and waste storage. These models aim to represent geological structures and estimate relevant physical and geochemical properties, while also incorporating and communicating uncertainty at appropriate spatial scales.

We present the current state of the Loop project (github.com/Loop3D), an open-source, interoperable, and probabilistic 3D geological modelling platform designed for efficient and reproducible construction of geologically realistic 3D models. Loop is built around three core Python libraries:

1. map2loop automates the extraction of geological information from regional maps and prepares structured inputs for 3D modelling.
2. LoopStructural constructs structurally consistent 3D geological models using a time-aware parameterization of geological events. Events such as faulting, folding, and intrusion are represented by structural frames — curvilinear coordinate systems defined by three scalar fields that approximate the principal directions of finite strain. These frames are fitted to structural data and combined according to geological history.
3. LoopResources uses the structural frames to guide geostatistical property estimation, enabling the integration of structural controls into resource modelling.

We demonstrate the complete Loop modelling workflow with a case study, from structural model construction through to resource estimation within the structural frame. By performing resource estimation using different structural frame geometries, we explicitly propagate geological uncertainty into the resource model and observe significantly greater variability in estimated resource volumes compared to traditional geostatistical simulations. This added variability reflects uncertainty in the geometry itself — highlighting which parts of the model most influence the outcome. Incorporating these structural uncertainties enables users to identify and prioritise areas where additional data could reduce risk, offering a more informed basis for exploration and development decisions.

Using the Loop platform to Build Structurally Complex 3D Geological Models: Towards Better Resource Models?

Laurent Ailleres[1,2,3], Lachlan Grose[1,2,3], Robin Armit[2]
[1]Monash University, Melbourne, Australia, [2]PGN Geoscience Pty Ltd, Melbourne, Australia, [3]Loop 3D Foundation Ltd, Melbourne, Australia

Loop is an innovative 3D geological modelling platform designed to build structurally constrained, data-driven, geological models. Loop integrates geological concepts with numerical methods, making it easier to create consistent, geologically realistic models. LoopStructural uses structural geology principles to interpolate surfaces and volumes, offering a flexible, open-source solution that can adapt to a wide range of complexly deformed geological settings. Loop also connects with other tools for geophysical inversion, uncertainty analysis, and visualisation, promoting a reproducible and scientific approach to geological modelling.

In LoopStructural, structural frames define a local coordinate system associated with a geological event, such as a folding event, a fault (or family of faults), an unconformity or an intrusion. Each frame consists of three orthogonal scalar fields related to the finite strain ellipsoid associated with each event; e.g., for fold, the principal shortening direction (perpendicular to the axial surface), the principal extension direction (usually perpendicular to the fold axis and parallel to the axial surface) and an intermediate direction (usually parallel to the fold axis) or for fault, a direction perpendicular to the fault surface, one direction parallel to the offset direction and the third direction being perpendicular to the other two. Structural frames allow Loop to model deformations histories in a time aware sense, meaning that the youngest event is be modelled first until the oldest foliations are modelled (usually bedding or lithological layering).

We demonstrate the capability for polydeformed terrane modelling in Bermagui, a well-known area for structural complexity and exhibiting multiple folding events. The model will be built progressively using field observations of overprinting relationships between dissolution cleavages in sandstones and crenulation cleavage in mudstone of the Ordovician-Silurian turbiditic sequence. Finally, we show how the frame could be easily and meaningfully applied to geostatistical methods and property simulation in complexly deformed geology.

Unlocking Deep-Time Prospectivity: Integration of Multiple Paleogeographic Models with Geochemical, Geochronological, and Thermochronological Big Data

Fabian Kohlmann[1], Wayne Noble[1], Moritz Theile[1], Romain Beucher[1], Xiaodong Qin[2], Brent McInnes[3], R. Dietmar Müller[2]
[1]Lithodat Pty Ltd, Melbourne, Australia, [2]The University of Sydney, Sydney, Australia, [3]Curtin University, Perth, Australia

Effective mineral exploration increasingly depends on robust spatial and temporal understanding of Earth's dynamic geological processes. LithoPlates, an innovative and highly versatile geoscience platform, uniquely addresses this by providing rapid access to multiple, interchangeable global plate tectonic reconstructions extending back to approximately 1.8 billion years. Users can seamlessly switch between different published plate models, enabling immediate comparison and critical evaluation of geological scenarios.

By integrating these plate tectonic reconstructions with comprehensive standardised big datasets, including major and trace element geochemistry, geochronology, thermochronology, and mineral deposit occurrences, LithoPlates delivers a powerful, interactive environment for visualising and interrogating geoscience data through deep time. Advanced dashboards and built-in interpolation and analytical tools facilitate rapid identification of critical relationships between tectonic events, magmatism, sedimentary processes, and mineral deposit formation.

A significant innovation of LithoPlates is its capability to reconstruct geoscience datasets palinspastically, preserving original spatial relationships and ensuring accurate temporal and spatial context for geological interpretation. This functionality is crucial for exploring mineralisation patterns and predicting prospective regions historically influenced by tectonic events, basin evolution, magmatic episodes, and hydrothermal activity.

Moreover, LithoPlates provides integrated Open REST APIs, allowing reconstructed geospatial datasets and rich associated metadata to be seamlessly consumed by external software platforms. This connectivity enables direct integration into advanced prospectivity analyses, including cutting-edge machine learning workflows and automated mineral targeting systems, significantly enhancing exploration efficiency and success rates.

LithoPlates thus represents a critical step forward in deep-time mineral systems exploration, combining flexible multi-model plate reconstructions with robust geoscience data integration and advanced analytical capabilities. This integrated approach empowers explorers to rapidly develop new geological insights, optimise exploration strategies, and improve predictions of mineralisation processes through geological time.

Unlocking the Potential of Core Photography Leveraging SOTA Open-Source Computer Vision Models

Ankita Singh[1], Alexey Khokhlenok[1], Harrison Button[1], Zubair Ahmed[1], Meghal Panchal[1], Maiko Sell[1]
[1]Rio Tinto, Perth, Australia

In mineral exploration, drilling for a target represents the ultimate test of our hypotheses about what lies beneath the Earth's surface. This process generates a wealth of sensor information, including core photography, which provides invaluable insights into subsurface geology. However, these photographic data are often collected in non-analytics formats, limiting their potential for advanced analysis and interpretation.

To address this challenge, Rio Tinto's exploration division has incorporated open-source AI models, such as Meta's Computer Vision models, to transform raw core photos into analytics-ready formats through human-in-the-loop prompts. Rio Tinto has built an end-to-end Python app-based workflow with cloud connections, allowing users to ingest and view core photography from recent or historical drilling projects. This workflow allows human prompts for core and depth markers through a cloud labelling template and invokes open-source models hosted within Rio Tinto's firewalls to carry out core segmentation and depth registration. Ultimately, this allows users to validate the quality of 'analytics-ready' core photography, which can later be used for geoscience-specific machine learning analysis.

By leveraging segmentation-by-prompting capabilities of these open-source models from leading technology and research sectors, mineral exploration can accelerate its analytics workflow. Building a machine learning model from scratch can be laborious and time-consuming, leaving little opportunity to influence the next drill targets based on insights from core photography.

Rio Tinto's approach defines the boundaries where open-source machine learning models intersect with geoscience datasets to create 'analytics-ready' datasets which can be utilized for more specialized geoscience-specific models. The combination of AI-assisted segmentation and human expertise allows for the rapid development of large, diverse datasets that can be used to train generic models for core image segmentation across various geological settings.

Structural Controls on a High-Grade BIF-Hosted Gold System: Umwelt Deposit Case Study, Back River Gold District, Nunavut, Canada

Cam Bartsch[1], Claudia Gordillo[1]
[1]B2 Gold, Vancouver, Canada

The Umwelt Deposit is the largest of five deposits that make up B2 Gold's Goose Project, part of the Back River Gold District (BRGD) located in the Slave Geological Province of the Canadian Shield. The BRGD is underlain by the Archean Beechey Lake Group, a thick turbidite succession in the greater Burwash Basin. Banded Iron Formation (BIF), comprising both oxide and silicate facies, occurs as thin but laterally continuous horizons interleaved in the sedimentary sequence.

The Slave Province underwent multiple contractional events between ca. 2.64 – 2.58 Ga, attributed to a series of collisions along the eastern margin of the Slave protocontinent. The most significant of these events was a protracted D2 shortening event between ca. 2.61 – 2.59 Ga, which resulted in the development of tight to isoclinal northwest-trending folds, attendant axial planar foliation (S2), and localized high strain zones. Syn-tectonic gold mineralization is concentrated in structural traps in the F2 fold architecture, and in veins and vein arrays genetically linked to the S2 fabric. Umwelt hosts combined open pit and underground reserves of over 6.4 million tonnes at > 8 g/t Au, preferentially occurring within the hinge zone and short limb of a moderately southeast-plunging asymmetric F2 fold.

With the advancement of Umwelt toward commercial production in 2025, infill drilling and underground development has provided some clarity on the controls on mineralization that were not fully realized during the early exploration phase. Key observations include: the presence of cryptic high-strain zones; the role of re-folding and second order folds on localizing gold; the influence of syn-kinematic QFP dykes on vein emplacement; and the identification of deformed extensional vein arrays. All help piece together the paragenetic history of the Umwelt deposit that will guide exploration of other BIF-hosted gold targets throughout the Back River Gold District.

Linking Geoscience Research to Orebody Knowledge at the Epizonal Fosterville Gold Deposits

Wessley Edgar[1]
[1]Agnico Eagle, Bendigo, Australia

The obvious economic value of the Swan Zone which evolved from within the Fosterville Gold Mine after 2015 has forever changed the international perceptions of the gold deposits in Victoria, Australia. More has been said about the mine's spectacular gold by promotional junior explorers and industry Brokers than from the mine itself which set to work mining whilst conducting in-house geoscience research. Now, with a majority of the Swan zone exhausted this orebody knowledge is successfully being applied into exploration and additional mining fronts some 5km north of Swan. A deposit that was discovered, drill-defined then mined all within a decade is Swan a singular Black Swan event or case of lightning striking twice? The deposits are an excellent place to build the knowledge to challenge traditional genesis models for orogenic systems – in an era open to novel gold mechanisms - amorphous silica fluids with colloidal metal species and nano-particle gold.

This presentation updates the geological features and in-mine observations, with research conducted since 2020 including deposit-wide multielement assay (4,495), hyperspectral alteration and detailed underground mapping with +370 samples used for +130 petrographic investigations including SEM, DSIMS, Raman Spectroscopic analyses upon a selected range of mineralization forms, host rocktypes and alteration. Sulfide separates were generated for Fission-Track geobarometry and Re-Os geochronology, plus mica Ar-Ar and SHRIMP Pb-U geochronological analysis.

This integrated data supports the Fosterville genesis model and is discussed relevant to the geological criteria which define the bonanza shear veins and sulfide halo, with some reference to other well-known Epizonal deposits in the Central Victorian goldfields. The key outcome is that lightning doesn't have to strike twice as separate hydrothermal events, but it may require progressively evolving fluids and particular local structural geometries and timing, to form the ideal large-scale versions as illustrated for the Swan shear vein.

"Fuchsite" and Chlorite in Carbonated Komatiites at the Kerr-Addison Deposit, Ontario, Canada: Genesis of Listvenites in Orogenic Gold Settings

Derek Leung[1], Andrew McDonald[1], Douglas Tinkham[1], Philip Lypaczewski[1,2]
[1]Laurentian University, Sudbury, Canada, [2]LithologIQ, Montreal, Canada

The association between Cr-bearing green micas (i.e., muscovite var. fuchsite and mariposite) and altered ultramafic rocks (listvenites) in orogenic gold settings is known, yet the genetic relationship between the two remain unclear. Altered komatiites from the 483 t Au Kerr-Addison deposit (Virginiatown, Ontario, Canada) were investigated to: (1) determine how muscovite var. fuchsite forms in ultramafic-hosted orogenic gold deposits (via petrography, SEM-EDS, and thermodynamic modelling); (2) apply mineral chemistry of muscovite and chlorite to elucidate hydrothermal processes and refine exploration techniques (using SWIR reflectance spectroscopy); and (3) relate trace-element mineral chemistry to lithogeochemistry for exploration in ultramafic-hosted terranes (using LA-ICP-MS and lithogeochemistry). Results show:

(1) The alteration assemblages of komatiitic protoliths (i.e., hydrated komatiite, regional carbonation, talc-chlorite, carbonate-chlorite, and carbonate-muscovite facies) and reaction textures (tremolite → dolomite, talc → magnesite + quartz, and chlorite → muscovite) are compatible with thermodynamic modelling, suggesting progressive carbonation in the presence of K+.

(2) The Al content in chlorite increases proximally via Tschermak substitution [[6](Mg,Fe2+) + [4]Si4+ ↔ [6]Al3+ + [4]Al3+], consistent with progressive carbonation of chlorite-bearing facies rocks. This zonation correlates with an increasing ~ 2250 nm peak position for chlorite in SWIR reflectance spectra.

(3) The Li content in chlorite increases proximally, from 55 to 400 ppm, interpreted to represent fluid contributions based on lithogeochemical additions in the carbonate-chlorite facies. Within the carbonate-muscovite facies, the muscovite does not show significant zonation in major to trace elements.

In conclusion, carbonation is a key process in the formation of listvenites in orogenic gold deposits. The extent of carbonation may be mapped on a deposit scale by interpreting the Al content in chlorite via SWIR reflectance spectroscopy. Given the lack of spatial variability recorded in muscovite chemistry at Kerr-Addison, muscovite chemistry may be useful for global-scale comparisons of ultramafic-hosted gold deposits over time and space.

Mafic Magmatic Origin of Orogenic Gold Systems

Daniel Wiemer[1], Steffen G. Hagemann[2], Anthony I.S. Kemp[2], Walter K. Witt[2], Jon Hronsky[3], Graham C. Begg[4], Trevor Ireland[5], Chris Fisher[2], Laure Martin[2], Nicholas Hayward[6], Carlos Villanes[7]
[1]Economic Geology Research Centre, James Cook University, Townsville, Australia, [2]Centre for Exploration Targeting, University of Western Australia, Perth, Australia, [3]Western Mining Services Pty Ltd, Perth, Australia, [4]LithoAtlas Pty Ltd, Perth, Australia, [5]School of the Environment, University of Queensland, Brisbane, Australia, [6]PredictOre Pty Ltd, Perth, Australia, [7]Compania Minera Poderosa S.A., Lima, Peru

Orogenic gold deposits are classically associated with metamorphic dehydration of crustal rocks, and over-pressurized fault valve mechanisms mainly at the mesothermal brittle-ductile transition. Granites, as alternative sources of Au-ore fluids are considered viable, if spatial and temporal relationships with gold mineralization are established. However, granites are not always observed in gold camps, or pre- or post-date mineralization. Furthermore, degassing of common granitic arc magma during crustal ascent precludes expulsion of typical orogenic aqueous-carbonic fluids (XH2O = 0.75-0.95 mol%) at meso- to hypozonal crustal levels (~2-6 kbar). Higher Au-solubility and metal-carrying capacity is shown for mafic magmas, especially for hydrous alkaline members. The latter, however, are volumetrically subordinate, questioning whether they can produce large deposits.

We report on low-preservation-potential incompatible element enriched hydrous mafic rocks as gold parental magmas. Petrological modelling demonstrates initial ascent of hydrous amphibole-rich magma mush stalling in the deeper mid-crust (>20 km; not exposed at mesozonal levels; common arc geotherm ~75 mW/m2). Near-solidus magmatic mineral assemblages (~660°C, 7kbar) match Ti-in-zircon temperatures spanning sub-solidus replacement (epidote) and thermal rejuvenation during magmatic recharge (amphibole breakdown). Antecrystic zircon confirm that the magmatic plumbing system evolved over >30Myrs through episodic (~5-Myr-duration) mafic recharge prior to final emplacement at mesozonal crustal levels. We show that volatile saturation and degassing of previously underestimated CO2-H2O contents during crustal passage can generate orogenic-like fluids. A mechanism for deep-seated fluid expulsion is recorded in observed amphibole breakdown during recharge. Hafnium-O in zircon isotopic data inform on a pre-enriched subduction-related metasomatized upper mantle source for the Au-parental mafic magma. For the first time, we present a power-law-like correlation between apparent trends for different Au-grades as a function of starting parental magma compositions (>5,500 data), showcasing a reduced grade-potential with increasing silica index confirming the higher capacity of mafic endmembers to produce higher-grade (and likely larger) deposits.

Red Hill: Unveiling the Geochemical and Hydrothermal Evolution of a Granitoid-Hosted Orogenic Gold System in the Yilgarn Craton, Western Australia

Renan De Souza[1], Guillermo Friera[1], Jim Sinclair[2], Carl Young[2]
[1]Northern Star Resources Ltd, Growth Kal Ops, Kalgoorlie, Australia, [2]Northern Star Resources Ltd, Geological Services, Perth, Australia

The Red Hill gold deposit, located in the Boorara Domain of the Eastern Goldfields Superterrane, represents an orogenic gold system hosted within the Red Hill granodiorite stock. Gold mineralisation is closely associated with sub-planar quartz-carbonate vein arrays. Geochemically, the granodiorite displays a calc-alkaline affinity and a metaluminous to weakly peraluminous character, with high Sr vs Na_2O + CaO (wt%) ratios characteristic of high-SiO_2 adakites. However, elevated Mg#, Ni, Cr, and incompatible element concentrations (Sr/Y, Cs, Ba, LREE), along with depleted HFSE levels (Nb, Ta, Zr, Hf, Ti), align more closely with a sanukitoid-like granodiorite composition.

The Red Hill granodiorite is hosted within the mafic-ultramafic clast-rich Ballarat Conglomerate and has an elongated geometry along the axis of a SW-plunging regional syncline. The granodiorite and surrounding stratigraphy are transected by NNW- and NE-trending shear zones. The deposit comprises multiple generations of quartz-carbonate veins with different wallrock alteration assemblages and variable gold content. An early barren halo of hematite-carbonate (dolomite to ankerite) alteration is overprinted by an inner albite-quartz-ankerite-sulphide assemblage, manifested as microcrystalline albite and quartz replacement of the groundmass. This later alteration phase is intimately associated with the formation of gold-bearing sulphide stringers and quartz-carbonate veins. In addition to these alteration phases, a white mica–dominated alteration phase is also present.

This study integrates datasets acquired from laser-induced breakdown spectroscopy (LIBS), computed tomography (CT) radiographs, and XRF analyses (Orexplore GeoCore X10) collected on half core, complemented by four-acid digest ICP multi-element analyses on pulverised half core samples. These datasets are used to characterise the mineralogical and geochemical evolution of hydrothermal alteration and vein development, which has implications for a revised genetic model. The results of the study provide a robust mineral system framework to support well-constrained resource and geometallurgical models.

Sunday Creek: High-Grade Gold-Antimony Discovery in Clonbinane, Victoria

Kenneth Bush[1], Michael Hudson[1]
[1]Southern Cross Gold Consolidated, Clonbinane, Australia

Sunday Creek is located within the Melbourne Structural Zone in the Lachlan Fold Belt. The regional host to the Sunday Creek mineralisation is an interbedded turbidite sequence of siltstones and minor sandstones metamorphosed to sub-greenschist facies and folded into a set of open NW trending folds.

Mineralisation is structurally controlled, with increased mineralisation associated with brittle-ductile shear veins that show quartz-stibnite extension veining, stibnite-gold-matrix breccias and disseminated mineralisation in the form of arsenian pyrite, pyrite and arsenopyrite. The main host for mineralisation is an east to north-east trending zone of intensely altered 'bleached' silica-sericite-carbonate siltstones, mudstones and sandstones and a silica-sericite-carbonate-fuchsite altered dyke rocks that ranges from 20 - 110 m wide. A large arsenic anomaly is associated with gold mineralisation, mostly represented by disseminated pyrite with proximal arsenopyrite bearing zones in close proximity to veining rounds out the larger visible alteration footprint.

Mineralised vein arrays cross the host structure on a predominate north-west orientation with individual vein sets defined over 600 m depth extent from surface to over 1,100 m below surface,2.5 m to 3.8 m wide (median widths) (and up to 10 m), and 20 m to 100 m in strike. Each of these zones repeats every 10 m to 20 m within the Apollo, Rising Sun, Golden Dyke and Apollo areas with over 70 vein sets currently defined to date. High-grade "cores" are observed within vein arrays and have much more complex multi-phase textures and high antimony grades (+5% Sb) and typically have visible gold or +20g/t Au up to 1000+ g/t Au values.

At Sunday Creek, and as is typical for epizonal deposits (for example Fosterville), visible gold becomes increasingly significant at depth below approximately 700-800 m. This represents the different temperatures and changes in structural regimes of formation of epizonal Au-Sb and Au dominant mineralisation.

Expanding the Chronological Evolution of Winu Cu-Au Deposit (Western Australia): Insights from Modern In Situ Titanite-Apatite-Calcite Petrochronology

Bruno Vieira Ribeiro[1], Jade Palmer, Hilke Dalstra, Christopher Lawrence Kirkland, Ana Despaigne Diaz
[1]Timescales of Mineral Systems Group, Curtin University, Perth, Australia

Mineral systems evolve through a protracted process with multiple stages of fluid activity and mineral alteration. Such scenario promotes mineral recrystallisation and fluid-temperature driven isotopic resetting of key geochronometers, challenging to establish the chronology of mineral systems and its genetic model. Modern petrochronology opens a new avenue to elucidate the chronological evolution of mineral systems. We present novel in situ titanite, apatite and calcite U-Pb dates from the concealed Winu Cu-Au deposit located in the Great Sandy Desert, Western Australia. Winu is hosted in metamorphosed massive sandstones and siltstones, with mineralisation associated with numerous thin, brittle veins and breccias. Although molybdenite Re-Os model ages suggest chalcopyrite precipitation between 658-655 Ma, the timing of vein formation is uncertain. Medium-to fine-grained anhedral titanite disseminated in polymineralic fine-grained matrix (quartz, alkali feldspar, micas, chlorite and zoisite) yield U-Pb ages of 668 ± 3 and 666 ± 3 Ma, whereas medium- to fine-grained anhedral apatite disseminated in the matrix yield U-Pb ages spanning 670–600 Ma. These data suggest that both phases record the regional high-temperature metamorphism and ductile deformation in the Paterson Orogen, which promoted early hydrothermal fluid activity as identified in the Nifty Cu deposit. Yet, fine-grained, disseminated calcite grains in polymineralic matrix yield the oldest U-Pb date of 522 ± 17 Ma, whereas calcite veins yield Carboniferous to Permian dates spanning 375–270 Ma. Although often associated with mineralised assemblages, the calcite U-Pb dates apparently record the timing of large-scale intracontinental reactivation of craton margins coeval with the development of the Canning Basin. The study reinforces a multi-stage evolution, with early hydrothermal activity associated wlth Neoproterozoic tectonism and later overprints coinciding with intracontinental extension events. These integrated datasets provide new constraints on the timing of structurally controlled mineralization at Winu, refining exploration models for orogenic-related ore systems in the region.

The Geology and Breccia Evolution of the Havieron Breccia-Hosted Au-Cu Deposit

Isaac Brown[1], Lejun Zhang[1], David R. Cooke[1], David Finn[2], Anthony C. Harris[2]
[1]University of Tasmania - CODES, Hobart, Australia, [2]Newmont Corporation, Melbourne, Australia

The Havieron breccia-hosted Au-Cu deposit was discovered in 2018 under approximately 420 m of post-mineralisation Permian glacial cover. It is located in the Paterson Province of Western Australia and hosted by lower greenschist metamorphosed sedimentary rocks of the Yeneena Basin, which also host the Telfer and Winu Au-Cu deposits. Pre-mining resources estimate that Havieron contains >7 Moz Au and 275 Kt of Cu. Mineralisation at Havieron is exclusively hosted in a large breccia complex that preserves evidence of the evolution of brecciation and mineralisation. The initial stages of brecciation evolved from quartz- and actinolite-cemented to actinolite-, biotite-, and carbonate-cemented breccias, locally with corroded clasts. Later stages of breccia formed through solution collapse and were cemented by carbonate. Rheological and chemical contrasts between the host meta-siltstone and meta-sandstone units containing variable carbonate content, and with dolerite sills localised breccia formation and high-grade mineralisation. Havieron has an unusual Au-Cu-Bi-Te-Ni-Co-W-REE-Sn metal association, reduced (pyrrhotite-stable) minerology, intimate connection to magmatism and is hosted by a large scale magmatic-hydrothermal breccia complex. Havieron is therefore interpreted to be a variation on the proximal style of intrusion-related Au-Cu deposit, that contrasts with the distal intrusion-related Au-Cu deposits that occur at Telfer and Winu. The recent discoveries of Winu and Havieron highlight the varied deposit styles and prospective nature of the Paterson Province, and the importance of effective exploration under post mineralisation cover.

What Sulfur Isotopes Reveal About Ore Formation in the Southwest Yilgarn Craton

Paul Duuring[1,3], Ignacio González-Álvarez[2,3]
[1]Geological Survey Of Western Australia, Perth, Australia, [2]CSIRO, Mineral Resources, Discovery Program, Perth, Australia, [3]Centre for Exploration Targeting, The University of Western Australia, Perth, Australia

Multiple sulfur isotopes ($\delta^{34}S$, $\Delta^{33}S$, $\Delta^{36}S$) provide critical insights into the evolution and origin of sulfur in mineralizing systems. This study uses in situ secondary ion mass spectrometry (SIMS) to analyse sulfide minerals from 13 mineral deposits across the southwest Yilgarn Craton, Western Australia, encompassing porphyry, volcanic-hosted massive sulfide (VHMS), and granulite-facies gold systems.

Porphyry and several VHMS deposits yield near-zero $\Delta^{33}S$ values and $\Delta^{36}S/\Delta^{33}S$ ratios indicative of mass-dependent fractionation, consistent with a dominantly mantle-derived sulfur source. In contrast, metamorphosed gold systems show broader $\Delta^{33}S$ ranges (−0.7 to +1.2‰) and mass-independent fractionation signatures, pointing to the involvement of Archean sedimentary sulfur reservoirs. Some deposits are isotopically homogeneous, while others record multiple, compositionally distinct fluid sources within a single mineralizing event.

These results demonstrate the retention of sulfur isotope heterogeneity across deposit types, even in high-grade metamorphic settings. The findings underscore the utility of multiple sulfur isotope analyses for tracing fluid sources, reconstructing ore-forming processes, and distinguishing crustal–mantle interactions in Archean terranes. This study provides a geochemical framework to refine metallogenic models and guide exploration in the southwest Yilgarn Craton.

Quantifying Rifting History of the Proterozoic Mt Isa Superbasin and Its Relationship to Base Metal Mineralisation

Edgar Leong[1], Mark Hoggard[1], Haining Chang[1], George Gibson[1], Karol Czarnota[2]
[1]Australian National University, Canberra, Australia, [2]Geoscience Australia, Canberra, Australia

A series of world-class sediment-hosted mineral systems are located in the Carpentaria Zinc Belt of northern Australia. Despite over a century of mining, the first-order geodynamic setting of these Proterozoic basins remains highly debated and hampers efforts to expand exploration into frontier regions, particularly those areas overlain by younger cover. Here, we combine field observations, seismic reflection profiles, and borehole data to reconstruct the depocenter of Mt Isa superbasin stratigraphy in the vicinity of the Lawn Hill Platform. We perform backstripping analysis to extract its subsidence history and undertake rift modelling to constrain the timing and intensity of extension. Our results support a link between active extension and formation of sediment-hosted mineral systems, while IOCG emplacement post-dates rifting events. The analysis provides a novel framework for assessing prospective time periods for mineralisation in frontier basins by using their respective subsidence histories.

Magma and Salt: The Discovery of the Oktyabrsky Supergiant and the Role of Salt-Magma Location on Genesis and Value

Steve Beresford[1]
[1]University Of Western Australia, Australia

The Oktyabrsky Cu-PGE-Ni deposit is the world's most valuable metal deposit of any type. The deposit was discovered in 1965 in the Noril'sk district, Russia, at a depth and complexity of exploration searchspace radically different from the original discovery of the smaller Noril'sk deposit. Oktyabrsky is worth over 50 times the insitu value of Noril'sk!

The discovery of Nori'lsk is well documented in Russian, but the discovery of Oktyabrsky is poorly documented. In this contribution we integrate work on the history of the discovery of the deposit, with new work on the role of magma intrusion interaction with evaporites, and in particular the relative location of magma emplacement with respect to salt position within the target sedimentary basin.

The location of autochthonous salt is considered the primary ingredient that controls both the deposit quality and discoverability. Targeting datasets thus shift away from magmatic processes towards basin studies with new prediction and detection tools. We integrate this basin analytical approach to new geochemical and mineralogical data on evaporite sub types.

Oktyabrsky represents a rare deposit style but with a highly salient deposit footprint linked to economic metrics. At a moment when timeline to discovery and capital intensity are considered key metrics, this deposit sub type represents a premium alternative source of Copper, let alone Ni, PGE, and semi-metals.

The results of this integrated basinal approach is applied to the Central Lapland Greenstone belt in Finland, and is the subject of numerous ongoing 'real world' exploration programs globally.

A Remote-Predictive Approach to the Geodynamic Evolution of the Coriolis Troughs

Elizabeth Xu[1], Melissa Anderson[1]
[1]University Of Toronto, Toronto, Canada

Seafloor massive sulfides (SMS) are a modern analogue of volcanogenic massive sulfide (VMS) deposits which provide a window into the metallogeny of these deposits. SMS deposits are formed via the circulation of hydrothermal fluids in the oceanic crust and have been previously found in extensional settings such as mid-ocean ridges and backarc basins. However, it is not clear at what stage large hydrothermal systems develop relative to backarc development. This knowledge gap will be addressed with a remote-predictive approach in the southern Vanuatu Coriolis Troughs—a nascent backarc rift that extends between ~17°41′S and 20°3′S. Additionally, the structural and magmatic evolution of the Coriolis Troughs may be influenced by the many tectonic processes it hosts, such as ridge collisions, hinge retreat, and slab tear. As such, this study aims to improve our understanding of the structural and magmatic evolution of nascent backarc rifts through a remote-predictive mapping approach. The structural and geological mapping was completed at a scale of 1:100,000 in ArcGIS Pro, using available ship-track bathymetry, satellite altimetry, and geophysical datasets. For optimal visualization of structures, the bathymetric data was transformed using geoprocessing tools. Geologic units are identified and classified based on the completed structural map, seafloor morphology, and geophysical datasets, including vertical gravity gradient (VGG), Earth Magnetic Anomaly Grid (EMAG), and side-scan sonar data. In seismically active regions, shallow centroid moment tensors (CMTs), will be compared with the mapped structures to interpret fault kinematics, their stress regimes, and how geodynamic events control the formation and development of new structures. The mapping outcomes of this study will help us to better understand the geodynamic evolution of nascent backarc rifts and their spatial and temporal relationship with metallogeny. This will help improve future exploration strategies.

Integrating Machine Learning and Conditional Simulation for Geometallurgical Block Modelling: A Case Study from the Resolution Copper Project

Cecilia Artica[1]
[1]University of Queensland, Brisbane, Australia, [2]First Quantum Minerals, Perth, Australia

Block models used for mineral resource estimation and grade control are fundamental to mine planning and design. Traditionally, these models focus on revenue-generating metals and follow well-established industry standards such as JORC and NI 43-101. However, as orebodies become more complex and lower in grade, the need to consider non-revenue-generating geological features in block models becomes increasingly important. These include deleterious elements (e.g., arsenic), minerals that impede processing (e.g., talc), and physical properties (e.g., rock hardness) that influence recovery, throughput, and energy consumption. Geometallurgy addresses these considerations by linking geological characteristics to metallurgical performance and overall project economics.

This study introduces an integrated geometallurgical modelling approach applied at the Resolution Copper Project, focusing on rock hardness as measured by Bond Work Index (BWi). To overcome the challenge of sparse BWi test data, machine learning techniques were used to predict BWi values from well-sampled geological variables. A Gradient Boosting Machine (GBM) was developed using six key geochemical predictors (Al, Be, Cs, Mg, Mo, Zn) and lithology. The model achieved strong predictive performance ($R^2 = 0.75$; RMSE = 1.2 kWh/t).

To account for spatial variability and uncertainty, geostatistical simulations were conducted. Lithology was simulated using the Hierarchical Truncated Pluri-Gaussian method, while continuous variables were transformed via Projection Pursuit and simulated using the Turning Bands method. These simulations preserved multivariate relationships and spatial continuity.

The trained GBM was applied to the simulated datasets, generating a geometallurgical block model of BWi that captures both spatial variability and uncertainty. This model enhances predictive capacity for mill throughput and supports better-informed mine planning decisions. The study demonstrates how integrating machine learning with geostatistics enables the creation of practical and predictive geometallurgical models, adding substantial value to resource development and operational planning.

A Machine-Learning Workflow for the Automated Stratigraphic Interpretation of Pilbara Iron Ore Stratigraphy

Elisabeth Scibiorski[1], Jack Maughan[1], Yasin Dagasan[1], Rian Dutch[1], Bhanu Voutharoja[1], Leam Gatehouse[2], Eric Hein[2], Patric Wilson[2]
[1]Datarock, Australia, [2]Rio Tinto Iron Ore, Australia

Accurately modelling subsurface stratigraphy is crucial for iron ore mining, but manual stratigraphic interpretation is a time-consuming process with varying consistency. We developed a machine learning (ML) workflow to automate interpretation at an iron ore deposit in the Pilbara, Western Australia, using a combination of conventional downhole wireline geophysics (gamma, density, magnetic susceptibility) and geochemical assay data.

The workflow comprises a nested XGBoost model architecture across five levels, designed to predict: detrital/bedrock boundary; mineralised/waste classification; weathering status; stratigraphic unit; and strand unit (a subdivision of stratigraphic unit). Predictions are generated for each 2m geological interval in turn, starting with the detrital/bedrock model. Selection and application of the next model in the sequence depends on the result of the previous model.

Post-prediction processing is used to reduce noise and improve geological continuity. Because the XGBoost models treat each interval in spatial isolation, the initial sequence of predictions may not be geologically reasonable; e.g., multiple alternating detrital and bedrock intervals within a single drillhole. The softmax class probabilities are used to smooth the initial predictions, with algorithms ranging from the enforcement of a single detrital/bedrock boundary at the most probable location, to the use of a Hidden Markov Model for the stratigraphy and strand models.

A robust flagging system is implemented to detect anomalies in ML predictions, both within individual drillholes and in a spatial context. Downhole flagging checks for internal consistency against constraints such as minimum unit thicknesses, and spatial flagging evaluates prediction continuity with neighbouring drillholes.

Overall, the automated workflow is able to produce geologically reasonable stratigraphic predictions, and the models perform well when training data is sufficiently representative of the target geology. These results demonstrate that automation via ML is a promising route to improve the consistency, reproducibility, and efficiency of stratigraphic predictions in iron ore deposits.

Challenges in Validating High-Resolution In-Hole Geoscience Data in Metalliferous Open Pit Mining

John Jackson[1], Camilo Riquelme[2], Mel Quigley[3]
[1]IMDEX Ltd, Brisbane, Australia, [2]IMDEX Ltd, Santiago, Chile, [3]IMDEX Ltd, Sydney, Australia

Recent technological advances now facilitate the routine acquisition of high-resolution, high-density in-hole geoscience data from blast and grade control holes. Emerging technologies such as FastGrade/OreSight, BLASTDOG, and Iberia complement traditional wireline logging; however, their adoption in metalliferous operations typically remains sporadic and project-specific.

Effective in-field validation of these data systems is critical both at initial deployment and throughout ongoing operations. Validation involves multiple layers—from sensor calibration and data integrity, site objectives to adoption—and each introduces distinct challenges. While deposits within similar geological frameworks may share many attributes, factors such as deposit geology, operational goals, workforce experience, and local site practices can significantly vary, complicating consistent validation approaches.

Here we present insights and practical learnings from validating the BLASTDOG system at porphyry copper and iron ore operations. The in-hole BLASTDOG data acquired at a 1 cm resolution is typically validated against traditional top-of-hole drill-chip sampling, introducing issues related to spatial mismatch, sample representativity, and analytical accuracy. Further complexity arises when comparing measured physical properties (e.g., magnetic susceptibility, gamma, conductivity) to derived top-of-hole parameters such as NIR-based mineralogy, geochemical assays, and geological logging. In an iron ore case study, outputs were validated against laboratory metallurgical test results highlighted additional interpretative and measurement uncertainties.

Moreover, successful technology adoption is influenced by the differences in the knowledge and experience of site personnel. This difference is reflected in variability of the validation data especially relating to the geological logging of drill chips, understanding of the relationships of alteration, veining and mineralisation in the validation areas. The level of knowledge and experience also impact in resistance to changes in process.
Ultimately, while high-resolution downhole tools significantly enhance orebody knowledge, their validation and adoption demand rigorous, site-specific strategies that consider both technological and human factors.

AI-Driven Core Logging and Drill Optimization Using High-Resolution Core Scanning: Enhancing Geological Models and Reducing Exploration Risk

Grant Sanden[1], Patrick Redmond[1], John Manchuk[1], Michelle Legat[1], Ana Chiquini[2]
[1]GeologicAI, Calgary, Canada, [2]Resource Modeling Solutions, Edmonton, Canada

The successful shift to clean energy and sustainable development hinges on timely access to large quantities of critical minerals. Yet, mineral discovery rates are falling, costs are rising, and the path from discovery to production is slower and riskier than ever. Many mines underperform in their early years, delaying returns and deepening the global supply gap. A key factor behind these challenges is inadequate geoscience data collection and analysis, which leads to poor deposit characterization resulting in unclear risk assessment, suboptimal mine design, and weakened decision-making.

Drill core data remains the most critical and costly component in evaluating a mineral deposit—yet it is rarely collected and utilized optimally. High-resolution core scanning is transforming how the exploration and mining industry acquires drill core data and generates core logs. Traditional logging is slow, subjective, and inconsistent. In contrast, dense multi-sensor datasets from core scanning, when integrated with geological knowledge and spatial context, enable auto-logging and auto-domaining through advanced machine learning algorithms. The resultant digital logs are consistent, spatially coherent, timely, and cost-effective. They provide a firm foundation for geological modelling, downstream risk quantification, and decision making. Digital logs feed directly into a centralized “digital core table” enabling remote access and more collaborative interpretation supported by AI-generated insights that improve efficiency, consistency, and data quality. Crucially, real-time scanning data can now inform drill hole optimization during active drilling campaigns, allowing for adaptive decision-making focused on zones of high geological uncertainty. This maximizes the value of each meter drilled while guiding drill hole spacing strategies that align with resource classification targets and budget constraints.

This presentation will feature real-world case studies that demonstrate the application of these integrated workflows, highlighting their impact on cost reduction, accelerated decision timelines, and improved engineering precision.

How confident are We in Our Models? Unidentified Dataset Dependencies may Inflate Machine Learning Performance Metrics

Taryn Scharf[1], Matthew Daggitt[2], Luc Doucet[1], Christopher Kirkland[1]
[1]Curtin University, Perth, Australia, [2]University of Western Australia, Perth, Australia

Machine learning algorithms are increasingly used to develop predictive models across Earth Science including mineral systems. The accuracy of reported model performance is important if models are to reliability inform decision making. However, unidentified data dependencies can contribute to 'data-leakage' between training and testing datasets. In effect such leakage promotes model overfitting. This may cause a model to have artificially higher performance on the test set and reduced generalisability.

Geological datasets are often hierarchical, with multiple analyses (e.g. assay data) drawn from a single sample (e.g. drill core), and multiple samples themselves relating to a single phenomenon (e.g. mineralised region). Here we demonstrate the effects of hierarchical data structures on algorithm performance by simulating a dataset of samples comprising multiple analyses that are related through latent variables (e.g. individual analyses linked to a common compositional sample). With this simulated dataset, we illustrate conditions that exacerbate the impact of data-leakage in hierarchical datasets. We then demonstrate this phenomenon in real-world datasets.

This work highlights that dependencies are often present in geological datasets. If these dependencies are unidentified and untreated, they may result in over-optimistic estimates of model performance.

A New Paradigm for Geoscientific Insights in the Digital Era: Large Scale Modelling of Orogenic Systems

Jean-Phillipe Paiement[1]
[1]Vrify Technology, Vancouver, Canada

We present a fully automated, data-centric workflow for mineral system exploration that converts heterogeneous geoscientific inputs into high-resolution, probabilistic prospectivity maps within a unified processing framework. Raw datasets—including surface and down-hole geochemistry, airborne and ground geophysics, structural lineament inventories, and 3D wireframe models—are ingested via schema-driven parsers and subjected to robust quality control (outlier detection, spatial–temporal alignment). Domain-aware feature engineering extracts exploration-relevant signals through computer vision decompositions of potential-field grids, tensor-based structural fabric descriptors, automated lineament detection and probabilistic lithological indices derived from geochemical ratios, spectral signatures and lithological labels. Furthermore, generative predictive models reconstruct sparse legacy data (e.g., lake sediment geochemistry, legacy gravity surveys or soil survey data) to augment under-sampled areas. A multi-task deep learning model, trained on >130 000 ground-truth samples across orogenic Au, porphyry Cu–Au–Mo, and LCT pegmatite systems, performs mineral system predictions and pixel-level property reconstruction. We show an example for a large-scale orogenic belt, where model interpretability is achieved via SHAP values enabling understanding of feature importance to predictive modelling results. Application to the Abitibi Greenstone Belt demonstrates accurate recovery of known gold corridors, identification of blind targets beneath cover, and new understanding of multi-scale controls on mineralization. Specifically, an interpolated absolute bouguer anomaly grid, decomposition filter of a reduce to pole magnetic intensity grid, predictive surface geochemical response grids, and simulated strike field grids were strong predictors for known Au systems. These insights demonstrate that AI tools can be used to map complex and subtle multiparameter patterns related to mineralized systems from public domain data for efficient exploration area prioritization. The presented workflow delivers rapid, reproducible, and transparent subsurface insights to guide exploration targeting.

Retrograde Overprint of the MEM2 Gold Deposit: Evidence for Polyphase Mineralization Along the Antimony Line, Murchison Greenstone Belt, South Africa

Luke Carlton[1], Alexander Kisters[2], Gary Stevens[2]
[1]Tect Geological Consulting, Cape Town, South Africa, [2]Stellenbosch University, Cape Town, South Africa

The Antimony Line is a 35 km-long shear zone in the central parts of the Murchison Greenstone Belt of South Africa. It exhibits both gold-only and antimony-gold polyphase mineralization, which is uncommon within most Archean greenstone belts. Despite this atypical mineral endowment, the conditions under which this polyphase mineralization developed are still poorly constrained, as previous studies have exclusively focused on its world-class antimony-gold mineralization.

This study characterizes the disseminated, gold-only mineralization of the MEM2 deposit, which is located along the eastern extent of the Antimony Line. The 3D modelling, structural mapping and petrographic analyses of the wallrocks and mineralization reveal that mineralization developed as vein-hosted gold-sulphide mineralization during higher-temperature (T >500 °C) amphibolite-facies conditions. Hosted within discrete muscovite-biotite-quartz-carbonate schists within a larger, unmineralized talc-chlorite-quartz-carbonate schist package, the vein-hosted gold mineralization has subsequently been deformed, transposed and overprinted by greenschist-facies shear bands which reflect the pervasive greenschist facies (T ~450-350 °C) retrogression of the host rocks of the Central Murchison Unit. The intensity of this structural overprint is such that, on a micro- to deposit-scale, gold mineralization has been completely remobilized and attenuated into the pervasive greenschist-facies subvertical foliation and steeply plunging lineation fabrics of the Antimony Line.

The recognition of this higher-temperature, amphibolite-facies gold-only mineralization and its subsequent greenschist-facies overprint, allows for a better temporal contextualization of the gold-only and antimony-gold polyphase mineralization along the Antimony Line. This study suggests that the gold-only mineralization along the Antimony Line occurred earlier within the metamorphic evolution of the Central Murchison Unit with antimony-gold mineralization developing during later, lower-temperature, greenschist-facies conditions.

IOCG, ISCG and Genetically Affiliated Deposits in the Cloncurry District, Queensland, Australia: Diverse Products of One Mineral System

Vladimir Lisitsin[1], Courteney Dhnaram[1], Elena Belousova[1], Alkis Kontonikas-Charos[1], Rhiannon Jones[1]
[1]Geological Survey of Queensland, Brisbane, Australia

The Cloncurry district in the eastern Mount Isa Province hosts over 50 Cu±Au±Co±REE Proterozoic mineral deposits commonly classified as Iron Oxide Copper-Gold (IOCG) or Iron Sulfide Copper-Gold (ISCG) deposits. The deposits have diverse mineralogical and geochemical signatures, with major implications for mineral exploration. Over the past decade, the Geological Survey of Queensland (GSQ), in collaboration with CSIRO and multiple other research and industry partners, have undertaken a program of systematic characterisation of these deposits, focusing on their mineralisation and alteration signatures, and distal footprints. The program covered the 15 largest deposits in the district (including Ernest Henry, SWAN, Osborne, Eloise, Jericho, Little Eva, Mt Dore, E1, Rocklands and Kalman) and numerous smaller deposits. Consistent data acquisition included continuous drill core scanning (HyLogger, Minalyzer), comprehensive multi-element geochemistry (68 elements), micro-XRF scanning (Maia Mapper, Bruker Tornado), BSE-EDS mineralogy (TESCAN TIMA), mineral chemistry (chalcopyrite, pyrite, magnetite, apatite, garnet) – including isotope geochemistry (Cu, Zn, Fe, Nd, S, C and O and geochronology of ore and alteration minerals – U-Pb, Re-Os, Ar-Ar, Rb-Sr). This data coverage, variety and consistency enable an objective district-scale metallogenic assessment of the Cloncurry district.

Copper-dominant deposits in the district have a common Cu-Au-Ag-S-Te-Se ore geochemistry, but there are several distinct geochemical signatures within the group. Ernest Henry and E1 have significantly enriched Te-Cu-Re-As-Au-Mo-Se-Bi-Co-W-Sb-Ag-Ba-F and proximal K-feldspar-magnetite alteration, whereas the other deposits have different (generally much simpler) geochemical signatures and predominantly Na-Ca-Fe, Fe-Ca and K-Ca proximal alteration. In addition to common magnetite-dominant IOCGs (K-feldspar-magnetite breccias, actinolite-carbonate-magnetite veins, 'IOCG Cu-REE skarns' – each with distinct ore geochemistry) and pyrrhotite-dominant ISCG vein deposits, the district also hosts genetically affiliated low-iron alkali-calcic Mo-Re±Cu, Fe, U, REE and Au deposits. The extreme metallogenic diversity in the Cloncurry district is the product of a long-lived (~100 Ma) Metasomatic Iron Alkali-Calcic (MIAC) mineral system.

Sulfur Isotope Fingerprints of Sediment-Hosted Copper Deposits in Mt Isa Western Succession, Australia

Jemi Ahnaf[1], John Mavrogenes[1]
[1]Research School of Earth Sciences, The Australian National University, Canberra, Australia

The genesis of sediment-hosted copper deposits in the Mt Isa Western Succession has been extensively debated, particularly regarding the syngenetic vs epigenetic models. One way to potentially clarify such models is to track the evolution of sulfur isotopic fingerprints of sulfide minerals. Here we present sulfur isotopes for the Mammoth sediment-hosted deposits from SHRIMP-SI, in addition to data from Lady Annie, Lady Loretta, and Mt Isa (from the literature). Based on our paragenetic study, pyrite formed at several stages, mainly during diagenesis (syn-diagenetic pyrite) and afterwards (epigenetic pyrite), two events distinguishable by sulfur isotopes. The median values of sulfur isotopes for Mammoth, Lady Annie, Lady Loretta, and Mt Isa syn-diagenetic pyrite are -4 ‰, 27‰, 11‰, and 5‰, respectively and median values of sulfur for Mammoth, Lady Annie, Lady Loretta, and Mt Isa epigenetic pyrite are -11‰, 20‰, 6‰, and 10‰, respectively. The ore-bearing copper sulfides such as chalcopyrite and bornite yield sulfur isotopes closer to the epigenetic pyrites than the syn-diagenetic pyrites, indicating ore precipitation under reducing conditions, which supports the epigenetic model. Syn-digenetic pyrite with enriched sulfur isotopes is interpretated to have formed from sulfate-reducing bacteria while epigenetic pyrite and copper sulfides with moderate to depleted sulfur isotopes are linked to copper-rich hydrothermal fluids. The distinctive depleted sulfur isotopes of both syn-diagenetic and epigenetic pyrite at Mammoth are interpreted as due to fluid-host-rock interaction, as Mammoth is the only copper deposit in the region hosted by silicious sandstone while others are hosted by dolomitic shales.

Orogenic Copper Systems: A Global Perspective and Exploration Framework

Matthew Hales[1,2,3], Christopher Lambert[1], Ian Basson[1]
[1]Tect Geological Consulting, CAPE TOWN, South Africa, [2]University of the Witwatersrand, Johannesburg, South Africa, [3]DSI-NRF Centre of Excellence for Integrated Mineral and Energy Resource Analysis, Johannesburg, South Africa

Copper mineralisation is found across a diverse range of tectonic settings, including porphyries found in collisional settings, to deposits such as Stratiform Sedimentary-hosted, Volcanogenic Massive Sulphide and Iron-Oxide-Copper-Gold (IOCG) systems typically formed in extensional settings. While these archetypal deposit types have proven to be (and remain) fruitful Cu sources, the increased demand in Cu due to the Just Energy Transition over the past five years means that the identification and understanding of alternative Cu sources and systems is becoming increasingly vital, especially in the context of tectonic settings that have not previously been considered.

The latter stages of the orogenic cycle is an example of an alternative tectonic setting that shows potential for Cu exploration, evidenced by several Cu deposits and districts within the Mt. Isa Inlier (Eloise, Tennent Creek, Ernest Henry), Mauritanides Orogenic Belt (Guelb Moghrein), Damara Orogenic Belt (Omitiomire, Onganja), Lufilian Arc (Kansanshi, Lumwana), the Itacaiúnas Belt (Sossego), Nautanen Deformation Zone (Nautanen), and the Aravalli–Delhi Fold Belt (Khetri). Copper ± (Au±U±REE±Mo±Co±Ni) mineralisation in the aforementioned systems was produced by fluids migrating along regional-scale structures related to late-orogenic exhumation processes. Sodium ± Ca±K alteration associated with mineralisation typically produced alteration assemblages that reflect cooling from Amphibolite or upper-Greenschist conditions during exhumation. Both mineralisation and alteration phases reflect chemical traps, which include redox buffers, pH buffers and/or P-T changes. Notably, many of the aforementioned deposits lack the voluminous Fe-rich phases common to IOCG deposits, thus distinguishing them from conventional IOCG systems (although some have been classified as orogenic IOCGs; Skirrow, 2022).

Using the characterisation of Orogenic Copper Systems described above, this research aims to outline exploration methodologies that use an integrated approach, combining deformation histories, structural controls on mineralisation, types and extents of alteration and changes in metamorphic facies, to effectively target Orogenic Copper Systems.

Geodynamics Control the Variability of Magmatic Ni-Cu-PGE Sulfide Mineral Systems Through Time

David Holwell[1], Daryl Blanks[2], Marco Fiorentini[3]
[1]University Of Leicester, United Kingdom, [2]BHP, United Kingdom, [3]University of Western Australia, Australia

Conventional models for the magmatic Ni-Cu-PGE sulfide mineral system emphasise the requirement for a large heat driver to sufficiently melt enough mantle peridotitic olivine to liberate Ni into the resultant melts. Recent work, however, has shown that hydrous, metasomatized mantle can also produce Ni-rich melts but at lower temperatures. Based on this understanding, we have identified evidence of diverse mantle source compositions and melting dynamics in the variable chalcophile metal signatures of magmatic Ni-Cu-PGE deposits through time. Through Earth history, the metal signature of magmatic sulfide systems evolves, with mostly high temperature melting events producing Ni- and PGE-dominant deposits >1.8 Ga, associated with large igneous provinces (LIPs). Conversely, from 1.8 Ga onwards, low temperature melting of hydrous mantle becomes more common, forming dominantly Ni- and Cu-dominant rich deposits that are characteristically poor in PGE, and have no spatial or temporal link with LIPs. We interpret these secular changes to be a function of: (1) a cooling mantle from the Archaean to the present; (2) varying compositions of mantle as a product of a spectrum of mantle metasomatism processes; (3) the mineralogical deportment of chalcophile metals in mantle rocks; and (4) the variety of thermal triggers to melt such assemblages. We provide a predictive capacity to assess the relative metal fertility of mafic-ultramafic systems through time.

Mafic-Ultramafic Ore Systems: From Genetic Insights to Exploration Tools

Margaux Le Vaillant[1], Giada Iacono-Marziano[2], Anja C. Slim[3], Stefano Caruso[1], William D. Smith[1], Louise E. Schoneveld[1], Stephen J. Barnes[1]
[1]CSIRO, Kensington, Australia, [2]ISTO - CNRS, Orleans, France, [3]Monash University, Clayton, Australia

Our research group focuses on translating recent advancements in our understanding of genetic processes at play within mafic-ultramafic magmatic ore systems into practical exploration tools and workflows.

At the terrane-to-camp scale, we are developing methods to utilize indicator mineral chemistry for assessing the prospectivity of intrusive complexes. Spinel, olivine, and pyroxene chemistries can inform: 1) the nature of the host rock, 2) the potential assimilation of crustal material by the melt, 3) segregation and interaction with immiscible sulfide melt, and 4) overall dynamism of the magmatic system. To enhance prospectivity analysis and target ranking, we have begun implementing machine learning models using olivine and spinel composition as a starting point, testing those models at the terrane scale on data from Geoscience Australia's heavy mineral map, and at the camp scale on samples from industry partner's exploration projects.

At the prospect scale, our approach integrates experimental high-P&T petrology and analogue fluid dynamic modelling to deepen our understanding of these ore systems. Through high-pressure, high-temperature experiments, we are refining the role of volatiles in genetic processes for magmatic sulfide ores. This is leading to the identification of potentially prospective mineral chemistries such as halogens and other trace element contents in apatite, indicative of the presence of volatiles, favorable to the generation of high tenor sulfides and high-grade systems. Analogue experiments provide insights into sulfide transport, coalescence, and accumulation. We are now identifying specific characteristics, such as sulfide droplet size distributions and distinct sulfide textures, linked with mineralising processes such as sulfide percolation, and used to vector towards massive sulfide accumulations.

Finally, one objective is developing distance-to-ore machine learning models at the deposit scale. These models integrate whole-rock geochemistry, modal mineral proportions, and mineral chemistry - potentially incorporating sulfide size distributions - to enhance exploration efficiency and improve targeting accuracy.

Controls on PGE Mineralization in the Lac des Iles Intrusive Suite: Insights from Trace Element Geochemistry and δ34S in Sulfides

James Tolley[1], William D. Smith[2], Wyatt Bain[3], Peter Hollings[1], Richard A. Stern[4]
[1]Lakehead University, Thunder Bay, Canada, [2]CSIRO Mineral Resources, Perth, Australia, [3]Western University, London, Canada, [4]University of Alberta, Edmonton, Canada

The Lac des Iles Intrusive Suite (LDI-IS) comprises a ~30 km ring-shaped array of Neoarchean mafic-ultramafic intrusions in northwestern Ontario, Canada. These intrusions host variable palladium-bearing sulfide mineralization, including the mine block intrusion of the world-class LDI mine. While the LDI intrusions have been extensively studied, the coeval mafic intrusions of the LDI-IS have received comparatively little attention. This study integrates mineralogical characterization, with in situ trace element chemistry (LA-ICP-MS) and sulfur isotope analysis (SIMS) of base metal sulfides to: (1) characterize ore mineral assemblages; and (2) understand the petrogenesis of individual intrusions and their context within the LDI-IS.

Sulfide assemblages of the LDI-IS show a continuum from primary magmatic ores of pyrrhotite-pentlandite-chalcopyrite-pyrite to weakly altered pyrite-chalcopyrite ± millerite. Palladium concentrations are typically <2 ppm, except for pentlandite from the Wakinoo Lake intrusion, which averages 726 ppm Pd. These values are comparable to mineralized LDI orebodies (e.g., the B2 zone, where pentlandite averages ~1000 ppm Pd). Sulfide S/Se ratios across the LDI-IS range from ~2000–4500, are within, or slightly lower than, mantle range (~2850–4350). The Dog River and Legris Lake intrusion sulfides display elevated S/Se ratios (22000 and 30500, respectively), suggesting an additional sulfur input. However, all $\delta^{34}S$ values from the LDI-IS lie predominantly within the mantle range ($\delta^{34}S = 0\pm1‰$).

Sulfides in the LDI-IS are magmatic with monosulfide generation playing a key role in Pd enrichment. The sulfide mineralization across the LDI-IS records a range of processes that collectively influence PGE mineralization. These processes include early sulfur saturation, high but variable R factors from fluctuating volumes of magma fluxing through local structures, and localized hydrothermal upgrading. This study of genetically related, yet variably mineralized intrusions provides important insight into the petrogenesis of the LDI-IS and the processes responsible for the exceptional Pd-PGE endowment at LDI.

Dual Horizon Ni-Cu-PGE Mineralisation in the Deep Platreef, Mogalakwena Mine Complex, South Africa: Insights into Formation of High PGE Tenors

Kate Canham[1], David Holwell[1], Iain McDonald[2], Hannah Hughes[3], Lara Du Preez[4], Erin Thompson[1], Laura Hepburn[5], Adrian Boyce[5], Nico Bompard[5], Andrew Tait[5], Andy Lloyd[5]
[1]Centre for Sustainable Resource Extraction, University of Leicester, Leicester, UK, [2]School of Earth and Environmental Sciences, Cardiff University, Cardiff, UK, [3]Camborne School of Mines, Department of Earth and Environmental Sciences, University of Exeter, Penryn, UK, [4]Anglo American South Africa, Johannesburg, SA, [5]Scottish Universities Environmental Research Centre, Stable Isotope Laboratory, East Kilbride, UK

Multiple mechanisms can generate high PGE tenors in magmatic sulfide deposits. However, for the Platreef Ni-Cu-PGE deposit in the Northern Bushveld Complex, South Africa, there is still no widely accepted mechanism. The down-dip extension of the Platreef at Sandsloot, Mogalakwena Mine Complex, hosts two distinct mineralised subunits: the Pt+Pd-rich PGE-Reef and, approximately 150 meters below, the Base Metal Zone (BM Zone), which is Fe-Ni-S-rich and newly recognised as being enriched in IPGE(Os-Ir-Ru)+Rh. The PGE-Reef contains Pt+Pd tenors up to 1900 ppm, while the underlying BM Zone contains IPGE+Rh tenors up to 1300 ppm – making this a prime area to evaluate mechanisms of high tenor formation.

Previous work has shown the BM Zone likely represents the monosulfide solid solution (MSS) portion of a sulfide liquid, whereas the PGE-Reef reflects a more typical sulfide liquid. In the PGE-Reef PGEs are predominately contained within Pd-rich pentlandite, symplectitic Pt-Fe PGM, and Pt±Pd-Bi-Te PGM. Interaction with country rock dolomite elevated fO_2 in the PGE-Reef, promoting the formation of high-temperature magmatic Pt-Fe alloys. Subsequent interaction with carbonate- and Pb-Zn±Cl-rich hydrothermal fluids modified the PGM and sulfide assemblage. $\delta^{34}S$ values (relative to VCDT) of +1.0 to +3.4 ‰ support a dominant magmatic origin to the sulfide liquid, with heavier values towards the footwall and around calc-silicate xenoliths suggesting contamination.

We propose the PGE-Reef and BM Zone represent two distinct generations of sulfide liquid. Their high PGE tenors are interpreted as primarily magmatic in origin, controlled by high R-factor and proximity to a feeder zone. The MSS-associated sulfides of the BM Zone preserve evidence of an early, ultra-high-tenor sulfide liquid, while fluid interaction modified sulfide and PGM in the PGE-Reef. These findings elucidate the complex interplay of magmatic and fluid processes that govern ore mineralogy and metal tenors in the deep Platreef.

The Thomson River Cu-Ni-PGE Deposit of the Southeast Lachlan Fold Belt: Implications for Cu-Ni-PGE Prospectivity in Silicic Dominated Magmatic Systems

Steven Boger[1], Reid Keays[2], Tobias Schlegel[3]
[1]The Geological Survey of Victoria, Melbourne, Australia, [2]Monash University, Melbourne, Australia, [3]CSIRO Mineral Resources, Perth, Australia

World-class Cu-Ni-PGE deposits are hosted by strongly differentiated ultramafic and mafic rocks crystallised in the plumbing systems of large igneous provinces (LIPs). LIPs are of intra-plate origin, comprised predominantly of mafic and ultramafic rocks and characterised by high-volume surface magmatism and extensive subvolcanic intrusive rocks.

The Thomson River Cu-Ni-PGE deposit crops out in the southeast Lachlan Fold Belt in central-east Victoria. Primary ore minerals are chalcopyrite, pyrite and pyrrhotite, with minor pentlandite-cubanite Platinum and palladium are the dominant PGE present and hosted by sperrylite and merenskyite, respectively. The Thomson River deposit is unusual in the respect that it is hosted mafic and ultramafic rocks that represent a relatively minor component of a late Devonian phase of voluminous silicic magmatism. The LIP origin of the Thomson River deposit can, nevertheless, be established via comparison with the Siberian Traps, the host rocks to the Norilsk-Talnakh Cu-Ni-PGE deposits of the Russian arctic. The mafic and ultramafic rocks from both regions: (1) compositionally overlap, (2) were emplaced coeval with crustal extension and terrestrial sedimentation, (3) occur over similar geographic extents and, (4) coincide temporally with major mass extinction events.

Whereas the mafic and ultramafic LIP magmas that formed the Siberian Traps were emplaced into the shallow crust or extruded to the surface, the compositionally equivalent rocks in Victoria mostly stalled at depth and formed an extensive magmatic underplate. The thermal effects of this underplate are reflected in the upper crust via voluminous shallow-level silicic granitoids and the coeval volcanism. The presence of Cu-Ni-PGE mineralisation in both regions highlights that processes in the mantle – in particular the formation of the sulphide undersaturated mantle partial melts - is more important for Cu-Ni-PGE prospectivity that the surface expression of magmatism.

Cumulates and the Orthomagmatic to Porphyry Transition: New Insights from the Lorraine Cu-Au-Ag Deposit, Quesnel Terrane, Central British Columbia

Matthew Manor[1], Nichole Moerhuis[1], Shaun Barker[1], Huaze (Derek) Xu[1], James Lang[2]
[1]Mineral Deposit Research Unit, University of British Columbia, Vancouver, Canada,
[2]NorthWest Copper Corp, Vancouver, Canada

Ultramafic-mafic arc cumulates offer rare glimpses into primitive melt evolution and transport through the middle to upper crust of accretionary orogens. In the Quesnel Terrane of British Columbia, Late Triassic to Middle Jurassic ultramafic suites occur as either Alaskan-type complexes (e.g., Tulameen, Polaris) or alkaline pyroxenite-syenite complexes (e.g., Duckling Creek syenite complex, Bootjack stock) that are commonly spatially and/or temporally related to Cu-Au porphyry deposits. The ca. 180 Ma Lorraine Cu-Au-Ag deposit is hosted in the Duckling Creek syenite complex, part of the Hogem Batholith. Host rocks form a crudely zoned clinopyroxenite-syenite complex comprising pre- to syn-mineral clinopyroxenite, melanocratic to leucocratic syenite and monzonite, rare pseudoleucite-bearing syenite, and late-mineral monzodiorite; all units are intruded by post-mineral quartz syenite and syenite dykes. The cumulates are mineralogically and geochemically linked to variable proportions of cumulus minerals (clinopyroxene-biotite-magnetite-apatite-hornblende-alkali feldspar-plagioclase-leucite). Orthomagmatic mineralization is present in three distinct zones – Lower Main, Upper Main, and Bishop – as disseminated to net-textured bornite-digenite, chalcopyrite-bornite, and pyrite-chalcopyrite (± magnetite ± sphalerite), with complex exsolution and symplectite textures associated with abundant Ag-tellurides (hessite), Pb-selenides (clausthalite?), and electrum. Locally, disrupted and partially resorbed sulfide textures occur with muscovite-epidote-chlorite-carbonate ± garnet alteration, and suggest a late, post-mineral fluid overprint. New LA-ICP-MS $^{206}Pb/^{238}U$ zircon dates show initial crystallization at ca. 182-175 Ma with later recrystallization between ca. 160-150 Ma. The younger zircon population displays irregular and disrupted zonation in CL, along with high U and Ce, and low Ti, which is indicative of a low-temperature, oxidized fluid overprint on primary igneous zircon grains. We interpret Lorraine as a rare example of a fractionated, oxidized, mid-crustal Cu-rich orthomagmatic deposit, analogous to Loch Borralan (Scotland), providing new constraints on arc magmatism, emplacement depth, and the orthomagmatic to porphyry Cu-Au transition in island arcs globally.

Go Small or Go Home… The Evolution of Using the Ultrafine Soil Fraction for Mineral Exploration

Ryan Noble[1], Anicia Henne[1], Ian Lau[1], Bobby Pejcic[1]
[1]Csiro, Australia

Ongoing discovery success requires innovation especially in sample collection and analysis. Soils are widely used as a geochemical sample medium and whilst large geochemical surveys conducted by mining companies are common, the suite of data we collect beyond standard soil chemistry has remained stagnant. In addition, the soils that host/adsorb the mobile element signature are the smallest size fractions, especially in transported cover, so we should consider concentrating the clay size fraction (<2 μm) as an improved sample medium. An innovative, new integration of analytical methods known as UltraFine+® extracts the "standard soil chemistry" of the <2 μm soil clay fraction, which is combined with soil spectral mineralogy proxies and physicochemical properties to improve interpretation of soil chemistry. Following the refinement of the technique, we commercialised and then applied the new UltraFine+® workflow to numerous orientation site studies, including reprocessed archived regional soil samples. A series of experimental research was conducted to demonstrate the value of using <2 μm fractions for exploration geochemistry to give industry confidence in the benefit of this technique including testing that international quarantine treatment procedures would not impinge upon results for soils collected outside Australia and the addition of REE, Br and Pd (that were not initially considered). Thousands of soils have now been tested in the vicinity of known mineral deposits (including background areas. Analysing fine fractions (<2 μm) generates reproducible, reliable results, with greater concentrations compared to standard techniques. Key benefits are the removal of nugget effects (for Au) and a marked decrease in censored results for Au, less variation with sampling depth in soil profiles, and value from (re-) assaying regional soil samples to generate new targets. To date the new process has been applied on all continents (except Antarctica) and is expected to become an industry standard in the future.

Using Automated pXRF Data Collection to Model Gold Deportment and Litho-Geochemistry through Regolith in the Yilgarn Craton, Western Australia

Aidan Kitchener[1], Simon Griffiths[1], Harry Brookes[1]
[1]Third Planet Exploration Services Ltd., Bristol, United Kingdom

The systematic deployment of portable X-ray fluorescence (pXRF) at mineral exploration projects provides robust multi-element geochemical data comparable with laboratory analytical methods. Recent innovative technologies such as detectORE™ and RokBot™ from Portable PPB facilitate rapid, on-site data collection of low-level gold (down to ~5ppb) and multi-element analyses using pXRF. Obtaining these data within days of sample collection enables shorter reaction times in mineral exploration, fast tracking discoveries, saving time and money.

At Mulgabbie, an orogenic Au deposit in the Yilgarn Craton, Western Australia, we used pXRF data from air core, diamond and reverse circulation drilling collected by detectORE™ and RokBot™ technology to outline gold deportment and generate a project specific litho-geochemistry model through regolith.

We validated detectORE™ gold determinations by comparing the values statistically and spatially with laboratory fire assay data. The base of regolith was defined using mobile element ratios. Assessing immobile elements, we developed a litho-geochemical classification that maps metavolcanic rock types through the regolith and into bedrock. This classification, in addition to gold analyses, interpreted sulphide mineralogy and pathfinder element distribution were then projected in cross-sections to provide a spatially coherent, 3D geochemical model.

Step by step analysis of the geochemical model resulted in identification of a clearly more favourable host rock for mineralisation at Mulgabbie, with gold, pyrite and pathfinders constrained to a dolerite unit. The model also highlights gold, arsenic and remnant sulphide distribution in the regolith overlying mineralisation, providing insight into dispersion in the weathered environment. A direct comparison of detectORE™ gold with fire assay gold shows a strong positive correlation.

Our work demonstrates that incorporation of new technologies in active exploration projects can rapidly improve decision making in near real time. In addition, demonstrating the potential to optimize the probability of discovery and reduce the time to discover and develop mineral deposits.

Field-Based Mineral Chemistry Using Portable Laser-Induced Breakdown Spectroscopy (pLIBS) for Exploration and Mineral Processing

Cassady Harraden[1], Shaun Barker[1], Andrew Somers[2], Morgan Jennings[2], Benjamin Vautour[2]
[1]Mineral Deposit Research Unit, University of British Columbia, Vancouver, Canada, [2]SciAps Inc., Andover, United States of America

There is a growing demand for real-time, cost-effective tools in mineral exploration and extraction. Accurately interpreting the trace element chemistry of alteration minerals can provide powerful vectors to mineralization, improving discovery efficiency and reducing exploration costs. In production settings, understanding trace element distributions in valuable by-products or deleterious elements is equally important. Traditional techniques such as laser ablation ICP-MS (LA-ICP-MS) and scanning electron microscopy (SEM) are lab-based, time-consuming, and require sample preparation, limiting their utility in field campaigns and mineral processing assessments. As a result, insufficient mineralogy and mineral chemistry often leads to missed or delayed opportunities in discovery and recovery. Laser-induced breakdown spectroscopy (LIBS) is an atomic emission technique that enables rapid, low-cost chemical analysis with minimal sample preparation. Portable LIBS (pLIBS) systems allow in-situ, high-resolution elemental mapping and can detect a wide range of elements, including low atomic number elements like Li, Na, and Mg which are difficult to analyze with most other portable instruments. Using a 16×16 analysis grid (~2.4×2.4 mm), pLIBS spectra are collected and processed mathematically to extract relative elemental intensities. These are compared against a user-defined mineral library using cosine similarity, providing the probability of mineral phases at each shot location. Mineral scores are spatially assigned, generating mm-scale maps that can be interrogated by mineral species to visualize intra-mineral trace element chemistry.

This approach enables real-time, field-based assessments of mineralogy and geochemistry, supporting vectoring toward mineralization and identification of by-products or deleterious elements. While the precision of pLIBS is lower than that of lab-based methods like LA-ICP-MS, its speed, portability, and broad elemental detection range make it ideal for rapid field deployment. This capability allows large volumes of mineral chemistry data to be generated in real time, supporting faster discovery and improved evaluation of recovery opportunities.

Utilising Indicator Minerals to Narrow the Search Space for Greenfields Exploration

Walid Salama[1], Louise Schoneveld[1], Catherine Spaggiari[1], Margaux LeVaillant[1]
[1]CSIRO Mineral Resources, Perth, Australia

Greenfields exploration beneath passive margin basins is challenging due to the thick cover, which requires deep drilling to sample the basement. The 'Seahorse' tenement package in Western Australia is one such region where the Proterozoic basement rocks are overlain by up to 500 m of basin cover. The Mesozoic Bight Basin is filled with early to late Cretaceous carbonaceous siliciclastics in a series of grabens and half grabens, which are unconformably overlain by Cenozoic carbonates of the Eucla Basin. The collaboration between CSIRO and BHP targeted the basement-Cretaceous and Cretaceous-Cenozoic paleosurfaces for analysis of indicator-heavy minerals to detect mechanical dispersion haloes at the Seahorse project, which included a strike length of 350 km along the Rodona Shear Zone tectonic boundary covering an area of approximately 13,000 km2. Although BHP initially targeted massive Ni-sulphides associated with magnetic-gravity anomalies in the basement, our research opens new avenues for additional styles of mineralisation. Four assemblages of indicator sulphide minerals were identified based on their texture and mineral chemistry. These include: 1) detrital, composite Cu-As-Co-Fe-Zn sulphides and Bi tellurides indicating epithermal or skarn sources; 2) detrital Ni-Co-Cu sulphides indicating a volcanogenic Ni sulphide source; 3) detrital Cu sulphides associated with titano-magnetite, ilmenite and apatite indicating IOCG-style deposits; and 4) diagenetic (sediment-hosted) Pb-Zn sulphides in the Madura Formation carbonaceous siltstones. The first two types are restricted mainly to the Cretaceous-Cenozoic unconformity, implying proximal prospective basement paleohighs as the main source of sulphides, which is supported by the subangular shape and the composite structures of the sulphides. The third type is restricted to the basement-Cretaceous cover, whereas the fourth type is likely formed by microbial and hydrothermal processes close to faults. Targeting indicator minerals close to paleosurfaces and linking them to paleotopographic processes can reduce the search space and deep drilling costs for greenfields exploration.

The Application of Radiogenic (Sr and Pb) and Stable Metal (Cu) Isotopes in Undercover Exploration, Northwest Queensland

Joseph Tang[1], Balz Kamber[2], David Murphy[2], Ryan Mathur[3], Brandon Mahan[4]
[1]Geological Survey Of Queensland, Brisbane, Australia, [2]Queensland University of Technology, Brisbane, Australia, [3]Juniata College, Pennsylvania, USA, [4]Melbourne Analytical Geochemistry, School of Geography Earth and Atmospheric Sciences, University of Melbourne, Melbourne, Australia

Ninety-five percent of mineral occurrences in northwest Queensland (NWQ) are discovered within the exposed Proterozoic rocks of the Mount Isa and Georgetown areas, with only a few, relatively recent finds (e.g., Ernest Henry, Cannington) concealed under Mesozoic cover rocks of the Carpentaria and Eromanga basins. Geophysics and geological mapping indicate prospective Proterozoic rocks extend under these Mesozoic basins, which hamper mineralisation signatures from reaching the surface.

Groundwater offers an effective alternative exploration methodology to costly geophysics and drilling. Pore waters inherit the chemical signatures of their host rock, and metal solutes can be transported up to 5 km from source. Solute concentrations are typically in parts-per-billion, are highly susceptible to dilution and evapotranspiration.

Metal isotopic ratios (206Pb/204Pb, 207Pb/204Pb, 65Cu/63Cu, 87Sr/86Sr) are largely unaffected by dilution or concentration factors and are robust geological tools for identifying source regions and sulphide weathering. By plotting Pb isotopes in groundwater against published Pb-Pb growth curves for the Mount Isa Inlier, potential Proterozoic source rocks can be identified, and hence, axiomatically enable informed decisions to be made regarding their mineral prospectivity. All major deposits in NWQ are Proterozoic and to be able to assign an anomalous water to the same mineralisation epoch will help explorers plan their undercover exploration programs. δ65Cu for chalcopyrite, pyrite and native Cu in the Swan, Eloise, Jericho, and Ernest Henry deposits, ranges from -1 to 2. Weathering of Cu ores preferentially concentrates heavy 65Cu relative to 63Cu and contouring the δ65Cu values will telescope them towards Cu-associated sources. The 87Sr/86Sr ratio has a strong relationship with source lithology. It clearly distinguishes Great Artesian Basin waters from waters of the Proterozoic Mount Isa and Georgetown Inliers.

Metal isotopes when used in conjunction with water chemistry, sound regional isotopic and geological knowledge, are therefore powerful tools for predicting mineral prospectivity.

LandScape+® - A new Software Providing Machine-Learned Landscape Context and Target Definition from your Exploration Surveys

Anicia Henne[1], Ryan Noble[1], Dave Cole[1], Morgan Williams[1], John Hille[1], Sam Bradley[1]
[1]CSIRO, Perth, Australia

It has been common practice in greenfields mineral exploration to use soil geochemistry with little regard for landform setting. Especially in varied cover, exploration success is often hindered by the lack of tools to understand and evaluate near-surface geochemical anomalies in context. Five years ago, the CSIRO, with the support of over 30 industry sponsors, set out to identify machine learning approaches to integrate soil geochemistry with spatial data layers (i.e., SRTM digital elevation radiometric and Sentinel-2 satellite data). The aim was to improve our ability to identify targets and understand the spatial variation and influence of regolith types on soil geochemistry. Since the conclusion of this research, the CSIRO has improved, tested and commercialised this research to provide a hands-on tool to the exploration industry; LandScape+®. This new software creates landscape models from publicly available spatial data layers almost anywhere in Australia and uses these to detect statistically relevant geochemical outliers from soil survey results. The user uploads their data to a user interface to tailor for the best fit for each survey area. The model is used to rapidly generate elemental outliers and multielemental trends from the dataset. The data can also be exported to a GIS software of choice. Overall, this process accelerates the review of geochemical data and allows the explorer to make decisions with more confidence. This talk will cover the background R&D that led to the development of the software and introduce relevant machine learning and regolith science concepts. We will show where this approach helps to better interpret geochemical results with examples from across Australia, highlight some practical limitations, and take a peek at the user interface to showcase some of the statistical and geospatial visualisation tools.

Data Before Algorithms: An Open-Source Approach to Legacy Geochemical Method Categorization

Sam Scher[1], Tom Carmichael[2]
[1]LKI Consulting Inc, Washington, United States, [2]Datarock, Adelaide, Australia

Legacy geochemical datasets frequently contain ambiguous analytical methods, undocumented modifications, and inconsistent reporting standards. These issues significantly impede data integration, interpretation, and, critically, the confidence placed in geochemical analyses by non-geochemist stakeholders, such as data scientists and machine learning specialists. The growing trend of multidisciplinary use of geochemical data has increased the risk of inappropriate inclusion or exclusion of these datasets due to misinterpretation or perceived complexity.

Currently, addressing analytical ambiguity predominantly relies on manual approaches, including subjective leveling and detailed univariate analyses, methods that are labor-intensive, inefficient, and potentially error-prone. To enhance objectivity and streamline this process, we developed a generalized, reproducible workflow implemented through open-source R code designed to automatically categorize instrumental geochemical data based solely on statistical distributions and patterns of missingness inherent to the datasets.

This workflow starts with exploratory data analysis to systematically detect distinct data subsets by evaluating precision, accuracy thresholds, and joint probability density functions. By identifying characteristic data distributions and missing data signatures, our methodology minimizes reliance on subjective judgment. However, acknowledging the inherent complexity of geochemical data, a final manual step of leveraging probability plots and spatial distributions ensures the robustness of method categorization.

Our approach addresses a critical technological gap by ensuring that interpretations derived from legacy datasets reflect accurate analytical contexts. By providing clarity to analytical origins, we mitigate common pitfalls associated with multidisciplinary usage of geochemical data, ultimately preventing flawed interpretations and enhancing the reliability of data-driven mineral exploration decisions. Thus, this method places appropriate emphasis on robust data characterization before applying advanced analytical technologies, avoiding scenarios where the technological cart precedes the geochemical horse.

Olympic Dam IOCG-U deposit: significant new additions to the mineral endowment

Kathy Ehrig[1,2], David Haddow[1], Max Shawcross[1], Nigel Cook[2], Cristiana Ciobanu[2]
[1]BHP Copper SA - Olympic Dam, Adelaide, Australia, [2]School of Chemical Engineering, The University of Adelaide, Adelaide, Australia

The Olympic Dam breccia hosted Fe-oxide Cu-U-Au-Ag (IOCG) deposit is without doubt one of the world's largest metalliferous deposits. Most of the publically declared Resource lies at depths of ~300-1200 m below the surface, with the deepest known areas in the SE part of the deposit. Recent mine-based diamond drilling has outlined mineralization extending for more than 2 km along strike and 1 km in depth below previously delineated mineralization (referred to as the Deeps). To date, drilling has yet to close off mineralization at depth and its lateral extents.

Mono- and polymictic-breccia textures are complicated to interpret, yet the ore and alteration mineralogy are systematically zoned across the deposit. Large syn-post mineralization faulting is followed by partial erosion during post-mineralization events with regional expression. Even though bornite-chalcocite-chalcopyrite-pyrite are present across the shallower parts of the deposit, chalcopyrite and pyrite are the dominant sulfides in the Deeps which also contains uraninite-coffinite-brannerite and Au-Ag minerals. The ratios of hematite to magnetite and siderite to Ca-Mg-Fe-carbonates decreases, while chlorite and carbonates increase with increasing depth. Likewise, the REE mineralogy changes from bastnäsite-synchysite-florencite to monazite-apatite-xenotime-allanite. Muscovite/illite, quartz, barite, fluorite and intensely altered K-feldspar are also present. Hydrothermal biotite has not been observed. The best indicator of transitioning out of mineralisation, even though the rock is significantly altered, is a rapid decrease in the concentrations of sulfides, in particular pyrite.

The distinct mineralogy and associated geochemical signatures across the deposit are used for vectoring to mineralization and recognizing when the limits of Fe-oxide alteration are nearby. The geochemical signature (Mo-As-Sb-Sn-W) of the Deeps is similar to the mineralized areas which overlie the Deeps. Hematite is the dominant mineralogical host of Mo-As-Sb-Sn-W. In fact, granitophile element (U-W-Mo-Sn) enriched hematite is ubiquitous to other IOCG deposits in the region surrounding Olympic Dam.

The Intersection of the Geoscientist with Business

Jim McCluskey[1]
[1]BHP Copper South Australia, Adelaide, Australia

Geoscience information is integral to estimating value in the resources sector. It is often described as the basis of value by geoscientists, but the vast majority of geoscientists cannot describe well how the information they generate is used in estimating value, and in decision making. Knowing what information matters most is not intuitive nor is it the same each time, the value of information is discrete per deposit. Data acquisition for decision making makes up the longest component of a project life cycle, exceeding often cited critical path items such as approvals, capital project study and seeking funding. The power of an individual through their actions to integrate with other disciplines and materially impact value, returns, and time to market is often under-estimated. The geoscientist with an understanding of decision making, valuation, modifying factors, investment risks and uncertainties, can make a profound difference by turning vague concepts such as in situ value or contained metal, into cashflows profiles and return on investment, ahead of their peers and at a competitive advantage for an organization. Such a geoscientist is a powerful asset for a resources company, changing strategic planning and materially improving project financials. While the defining quality of the geoscientist that lives this is curiosity, the interdisciplinary understanding can be deliberately built into people by an organization, or by the direct development planning of the individual. Quantifying the value of geoscientific information will be demonstrated with tangible examples to illustrate the critical intersection of geoscience with business strategy.

Structural Insights of the Olympic Dam Cu-U-Au-Ag Deposit, South Australia Provided by Hardrock 3D Seismic Data

David Haddow[1], Matthew Goldman[1], Kathy Ehrig[1], Jared Townsend[2], Carolina Pimentel[3], Ian James[3], Erwann Lebrun[3]
[1]BHP Copper South Australia Geoscience, Adelaide, Australia, [2]BHP Innovation, Perth, Australia, [3]BHP Resource Centre of Excellence, Perth, Australia

The supergiant Olympic Dam breccia hosted Fe-oxide Cu-U-Au-Ag deposit is located in the Olympic Cu-Au province, on the eastern margin of the Gawler Craton. The deposit is hosted in a hydrothermal breccia system known as the Olympic Dam Breccia Complex which occurs within the Roxby Downs Granite and is unconformably overlain by ~350 metres of undeformed Neoproterozoic-Cambrian sedimentary successions. The deposit has undergone a complex tectonic, magmatic, and hydrothermal evolution, with much of the early structural architecture overprinted by physical, chemical, and volcanic brecciation mechanisms. The deepest part of the deposit (Olympic Dam Deeps) extends to >2 kilometres depth southeast of the Mashers Fault Zone.

The Olympic Dam Deeps was drilled extensively prior to receiving fully processed hardrock 3D seismic data. This drilling has helped characterise features observed in the seismic data including the extents of the Deeps Fe-oxide breccias and associated structural architecture. High seismic reflectivity coincides with drilled Fe-oxide breccias. The seismic reflectors are discontinuous and comprise high and low amplitude reflectors as expected for heterolithic breccias. These breccias are structurally bound by northeast-southwest and northwest-southeast faults including the Mashers Fault Zone (western extent) and Woodall Junior Fault (southern extent).

Continuous linear seismic reflectors also exist in the seismic data across the deposit. These reflectors are interpreted to be mafic and felsic dykes and sill intrusions, associated with high or low amplitude reflectors depending on primary composition and subsequent alteration. The steeply dipping northwest-southeast trending Gairdner Dolerite dyke swarm is the dominant feature imaged by the seismic data.

Other interpreted intrusions are also imaged in the seismic data including moderately dipping north-south trending dyke intrusions, shallow dipping dykes and sill intrusions. These dykes and sills have been identified in areas surrounding the deposit but are not observed in the deposit suggesting they may have formed pre- or syn-mineralisation.

A Temporal Framework for the Genesis of the Carrapateena Iron Oxide Copper-Gold (IOCG) Deposit, Northern Gawler Craton, South Australia

Bradley Cave[1], Martin Hand[1], Laura Klingberg[2], Dana Leslie[2], Shaun Light[2], **Richard Lilly**[1], Brooke North[1], Jo Park[2], Jan Varga[2]
[1]University Of Adelaide, Adelaide, Australia, [2]BHP, Adelaide, Australia

The Carrapateena Iron Oxide Copper-Gold (IOCG) deposit is located on the eastern margin of the Gawler Craton, approximately 475 km NE of Adelaide, South Australia. The deposit is hosted within the ca. 1860 Ma Donington Suite granite and encompasses polylithic breccias and various volcanic units. The Carrapateena deposit contains multiple stages of alteration/mineralisation, which in paragenetic order includes: (1) pre-mineralisation alteration; (2) hematite I; (3) Cu-Au mineralisation; (4) hematite II & REE mineralisation; (5) uranium mineralisation; (6) post-mineralisation alteration/veining. In order to establish a temporal framework of the deposit, multi-mineral U–Pb geochronology was conducted on a variety of lithologies within the deposit, as well as the various stages of alteration/mineralisation. Zircons from a porphyritic volcanic unit contain an age population with 207Pb/206Pb mean weighted ages of 1591 ± 19 Ma (MSWD = 0.58). These ages correlate with age populations from zircons from two samples composed of polylithic breccias, which produce 207Pb/206Pb mean weighted ages of 1595 ± 18 Ma (MSWD = 0.01) and 1593 ± 19 Ma (MSWD = 0.72). In both instances, the ca. 1590 Ma population is interpreted to represent the age of porphyritic volcanic crystallisation and polylithic breccia formation, correlating with a period of widespread magmatism throughout the Gawler Craton. U–Pb geochronology conducted on apatite, hematite and xenotime associated with various stages of mineralisation all produce ages of ca. 1585 Ma, indicating a major metallogenic event throughout the Gawler Craton. U–Pb geochronology performed on uraninite, apatite within a paragenetically late vein and hematite that displays a distinct “vuggy” texture records a second distinct hydrothermal event from ca. 600–500 Ma. This second event is interpreted to reflect the age of deposit-wide fluid circulation during the Delamerian Orogeny (520–470 Ma), resulting in the deposition and/or dissolution-reprecipitation of uraninite and post-mineralisation alteration/veining.

New Tectono-Stratigraphic Model with Revitalised Discovery Potential for Olympic Dam-type Haematitic Iron Oxide Copper Gold (HIOCG) Deposits

John Anderson[1]
[1]Austrike Resources Pty Ltd, Glenalta, Australia

A review of public data addressed questions raised by decades of minerals exploration on the Gawler Craton, including whether a spectrum of deposit styles for the Olympic Metallogenic Event (OME) was connected by a common stratigraphic marker and universal targeting vectors.

Discovery of the Paris Ag (Pb) epithermal, Helen Cu Au Ag skarn, and Nankivel Cu (Au) porphyry system and subsequent research indicated the system formed coevally with the HIOCG breccia pipes on the opposite side of the massif of Gawler Range Volcanics (GRV). Collaborative and public precise zircon dating established Olympic Dam (OD), Paris and Nankivel were deposited in the 5-million-year interval between the Lower and Upper GRV. Researchers had recognised remnants of an absent sedimentary basin within the OD pipe. This raised the question of how the basin was formed and vigorously eroded off in such a short period to the same upper level of multiple and widespread HIOCG pipes.

In this study, the stratigraphy hosting the Emmie Bluff "stratiform" HIOCG is correlated with the Lower Pandurra Formation. Unit PF2 Red Shale represents the regional marker of the OME "Big Bang". Sedimentary inclusions in OD Deeps and Oak Dam West pipe are also correlated with the Lower Pandurra. The missing basin was hidden in plain sight.

The new model proposes that cessation of double subduction caused a rapid change to extension and sourcing of metals from two differently prepared mantles with precipitation at the palaeosurface. A waning phase of PF3 haematitic siltstone may be correlated with the Upper GRV that filled the southern part of a collapsed mega-caldera. Although the HIOCG potential is restricted to the Olympic Basin mapped by Lower Pandurra, the model provides new far-seeing geochemical and sedimentological vectors to prospective brownfields and greenfields targets.

Critical Ingredients of IOCG Mineral Systems: Insights from a New National Mineral Potential Model of Australia Using a Hybrid Approach

Jonathan Cloutier[1], Arianne Ford[1], David Huston[1,2], Marcus Haynes[1], Anthonyy Schofield[1], Michael Doublier[1], Guillaume Sanchez[1], Jingming Duan[1], James Goodwin[1], Eloise Beyer[1], Geoff Fraser[1], Yanbo Cheng[1], Kathryn Waltenberg[1], Antony Burnham[1], Karol Czarnota[1]
[1]Geoscience Australia, Canberra, Australia, [2]Australian National University, Canberra, Australia

Iron oxide copper-gold (IOCG) deposits are part of a contentious class of deposits wherein key formation processes are still highly debated. Resolving key processes necessary for the formation of IOCG deposits is critical as they contain many elements that are required for the transition to a low carbon economy. This study focuses on robust statistical analysis of new and updated mappable criteria used to represent as spatial proxies for IOCG mineral systems processes. Using a hybrid data- and knowledge-driven approach, a national-scale assessment of the mineral potential for IOCG mineral systems in Australia has been undertaken. A total of 149 mappable criteria were statistically assessed using the Kolmogorov-Smirnov test. From these, 14 criteria were found to show a satisfactory statistical relationship with known IOCG mineralization and were retained for the model. The statistical approach used in this study permits the assessment of the relative importance of each mappable criteria. Overall, the best predictors for IOCG mineral system are mappable criteria representing energy sources followed by ore depositional gradients, fluid pathways and architecture and sources of metals, fluids and ligands.

The mineral potential model successfully predicts the location of 91.7% of known IOCG deposits and occurrences in 8.3% of the area, reducing the exploration search space by 91.7% and opening new exploration search space in under-explored regions of Australia. Underexplored regions for IOCG that demonstrate favorable geological evolution and high prospectivity include the Delamerian Orogen, the Lamboo Province and Tanami Orogen, the western Gawler Craton, the Hooper Province and the Musgrave Province.

This study highlights how precompetitive geoscience data can be evaluated and utilized within a geologically and statistically robust framework to produce improved mineral system formation models and mineral potential models with strong predictive performance.

Giant IOCG Deposits Formed in Late Paleoproterozoic to Early Mesoproterozoic Rift Basins After the Assembly of the Columbia Supercontinent

Xin-Fu Zhao[1], Zhi-Kun Sun[1], Xue-Qing Yin[1], Mei-Fu Zhou[1]
[1]China University of Geosciences (Wuhan), Wuhan, China

The late Paleoproterozoic to early Mesoproterozoic is often regarded as the onset of the "boring billion," characterized by tectonic quiescence and subdued magmatism. However, this period is also characterized by the formation of giant iron oxide-copper-gold (IOCG) deposits globally. In this study, we document multiple 1.66–1.65 Ga IOCG deposits in the Kangdian region of South China, where mineralization is spatially and temporally associated with 1.69–1.65 Ga diabase intrusions and variable-sized hydrothermal breccia bodies. The paragenetic sequence of these deposits typically comprises: (I) pre-ore Na-(Ca) alteration dominated by albite (with local amphibole); (II) Fe-(REE) mineralization marked by magnetite, siderite, and subordinate REE-enriched apatite; and (III) Cu-(Au-REE) mineralization featuring chalcopyrite, ankerite, biotite, K-feldspar, sericite, chlorite, and locally bornite and LREE minerals. The deposits underwent multiple tectonic and magmatic events
These deposits are hosted within the 1.74–1.68 Ga metavolcanic–metasedimentary sequences of the Dahongshan, Hekou, and Dongchuan Groups, which represent fluvial to intertidal sedimentary–volcanic successions deposited in a late Paleoproterozoic rift basin on the western margin of the Yangtze Block. The stratigraphy transitions from basal conglomerates and sandstones with minor tuffaceous and mafic volcanic rocks to interbedded carbonates. Notably, this rift succession in the southwestern Yangtze Block exhibits striking similarities to contemporaneous sequences in north-eastern Australia and the Wernecke supergroup in north-western Laurentia. Given the coeval occurrence of IOCG deposits across these continents, we propose that these deposits formed by the conversion of continental rift basins following the assembly of the Columbia supercontinent. We further suggest that the formation of giant IOCG deposits requires a combination of critical factors: (1) thick sedimentary-volcanic sequences, (2) upwelling mafic magmas derived from the asthenospheric mantle, and (3) large-scale hydrothermal fluid circulation under elevated geothermal conditions.

Deciphering the formation of Andean Iron Oxide-Copper-Gold Deposits: New Advances in their Geology and Exploration

Irene Del Real[1]
[1]Pontifica Universidad Católica De Chile, Santiago, Chile

Andean IOCG deposits form part of the Andean-Iron-Belt which extends from central Peru to central Chile and hosts world-class deposits such as Candelaria, Mantoverde, Mina Justa and Raúl-Condestable. Andean IOCG deposits formed between the Upper-Jurassic and Mid-Cretaceous at a convergent plate boundary at the end or just after back-arc extension related to steep subduction. These deposits correlate in time and space with IOA and stratabound Cu-Ag deposits and are mostly hosted in volcanic/volcanosedimentary sequences, although sometimes they can be hosted in shear zones within intrusions. Andean IOCG deposits have a strong structural control with both transtensional and transpressional environments, and in the Chilean section of the belt they have a close relationship with the Atacama Fault System. Mineralizing hydrothermal fluids are mostly characterized as magmatic-hydrothermal, although external fluids potentially from evaporites can occur at later stages of mineralization. Alteration facies in IOCG deposits are characterized by an early and widespread Na-Ca alteration, followed by Ca-Fe alteration that can form IOA mineralization in some cases, high-temperature Fe-K alteration that can form magnetite dominant IOCG mineralization (e.g. Candelaria) or low-temperature Fe-K that can form hematite dominant IOCG mineralization (e.g. Mantoverde), and later lower-temperature K-Si-Al ± Fe-Ba alteration. Mineralization can be hosted in structures and in favorable lithologies, forming structurally controlled and/or lithologically controlled ore bodies. Copper mineralization is predominantly found as chalcopyrite, although in some cases bornite is also significant (e.g.Mina Justa/Raúl Condestable). The relationship between IOCG and magmatism is not direct, as these deposits are not hosted in intrusions that sourced mineralization, although magmatism coeval with IOCG mineralization is documented to show more oxidized conditions favorable for Cu mineralization. In contrast, magmatism coeval with early Ca-Fe alteration and IOA mineralization is documented to show more reduced conditions suggesting that overall conditions to form one deposit over the other are different.

The Tujuh Bukit District: An Emerging Cluster of Cu-Au Porphyry Systems in East Java, Indonesia

Alexsei Robert Taube[1], Mochammad Hafid Rahadi[1], Arief Bastian[1], James William Sweeney[1]
[1]Merdeka Copper Gold, Banyuwangi, Indonesia

The Tujuh Bukit district in East Java, Indonesia, represents an emerging Cu-Au porphyry cluster formed in a subduction-related island arc setting with the mineralising event occurring around 5–4 Ma. The main Tujuh Bukit porphyry system, discovered in 2012, and is hosted by dioritic to tonalitic intrusions emplaced into volcanic-sedimentary sequences. It contains a current resource of approximately 1.7 billion tonnes at 0.5% Cu and 0.5 g/t Au. Recent drilling has confirmed five satellite porphyry systems: Candrian, Gua Macan, Katak, Tujuh Bukit North, and Tujuh Bukit South, redefining the area as a district-scale porphyry cluster. Several of these discoveries will have maiden Mineral Resource Estimates released in 2025.

Hydrothermal alteration is dominated by prograde potassic assemblages (biotite ± K-feldspar ± magnetite), overprinted by minor phyllic, argillic, and localised advanced argillic alteration. Mineralisation is primarily chalcopyrite with subordinate bornite, associated with quartz-magnetite stockwork veining that controls the distribution of Cu-Au grades. The satellite systems are slightly older and exhibit higher Au/Cu ratios than the main Tujuh Bukit porphyry, suggesting a more Au-rich character.

District-scale structural architecture is characterised by NW-SE trending corridors with N-S opening structures that facilitated magmatic and hydrothermal fluid ascent. Vertical mineralisation extends over 1,000 meters in the main system, with potential for similar profiles in the satellite centres. Ongoing exploration continues to test numerous targets, including Gunung Manis, Lompongan, and Salakan, while untested airborne magnetic geophysical bullseye and geochemical anomalies highlight additional potential across the district.

The Tujuh Bukit district displays the hallmark features of a fertile porphyry province, with multiple mineralised centres likely derived from a single magmatic-hydrothermal event. The evolving geological model underscores significant upside for further discoveries and resource expansion in this underexplored yet highly prospective region.

This paper describes the newly discovered deposits and the implications for expansion in the Tujuh Bukit District.

Geology and Geochemistry of an Emerging Copper – Gold Discovery in the Autonomous Region of Bougainville, Papua New Guinea

Patrick Highsmith[1], Antonio Arribas[2], Andrias Kristianto[1], Antonius Yuniarto[1], Joseph Espi[4], Regis Neroni[3]
[1]Island Passage Exploration Ltd, NASHVILLE, United States, [2]The University of Texas at El Paso, Earth, Environmental and Resource Sciences, El Paso, United States, [3]NewGen Geo Pty Ltd, Perth, Australia, [4]University of Papua New Guinea Earth Sciences Division, Port Moresby, Papua New Guinea

Operations at the world-class Panguna porphyry copper-gold mine on Bougainville Island were tragically halted by civil war in 1989; a moratorium on mineral exploration ensued until 2015. EL02, the first metals exploration license under that act, was granted to a landowner company called Isina Resource Holdings (IRHL). Island Passage Exploration (IPX) is IRHL's technical and finance partner.

IPX and IRHL began exploration in March 2024. The team utilizes prospecting, mapping, geochemistry, geophysics, infrared spectroscopy, and micro-XRF to explore this highly prospective new district along the Bismark Volcanic Arc. This approach has identified numerous drill targets for vein-controlled and porphyry copper-gold mineralization. Early results are extraordinary, including channel samples of 8.0 metres averaging 57 g/t gold and 1.4% copper.

The license is partially underlain by the Isina Monzonite complex (4.82 – 7.43 Ma). Most of the remaining area is within the Miocene Toniva formation, a composite of volcanic sediments, lavas, and limestones.

Stream sediment sampling has been instrumental in target delineation. Local social awareness specialists lead geologists into remote areas, sometimes encountering artisanal vein mining. Quartz-sulphide veins occupy persistent structures (>3 km) in the Toniva formation while mineralized crossing structures enhance width and grade. Vein-related silicification and argillization overprint regional propylitic alteration. The dominant pathfinder elements with the precious metals and copper are bismuth and tellurium.

Explorers have also developed drill targets within the intrusive complex, where strong copper, gold, and molybdenum occur in rocks and soils. Intrusive rocks of varying composition exhibit widespread propylitic alteration with localized phyllic, argillic, and advanced argillic alteration. Magnetic and radiometric data correlate with geochemical targets near the margin of the intrusive complex. Some magnetic intrusions resemble the Panguna porphyry in scale and character.

The high-grade veins and copper-molybdenum-gold targets of porphyry-scale provide for some of the most outstanding drill targets these authors have encountered.

The Significance of the Mineralising Magma Composition on the Diagnostic Exploration Criteria for Porphyry Copper Deposits

Ross Large[1], Imants Kaverlieris[2], David Royle[3], Terry Hoschke[4]
[1]University Of Tasmania, Hobart, Australia, [2]Plus Minerals LLC, Ulaanbaatar, Mongolia, [3]Consultant, Brisbane, Australia, [4]Consultant, Perth, Australia

Recent research on alteration assemblages of porphyry copper deposits using large multielement datasets has led to the recognition of three groups of porphyry copper deposits: diorite group, granodiorite group and monzonite group (Large, 2025). These groups have different alteration mineral assemblages, Au/Cu ratios, tonnage-grade characteristics and geophysical signatures.

The diorite group, having the lowest K and highest Ca & Fe content of the three groups, constitute around 40% of all porphyries, commonly have an albite-rich copper core zone with biotite but less K-feldspar, and more elevated chlorite, magnetite, pyrite, epidote and actinolite in the copper core than the other two groups. They are the most Au-rich and Cu poor, with about 70% of the global group of Au-rich porphyries (Au > 0.4 g/t), and have the lowest mean tonnage of 735 Mt.

The granodiorite group, with higher K and lower Ca & Fe than the diorite group, also constitute about 40% of all porphyries, display moderate K-feldspar alteration in the copper-zone, with outboard sericitic alteration. They have lower mean Au content than the other groups comprising only 10% of the global group of gold-rich porphyries. However, they form 47% of the Cu-rich porphyries (Cu > 0.5%), all the Cu-Mo porphyries, commonly lack a magnetite-bearing core and have the greatest mean size of 1001 Mt.

The monzonite group (commonly called alkalic) constitute about 10% of all world-wide porphyry copper deposits and display intense potassic alteration in the Cu-Au-bearing core. They comprise about 20% of the global group of gold-rich porphyries (Au > 0.4 g/t), have a mean tonnage of 790 Mt, low pyrite content and elevated hematite/magnetite ratio in the copper core.

Ref: Large, RR., in press, Simple graphical tools to understand the relationship between porphyry composition, hydrothermal alteration, mineralogy and copper-gold grades in porphyry copper deposits: Ore geology Reviews

Synchronous Carbonate-Base Metal Gold and Skarn Copper-Gold Mineralization New York-Wellington Area, Ok Tedi District, PNG: Implications for Genetic Models

Peter Pollard[1], Holly Stein[2], C. Mark Fanning[3], Peter Olubas[4], Benjamin Peni[4], Moss Batap[4], Mike Erceg[5]
[1]Pollard Geological Services Pty. Ltd., The Gap, Australia, [2]AIRIE, Fort Collins, USA, [3]Arise Geosciences Pty. Ltd., Canberra, Australia, [4]Ok Tedi Mining Limited, Tabubil, Papua New Guinea, [5]Canterbury Resources Limited, Pyrmont, Australia

The New York-Wellington area, approximately 1 km northeast of the Ok Tedi Cu-Au mine contains Jurassic to Neogene sediments intruded by Pleistocene monzodiorites. This area hosts several different styles of copper and gold mineralization that formed in the period 1.54-1.41 Ma, prior to mineralization at Ok Tedi. The main intrusions in this area include the New York, Wellington and East Cheam Monzodiorites which have SHRIMP U-Pb zircon ages of 1.535±0.073 Ma, 1.551±0.080 Ma and 1.359±0.068 Ma respectively.

Sparse porphyry-style D veins in the New York Monzodiorite have a Re-Os molybdenite age of 1.5403±0.0048 Ma, while B veins in the New York and Wellington Monzodiorites have Re-Os molybdenite ages of 1.5055±0.0031 Ma to 1.4956±0.0042 Ma (n=4). The monzodiorite- and siltstone-hosted New York Breccia consists of early magmatic breccia overprinted by carbonate-base metal gold mineralization which has an age of 1.412±0.047 Ma based on the U-Pb zircon age of an inter-mineral dyke. This is contemporaneous with the Wellington Cu-Au skarn and underlying calc-silicate breccia with Re-Os molybdenite ages of 1.413±0.003 Ma and 1.418±0.005 Ma respectively. Late gold-rich intermediate-sulphidation epithermal veins with quartz-rhodochrosite-sphalerite-galena have not yet been dated.

Stratigraphic reconstruction and the lack of a high-sulphidation lithocap at Ok Tedi suggests 1-2 km of erosion since formation of porphyry and skarn mineralization at 1.32-1.14 Ma. The pre-mining elevation of Mt Fubilan was 2084 m compared to 1500 m for the top of the Wellington skarn and 1400 m for the New York Breccia. The New York Breccia has been drilled to a depth of >1200 m without any sign of transitioning to porphyry-style mineralization, suggesting that carbonate-base metal gold mineralization formed 3-4 km below the paleosurface and at considerably deeper levels than the later porphyry and skarn Cu-Au mineralization at Ok Tedi.

An Updated Appraisal of the Mineral Prospectivity of the Arabian Shield

Jamie Price, James Davey[1], Laura Ward[1], Stephen Johnson[2], Mark Campodonic[1], Peter Johnson[1], Jon Hronsky[1], Bruce Eglington[1], Abdulrahman Al Ghamdi[3], Ozgur Sapanci[3], Thiago Schlicta[3]
[1]SRK Consulting (UK) Ltd., Cardiff, United Kingdom, [2]SRK Consulting Australasia Pty Ltd, West Perth, Australia, [3]Ma'aden (Saudi Arabian Mining Company), Riyadh, Saudi Arabia

The Arabian Shield forms the eastern component of the Arabian-Nubian Shield (ANS), which straddles the Red Sea and represents one of the largest portions of exposed, juvenile Neoproterozoic crust on Earth. The Arabian Shield has been subject to significant historic academic study in parallel with geological survey-driven mineral exploration, resulting in tectonic models of formation with broad consensus. However, renewed interest in the region by mineral explorers, combined with newly available high-resolution datasets, has highlighted the uncertainty of historic geodynamic models, as well as resultant knowledge gaps regarding the geological prospectivity of some mineral systems in the shield.

This study builds upon a revised geodynamic framework developed by SRK and Ma'aden for the Arabian Shield and focuses on mineral prospectivity mapping and target generation for a wide range of mineral systems.

The results demonstrate a broad range of metallogenic potential that is temporally, spatially and genetically diverse, reflecting a protracted history of tectonism and magmatism across the Arabian Shield. Some of the oldest metallogenic events (ca. 870 Ma) comprise primitive oceanic crust development and ophiolite obduction, preserving podiform chromite mineralisation along the Yanbu and Bi'r Umq suture zones. Bi-modal magmatism emplaced during the accretion of juvenile arc terranes during a period of plate reorganisation at ca. 700 Ma, prior to Nabitah orogenesis, are prospective for VMS, epithermal and magmatic Ni-Cu-PGE systems. Suturing of the western and eastern halves of the Arabian Shield (ca. 660-620 Ma) resulted in the development of the dominantly N-S trending Nabitah Orogenic Belt that is highly prospective for orogenic gold mineralisation, extending undercover along a newly interpreted arcuate geometry that trends northeast of the exposed Arabian Shield. Voluminous post-orogenic felsic magmatism (ca. <620 Ma) exploited reactivated NE-SW trending pre-Nabitah translithospheric structures, resulting in localised occurrences of highly evolved peraluminous and peralkaline granitic complexes prospective for Sn-W-Li-Cs-Ta and REE-Zr-Nb-Ta mineralisation, respectively. A series of volcanic arcs accreted to the ANS throughout pre- to post-Nabitah orogenesis demonstrate potential fertility for porphyry and related epithermal mineralisation, ranging from the Jiddah and Birak arcs in the west, to the Ar Rayn arc in the east of the exposed ANS.

The results of this study provide an updated understanding of the geological history of the Arabian Shield based on the integration of a wide range of geoscientific datasets and related expertise. The updated geodynamic framework, along with mineral system models for eight geologically permissible deposit styles, form the basis of the widest range of prospectivity maps and associated exploration targets generated to date in the Kingdom of Saudi Arabia.

Sulphur flushing as a driver of copper accumulation in magmatic arcs

Teresa Ubide[1], Fidel Costa[2], Mark Kendrick[1], Andrew Berry[3], Henny Cathey[4], Andy Tomkins[5]
[1]The University Of Queensland, Brisbane, Australia, [2]Institut de Physique du Globe de Paris, Paris, France, [3]Australian National University, Canberra, Australia, [4]Queesland University of Technology, Brisbane, Australia, [5]Monash University, Melbourne, Australia

Porphyry copper ore deposits in magmatic arcs supply ~75% of the world's copper. Because copper is a critical element in our race to decarbonise, there is great interest in the possible causes of copper enrichment in parental arc magmas, and copper-sulphide precipitation mechanisms, which may involve external gas fluxes as well as hydrothermal processes. Here we combine volcanic SO2 degassing fluxes with the petrological record of Cu and S enrichment in associated lavas. We focus on recent eruptions of Mount Mayon open-vent volcano in the Philippines, which had notably different explosivity and SO2 flux. We apply high-resolution crystal-scale element mapping to explore copper sulphide precipitation in the subvolcanic environment immediately preceding these different eruptions.

Melt inclusions preserved in sieve-textured plagioclase phenocrysts erupted between 2006-2009, after an elevated SO2 degassing flux, are extensively crystallised and contain clinopyroxene and magnetite together with droplets of copper sulphide (digenite). The bulk composition of the melt inclusions (including crystals) is andesite-dacite, similar to typical arc lavas but with exceptional Cu enrichment. The residual melt is dacitic and contains up to 1 wt.% Cl (~3× typical arc magma values). Similar melt inclusions in plagioclases from the large 2000 eruption, before the rise of SO2 degassing flux, lack equivalent copper-enrichment.

These observations suggest that external SO2 fluxes may exert a critical control on copper localisation in arc magmas. Flushing of shallow plagioclase mushes with sulphur-rich vapours as well as Cu-bearing magmatic brines might be a critical step in the enrichment of arc magmas necessary for the production of large porphyry copper deposits.

Advanced Mineralogical, Spectral, and Geochemical Tools for Locating Upflow Zones and Mineralization in the Lithocap Environment

Lejun Zhang[1], David Cooke[1], Noel White[1,2], Evan Orovan[1,3], Xin Ni Seow[1], Yi Sun[1], Jaime Osorio[1]
[1]CODES, University of Tasmania, Hobart, Australia, [2]Hefei University of Technology, Hefei, China, [3]Thesis Gold, Vancouver, Canada

Lithocaps are extensive domains of hypogene silicic, advanced argillic, and argillic altered rocks that develop between the paleosurface and shallow crustal intrusive complexes. They may host high sulfidation mineralization in breccias, veins, or as disseminations in silicic and advanced argillic alteration zones above porphyry intrusions. These high sulfidation mineralization can also be superimposed on earlier-formed porphyries in regions affected by rapid uplift and exhumation. However, their large volumes, complex mineralogy, and poorly defined alteration zonation make exploration particularly challenging.

Previous research (e.g., Chang et al., 2011) developed geological and geochemical vectoring tools based on field geology, whole-rock geochemistry, alunite spectral features, and alunite chemistry, to assist in identifying upflow zones and/or mineralized centers. Despite these advances, significant challenges remain in tracing hydrothermal fluid pathways and estimating the depth to underlying porphyry intrusions.

Recent work, using a combination of high-resolution short-wave infrared (SWIR) spectroscopy, scanning electron microscopy (SEM), cathodoluminescence (CL), and laser ablation inductively coupled plasma mass spectrometry (LA-ICP-MS) mineral chemistry, distinct mineralogical assemblages and textural features, we have identified, documented, and characterized upflow zones in several deposits worldwide. These zones are characterized by high-temperature advanced argillic alteration assemblages, containing abundant zunyite, topaz, diaspore, pyrophyllite, white mica, aluminium-phosphate-sulfate (APS) minerals, and rutile. Their mineralogical zoning, textures, and mineral chemistry provide key insights into fluid directionality and proximity to porphyry centers, and have proven to be reliable indicators of hydrothermal fluid feeder zones. By integrating these datasets, we have developed and refined new exploration tools that enable the identification and prioritization of epithermal and porphyry targets both within and beneath lithocaps. These tools offer a more predictive framework for mineral exploration in complex terrains and improve the likelihood of discovering concealed porphyry mineralization.

Close-Range Hyperspectral Sensing of Mineralized Outcrops in Open-Pit Mines: Analysis of the Lithocaps of the Allumiere-Tolfa Epithermal System (Central Italy)

Anna Sorrentino[1]
[1]University of Naples "Federico II", Naples, Italy

This study aims to demonstrate how the new generation of portable spectral cameras can significantly enhance geological mapping, enabling rapid acquisition of detailed mineralogical information directly in the field. We showcase the Allumiere-Tolfa mining district (central Italy), where lithocaps associated with a high-sulfidation epithermal system are characterized by alunite and polymetallic mineralizations in addition to a widespread, structurally-controlled and zoned hydrolytic alteration of the hosting volcanic rocks.

High spatial and spectral resolution reflectance VNIR-SWIR imaging was acquired on four outcrops in two open-pit mines using tripod-mounted Headwall Photonics cameras, covering the 400-2500 nm spectral range, with 270 (VNIR) to 166 (SWIR) spectral bands. These spectral images were complemented with 3D modelling obtained via drone imagery and Structure from Motion reconstruction. Hyperspectral data were validated through laboratory spectroscopy and mineralogical analysis on twenty-nine ground samples. Identification and characterization of optically active minerals were performed using minimum wavelength mapping and band ratios. Distinct hydrothermal alteration zones were mapped by using absorption features of smectites, kaolin-group, and alunite-group minerals at ~2160, 2200, and 2290 nm in the SWIR range. Kaolinite polytypes and alunite composition were investigated through the wavelength position of the 1400 nm doublet and sulfate-related absorptions between 1400-1800 nm, enabling estimation of the physicochemical conditions under which the hydrothermal alteration developed. The Fe-oxy-hydroxides signature at ~900 nm allowed mapping the supergene alteration. The hyperspectral images of the Allumiere quarry were finally draped onto a 300-m wide 3D model, for improving the spatial visualization of the alteration facies and their relationship in 3D with tectonic structures and other geological features.

This high-resolution, multi-scale approach, combined with 3D modelling, provided key insights into the geometry of alteration mineral assemblages, serving as a powerful tool for rapid, cost-effective, detailed characterization of exposed vertical wall-rocks, applicable to other active and/or inactive mining areas.

Application of Satellite and Proximal Hyperspectral Sensing to Target Li Mineralization in Volcano-Sedimentary Deposits: The McDermitt Caldera (USA) Case Study

Francesca Corrado[1], Francesco Putzolu[2], Robin Armstrong[2], Nicola Mondillo[1,2], Rita Chirico[3], Bruno Casarotto[3], Matteo Massironi[3], Danielle Fuller[4], Rebecca Ball[5], Richard Herrington[2]
[1]Department of Earth, Environment and Resources Science, University of Naples "Federico II", Naples, Italy, [2]Resourcing the Green Economy Theme, Natural History Museum, London, United Kingdom, [3]Department of Geoscience, University of Padua, Padua, Italy, [4]Image and Analysis Centre, Natural History Museum, London, United Kingdom, [5]Jindalee Lithium Ltd, Perth, Australia

This study provides satellite and proximal hyperspectral analysis of lithium (Li)-bearing volcano-sedimentary environments aimed at determining the target absorption features of alteration assemblages to be used as exploration vectors towards analogous Li-mineralized systems. The approach was applied to the McDermitt caldera (USA), where Li mineralization occurs in the form of clay minerals forming from the alteration of glass-rich extrusive igneous rocks in endorheic lacustrine basins.

The surface-exposed areas of the caldera were investigated using hyperspectral imagery from the German Environmental Mapping and Analysis Program (EnMAP). Satellite data were validated by analyzing samples from the Jindalee McDermitt Li deposit, using hyperspectral imaging with Headwall Photonics cameras. Complementary mineralogical and geochemical analyses were also conducted on the same samples to further support spectral interpretations.

The creation of hyperspectral feature distribution maps was achieved using band ratio analysis and minimum wavelength techniques, implemented in ENVI 5.6 and the open-source Python package “hylite”. This methodology enabled the detection of various minerals, including amorphous silica, Fe-smectite, Mg(Li)-smectite (hectorite), calcite, and zeolites. The samples, which predominantly contain a Mg(Li)-smectite + amorphous silica assemblage, show distinct absorption features at 2200 nm and 2306 nm, detectable within the EnMAP spectral range. Analysis of the corresponding hyperspectral feature distribution maps and the comparison with ground control samples, compared with ground data, confirmed that the 2200 nm and 2306 nm absorption features can be effectively used as mineralogical-hyperspectral vectors for identifying Li-prospective areas on a caldera-scale. This spectral footprint was applied to the High Rock Caldera Complex, a similar system without Li mineralization. The analysis revealed the absence of the characteristic 2200+2306 nm feature, indicating that these spectral features could be used to target lacustrine sequences with mineralogical characteristics similar to those found in Li-mineralized deposits.

Geology and Geochronology of the Mount Marion Lithium Pegmatite Deposit, Yilgarn Craton, Australia

James Porter[1], William Candish[1], Seth Morris[1], Victoria Peterson, Anthony Kemp[2], James Porter[1], Ashok Doorgapershad[1], Jack Stirling[2]
[1]Mineral Resources Limited, Perth, Australia, [2]University of Western Australia, Perth, Australia

The Mount Marion lithium deposit is a globally significant hardrock lithium mine located in the Eastern Yilgarn Craton of Western Australia, a well-endowed region for lithium as well as gold and nickel. Although Mount Marion was discovered in 1913, it received relatively little attention nor research for almost a century. Over the last decade, multiple drilling campaigns have delineated a sizable resource of 66.1Mt @ 1.36% Li_2O and have helped greatly improve our understanding of the deposit.

Lithium mineralisation at Mount Marion is hosted in spodumene-bearing pegmatites that intrude the NW to NNW striking Archaean Saddle Hills Greenstone belt. The deposit is made up of around 100 pegmatite bodies across an ~8km trend. The pegmatites lie sub-parallel, trending roughly northeast and dip at ~10° - 30° to the northwest, as well as some less common steeper dyke-like orientations. Individual pegmatites vary significantly in size, from <2m up to 35m in thickness and up to ~1500m in strike length.

Pegmatite mineralogy is dominated by quartz-albite-spodumene-muscovite with minor garnet, microcline, apatite and tourmaline. Typical pegmatite accessory minerals such as columbite-tantalite and cassiterite are present, but relatively rare, and occur sporadically throughout the deposit.

New U-Pb geochronological studies of apatite from two samples have yielded Tera-Wasserburg lower intercept ages of 2617 ± 14 Ma (MSWD = 0.64, n = 54) and 2629 ± 34 Ma (MSWD = 0.48, n = 53), providing the first direct age constraint for the lithium-bearing pegmatites at Mount Marion. This age is consistent with other lithium-caesium-tantalum (LCT) pegmatites throughout the Yilgarn Craton. The pegmatite age is also coincident with widespread low-Ca granite emplacement, suggesting that the regional tectonic regime provided pressure-temperature conditions conducive to fractionated melt production and subsequent LCT pegmatite formation.

Isotopic Constraints on Tungsten-Beryllium Mineralisation at Mahuaping, SW China

Bin Fu[1], Yanqin Fan[1], Maochun Zhao[2], Bihong Zhang[3]
[1]State Key Laboratory of Critical Mineral Research and Exploration, Institute of Geochemistry, Chinese Academy of Sciences, Guiyang, China, [2]Yunnan Institute of Geology and Mineral Resource Exploration, Kuming, China, [3]Yunnan Institute of Geological Sciences, Kunming, China

Beryllium deposits in China are largely associated with peraluminous igneous rocks. They can be hosted within pegmatite, granite, quartz vein, skarn, greisen or volcanic rocks [1]. Late Paleogene pegmatite-type beryllium mineralisation commonly occurs in the Tibetan-Sanjiang Tethyan metallogenic belt of western Yunnan Province, SW China [2]. However, intrusive rocks have yet been identified around the large Mahuaping tungsten-beryllium-fluorite deposit in North Lijiang, at the junction of the Yangtze Block, the Changdu-Simao Block, and the Songpan-Ganzi accretionary complex [3]. Contrasting mineralisation is hosted within meta-sedimentary rocks on both sides of the lithological interface: scheelite-beryl-euclase-calcite±galena-muscovite-fluorite veins in Middle and Upper Devonian marble; wolframite-beryl-pyrite±fluorite±cassiterite-muscovite-quartz veins in Lower Devonian slate and phyllite.

Fluid inclusions in vein quartz are either mixed CO_2 – low-salinity aqueous inclusions with total homogenisation (Th) of 270-340°C, or low-salinity aqueous inclusions (ca. 6 NaCl wt.% eq. and Th=180-260°C). Oxygen isotope compositions of quartz ($\delta^{18}O$=11.5–13.8‰, VSMOW), cassiterite (1.7–2.7‰), wolframite (1.0–8.7‰), beryl (10.4–11.7‰), and scheelite (2.7–4.8‰) from several samples analysed by using secondary ion mass spectrometry suggest that mineralising fluids are of multiple origins.

Both muscovite Ar-Ar and wolframite and scheelite U-Pb dating gave a comparable age of 35–30 Ma, indicating that the tungsten-beryllium-fluorite mineralisation formed possibly by unidentified Oligocene magmatism at depth as well as compression, strike-slip and thrust nappe structures of the Yangtse Block in North Lijiang during the Himalayan Orogeny.

Similar tungsten-beryllium mineralisation also occurs in Dabashan of the Qinling orogenic belt, Central China, which is related to emplacement of felsic intrusions in vanadium-rich meta-sedimentary rocks including carbonaceous slate, phlogopite schist and marble [4].

[1] D.H. Wang, et al. (2017) Geol. Survey China 4(5): 1-8.
[2] C.Y. Zhang, et al. (2021) Acta Petrol Mineral 40: 452-464.
[3] W.C. Li, et al. (2025) Ore Geol. Rev. 179: 106553.
[4] H.Z. Dai, et al. (2018) Rock Mineral Anal 37(3): 336-345.

Melt-Melt Immiscibility as a Key Mechanism for Spodumene-Quartz Mineralisation in Pegmatites, Insights on a Novel Concept from Andover, Western Australia

Joshua Combs[1], **Richard Cubitt**[1]
[1]Azure Minerals, Perth, Australia

Since discovery of the spodumene-bearing pegmatites hosted in the Andover Intrusion ~5km SE of Roebourne, Azure Minerals has compiled a world-class geochemical, structural, geological (lithological, mineralogical, textural) and spectral dataset. This dataset provides a robust framework for detailed characterisation of the Andover pegmatite swarm and offers key insights into the spatial and temporal evolution of this lithium-mineralised system and points towards a unique origin story.

An anatectic origin (as originally proposed by Stewart, 1978) for the Andover pegmatites is proposed, aligning with a recent notable shift in published literature (e.g. Koopmans et al., 2023). The lithium-enriched pegmatitic melt was structurally emplaced within the Andover Mafic-Ultramafic complex. The pegmatite underwent a significant degree of fractionation within the structural network during emplacement, preserved as systematic variations in whole rock geochemistry within the pegmatite swarm. This open system behaviour facilitates a more extreme degree of fractionation to a pure spodumene-quartz melt, and in much higher volumes than would be plausible with closed system fractionation.

Andover provides compelling evidence for the role of melt–melt immiscibility as a critical ore-forming process - a mechanism that has not received any attention in the context of lithium pegmatites to date. This mechanism facilitates the crystallisation of high quality spodumene, high in lithium content and low in deleterious elements such as iron.

This revised genetic framework has wide-ranging implications from deposit-scale resource modelling and geometallurgical domain definition, through to regional exploration strategies targeting lithium pegmatites within the northwest Pilbara Craton and analogous geological settings globally. The recognition of high-quality immiscibility-driven spodumene mineralisation opens new conceptual frontiers in lithium pegmatite ore forming systems.

Innovating Discovery: Field Sensors and Breakthrough Technology at the Sinclair Cesium Mine, Western Australia.

Nigel Brand[1,2], David Crook[3]
[1]Geochemical Services Pty Ltd, West Perth, Australia, [2]CET, University of Western Australia, Crawley, Australia, [3]Ore Source Pty Ltd, West Perth, Australia

Field-portable sensors were instrumental in both the discovery and development of the Sinclair Cesium Deposit in Western Australia—Australia's first commercially extracted Cesium mine and only the third pollucite deposit mined globally.

The deposit was discovered and delineated through the innovative application of field-portable X-ray fluorescence (pXRF) to an extremely fractionated lithium–Cesium–tantalum (LCT) pegmatite system. A key adaptation was the development of a Lithium Index algorithm integrated into pXRF software, enabling rapid identification of coherent soil anomalies suggestive of fertile pegmatites within the Pioneer Dome. One such anomaly ("Anomaly 8") led directly to the first drill intersection of Cesium-rich mineralization in August 2016.
A robust workflow integrating real-time field data, laboratory confirmation, and QA/QC protocols enabled precise Cesium quantification during exploration, drill-and-blast, stockpiling, and shipping. The turnaround speed of in-field analysis proved critical to the success of the tightly constrained three-month mining operation, far outpacing the responsiveness of commercial laboratories.

To support ore-block definition and mineral identification, an extensive spectral reference library of LCT pegmatite minerals was compiled, validated, and deployed using portable Raman spectroscopy. This facilitated rapid in-situ mineral confirmation and strengthened confidence in grade control and extraction planning.

Following Sinclair, continued regional exploration using these portable technologies led directly to the discovery of spodumene mineralization at Dome North. The Sinclair project proved that field-adapted technologies can transform mineral discovery from slow, laboratory-dependent processes into fast, precise, real-time decision-making— bridging the gap between geoscience innovation and operational impact in the field.

Multiple Mineralisation Events Demonstrated in Complex Li-Cs-Ta Pegmatites at Buchanan Creek, Queensland

Benjamin Hines[1], Lisa Kearney[1,2], Daniel Killen[1], Andrew Christy[2,3], Mitchell de Bruyn[3], Renjie Zhou[4]

[1]Geological Survey of Queensland, Brisbane, Australia, [2]School of Earth and Atmospheric Sciences, Queensland University of Technology, Brisbane, Australia, [3]Central Analytical Research Facility, Queensland University of Technology, Brisbane, Australia, [4]School of the Environment, University of Queensland, Brisbane, Australia

Characterisation of pegmatite dykes from Buchanan Creek, located in the Georgetown Inlier, Northeast Queensland, has produced a detailed story of polyphase deformation within the sampled Li-Cs-Ta (LCT) pegmatite dykes. The primary lithium ore mineral spodumene has recently been identified in this system, in addition to other Li-bearing phases, including amblygonite, lepidolite, and rare elbaite. Caesium and rubidium mineralisation is strongly associated with micas, including the first reported instance of sokolovaite (Cs-polylithionite) in Australia. A multidisciplinary approach encompassing regional mapping, bulk rock geochemical sampling, visual and hyperspectral logging of drill core, and microanalysis (micro-XRF, LIBS, TIMA, Raman, LA-ICPMS) have been applied to characterising the Buchanan Creek LCT pegmatite system. The ca. 1550 Ma Buchanan Creek pegmatites are hosted within tightly-folded, lower greenschist facies Paleoproterozoic dolerite and metasedimentary rocks. Complex pegmatites of the Buchanan Creek area are lenticular in shape, and are typically oriented northwest, following the strong regional S1 foliation. Dykes display strong internal zonation, commonly having a quartz core, with a lepidolite dominated (± quartz, albite, amblygonite) intermediate zone, and an albite-muscovite dominated wall zone. Where present, spodumene and amblygonite are most abundant in the core to intermediate zones of the studied dykes. Cassiterite, tantalite, and microlite are late in the mineralisation sequence, and are commonly associated with albite, and late crosscutting aplitic dykes. Spodumene appears to have formed early in the paragenetic sequence, and subsequently has been largely replaced by amblygonite. Further deformation and remineralisation of incompatible elements (Li, Cs, Rb) is indicated by common brecciation of amblygonite, infilled by early trilithionite, and later polylithionite. Collectively, these observations demonstrate multiple fluid-flow and deformation events affecting the distribution of critical metals and the mineralogy of this system, providing key insights for the future exploration of this system.

The Role of Iron-Rich Hydrosaline Liquids in the Formation of Kiruna-Type Iron Oxide–Apatite Deposits: Evidence from Fluid Inclusions

Li-Ping Zeng[1], Xin-Fu Zhao[1], Carl Spandler[2]
[1]State Key Laboratory of Geological Processes and Mineral Resources, and School of Earth Resources, China University of Geosciences, Wuhan, China, [2]Department of Earth Sciences, The University of Adelaide, Adelaide, Australia

Kiruna-type iron oxide-apatite (IOA) deposits, an important source of iron, show close associations with andesitic subvolcanic intrusions. However, the processes of ore formation and the mechanism controlling iron concentration remain uncertain. Several genetic models have been proposed, including iron oxide or sulfate (- carbonate) melts formed via magma immiscibility, flotation of igneous magnetite micro-aggregates, and replacement by magmatic hydrothermal fluids. Recent studies have focused on high-temperature melt/fluid inclusions, as these provide critical clues for understanding the nature and origin of the initial ore-forming fluids/liquids or iron-rich melts.

Here, we report the widespread presence of high-temperature, water-poor, multi-phase hydrosaline liquid inclusions in pre- and syn-ore minerals (including albitized zircon, garnet, diopside, and magnetite) from representative IOA deposits in eastern China. These inclusions consistently homogenize to a liquid phase upon vapor disappearance during heating at temperatures ranging from 787 to 928°C (average 865°C). These homogenization temperatures roughly represent the entrapment temperatures of the hydrosaline liquids, supported by other independent geothermometric data. Additionally, the hydrosaline liquids can be stable at temperatures and pressures of subvolcanic settings in the simplified H2O-NaCl phase diagram. LA-ICP-MS analyses show these inclusions predominantly contain 3 to 10 wt.% Fe, highlighting their significant iron transport capacity. Furthermore, elemental ratios (e.g., K/Na and Ca/K) of the hydrosaline liquids and B-Sr-O isotopic data from associated minerals suggest that such hydrosaline liquids were likely immiscible from the dioritic magmas with high Cl/H2O in subvolcanic settings, typically at depths ≤2.5 km.

Therefore, we propose a novel genetic model in which high-temperature alteration and magnetite ores were formed simultaneously through host-rock interaction with, and/or decompression and cooling of, the hydrosaline liquids. This model provides important insights into the distinctive characteristics of IOA deposits in shallow magmatic-hydrothermal systems.

Experiments Supporting IOA Deposit Formation by Interaction of Evaporites and Silicate Magmas

Shengchao Yan[2], **Michael Anenburg**[1], Bo Wan[2], John Mavrogenes[1]
[1]Australian National University, Canberra, Australia, [2]Institute of Geology and Geophysics, Chinese Academy of Sciences, Beijing, China

The key processes that lead to iron-oxide apatite deposit formation are hotly debated. Several genetic processes have been previously suggested, including immiscibility between iron-phosphate and silicate magmas, flotation of magnetite by adherence to hydrothermal fluid bubbles, metasomatism by Fe-rich saline brines, and crystallisation alongside molten salts. Although each is a viable model supported by both natural observations and experiments, a common link between them is still missing.

Here we show high pressure and temperature experiments inspired by the El Laco IOA deposit. We ran layers of silicate magma, sulfate–chloride evaporites, ironstones, and phosphorites at 0.4 GPa and 1000 to 1200 °C. We see formation of emulsion consisting of immiscible silicate and iron-phosphate melts, a third liquid of Fe-rich sulfate–chloride composition, and magnetite covered with phosphate-rich fluid bubbles. Upon quench, dendritic magnetite crystallised out of the silicate magma.

Our results show that when regular andesitic magmas assimilate evaporites, iron and phosphate solubility in the silicate increases substantially. The heat causes melting of the evaporites themselves. These liquid salts are highly corrosive and can effectively extract iron and phosphate from surrounding rocks, and maintain a high equilibrium level of these components in the silicate. Fluid bubbles will float the voluminous magnetite that forms during cooling, as well as transport phosphate within the fluid. Finally, the saline fluids have high iron-carrying capacity and will readily metasomatise existing rocks, forming additional magnetite.

This shows that all the previously-suggested IOA formation models are correct, and occur contemporaneously. This is triggered when arc-related silicate magmas intrude sedimentary sequences that contain evaporites and iron–phosphate rich rocks.

Alteration-Derived Volume Changes Control the Location of Iron Oxide-Cu-Au Mineralization

Tobias Schlegel[1], James Austin[1], Weihua Liu[1], Renee Birchall[1], Belinda Godel[1]
[1]CSIRO Mineral Resources, Australia

Iron oxide-copper-gold (IOCG) deposits are a key source of copper crucial for the energy transition. The Ernest Henry deposit in Queensland, Australia, is the region's largest IOCG system. Early deposit formation models emphasized hydrothermal brecciation and magnetite–chalcopyrite infill with limited overprinting, contrasting with tectonic brecciation and alteration-related mineralization models but aligning with the role of structures in focusing fluid flow and Cu deposition.

This study synthesises published data from >200 samples, including modal mineralogy, inferred magnetic fabrics, and combines them with thermodynamic modelling to evaluate alteration and porosity evolution at Ernest Henry. Alteration halos, marked by systematic changes in mineralogy, geochemistry, porosity, and petrofabric intensity, are gradational and cross lithological boundaries and structures. High-grade Cu–Au mineralization is associated with a recognizable hydrolytic alteration overprint of pre-Cu Na–Fe and syn-Cu K–Fe alteration assemblages.

The thermodynamic modelling of sequential Fe–K alteration (stage I) and hydrolytic fluid-rock interaction (stage II) involved two fluids, both approximating fluid compositions published from IOCG deposits. Stage I interaction of Na–Fe–K–S–CO_2–Cl fluids (200 ppm Cu) with albite + magnetite host rock at 300–400 °C yields, depending on the fluid/rock ratio, dominantly K-feldspar + magnetite with un- to subeconomic chalcopyrite grades. Stage II alteration by acidic, low Cl-salinity Na–Mg–SO_4 fluids (500 ppm Cu) at 225–250 °C enhances Cu deposition to the observed chalcopyrite grades, reproduces key mineral assemblages, their concentrations and distribution determined.

High-grade chalcopyrite distribution reflects mineralogical and porosity changes caused by K–Fe and hydrolytic alteration, including partial destruction of magnetite and K-feldspar. This generates porosity in ore zones, enhancing permeability, while porosity decreases in alteration halos nearby.

These findings, along with subtle changes in magnetic fabrics, show that structural preparation and alteration-induced porosity–permeability changes jointly control high-grade Cu mineralization.

What Controls the Geometry of Ernest Henry Ore Shoot?

Adrian Corvino[1], Jonathan Standing[1], Scott Stephens[2], Ned Howard[2]
[1]Model Earth Pty Ltd, Unit 2/80 Colin St, Australia, [2]Evolution Mining Pty Ltd, 175 Liverpool Street, Australia

Ernest Henry Cu-Au deposit represents a blind IOCG breccia pipe plunging south. Here, we recognise boudinage, stretching lineation and folding as important structures governing its linear, pipe-like development. Elongation axes of these structures are consistently parallel to the ore shoot. They underscore a fundamental geometric relationship that must be reconciled within breccia-based models.

Boudinage features include neck veins, foliation boudinage, disaggregation breccia, and small ragged ductile fractures called birdswing veins. Elongate amygdules, phenocrysts and breccia clasts define a stretching lineation aligned with boudin extension axes. High strain zones show streaky elongation banding and L or L-S fabrics. Strain is constrictional. Elongations are up to 400% or more.

Foliation bends from SE-dipping near surface to steeply W-dipping and overturned at depth. Continuous bending defines a broad fold hinge along which the ore shoot is aligned. Ore breccia itself is not a discordant pipe but seems stratigraphically conformable and has a twisted curviplanar shape mimicking the fold hinge. The fold axis is subparallel to boudinage and may suggest transtensional folding with elongation and dilation along the hinge zone.

Deeper down, ore breccia resembles volcaniclastic rock, and the textural distinction between these rocks is blurred. New 3D modelling shows that they occupy the same folded stratum in the volcanosedimentary pile.

In our model, ore breccia development overlapped with syn-boudinage disaggregation and involved metasomatic replacement of 'apparent volcaniclastics' along the fold hinge zone. Evolving rheology contrasts influenced strain and fluid partitioning. Despite alteration complexity, there is strong evidence for geometric continuity of elongation fabrics across the ore breccia. Thus, boudinage and pseudobreccia processes help explain ore breccia, whereas a breccia-pipe per se does not explain the boudinage or fabric continuity.

Our observations are consistent with a fold or flexure-controlled replacement model for Ernest Henry. This draws comparison to neighbouring deposits E1 and Monakoff.

Exploration success and refinement of mineralisation models in the Western Forelands of the Democratic Republic of Congo.

Amanda Stoltze

Ivanhoe staked a 50,000 square kilometres land package in the Democratic Republic of Congo in the late 1990s consisting of ground peripheral to the sought-after Congo Copper Belt (CCB). Systematic high-quality geophysics and geochemistry was collected resulting in the discovery of Kamoa in 2008. Kamoa has been followed by further discoveries in the region, most notably the high-grade Kakula deposit in 2016, with a total of 1 901 Mt at 2.5% Cu identified to date.

Sediment-hosted Cu at Kamoa-Kakula in the Western Forelands (WF) of the Central African Copper Belt occurs in a setting not previously thought to be prospective. The ore host to the historically known deposits on the CCB, the Mine Series, is not present and there is no evidence of salt deposition on the WF. These factors were widely believed to limit the prospectively of the WF and were the reason for limited exploration prior to 2000. Mineralisation at Kamoa-Kakula is hosted in pyritic siltstones and diamictites of the Grand Conglomerate at the base of the Nguba Group, and is underlain by a thick sequence of "red bed" clastics which allowed the circulation of mineralising basinal fluids.

Ivanhoe has continued to invest in exploration and this persistency has resulted in the recent discovery of the Makoko and Kitoko orebodies to the west of Kamoa-Kakula which has further modified the regions mineralisation model. The mineralisation is hosted in a pyritic siltstone reductant similar to Kamoa-Kakula but with the Grand Conglomerate directly overlying basement without a proximal sandstone aquifer. Basement proximal to mineralisation is pervasively hematite altered with a high proportion of fractures and veins in comparison to basement intersected in areas distal to mineralisation. This change in the basement associated with mineralisation suggests fluids may have circulated in the basement without the need of a sandstone aquifer.

Building Bridges in Geoscience: EuroGeoSurveys in Fostering International Resource Collaboration

Julie Hollis[1], Francesco Pizzocolo[2], Patrick Wall[1]
[1]EuroGeoSurveys, Brussels, Belgium, [2]TNO, The Geological Survey of the Netherlands, Eindhoven, The Netherlands

The quest for resources to drive the energy transition requires robust international collaboration in geoscience. Europe aims to strengthen its access to critical minerals – domestically and through strategic international partnerships – through initiatives such as the Critical Raw Materials Act and, more broadly, the Clean Industrial Deal.

EuroGeoSurveys – a not-for-profit association representing the Geological Surveys of Europe – actively supports these efforts through initiatives that address resource challenges at both European and global scales. EuroGeoSurveys flagship project is the Geological Service for Europe project, which is delivering pan-European harmonized geological data through the European Geological Data Infrastructure. This project supports informed decision-making on resource management and sustainable development at European scale. EuroGeoSurveys is also active internationally, engaging with organizations worldwide to share best practices, harmonize data standards, and promote responsible resource governance. This includes delivering geoscience support to a number of EU bilateral strategic international partnerships, including the EU-Australia partnership. In Africa specifically, the work of EuroGeoSurveys has – for the past 8 years – been central to building the PanAfGeo project, which focuses on enhancing geoscientific capacity across Africa through training of geologists from the African geological surveys, in partnership with the Organisation of African Geological Surveys. This capacity-building initiative has not only enhanced Africa's ability to sustainably manage its geological resources but has also built a vibrant, engaged geoscience community spanning two continents, with institutional impacts at national and continental scale. This paves the way for a much expanded and minerals-focused programme – PanAfGeo+ – which launched this year. These and other EuroGeoSurveys collaborative activities, projects, and initiatives highlight that while harmonised data, knowledge sharing, and collaborative projects are vital, collaboration itself has inherent value. As a very human endeavor, collaboration can deliver fundamental rewards that extend well beyond finite project goals.

The Critical Minerals Geoscience and Data (CMGD) Initiative of the Geological Survey of Canada

Eric Potter[1], **Geneviève Marquis**[1], Victoria Tschirhart[1]
[1]Geological Survey of Canada, Ottawa, Canada

The Geological Survey of Canada's (GSC) Critical Minerals Geoscience and Data (CMGD) initiative is one component of the Government of Canada's $3.8B Critical Minerals Strategy that is supporting the transition to a green economy and strengthening supply chains. While CMGD funding is primarily for federal government-led research, the initiative also includes funding opportunities for Canadian provincial and territorial surveys over the four-year lifespan of the program. CMGD research is focused on six priority minerals from Canada's critical minerals list: cobalt, copper, graphite, lithium, nickel, and rare earth elements. These minerals are essential for sustainable economic development yet face global supply risks, limited and/or declining domestic production. To address these challenges, CMGD is structured around four integrated solution streams: 1) knowledge base; 2) ore system studies; 3) advanced predictive geoscience; and 4) market assessments.

The critical minerals knowledge base is a national sampling and analytical program to characterize critical mineral contents and associations in mineral deposits and derivative materials (e.g., ores, concentrates, tailings, and waste). Critical mineral ore system studies involve thematic geoscience research targeting both conventional and emerging sources of critical minerals, developing models and knowledge to support exploration. The advanced predictive geoscience solution builds on solutions 1 and 2, using advanced analytics to accelerate and guide responsible exploration. This includes creation of national mineral potential maps that consider not only geological data, but environmental, social and governance (ESG) factors to streamline exploration and reduce project development delays. Finally, the market assessments solution is examining critical mineral supply and demand to inform Canada's position in global markets and support our domestic critical mineral value chains. Together, these solution streams aim to de-risk mineral exploration and development in Canada by delivering high-quality, public geoscience data and knowledge to guide responsible resource development, land use decisions, and policymaking.

Environmentally Sustainable Production of Critical Metals in Tasmania: A Collaborative Multidisciplinary Initiative

Owen Missen[1], Yamila Cajal[1], David Cooke[1], Ivan Belousov[1], Lejun Zhang[1], Julie Hunt[1], Sharon Fraser[2], Michael Roach[1], Mohammadbagher Fathi[1], Wei Hong[1], Sheree Armistead[1], Kim Beasy[2], Angela Santos Costa[1], Vinicius da Cruz[1], Jose Barillas Dias[1], Emrecan Yurdakul[1], Javier Gil Rodriguez[1], Nelao Naimbale[1], Christopher Allen[1], Alfredtina Appiah[1], Emmanuel Musa[1], Pratichee Mondal[1], Jane Hall-Dadson[2]
[1]Centre for Ore Deposit and Earth Sciences (CODES), University of Tasmania, Hobart, Australia, [2]School of Education, University of Tasmania, Hobart, Australia

Tasmania hosts important resources of critical and strategic metals, with: mining operations currently producing tungsten and tin; advanced exploration projects defining rare earth element and magnesium prospects; and critical metals such as cobalt identified as potential by-products from active base and ferrous metal mines, and potentially from mine wastes. To support the growth of the critical minerals industry in Tasmania and foster additional development, an extensive collaboration between academia, industry, and government organizations has been developed to provide up-to-date geological and geometallurgical data. The collaboration is funded by the Australian Government Department of Education and led by CODES, in conjunction with the School of Education at the University of Tasmania. It includes collaboration with ten industry partners, Mineral Resources Tasmania, the Tasmanian Mining Manufacturing and Energy Council, and other national and international university partners. The research includes three main components: (1) orebody knowledge and characterization of critical metals, (2) pathways to production, including minerals processing and environmental geology, and (3) education and engagement. Eleven doctoral students are being trained in the collaborative project, with most combining field research and analytical work to link the macro and micro scales of critical metal deportment. Our Education team is investigating barriers for young people entering into the mining industry, particularly focusing on Tasmania's mining-dominated but isolated west coast region.

The three components of the project have allowed the team to produce research outcomes via our innovative multidisciplinary collaboration between natural and social scientists. Our research results have implications for exploration targeting and processing of critical metals and on regional community engagement and barriers to STEM education, both in Tasmania and globally. We highly recommend that future multidisciplinary projects collaborate with departments such as education and social sciences to expand the reach and impact of geoscience research.

The Global Critical Minerals in Ores (CMiO) Database – A Review and Update

Evgeniy Bastrakov[1], George Case[2], Michael Gadd[3], Garth Graham[2], Kristy Guerin[1], Steph Hawkins[1], Albert Hofstra[2], David Huston[1], Christopher Lawley Christopher Lawley[3], Vladimir Lisitsin[4], Simon VanDerWielen[1], Bronwen Wang[2]
[1]Geoscience Australia, Canberra, Australia, [2]United States Geological Survey, USA, [3]Geological Survey of Canada, Canada, [4]Geological Survey of Queensland

Critical minerals are essential to modern technologies and national security yet supply chains remain vulnerable to disruption. Identifying and assessing these resources is challenging, particularly because public data on their concentrations is often limited, especially for byproducts in ores mined primarily for other commodities. To address this, the Critical Minerals Mapping Initiative (CMMI), a collaboration among the U.S. Geological Survey, Geological Survey of Canada, and Geoscience Australia, was established in 2019.

A key output of the CMMI is the Critical Minerals in Ores (CMiO) database, a comprehensive resource designed to enhance understanding of global critical mineral distributions. As of May 2025, this database includes a compilation of over 20,000 high-quality multielement geochemical analyses of ore and related samples, representing over 100 deposit types across 10 distinct deposit environments. This database employs a novel, consensus-based classification scheme developed by CMMI (https://pubs.usgs.gov/publication/ofr20211049), enabling robust comparisons of geochemical signatures from ore deposits globally.

The CMiO database supports the assessment of both primary and byproduct critical mineral potential. For example, analysis of data from porphyry copper-molybdenum-gold deposits allows for quantification of potential byproduct recovery of elements like tellurium, selenium, and rhenium. Similarly, the database aids in understanding differential enrichment of critical minerals, such as cobalt and rare earth elements, within various types of iron-oxide-copper-gold (IOCG) and iron-oxide-apatite (IOA) groups of deposits, informing exploration and resource assessment efforts.

The CMiO database is a vital, steadily expanding data source for researchers, industry, and governments seeking to understand and secure future critical mineral supplies. An interactive map, the downloadable dataset (https://portal.ga.gov.au/persona/cmmi), and a set of data analysis and visualisation tools are publicly available. CMMI invites submission of high-quality geochemical data for well-characterized ore samples to support the continued growth of the CMiO database. A submission guide is available at: https://doi.org/10.26186/149408.

Disruptive Thinking and Transformational Research to Shape the Future Resource Workforce

Marco Fiorentini[1], Anita Parbhakar-Fox[2], Carl Spandler[3], Tony Kemp[1], John Mavrogenes[4], Andrew Berry[4], Wei Liu[1], Natasha Bartlett[1]
[1]University of Western Australia, Perth, Australia, [2]University of Queensland, Brisbane, Australia, [3]University of Adelaide, Adelaide, Australia, [4]Australian National University, Canberra, Australia

The global transition to a high-tech, 'green' energy future requires a fundamental shift towards commodities characterised by high risk of supply chain disruption. This pivot is not easy since it requires a change in the way we explore for and process minerals and requires a work force with a future-focused skillset. Accordingly, there is growing recognition for the importance of skills training, as highlighted in the respective critical mineral strategies of many nations and nation blocks globally. Optimal skills training demands much closer collaboration between the higher education, government, public and the resources sectors. Established in November 2024 in Australia, this centre is unique in its kind as the term 'resource' encompasses both the mineral commodities and the specialist human resource. Our centre represents an unprecedented partnership across the minerals value chain. It strategically brings together four major Australian universities, with significant funding from the Australian Research Council, major government agencies and state/territory geological surveys, as well as mining companies and technology firms advancing mineral discovery and extraction.

The centre will revolutionize the path for training in the geosciences and will drive transformational outcomes for industry and government by translating cutting-edge skillsets and knowledge into applied capacity. This centre aims to leverage complementary strengths with other centres operating in the same framework, providing tailored skill development solutions, fostering collaborative events, co-designing and marketing training courses, and providing broader opportunities for industry internships and placements.
This centre will bridge the gap between mineral systems science, mineral exploration protocols and ore processing/metallurgical extraction. It will train the next generation of geoscientists with an essential understanding of the whole value chain of the critical resources of the future, which are needed to unlock the critical mineral potential of any geological environment and to enable the transition to a high-tech, clean energy society.

Collaboration in Action: Advancing Geoscience through Government, Research and Industry Partnerships

Janelle Kerr[1]
[1]Geological Survey Of Queensland, Brisbane, Australia

Collaboration is integral to the Geological Survey of Queensland's (GSQ) operations, supporting a range of projects across geoscience and related fields. GSQ partners with universities, government organisations, and industry stakeholders to combine expertise and resources, enabling the generation of insights that would not be achievable independently. Many projects are closely aligned with industry needs, ensuring that outputs are targeted and fit for purpose. For example, the deposit characterisation project has over the last 7 years produced extensive data and knowledge on Queensland's mineral deposits, offering new insight into how these deposits formed and how to target for them.

Collaborations with government-funded organisations facilitate standardisation of data and workflows, and enable participation in national programs, such as those led by Geoscience Australia. This approach has been particularly impactful in geophysics, where consistent data quality, technical requirements and formats improve accessibility for industry users. University partnerships provide access to specialised knowledge, equipment, and research capacity. These collaborations also support student and early-career researcher development, fostering long-term workforce development.

The Collaborative Exploration Initiative (CEI) and Collaborative Development Program (CDP) are the most direct means of support for industry offered by GSQ. These programs also allow us to incentivise novel, innovative and collaborative projects being undertaken by companies, especially in the upstream exploration space. The recent addition of CDP allows us to incentivise companies further down the value chain to consider whether it is worth taking a second look at materials previously considered waste through a new lens.

In addition to traditional geoscience, GSQ participates in non-traditional research collaborations, including work on circular economy principles. These initiatives focus on improving resource efficiency and quality, aligning with broader goals of sustainable resource development. GSQ's collaborative efforts aim to address key scientific and industry challenges, contributing to the responsible management and development of Queensland's resources.

Identification and Measurement of Minerals on Thin Sections of Intrusive Igneous Rocks Using Deep Learning

Stalyn Paucar[1]
[1]Central University of Ecuador, Quito, Ecuador

Artificial Intelligence has gone beyond theory, and it is used in everyday life, including politics, economics, medicine, social networks, web browsers, academic research, etc., and some consider it as the Fourth Industrial Revolution. Since the beginning of the 21st century, research on deep learning has been applied to different branches of the Earth Sciences. On this research I used the Facebook AI Research's Detectron2-v0.6 model to identify eight minerals of intrusive igneous rocks on thin sections, and the image processing library scikit-image to obtain quantitative measurements of each identified mineral.

For this task, the starting point is the workflow used in the development of AI models. 400 images were obtained from the British Geological Survey and The Open University database, and 62178 labels were prepared in the Roboflow platform. This was used to generate four datasets for training and validation, along with an external evaluation dataset of 20 images for the Detectron2-v0.6 model. This process was carried out in Google Colab.

The best result was obtained in the XPL-10X image dataset, achieving a total loss of 1.15, an average precision (AP, IoU = 0.5-0.95) of 15.7% and an average recall (AR, IoU = 0.5-0.95) of 26.3%. Scikit-image was also evaluated, and the relative error results were between 0-3%, which is appropriate for this task.

An application was designed and implemented, available in Google Colab and Hugging Face-Streamlit. This mineral identification and measurement tool allows the user to upload their own images and generate a PDF report useful in the geological field, which includes the results of mineral segmentation and identification, mineralogical analysis, and crystal size distribution analysis.

AI models show great potential for automating petrographic analysis, accelerating processes that have traditionally been manual and highly specialized. Their adaptability makes them valuable for the microscopic analysis of ore minerals.

Exploring the Digital Terrain: The Power of Geospatial Technology and GeoAI in Modern Mining

Aune Kamosho[1]
[1]QKR Navachab Gold Mine, Namibia University of Science and Technology, Windhoek, Namibia

The mining industry, traditionally rooted in fieldwork and manual data analysis, is undergoing a profound transformation through the integration of advanced geospatial technologies and artificial intelligence. GeoAI, the fusion of geospatial science and AI, is empowering mining companies to navigate the complexities of mineral exploration with unprecedented accuracy, efficiency, and sustainability.

At the core of this technological shift is geospatial technology, which provides the tools to map, analyze, and monitor the Earth's surface and subsurface in real-time. Satellite imagery, drone-based data collection, LiDAR scanning, and hyperspectral imaging are now integral to the exploration process. These technologies offer high-resolution mapping and allow for detailed analysis of vast, often remote, areas. By combining these data sources, geospatial technology creates a comprehensive digital model of the terrain, enabling geologists to identify promising mineral-rich zones more efficiently and with greater precision than ever before.

In parallel, GeoAI is revolutionizing how this data is interpreted. Artificial intelligence algorithms, trained on historical and real-time exploration data, are capable of detecting hidden patterns in geological formations, predicting the likelihood of mineral deposits, and suggesting optimal drilling locations. Machine learning enhances the decision-making process by automating data analysis and reducing human error, while also enabling more accurate predictions of mineral behavior and resource volumes.

These innovations are not only transforming mineral discovery but also improving sustainability efforts in mining. The combination of real-time geospatial monitoring, predictive modeling, and AI-driven analysis is helping to reduce environmental impact, optimize resource extraction, and minimize operational costs. Through smarter, more data-driven methods, the industry can address the growing demand for minerals while balancing the need for responsible, ethical, and efficient mining practices.

This presentation will explore the evolving role of GeoAI and geospatial technologies in the mining sector, reshaping the entire lifecycle of mineral exploration.

The Rock Whisperer: Integrating Automated EDA Workflows with LLM

Putra Sadikin[1], Steve Prager, Arthur Sedek, Pouria Mehrani
[1]Imdex, Melbourne, Australia

Traditional exploratory data analysis (EDA) in geoscience benefits from visual computation of large, multivariate datasets. Augmenting EDA with dimension-reducing workflows helps geoscientists derive insights and identify structures, such as correlations or similarities. This process integrates data interpretation with geological context, enhancing applications like exploration vectoring and rock characterization, which are crucial for ore body knowledge development.

Further enhancement involves the templatization of exploratory data analysis through recordable workflows. Recent developments have enabled these workflows in Headless GAS, a headless cloud-native environment of IMDEX ioGAS, allowing for automation and standardisation of complex geochemical data analysis processes, ensuring consistency and reproducibility. Headless GAS retains all the functionality of ioGAS, enabling cloud-based workflow processing and handling unlimited data sizes, ideal for "big data" from real-time sensors and scanning platforms.

Headless GAS is limited to recording workflows within the program, restricting workflow creation and dissemination to SMEs. This presentation explores integrating a headless cloud-native environment with LLMs to automate and scale workflow creation. Natural language requests replace traditional workflow recording, dynamically translating into actionable workflows. Domain knowledge and workflow paths are captured in knowledge graphs for LLM fine-tuning. This simplifies workflow recording, leveraging a domain-trained LLM for sophisticated analytics in ioGAS. Geoscientists describe desired outcomes, with the LLM determining optimal paths, shifting focus to goal-oriented analysis. Embedding an LLM enriches result context, interpreting visualizations and recommending further analysis steps to achieve desired outcomes.

Another key advantage is the potential for creating workflows without a traditional interface, which is particularly beneficial in coreshed applications where hands-free operation is crucial for safety. This method ensures reproducibility and scalability, with the LLM acting as a cognitive layer that bridges human-centric commands with the functionalities provided in the headless GAS environment. This foundational step paves the way for natural conversational workflows to interact with traditionally complex desktop applications.

Characterisation of Ore Deposits: Overcoming Cross- and Multi-Scale Challenges

Elena Belousova[1], Vladimir Lisitsin[1], Alkis Kontonikas-Charos[1], Al-Tamini Tapu[1], Rhiannon Jones[1], Friedrich von Gnielinski[1], Suraj Gopalakrishnan[1], Daniel Killen[1], Lisa Kearney[2]
[1]Geological Survey of Queensland, Brisbane, Australia, [2]School of Earth & Atmospheric Sciences, Queensland University of Technology, Brisbane, Australia

Economic geology research commonly focuses on individual mineral deposits, emphasising their genesis and deposit type classification. However, distal alteration footprints of individual deposits and their connections to regional mineral systems remain poorly understood and under-represented in publicly accessible geological sample and data collections. The Geological Survey of Queensland (GSQ) have achieved significant success in closing this gap in Queensland. GSQ built up an extensive reference collection of drill cores and ore and alteration samples from 110 mineral deposits representing most important mineral systems across the state and employed multiple advanced analytical methods for characterisation of deposits and their alteration footprints at various scales.

The most extensive data acquisition has been completed for mineral systems in the Mount Isa Province (IOCG, sediment-hosted Cu and Zn-Pb-Ag deposits) and intrusion-related mineral systems in NE Queensland, with multiple deposits investigated for each distinct mineral system. Each target deposit is characterised based on 3 to 6 representative drill cores. Continuous hyperspectral scanning is used first for fast mineralogical logging and identification of common rock-forming and alteration minerals, followed by other non-destructive core scanning, including continuous XRF and petrophysical data acquisition. Representative samples are then selected for comprehensive whole-rock geochemistry and detailed mineralogy and geochemistry using micro-XRF, LIBS and TIMA instruments. Selected well-characterised samples from each deposit are then further investigated to document mineral chemistry of key ore and alteration minerals, including isotope geochemistry.

Integrating results of these multiple complementary methods applied across an individual mineral deposit allows comprehensive characterisation of its mineralisation signature and alteration footprints. When systematically applied to characterise multiple genetically related deposits across a mineral district, and integrated into a regional geological context, this approach generates an invaluable dataset to successfully identify, map and navigate alteration footprints of a mineral system from deposit to district scale, ultimately assisting and facilitating mineral exploration.

Multi-Scale Characterisation of Intrusion-Related Gold Systems in NE Queensland: Insights from the Ravenswood Deposit

Al-tamini Tapu[1], Vladimir Lisitsin[1], Elena Belousova[1], Courteney Dhnaram[1], Friedrich von Gnielinski[1], Alkis Kontonikas-Charos[1], Daniel Killen[1], Suraj Gopalakrishnan[1], Lisa Kearney[2], Ben Hines[1], Rhiannon Jones[1]
[1]Geological Survey of Queensland, Brisbane, Australia, [2]School of Earth & Atmospheric Sciences, Queensland University of Technology, Brisbane, Australia

North-east Queensland hosts multiple gold deposits produced by late Palaeozoic intrusion-related gold systems (IRGS), including the Ravenswood, Mt Leyshon, and Kidston deposits. The Geological Survey of Queensland (GSQ) implemented an integrated, multi-scale approach to characterise mineralisation signatures and alteration footprints of IRGS in the region, using the Ravenswood camp as a key case study area. This approach involves consistent and systematic characterisation of individual gold deposits and their camps, combining continuous hyperspectral (HyLogger) and XRF (Minalyzer) drill core scanning, comprehensive bulk geochemistry and multiple micro-analytical techniques. Bulk macro-scale methods allow identification of overall mineralisation and alteration signatures, as well as delineation of alteration zoning, while micro-analytical tools document micro-scale mineral phase relationships and textures. GSQ employed micro-XRF (M4 TORNADO PLUS) and TESCAN TIMA microanalysis, which successfully identified free gold and electrum within quartz – pyrite – galena ± chalcopyrite assemblages. High-resolution elemental mapping (≤60 μm) also highlighted distinct Au–Pb–Bi correlations. In addition, this study documented systematic alteration of dioritic host rocks to muscovite-chlorite assemblages, locally overprinted by biotite, reflecting evolving hydrothermal conditions. This integrated approach addresses the geochemical, mineralogical, and metallurgical complexities at Ravenswood — a common challenge in Queensland's IRGS — and offers a transferable framework for exploration targeting in analogous mineral systems across the region.

Mapping Alteration Footprints of IOCG Systems: A Case Study of the Ernest Henry Camp, Cloncurry District, Queensland

Alkis Kontonikas-Charos[1], Vladimir Lisitsin[1], Elena Belousova[1], Courteney Dhnaram[1], Xin Wang[2], Lisa Kearney[3], Rhiannon Jones[1], Al-Tamini Tapu[1]
[1]Geological Survey Of Queensland, Brisbane, Australia, [2]School of the Environment, The University of Queensland, Brisbane, Queensland 4072, Australia, [3]School of Earth and Atmospheric Sciences, Queensland University of Technology, Brisbane, Queensland 4000, Australia

Vectoring towards ore deposits using a combination of geochemical and mineralogical indicators has been successfully applied in exploration for diamonds and porphyry copper deposits. However, practical applications of such studies are often limited by availability of suitable samples and/or analytical techniques. Furthermore, alteration footprints of porphyry deposits are typically <2 km. Conversely, IOCG systems are typified by large alteration halos (>5 km), comprising regional and multiple local stages of alteration, including Na-Ca, Ca-Fe, K-Fe, and/or hydrolytic. The Cloncurry district, Queensland, contains numerous economically important IOCG deposits, of which the Ernest Henry deposit is the largest and most well-studied. Here we apply our analytical workflow to track changes in geochemical and mineralogical footprints of large IOCG systems, using the Ernest Henry camp (including the E1 deposit) as a case study.

The Ernest Henry and E1 IOCG deposits are hosted within Proterozoic metavolcanic and metasedimentary rocks of the eastern Mount Isa Province. A challenge of navigating alteration footprints in the Ernest Henry camp is the presence of multiple generations of alteration phases occurring in mineral assemblages formed at the regional metasomatic (Na-Ca, albite-actinolite-titanite-carbonate), pre-ore (K-feldspar-biotite-magnetite-titanite), syn- (magnetite-chalcopyrite) and post-ore (carbonate-fluorite-barite). We selected over 200 samples from drill cores within and spanning 3-4 km outside the Ernest Henry and E1 deposits to determine how geochemical and mineralogical footprints change with distance to ore. Using multiple analytical techniques (multielement geochemistry, HyLogger, µXRF, SEM-TIMA, EPMA and LA-ICP-MS) to characterise the samples and target suitable phases, we build on existing works by taking a holistic mineral chemistry approach by analysing multiple generations of actinolite, biotite, chlorite, apatite, titanite, feldspars and sulphides to map integrated mineral system footprints using trace element and stable isotope geochemistry (S, Cu, Fe, O) that could be used for vectoring towards ore in other camps of the Cloncurry district and in similar terranes.

Structural Architecture of the Lufilian Arc: Arc-Scale Structural Controls on Mineralization Based on a New GIS Compilation

Ian Basson[1,2], Michael-John McCall[1], Corne Koegelenberg[1,2], Louis van Eeden[1]
[1]Tect Geological Consulting, Somerset West, South Africa, [2]Department of Earth Sciences, University of Stellenbosch, Stellenbosch, South Africa

A GIS-based structural overview of the Lufilian Arc, based on 65 georeferenced maps from research, technical reports and geophysical surveys, allows for a review of its structural evolution and the development of Cu/Co-mineralized zones. Regional-scale "bedding boudinage" or "foliation boudinage", depending on the grade of metamorphism and position within the arc, concomitant with the arc's flexure, created dilational zones into which mineralized fluids were channelled. These zones acted as "sumps" for detritus (olistostromes) shed from and pushed ahead of advancing thrust units. The base of these were poorly supported probably saturated, Roan Breccia-filled dilational zones, resulting in local delamination of disharmonically-folded, highly-fractured, pervasively-veined and mineralized lower portions of the Mines Subgroup (écaille or fragments), which slumped into these zones at various attitudes and positions. The new model does not exclude the role of halite-rich fluids in transporting mineralization or in facilitating thrusting, but proposes deformation mechanisms that do not require physical, en masse fragment transportation by salt. The new model accounts for the absence, in the Lufilian Arc, of features typically associated with salt diapirism, such as diapiric uplift, radial fracturing and drainage patterns, salt extrusion, salt mines, salt playas, brine or sulphur springs, and changes in vegetation or carapaces.

Skarn-Porphyry Transition

Zhaoshan Chang[1], Shiqiang Huang[1], James Cannell[2]
[1]Colorado School Of Mines, Golden, United States, [2]Minera Las Bambas, Lima, Peru

Skarn and porphyry styles of mineralization may occur in the same deposits hosted in carbonates, in various proportions from almost pure skarn to almost pure porphyry. This study attempts to address the industry need of using skarns to find porphyry deposits, and the need of predicting the porphyry potential of an existing skarn. The skarn-porphyry transition at various skarn-porphyry Cu and/or Au deposits are examined in this study. It is found that the transition occurs inside fertile intrusions and starts from massive to semi-massive endoskarn at the intrusion margin. Such endoskarns grade to endoskarn veins or disseminated skarn minerals inward in the intrusions, followed by a zone with mixed endoskarn vein and porphyry-style veins with K-alteration halos, and eventually to only high temperature porphyry veins in the interior of the intrusions. The endoskarn features garnet with red to dark red color, which indicates high formation temperatures. Other endoskarn minerals include pyroxene, minor wollastonite, and rare scapolite, plus various amounts of the typical retrograde minerals such as the early retrograde assemblage of epidote-amphibole-vesuvianite-quartz-Fe oxides, and later retrograde assemblage of quartz-carbonate-chlorite-sulfides. This transition is proposed to be controlled by the supply of Ca, mainly from the dissolution of carbonate wall rocks by weakly acidic magmatic hydrothermal fluids, and high temperatures. The magmatic hydrothermal fluids are typically saturated with SiO2, and contain abundant Fe, evidenced by Fe-Cl daughter minerals in fluid inclusions, and up to ~22 wt% Fe measured in single fluid inclusions. With the supply of Ca at intrusion margins, skarn minerals have to form at high temperatures (e.g., >500 °C) and relatively low CO2 fugacity. When the Ca supply diminishes inward, quartz-dominant veins of typical porphyry style (e.g., EDM – early dark mica veins; quartz-biotite-magnetite veins) will form. The implications to exploration will be discussed based on the understanding.

Carbon and Oxygen Isotope Analyses with Paired Geochemical Data in Exploration for Cu-Zn Skarn Mineralization at Antamayo, Central Peru

Lucas Marshall[1], Jimy Diaz Villafuerte[2], Miguel Alfaro[1]
[1]Teck Resources Limited, Vancouver, Canada, [2]Teck Peru SA, Lima, Peru

Skarns can host significant endowment of base and precious metals, commonly at grades that can support economic open pit and/or underground mining, and as such can comprise attractive exploration targets. However, along with other carbonate-hosted ore deposits, skarns are often marked by narrow and irregular alteration haloes, presenting a clear challenge for mineral exploration, as well as an opportunity, given potential for undiscovered ore systems.

We present geochemical data and carbon and oxygen isotopic analyses from skarn-focused exploration programs at the Antamayo prospect in the Miocene magmatic belt of central Peru, including both selective samples targeting visible alteration and mineralization, and systematic grid based samples. Isotopic data is compiled with analyses from other occurrences in central Peru, providing valuable context for defining the main isotopic reservoirs in the area, including calcite- and dolomite-bearing carbonate host rocks, and hydrothermal carbonate in equilibrium with two separate fluids, inferred to reflect magmatic-hydrothermal (skarn, hornfels and vein samples) and meteoric fluids (vein samples). Most data can be explained by equilibration of pure and impure carbonate host-rocks, with a magmatic-hydrothermal fluid with up to ~10% CO2, with input of meteoric fluids only locally evident. Our findings suggest that in exploration settings, carbon and oxygen isotopes are best used in tandem with geochemical analyses and geologic observations, and collectively can increase the confidence in identified anomalies, such that wide-spaced (e.g. 500 x 500m) systematic sample grids can be a cost-effective approach for early exploration targeting. While isotopic results at Antamayo do not delineate a materially larger footprint that evident from geochemical data, isotopic data appears to be less prone to impacts of selective sampling. Results speak to the importance of systematic sampling in delineation of subtle alteration footprints for skarns and other carbonate-hosted ore deposits in mineral exploration.

The Timing and Geochemistry of Dolphin W Skarn, King Island, Tasmania: A Micro-Scale Study of Garnet and Scheelite

Angela Isaura Santos Costa[1], David R. Cooke[1], Lejun Zhang[1], Yamila Cajal[1], Owen Missen[1], Carlos Carrasco[1]
[1]CODES - University of Tasmania, Hobart, Australia

The development of critical metals, such as tungsten (W), is crucial for sustainable technologies. Australia hosts 12% of global W resources with ~26% in Tasmania, including the poorly understood Dolphin W skarn in King Island. This study integrates macro- and micro-scale approaches to enhance the understanding of the timing and chemical characteristics of Dolphin. We conducted mine-scale mapping, 3D modeling, HyLogger3 spectral analysis, and SEM-CL, LA-ICP-MS, and state-of-art synchrotron XFM analyses of scheelite and garnet.

Our data, integrated with previous studies, identify Dolphin as a calcic skarn, characterized by two mineralization lenses, formed by the replacement of carbonate-rich units of the Neoproterozoic Grassy Group by magmatic-hydrothermal fluids from post-Tabberabberan granodiorites. Mineralization is structurally controlled, concentrated near east–west faults striking sub-parallel to the Sandblow Granodiorite contact and dipping 30–60° toward the intrusion. HyLogger3 mapping reveals a wavelength shift from grossular (~11,500-12,000 nm) to andradite (~11,800-12,400 nm) near mineralized intervals.

We explored micro-scale analyses of garnet and scheelite to investigate the temporal evolution of the skarn system, from high-temperature metamorphic-magmatic to magmatic-hydrothermal conditions. Our results show multiple stages of garnet and scheelite precipitation, with complex trace element zoning. Prograde garnet features Ca-Fe rich cores with HREE+Y halos, while the rims are Fe-Mn rich with higher Nb. Early-stage scheelite precipitation occurs as disseminations within Mn-rich garnet rims, exhibiting CL oscillatory zoning, Mo contents up to 12%, and elevated As. Retrograde-stage scheelite precipitates interstitially to garnet with later quartz-calcite infill, displaying CL-homogeneous textures, lower Mo (<2%), and higher Nb, Sr, and REE. U-Pb dating of garnet provides Late Devonian-Early Carboniferous ages (363.6 ± 2.6 Ma to 355.9 ± 2.4 Ma), predating the emplacement of the Sandblow Granodiorite (350.8 ± 1.8 Ma). These findings provide new key geochronological and geochemical insights into the Dolphin skarn, highlighting its significance for critical metal exploration in Tasmania.

Subtle Potassic Alteration as a Vector for Mineralization in the Silvertip CRD, Northern British Columbia

Max Hohl[1], Shaun L.L. Barker[1], Cody N. Delgado[1]
[1]Mineral Deposit Research Unit, Vancouver, Canada

Carbonate replacement deposits (CRDs) are significant sources for Pb, Zn, Ag, and potentially In. CRDs represent the distal expression of magmatic-hydrothermal alteration systems where the causative igneous intrusions remain often undiscovered. Exploration efforts are often hampered due to a lack of visible alteration phases in the carbonates hosting the ore lodes and rely primarily on systematic mapping of variations in δ13C and δ18O isotopic compositions. Isotopic variations can extend well beyond calc-silicate skarn alteration due to the relatively fast propagation velocity of isotopes compared to silicate- or sulfide-forming reactions. In many CRDs, siliciclastic shales form the hanging wall, representing an impermeability layer to the mineralizing fluid, promoting lateral fluid migration and formation of subhorizontal, tabular ore bodies (mantos). Mantos often occur along the contact between carbonates and siliciclastics. Locally significant mineralization can occur in the hanging wall, usually as high-angle veins.

The Ag-Pb-Zn±Cu Silvertip deposit in northern British Columbia, Canada, shows near-vertical, fault-related (chimneys), and manto-style mineralization hosted by the Middle Devonian McDame limestone. The Late Devonian Earn Group shales at Silvertip overlie the McDame Group. We show that subtle potassic alteration can be tracked in the Earn Group when carefully accounting for base metal enrichment due to redox processes in black shales. The potassic alteration correlates spatially with δ34S values in CRD-related sulfides centering around 0 ‰ in the Earn group, δ18O values from carbonate veinlets approximating 11 ‰, and mineralization containing increasing amounts of chalcopyrite and bursaite ($Pb_5Bi_4S_{11}$(Ag)), the latter exsolving native bismuth.

These results indicate a progressively dominant magmatic fluid and an increasing water/rock ratio, likely linked to closer proximity to the fluid-releasing intrusion. In the absence of volcanic or intrusive rocks, subtle chemical variations detectable with a p-XRF handgun can help identify potassic alteration and define fluid-flow paths toward a magmatic-hydrothermal fluid source.

Unlocking the Timing of Au-Sb-Hg Mineralization in the Kyrgyz South Tien Shan: Constraints from In-Situ U–Pb Dating of Hydrothermal Calcite

Aizat Zhaanbaeva[1], Ruizhong Hu[1], Douglas Kirwin[2], Rosalia Djenchurarva[3], Wei Gao[1]
[1]Institute Of Geochemistry, Chinese Academy Of Sciences, Guiyang, China, [2]Independent geologist, Bangkok, Thailand, [3]Institute of Geology, National Academy of Sciences of Kyrgyzstan, Bishkek, Kyrgyzstan

The Turkestan-Alay segment of the Kyrgyz South Tien Shan is known for hosting many sedimentary rock-hosted low-temperature Au-Sb-Hg ore deposits, e.g., the Khaidarkan and Chauvai. Recent studies have highlighted that these deposits exhibit similar characteristics to Carlin-type gold deposits (CTGDs) in Nevada, USA. Resolving mineralization age and its relationship to geodynamics is essential for understanding the ore genesis. However, this constraint is particularly difficult to achieve for CTGDs, due to the lack of suitable minerals for radiometric dating. Calcite is the most important hydrothermal mineral in CTGDs. Its crystallographic structure can accommodate U and Th in μg/g concentrations. With the advancements of laser ablation inductively coupled plasma mass spectrometry (LA-ICP-MS), the feasibility of in situ U-Pb dating on calcite has been successfully demonstrated for some hydrothermal gold deposits. Here, three calcite generations, Cal-1, Cal-2, and Cal-3, have been identified based on precipitation mode, mineral associations, and cathodoluminescence (CL) features from the Chauvai and Khaidarkan deposits. Among these, the Cal-2 is characterized by intimate intergrowth with cinnabar containing higher concentrations of Carlin-pathfinder elements (e.g., Au, As, Sb, Tl, and Te), distinct CL zoning, and MREE-enriched composition patterns, indicating that it is clearly associated and coeval with gold mineralization. The LA-ICP-MS dating of Cal-2 yielded U-Pb ages of 231.1 ± 6.2 Ma for Chauvai and 227.5 ± 5.3 Ma for Khaidarkan, representing the timing of Au-Sb-Hg mineralization in the Turkestan-Alay belt. Our new ages suggest that the Au-Sb-Hg deposits in the Turkestan-Alay segment formed in a post-collisional taphrogenic relaxation followed by an intraplate setting. Our case story suggests that hydrothermal calcite could be a robust chronometer for age dating of sedimentary rock-hosted low-temperature gold deposits worldwide.

Government Geoscience Accelerating Resource Discovery and Development

Marina Costelloe[1]
[1]Geoscience Australia, Canberra, Australia

Collaborative government geoscience programs deliver data and information on Australia's natural resources. Precompetitive geoscience is publicly available data and information that is shared freely by government for the benefit of all stakeholders. For the resources industry, precompetitive geoscience helps reduce investment risk and increases the odds of making a successful discovery during exploration. It comes before private sector exploration and investment, not instead of it.

Governments deliver precompetitive geoscience through integrated and complementary regional- to national-scale data acquisition, analysis, and interpretation studies of mineral and energy resource potential. Geoscience Australia works to support many stakeholders from the Australian Government, State and Territory Geological Surveys, industry, academia and communities. This ensures that reliable information is available to make informed decisions, and to establish and maintain a pipeline of exploration and development projects that can provide the raw materials needed for the production of renewable energy technologies (such as solar panels, wind turbines), electric vehicles, advanced manufacturing and defence technologies and capabilities.

Geoscience Australia's current program has several initiatives specifically supporting the minerals sector. Resourcing Australia's Prosperity aims to comprehensively map Australia's natural resources and accelerate the discovery of critical minerals and is a major federal government initiative. The Critical Minerals Research and Development Hub, which concludes in June 2026, is focused on industry prioritised research in collaboration with CSIRO and ANSTO. The Critical Minerals Mapping Initiative leverages international expertise to improve mineral prospectivity techniques and will feed into the Resourcing Australia's Prosperity initiative.

The outcome of this work provides trusted scientific evidence and advice that directly supports the Australian Government's policies and strategic priorities as well as mapping the nations mineral potential and other resources needed to support the energy transition. A thriving and sustainable resources exploration industry accelerates investment, resource development and a stronger economy.

Academia-Industry-Government Collaborative Projects to Explore the Proterozoic Basins in Australia and their Sediment-Hosted Resources

Juraj Farkas[1], Alan Collins[1], Morgan Blades[1], Stefan Loehr[1], Darwinaji Subarkah[1], Carl Spandler[1], Rosalind King[1], Sarah Gilbert[1], Simon Holford[1]
[1]University of Adelaide, Australia

The Proterozoic depositional systems in Australia not only provide unique sedimentary archives of the Earth system and environmental/biological evolution on our planet in deep time, but these systems or basins also host vast and still relatively underexplored sedimentary-hosed mineral and energy resources. This contribution will present key outcomes and research highlights from the recently completed and/or ongoing collaborative ARC Linkage projects between academia, industry and government, focused on novel analytical techniques (e.g. in-situ geochronology, metal-isotope chemostratigraphy, etc) applied to problems relevant to improved basin exploration, including sediment-hosted resources in the greater McArthur Basin (NT) and the Centralian Superbasin (NT, WA and QLD).

Specifically, we will introduce and discuss the recently developed and relatively rapid and high-throughput laser-based geochronology techniques, which were applied to the above basins for direct dating of the Precambrian sedimentary rocks and their depositional and diagenetic histories (via in-situ Rb-Sr dating of shales/illite, glauconite, detrital feldspar, etc). These novel dating techniques are thus highly complementary to more established and popular geochronology tools, such as U-Pb zircon and carbonate dating, used for basin exploration. In addition, we will also illustrate the application of novel metal-isotope chemostratigraphy (e.g. Ca, Cr, Cd isotopes) in marine carbonates and organic-rich shales, along with more routine chemostratigraphy tools (e.g. C and Sr isotopes, elemental ratios), and how these various tracers and isotope proxies might get reset via fluid-rock interactions related to the formation of sediment-hosted ore deposits in the studied basins. Overall, this contribution will provide latest updates on the application of novel geochemical/isotope and geochronology tools to de-risk resource exploration in the Proterozoic Basins in Australia, while illustrating the benefits of highly collaborative research developed via ARC Linkage projects between the Universities, Geological Surveys, CSIRO, resource exploration companies, and manufacturers of analytical instruments.

Assessing Proximity to Cu and Au Deposits Using Resistate Mineral Chemistry

Caroline Tiddy[1], Adrienne Brotodewo[1], Dave Giles[1], Jennifer Porter[1], Travis Batch[1], Luke Tylkowski[1], Ernest Opoku[1], Anthony Budd[2], Ross Cayley[3], Rob Thorne[4], Vaclav Metelka[5], Mike Taylor[5]
[1]Future Industries Institute, University Of South Australia; Mineral Exploration Cooperative Research Centre (MinEx CRC), Mawson Lakes, Australia, [2]Minerals, Energy and Groundwater Division, Geoscience Australia; Mineral Exploration Cooperative Research Centre (MinEx CRC), Canberra, Australia, [3]Geological Survey of Victoria, Melbourne, Australia, [4]CSIRO Mineral Resources; Mineral Exploration Cooperative Research Centre (MinEx CRC), Bentley, 6102, [5]AIC Mines, Subiaco, 6008

This presentation shows outcomes from a coordinated research effort within MinEx CRC investigating resistate mineral chemistry for assessing proximity to Cu and Au deposits. Resistate minerals are capable of withstanding weathering, erosion, and transport processes, and often survive in young, barren cover sediments that obscure older, prospective basement terranes. This is particularly relevant in Australia, where ~80% of basement rocks lie beneath younger cover, thereby obscuring economically significant mineral systems.

Our research has used iron oxide-copper-gold (IOCG) and iron sulfide-copper-gold (ISCG) systems in the Gawler Craton (South Australia) and Cloncurry District (Queensland) as well as orogenic Au systems in the central Victorian goldfields, southeastern Australia. We demonstrate that LREE, Y, and Th concentrations in monazite can indicate proximity to LREE-rich IOCG deposits, while Ca, S, and Th content are more relevant for ISCG systems (Tiddy et al., 2021: Minerals; Batch et al., 2024: J. Geochem. Expl.). Zircon from IOCG settings shows distinct heavy REE (HREE) fractionation and europium (Eu) anomalies (Brotodewo et al., 2021: Minerals). In orogenic Au systems, apatite displays distinct concave REE + Y profiles, along with characteristic Mn, Fe, Sr contents, Y/Ho ratios, and Eu anomalies (e.g., Tylkowski et al., in review: Gond. Res.).

These geochemical insights provide a basis for targeted mineral exploration sampling in covered terrains. Monazite and zircon retain their indicative chemistry into transported sediments, while apatite, although susceptible to acid weathering, preserves its original chemistry when intact. Sampling strategies that incorporate paleolandscape and paleoenvironment understanding can effectively use these resistate minerals to assess proximity to buried Cu-Au mineralisation. Future research includes using apatite for IOCG systems (for example) to add to the toolkit of mineral chemistry across multiple mineralising systems, effectively enhancing the ability to detect buried ore systems by expanding their geochemical footprint.

A Continental-Scale Heavy Mineral Reference Framework for Mineral System Detection

Brent McInnes[1], Patrice de Caritat[1,2], Alex Walker[1], Evgeniy Bastrakov[2]
[1]Curtin University, Perth, Australia, [2]Geoscience Australia, Canberra, Australia

The Heavy Mineral Map of Australia (HMMA) project provides the first continental-scale, internally consistent reference dataset for crustal heavy mineral assemblages across Australia (Walker et al., 2024; de Caritat et al., 2023). Conducted as a collaboration between Geoscience Australia and Curtin University under the Exploring for the Future (EFTF) program, this study applied automated mineralogical analysis to over 1,300 floodplain sediment samples collected during the National Geochemical Survey of Australia (de Caritat 2022).

HMMA samples were sieved to a 75–425 μm grain size and subjected to dense media separation. Heavy minerals (specific gravity >2.9 g/cm^3) were mounted in epoxy and analysed using a TESCAN TIMA system. The resulting dataset comprises over 145 million individual mineral grain observations and includes more than 160 mineral species. Because the samples represent basement-derived sediments collected at catchment outlets, the resulting mineral assemblages are interpreted to reflect the integrated makeup of high-density minerals in the Earth's upper crust.

We use a weighted grain count approach to quantify the relative abundance of heavy minerals in Australian catchment sediments and propose this as a continental-scale reference framework for mineral system detection.

Through case studies using heavy mineral data collected from greenfield and brownfield terrains, we demonstrate how this mineralogical baseline can assist in identifying anomalous heavy mineral assemblages in exploration samples - particularly in weathered environments where geochemical signals are often attenuated.

This method holds promise for the rapid, cost-effective detection of mineral system signatures by integrating routine stream sediment sampling and automated mineralogy techniques. The reference framework is transparent, scalable, and compatible with other mineralogical and geochemical datasets, supporting a wide range of applications in sediment provenance studies, mineral system fingerprinting and exploration targeting.

MinEx CRC's Model for Collaboration in Exploration Geoscience

David Giles[1,2], Andrew Bailey[1]
[1]Minex CRC, Perth, Australia, [2]UniSA, Adelaide, Australia

We established MinEx in 2018 just one year after publication of the Uncover Roadmap with its call for a ~15 year, > billion-dollar investment in a continent-wide collaboration in exploration geoscience. Uncover provided important context for MinEx in our start-up phase. You can see the shadow of Uncover in our mission statement: "To deliver market-ready technology, work practices and knowledge designed to increase the likelihood of mineral discovery [in frontier provinces] and facilitate the transition from discovery to mineral production." "Uncover Lite" is a description of our collaboration that we have heard on more than one occasion. In truth we don't mind the association, but it is not entirely fair to either MinEx or Uncover. We'd be more comfortable with "Uncover Previewed" (because MinEx has provided a template for the type of collaborations that could work on a larger scale and over a longer time) or "Uncover Under Cover" (because MinEx and our collaborators have quietly played our part in promoting many of the goals of Uncover without the formal recognition or funding of Uncover as its own entity). In this talk we will outline the achievements of MinEx through a selection of case studies from our research portfolio. We think these achievements speak for themselves. But there is a deeper story behind them, which is as much to do with the logistics, practicalities, people and relationships as it is to do with the science and engineering. Collaboration at scale can be complex, expensive, energy-consuming and at times frustrating. We won't shy away from presenting those challenges here. But collaboration can also be (and has been!) hugely rewarding. We hope that some of our learnings from MinEx CRC will help inform and promote the continued national (and international) collaborations that we will need to meet future mineral resource requirements.

Mineral Systems Modeling: New Technologies for Exploration Vectoring

Daniel Palmowski[1], Thomas Hantschel[1], Adrian Kleine[1]
[1]Terranta GmbH, Aachen, Germany

The rapid and large-scale increase in critical metal production required for the global energy transition demands innovative exploration technologies and predictive models to support informed investment decisions and reduce geological risk. Geological process modeling provides a framework for simulating the dynamic evolution of mineral systems, enabling a better understanding of ore-forming processes and guiding the definition of exploration vectors. Unlike petroleum systems, mineral systems are characterized by greater structural, hydrological, and geochemical complexity, necessitating more sophisticated and versatile simulation approaches. This study presents a generic geological process modeling workflow, demonstrated with a detailed case study of the Kupferschiefer sedimentary copper system.

We extended the classical Darcy flow formulation with advective and diffusive transport to enable accurate modeling of convection cells in complex and heterogeneous flow regimes. High-dimensional finite elements, unstructured grids, and adaptive local grid refinement techniques are used to resolve fluid transport at multiple scales, including advective, diffusive, and reactive transport. Chemical processes involving mineral, aqueous, and gas phases are modeled using Gibbs energy minimization solvers. To maintain computational efficiency, macro-models reduce solver runtimes from milliseconds to microseconds per call, making full-basin, multi-million-cell 3D simulations feasible.

The Kupferschiefer system, a world-class stratiform copper deposit in Central Europe, is modeled to understand basin-scale fluid-rock interaction, metal transport, and precipitation. Results highlight two critical ore-forming mechanisms: (1) focused fluid flow that channels oxidized, metal-bearing fluids into structurally and chemically favorable horizons, and (2) an advancing redox front that triggers copper precipitation, producing characteristic mineral zonation. The model reproduces observed vertical and lateral zoning patterns, including enrichment in chalcocite and bornite in relation to the advancing reduction front. These patterns serve as calibration targets and provide predictive criteria for exploration. The case study demonstrates how 3D geological process modeling enhances understanding of complex mineral systems and supports strategic exploration in sedimentary basins.

The Architecture and Dynamics of Coupled Failure and Fluid Flow in Magmatic-Hydrothermal Systems: Insights from Swarm Seismicity, Hakone Caldera, Japan

Stephen Cox[1], Yohei Yukutake[2]
[1]Research School of Earth Sciences, Australian National University, Canberra, Australia,
[2]Earthquake Research Institute, The University of Tokyo, 1-1-1 Yahoi Bunkyo-ku, Tokyo, Japan

A seismically very active hydrothermal system at depth beneath the Hakone caldera (Japan) provides new insights about the controls on permeability enhancement, fluid migration pathways and flow rates in fracture-controlled, magmatic-hydrothermal systems.

Seismicity at Hakone is dominated by migratory swarm activity that occurs when failure, permeability generation and fluid migration are driven by fluid pressurization within an overpressured, high fluid flux regime. From 1995 to 2019 eight major injection-driven earthquake swarms occurred within an area of 70 km^2 that extends to depths up to 7 km. Seismicity in this period involved 29,000 events in the magnitude range $-0.6 < Mw < 3.7$, driven by injection of approximately 500,000 m^3 of fluid. Recorded ruptures have diameters ranging from < 10 m to 500 m. Swarms typically last between several days and 250 days and involve cascades of many 1000s of ruptures that illuminate well-defined faults. Individual swarms exhibit diffusion-like migration of seismicity fronts and provide insights about fluid pathways, migration rates and flow directions. The seismically quiescent intervals between swarms last up to several years.

Seismicity occurs within a fault complex comprised mostly of small, conjugate, strike-slip faults that are nearly optimally-oriented relative to the NW-SE trending maximum principal stress. Many swarms activate just a few small fault segments, each having areas up to about 1 km^2. However, the largest swarms activate many hundreds of small faults that are distributed across the entire Hakone system. These are associated with injection of up to 100,000 m^3 of fluid at injection rates transiently as high as several m^3/s.

Injection-driven swarms cause episodic, large fluctuations in fluid pressures, fluid pressure gradients and flow rates during rapidly recurring slip events. This dynamic fluid regime has implications for processes and rates of ore formation that involve thousands of swarms over timescales of $10^4 - 10^5$ years.

Identifying Mineralisation Within the Sunrise Dam Gold Mine Using Geomechanical Targeting Technology

John Mclellan[1,2], Christine Shore[3], Perkins David[3]
[1]Director, GMEX (Geological Modelling for Exploration), Townsville, Australia, [2]Senior Research Fellow, Adjunct, James Cook University, Townsville, Australia, [3]AngloGold Ashanti Australia, Perth, Australia

Drill targeting within mine sequences can be difficult, particularly when mineralisation is located in narrow structural corridors, shear-zones, fracture systems and/or veins. Structural analysis and interpretations can be very useful in aiding the understanding of the geology, building a three-dimensional model and its deformational history. However, even with this knowledge targeting is always difficult. Understanding how deformation events have affected the rock units and consequent structure development, is key to highlighting structural targets for investigation.

Here, we use the currently accepted geological model for the Sunrise Dam Gold Mine (SDGM) in Western Australia, to develop a three-dimensional mesh representing the main architecture of the deposit. A 3D hexahedral model was developed and subjected to boundary conditions to simulate the main D4 deformation event (NE compression), most likely to have been responsible for the high grade steeply orientated mineralisation domains. Initial models were run for sensitivity analysis, and the known positions of ore lodes were used as validation for several input parameters. The parameter validation strongly suggested that mineralisation was a) closely related to areas of increased shear strain in the adjacent shear zones, and b) failure state (both shear and tensile) within the volcaniclastic packages and more competent intrusives, which often displayed volumes of increased pore fluid pressure.

The geomechanical modelling of the system has highlighted several key target areas that localise shear strain, increased pore fluid pressures and tensile failure. Shear strain localisation in and around the more competent intrusive rocks appear to be important, even from early stages of the modelling. The structural architecture accommodates most of the movement within the system, however, this has accommodated partitioning of strain into lithological units such as the volcaniclastic units and intrusive bodies, where there is a good correlation with areas of tensile failure, high pore pressures and known ore bodies.

Suspension-Driven Transport of Noble Metals in Hydrothermal Fluids

Nestor Alfredo Cano Hernandez[1], José María González Jiménez[2], Antoni Camprubí[1], Eric Morales-Casique[1], Eduardo González-Partida[3]
[1]Instituto de Geología (UNAM), Mexico City, Mexico, [2]Instituto Andaluz de Ciencias de la Tierra (IACT-CSIC), Granada, Spain, [3]Instituto de Geociencias, Juriquilla, Mexico

The transport of noble metals (Au, Ag) by metal-rich melts in hydrothermal ore systems has been increasingly documented over the past decade. This challenges the traditional view that ligands (e.g., chloride- and bisulfide-bearing) are the only effective metal carriers in hydrothermal solutions. Nonetheless, the conditions under which soluble and non-soluble metal transport coexist—and whether metal-rich melts can be physically mobilized by hydrothermal fluids—remain uncertain. Here we present evidence for the suspension-like transport of nano- to micron-sized sulfide-sulfosalt melts within hydrothermal fluids at temperatures below 400 °C. These melts crystallize into irregular, bleb-shaped polymineral inclusions composed of Ag sulfosalts, electrum, and base metal sulfides. Polymineral inclusions range from 5 nm to 40 µm, and are cogenetic with liquid + vapor fluid inclusions hosted in quartz. Numerical modeling using fluidization and settling dynamics demonstrates that hydrothermal fluids can mechanically carry these metal-rich nano- to micromelts at flow rates below 0.1 m/s. The chemical consistency between the nano- and micron-scale inclusions supports a model in which nanomelt precursors coalesce during episodic transport (or upon precipitation), contributing significantly to the formation of high-grade noble metal deposits.

Electrical Processes in Charge of Ore Formation

Chris Voisey[1], Andrew Tomkins[1], Joël Brugger[1]
[1]Monash University, Melbourne, Australia

Electrical phenomena may play a more fundamental role in geological processes than previously recognised. Previous work has revealed that gold can precipitate within quartz veins during seismic events via quartz's piezoelectric properties—where mechanical stress generates localized electric fields that facilitate ore-forming reactions, a process known as piezocatalysis. This mechanism introduces a new perspective on metal transport and concentration, with implications extending beyond gold mineralisation.

This presentation explores the intersection of mineral electrical properties, electrochemical reaction pathways, and piezoelectricity in ore-forming environments. We examine how mineralisation processes such as galvanic redox reactions at conductive mineral junctions, and factors governing metal nanoparticle stability (i.e. zeta potential, electrostatic attraction, and surface potential) are affected by the presence of piezoelectric potentials.

Experimental results demonstrate that electric fields generated by stressed quartz can significantly enhance metal deposition onto conductive minerals, particularly sulphides. We propose that quartz piezoelectricity promotes electrochemical deposition (i.e., 'electroplating') over spontaneous galvanic reactions and drives nanoparticle movement via electrophoresis rather than electrostatic attraction. We hypothesize that the directionality of applied voltages may govern the galvanic corrosion or protection of sulphides, influencing ore preservation or remobilisation during deposit formation.

Colloidal Flocculation and the Mechanical Transport of Gold: Causes of the 'Nugget Effect' in Hydrothermal Ore Deposits

Duncan McLeish[1,2], Anthony Williams-Jones[1], James Clark[1], Richard Stern[3]
[1]McGill University, Montreal, Canada, [2]Newmont Corporation, Vancouver, Canada, [3]University of Alberta, Edmonton, Canada

The nugget effect, a phenomenon that results in a strong skewing of the precious metal distribution in vein-hosted hydrothermal systems towards spatially discrete occurrences, or nuggets, poses a major challenge for exploring, modelling, and mining many types of ores. Although researchers have long investigated ways to mitigate the products of the nugget effect, namely the extraordinary grade variability seen in many bonanza-type deposits, relatively little progress has been made in explaining its fundamental causes. Recent recognition of the role that the mechanical transport and flocculation of metallic nanoparticles plays in the genesis of high-grade hydrothermal gold deposits, however, offers new insight into the formation of very coarse grained or nuggety gold. Here, we present a synthesis of nano- to micro-scale textural and chemical TEM observations from the Brucejack bonanza-type epithermal Au-Ag deposit in northwestern British Columbia, Canada. These observations demonstrate that flocculation (i.e., aggregation) of gold nanoparticles suspended in the ore fluid formed spectacularly high-grade (e.g., up to 42,100 g/t Au over 0.5 m core intervals) yet highly discontinuous, nuggety mineralization. We go on to show, using data from EMPA, high-resolution in situ SIMS, and LA-ICP-MS analyses of pre- to syn-mineralisation pyrite, that three distinct processes operated on gold transported as a colloid (i.e., a suspension of negatively-charged nanoparticles in an electrolyte solution). These processes drove flocculation through: (1) the electrochemical destabilisation of colloids by interaction with semiconductive p-type arsenian pyrite; (2) surface-charge-bridging of colloids by cationic flocculants added to the ore fluid through mixing with seawater; and (3) cooling of the ore fluid by boiling and/or seawater mixing. We conclude that these processes, which are common in many near-surface hydrothermal systems, can account for the genesis of nuggety gold in many hydrothermal deposit types and, by extension, help predict their occurrence at the deposit scale.

What is the Value of Ore Body Knowledge for the Mining Industry?

Shaun Barker[1], Cassady Harraden[1], Carolina Marin Saurez[1], Andrea Rutley[1], Warren Wegener[1], John Steen[2], Dave Lawie[3]
[1]Mineral Deposit Research Unit, University Of British Columbia, Vancouver, Canada, [2]Bradshaw Research Institute for Minerals and Mining, University Of British Columbia, Vancouver, Canada, [3]Imdex Limited, Perth, Australia

Despite decades of technological progress and growing access to high-resolution geoscientific data, many mining operations continue to fall short of their forecasted production targets and fail to deliver on projected financial returns. A key contributor to this systemic underperformance is insufficient or poorly integrated ore body knowledge (OBK), rooted in technical shortcomings during feasibility studies (e.g., McCarthy, 2003). Recent analysis by Gillis et al. (2024) reinforces this concern, identifying inadequate OBK as a leading cause of financial impairment in the Canadian mining sector, with technical errors contributing to over USD $1 billion in global annual losses. These issues also carry significant environmental consequences through inefficient resource extraction and increased waste.

Paradoxically, mining companies now have access to more subsurface data than ever before—thanks to advances in sensors, remote sensing, and AI-driven analytics—yet the conversion of this data into actionable, economically relevant knowledge remains a major bottleneck. We argue that the central challenge is no longer data scarcity, but rather the integration and contextualization of data into meaningful, decision-ready insights that can support technical, engineering, and financial assessments.

Addressing this challenge demands a fundamental shift in how ore body knowledge is generated, shared, and utilized across the mining value chain—from exploration through to mine planning, production, and processing. We propose an "exploration-to-mill" framework that promotes cross-disciplinary collaboration, enhanced data integration and visualization platforms, and the adoption of robust, transparent value-of-information (VOI) methodologies. By explicitly quantifying the financial impact of improved OBK, this approach offers a clear pathway to reduce risk, enhance predictability, and drive more sustainable and profitable mining outcomes.

Teaching AI the Language of Rocks: Transforming Orebody Knowledge

Dave Lawie

Orebody Knowledge (OBK) is increasingly recognised as the critical foundation for value creation across the mining value chain. Yet our most fundamental tool for building OBK — the rock microscope — has changed little since the 1860s. Even with digital imagery, the work has remained manual, siloed and dependent on skills that are literally retiring out of the industry.

RockStein is changing this. For the first time, automated multi-modal capture of thin and polished sections is being combined with petabyte-scale cloud storage and physics-informed artificial intelligence. Over the next three years, more than one million rock sections will be digitised into a unified, queryable, cloud-native library. Critically, this is not "AI for AI's sake": RockStein embeds the physics of mineral equilibria, disequilibria, pressure–temperature paths, prograde and retrograde metamorphism — allowing the system to reason about Earth processes, not just recognise pixels.

The result is a platform that geoscientists can use without friction: the same workflows, the same samples, but with orders-of-magnitude more insight, speed and reproducibility. Old datasets gain new life, new datasets scale globally, and together they create an enduring resource for exploration, mining, and academic research.

This is not AI being imposed on us by outsiders who don't understand our industry. This is geoscience bringing AI to itself — by geologists, for geologists, grounded in our science and our workflows. RockStein is designed to be transformative: turning the sunk cost of generations of microscope work into the foundation of a new era of OBK.

The Flow of Geological Information in the Mining Value Chain and Its Use for Operational Optimization

Carolina Marin Suarez[1], Shaun Barker[1], Cassady Harraden[1], Lindsey Heagy[1], Andrea Rutley[1], Dave Lawie[2]
[1]The University of British Columbia, Vancouver, Canada, [2]IMDEX, Perth, Australia

A comprehensive understanding of the orebody and uncertainties associated with geology is fundamental to the success of mining operations. Orebody knowledge (OBK), which includes insights into grade, mineralogy, and alteration, supports critical decision-making across the entire Mining Value Chain (MVC), from exploration to marketing and sales. Despite its intrinsic value, geological information is often overlooked or disconnected from downstream processes. Identifying and addressing this disconnect is essential for improving decision-making and optimizing operations. This research integrates insights from literature, professional experience, and expert input from across the mining value chain, from exploration to closure, to construct a flow diagram that traces the path of geological information throughout an open-pit porphyry copper operation. The research emphasizes the interconnectedness between geological data, and it's use in direct and indirect forms, to enable decision making across the MVC including geotechnical, mine planning, and metallurgy others.

The number of connections between datasets and the use of empirical and absolute data in resulting models have been quantified, providing insights into which models are more robust, and which rely more heavily on empirical information. As example of the utility of different types of information collected during exploration we highlight VNIR-SWIR (Visible to Near and Shortwave Infrared) hyperspectral data. This data enables the identification of clay minerals, informing vectors to mineralization, geotechnical and mineral processing. and indicators of environmental concerns such as acid mine drainage. These insights are essential for exploration, extraction, re-mining, waste management, and rehabilitation. Finally, to illustrate the integration of geological data in decision-making, a preliminary case study is presented. It demonstrates the use of a geological proxy related to feldspar destruction, CCNK (calcium, carbon, sodium, and potassium), as a predictor of throughput. The study explores how machine learning models, trained on hyperspectral data (SWIR), can predict this proxy and support operational decisions.

Magnetite Ores and the Green Steel Revolution: A Global Outlook on Resources and Future Prospects

Yoram Teitler[1], Erick Ramanaidou[1], Hayley McGillivray[2], Mitchell Bockmann[3], Carmen Krapf[3], Adrian Fabris[3], Anil Subramanya[2]
[1]CSIRO Mineral Resources, Kensington 6151, Australia, [2]Amira Global, Perth 6000, Australia, [3]Geological Survey SA | DEM - SPD – GSSA, Adelaide 5001, Australia

The drive to decarbonise the steel sector has led to renewed focus on high-grade magnetite ores (≥68% Fe) as a suitable feedstock for Direct Reduced Iron (DRI) production. These ores present a viable pathway to lower-emission steelmaking. However, their economic competitiveness hinges on several key factors, including deposit tonnage, head grade, recovery potential, mineralogical properties, comminution performance, and beneficiation requirements.

This study compiles and analyses global data on magnetite ore deposits to evaluate their suitability and constraints in the context of low-carbon steel production. Banded Iron Formation (BIF)-hosted magnetite systems represent the principal global source, with major operations across Canada, Western Australia, and South Australia. These deposits are typically fine grained and require substantial grinding to achieve mineral liberation—an energy-intensive step that elevates processing costs. However, certain BIFs that underwent high-grade metamorphism, such as Scully (Canada) and Warramboo (South Australia), contain coarser-grained magnetite, improving comminution efficiency and reducing energy consumption.

Other important sources include Iron Oxide-Apatite (IOA) deposits (e.g., Kiruna, Sweden; El Laco, Chile), with high Fe grades but requiring phosphorus removal, and Iron Oxide-Copper-Gold (IOCG) deposits (e.g., Marcona, Peru) associated with sulphur-rich gangue. Skarn-hosted deposits, such as Savage River (Tasmania), produce high-grade magnetite concentrates despite smaller scale. Layered mafic-ultramafic intrusions (e.g., Balla Balla, Western Australia) provide coarse-grained magnetite with potential for vanadium and titanium by-product recovery.

Processing magnetite ores involves energy-intensive crushing, grinding, and magnetic separation, with grain size critical to beneficiation efficiency. High-grade metamorphosed BIFs hold a significant advantage due to coarser grains, larger resource volumes, and lower deleterious element contents. While magnetic separation remains the main beneficiation method, selective flotation is sometimes necessary to meet concentrate quality targets. Optimising geological understanding and processing strategies is crucial to ensure the economic and environmental sustainability of magnetite ores in supporting the transition to green steel.

The Application of Data Fusion Algorithms and Machine Learning to Increase Orebody Knowledge

Penny Stewart[1]
[1]Petra Data Science, Brisbane, Australia

Data fusion ore tracking algorithms fuse geospatial geological data with historical downstream production data to develop site-specific geometallurgy, product quality and material handing models using supervised machine learning. The term "orebody learning" has been defined to describe the process of developing site-specific machine learning models from production data from across the mine value chain. The paper will demonstrate the application of orebody learning through a series of six case studies from a wide range of deposits from around the world. Over the past five years mining operations have extended the application of orebody learning to include optimisation of blasting designs and set points for digging, crushing, grinding and flotation recovery for a range of orebodies including; haematite, magnetite, epithermal gold, and coal overburden removal. The paper will also explore the potential for incorporating optimised designs and set points into the geology model to further expand orebody knowledge and capitalise on geological data investments for geology driven mine value chain optimisation.

The Critical Role of Ore Body Knowledge in Mining Success: A Systems Leadership Perspective

Andrea Rutley[1], Shaun Barker[1], Cassady Harraden[1]
[1]University of British Columbia - MDRU, Vancouver, Canada

Incomplete, inadequate, or poorly communicated ore body knowledge has led to catastrophic safety incidents and significant financial losses across the mining sector—consequences that transcend company size or operational purpose. A study of the Canadian mining industry estimates the global financial impact of these failures exceeds USD $1 billion, largely due to ineffective communication of technical risks.

Mining risks manifest as either internal (stemming from operational challenges such as geotechnical failures) or external (driven by factors such as market fluctuations and regulatory changes). Successfully navigating these risks requires an integrated approach that balances technical expertise, commercial strategy, and social responsibility. At the core of this approach is the concept of systems leadership, which offers a framework for fostering consistent, organization-wide alignment. This alignment enhances operational resilience by promoting an integrated understanding of the complexities inherent in the various facets of the minerals industry.

Effective systems leadership establishes coherent processes that account for the interplay among various operating units, promoting repeatable success and embedding robust communication pathways throughout the organization. A company's purpose statement should not be aspirational rhetoric—it must actively shape organizational behavior and decision-making. Yet, a review of the core values of the top ten global mining firms reveals that only two purpose statements demonstrate alignment across technical, commercial, and social dimensions.

There are a number of real-world operations failures (such as the geotechnical incidents at Quellaveco, Highland Valley Copper, and Bingham Canyon, or the geometallurgical challenges at Lihir Island Gold Mine and Copper Mountain Mine) which could have been avoided through systems leadership by fostering stronger integration between technical teams, commercial partners, and key decision-makers. This study underscores the necessity of consistent, structured risk communication and advocates systems leadership as a critical enabler in building resilient, high-performing, optimized mining organizations.

Modelling the Future of Copper Production in Queensland's NW Minerals Province: A New Foundation for Policy Development and Infrastructure Planning

Tom Evans[1], Juan Soto-Diaz[1,2], Pascal Bolz[1], Liza Forbes[1], Mark Noppé[1,3], Steven Micklethwaite[1], Louisa Rochford[1]
[1]Sustainable Minerals Institute, University of Queensland, Indooroopilly, Australia, [2]School of Economics, University of Queensland, Saint Lucia, Australia, [3]JKTech, Indooroopilly, Australia

The closure of the Mount Isa Copper Mine in 2025 highlights the importance of future copper production in the region. We have developed an economic model of Queensland's NW Minerals Province copper supply chain that forecasts this critical metal's production through to 2050. This will be used to estimate the impact of market variables and government interventions on the future productivity of the region.

We have compiled data on all significant copper producers and near-term projects in the province, including mining and processing capacity, copper, by-product and waste production, power and water consumption, capital and labour costs, and transport infrastructure utilisation. The pathway from in-ground resource to saleable copper is mapped out for each producer using a structural economic model that features a rich internal organisation of production and a detailed transport, power and water infrastructure network. This model defines the functions that characterise the processes and cost structures of each production stage, while integrating the infrastructure network that supports it. The model allows for a diverse range of variables to be incorporated at each node of production (mine, concentrator, smelter) and connector in the supply chain (highway, rail, port). Each network segment, consisting of multiple nodes and connectors can be assessed for its financial viability, allowing the model to be used to investigate which variables are the principal drivers of productivity, or conversely, represent the most risk to the supply chain.

The model presented here is a preliminary base-case study; however, this framework allows us to construct counterfactual scenarios for analysing the organization of the copper supply chain and its adaptation to future demand. Quantitative modelling of this complex system is a useful tool for assessing the effectiveness of government intervention in resources supply chains, and assessing the vulnerabilities and emergent opportunities arising due to changes in external variables.

Delamerian Porphyry Mineral Systems in South Australia: What Can Magmatic Mineral Chemistry Deliver on Fertility and Exploration?

Wei Hong[1,2,3,4], Adrian Fabris[3,4], Tom Wise[3,4], Alan Collins[1,4], Sarah Gilbert[5], Stacey Curtis[3,4,6]
[1]Department of Earth Sciences, School of Physical Sciences, The University of Adelaide, Adelaide, Australia, [2]Centre for Ore Deposit and Earth Sciences (CODES), University of Tasmania, Hobart, Australia, [3]Geological Survey of South Australia , Adelaide, Australia, [4]Mineral Exploration Cooperative Research Centre (MinEx CRC), Bentley, Australia, [5]Adelaide Microscopy, The University of Adelaide, Adelaide, Australia, [6]STEM, University of South Australia, Mawson Lakes, Australia

Paleozoic porphyry-style hydrothermal alteration was previously recognized in the Delamerian Orogen, South Australia in late 1960s-1970s, where porphyry prospects include Anabama Hill, Netley Hill, and Bendigo. However, limited exploration due in part to thick postmineralization cover hindered the understanding of the temporal context, metallogenic setting, and mineral potential of the porphyry systems along the Proterozoic continental margin of Australia. We have characterized the hydrothermal alteration and mineralization of these porphyry occurrences, and conducted microanalyses of zircon, apatite, molybdenite, and white mica. Zircon U-Pb analyses reveal that the intrusive rocks were emplaced mostly between 490 and 465 Ma. Molybdenite Re-Os dating identifies two Cu-Mo mineralization events at 480 and 470–460 Ma, respectively. In situ white mica Rb-Sr LA-ICP-MS/MS analyses return two age ranges: 470–460 and 455–435 Ma for phyllic alteration at these prospects. Zircon grains are characterized by heavy REEs enrichment relative to light REEs, high (Ce/Nd)N (1.3–45), and weak to moderate negative Eu/Eu* (0.2–0.78). Apatite crystals have Mn contents of 110–10,900 ppm, Y of 130–5,350 ppm, light REEs of 100–780 ppm, heavy REEs of 10–40 ppm, low Mg (< 670 ppm), and Sr/Y ratios (<5). The zircon and apatite chemistry results suggest that both Anabama and Bendigo complexes experienced prevalent (garnet-) amphibole crystallization from hydrous melts that have moderately high oxidation state (ΔFMQ +1 to +3) and elevated magmatic sulfur-chlorine components. The Delamerian porphyry systems are interpreted to have post-dated subduction-related magmatism in the region (514–490 Ma), forming within an inverted back arc regime. Porphyritic stocks, dikes, and aplites with ages of 470–460 Ma are the most likely hosts to porphyry-style mineralization. These results highlight the significance and potential of Early-Middle Ordovician intrusive systems to produce magmatic-hydrothermal mineralization in the Delamerian Orogen.

Zircon and Whole-Rock Trace Element Indicators of Magmatic Hydration State and Oxidation State Discriminate Copper Ore-Forming from Barren Arc Magmas

Gonzalo J. Henriquez[1], Robert R. Loucks[1], Marco L. Fiorentini[1,2], Charlotte M. Allen[3], T. Campbell McCuaig[4,1]
[1]Centre for Exploration Targeting, The University of Western Australia, Crawley, Australia, [2]ARC Centre in Critical Resources for the Future, School of Earth and Oceans, The University of Western Australia, Crawley, Australia, [3]Central Analytical Research Facility, Queensland University of Technology, Brisbane, Australia, [4]BHP, 125 St Georges Terrace, Perth, Australia

To meet surging requirements of copper for the green energy revolution, mineable resources subequal to all copper production in history must be found in the next two decades. We show that trace elements in zircon and whole-rock samples that are diagnostic of unusually high-pressure magmatic differentiation and high hydration state and oxidation state of their parent silicate melt are effective for discriminating copper sulphide-ore-productive arc magmas from infertile arc magmas. Tests on our database of 5,777 zircons from 80 igneous complexes, including 2,220 zircons from ore-generative intrusions in 37 major porphyry and high-sulphidation epithermal Cu(-Au-Mo) deposits worldwide, demonstrate that our magmatic copper fertility discriminants perform equally well in intraoceanic arcs, continental margin arcs, and continental collision orogens of Ordovician to Quaternary age. The ratio $Ce/\sqrt{(U \times Ti)}$ in zircon is a quantitative indicator of the relative oxygen fugacity of the silicate melt and its sulphur-carrying capacity. The ratio of the europium anomaly to ytterbium in granitoid melts and zircon is an empirically useful indicator of the melt's hydration state and ability to provide chloride-complexed metals to exsolving hydrothermal fluids. Arrays of cognate zircons on $(Eu_N/Eu^*)/Yb_N$ vs. $Ce/\sqrt{(U \times Ti)}$ plots have slopes that vary with pressure-dependent chlorinity of exsolving fluid and its efficacy in scavenging CuCl from the melt. We formulated a composite zircon copper fertility index (ZCFI) that can be applied to each microbeam spot analysis—$ZCFI = 104\ (Eu_N/Eu^*)/Yb_N + 5\ Ce/\sqrt{(U \times Ti)}$—and substantially decreases the number of zircon analyses needed for reliable prospectivity assessment, discriminating igneous complexes, arc segments, and time intervals within them that are likely to host magmatic-hydrothermal Cu(-Au-Mo) ore deposits. Our zircon indicators of Cu metallogenic fertility are equally applicable to in situ as well as detrital zircons, thereby making regional or watershed-scale exploration tool cost-competitive with other methods of geochemical exploration.

Redox-Sensitive Trace Elements in Apatite: Insights into Porphyry Copper Mineralization Environments

Maria Alejandra Rodriguez Mustafa[1], Shaun L. L. Barker[1], Matthew J. Manor[1], Brian McNulty[1], Jackson Partridge[1]
[1]Mineral Deposit Research Unit. University of British Columbia., Vancouver, Canada

Apatite readily incorporates numerous ions in its crystal structure, enabling it to record physicochemical conditions and trace the composition of parent melts or fluids. Its ubiquity in ore deposits makes apatite an ideal indicator mineral, particularly in porphyry systems that lack zircon.

We studied apatite from six porphyry deposits in British Columbia (BC), from silica-undersaturated Cu-Au Galore Creek to silica-saturated Cu-Mo Camp Creek. Apatite from the unmineralized silica-undersaturated Zippa Mountain suite was also analysed as a baseline. Apatite grains were identified and texturally characterised using petrography, micro-XRF, SEM-EDS, and cold cathodoluminescence (CL). Major and trace element compositions were quantified using EPMA and LA-ICP-MS.

Euhedral to anhedral apatite occurs in the groundmass and as inclusions in mafic phenocrysts, exhibiting complex cold CL zoning in brown, yellow, green and purple hues. Most grains are fluorapatite (0.6-0.8 F apfu) with < 0.05 Cl apfu. Apatite from Galore Creek has the highest average concentrations of redox-sensitive elements (~3000 ppm S, ~250 ppm V, ~150 ppm As). Similar enrichments and elevated Sr concentrations (>1000 ppm) occur in other deposits in BC that are associated with silica-undersaturated rocks, which also display heavy rare-earth elements depletion relative to deposits associated with silica-saturated rocks. In contrast, apatite from the unmineralized Zippa Mountain suite lacks enrichment in V, S, and As but still contains elevated Sr concentrations (up to 8 wt%).

These data indicate that the high concentrations of redox-sensitive elements in apatite reflect the oxidized nature of the mineralizing environment. Moreover, the anomalous chemistry of apatite from silica-undersaturated deposits appears more strongly influenced by magmatic-hydrothermal processes rather than primary melt composition. Apatite with such distinctive enrichments thus represents a useful indicator of silica-undersaturated Cu-Au porphyry mineralization.

Zircon Trace-Element Indicators of Magmatic Gold Fertility

Robert Loucks[1], Carolina Nogueira Mafra[1], Charlotte Allen[2], Marco Fiorentini[1,3]
[1]Centre for Exploration Targeting, School of Earth Sciences, University of Western Australia, Crawley, Australia, [2]Central Analytical Research Facility, Queensland University of Technology, Brisbane, Australia, [3]ARC Centre in Critical Resources for the Future, School of Earth and Oceans, The University of Western Australia, Crawley, Australia

With the aim of identifying trace-element characteristics of zircon that are diagnostic of magmatic fertility for generating gold-rich magmatic-hydrothermal deposits, we have analyzed by LA-ICPMS 1067 zircon grains in intrusions that exsolved gold-ore-forming magmatic-hydrothermal fluids in 12 major deposits, and we compiled published analyses of 1078 zircons in gold-ore-forming intrusions in another 28 major porphyry-, skarn-, high-sulfidation-epithermal, and Carlin-type gold-dominant deposits at Phanerozoic convergent plate margins worldwide. We compared those trace-element compositions of zircons in gold-ore-forming intrusions with 2044 zircon analyses from granitoid intrusions in unmineralized Phanerozoic arc segments and time intervals worldwide. The comparisons show that zircons from gold-ore-forming intrusions tend to have higher Ui/Yb and Nb/Y depending on the metallogenic region. Zircons in gold-ore-forming intrusions have consistently higher ratios of the composite ratio 104 × (Eu/Eu*)/YbN + 5 × Ce/√(Ui×Ti), wherein (Eu/Eu*)/YbN is an indicator of silicate melt hydration state; Ce/√(Ui×Ti) measures silicate melt oxidation state in which Ui represents radioactive-decay-corrected initial ppm U at the time that dated zircons crystallized. We applied the zircon (Eu/Eu*)/YbN hygrometer and Ce/√(Ui×Ti) oxybarometer to a 6-Myr time series of pre-ore to late-mineralization samples from the Yanacocha igneous complex to track the temporal evolution of melt hydration state and oxidation state. Our results indicate that pre-ore intrusions were less hydrous and less oxidized than ore-stage melts. In orogenic stress regimes, replenished magma chambers residual granitoid melts inherit an accumulation of volatiles and oxidized members of redox couples, which contribute to enhanced Au-Cu metallogenic fertility. Due to the mechanical and chemical robustness of zircon, these discriminants of magmatic gold fertility are applicable to detrital zircons to identify watersheds containing gold-fertile igneous complexes upstream from the sediment sampling site.

Zircon and Melt Inclusion Geochemistry from the 1.1 Ga Jogran Porphyry, Canada: Implications for Porphyry Formation in a Non-Arc Setting

Pete Hollings[1], James Tolley[1], Jacob Hanley[2], James Crowley[3], Sasha Tsay[4], Zoltan Zajacz[4]
[1]Lakehead University, Thunder Bay, Canada, [2]St Mary's University, Halifax, Canada, [3]Boise State University, Boise, USA, [4]University of Geneva, Geneva, Switzerland

The Jogran quartz-monzonite porphyry (20 million tonnes @ 0.19%Cu), is one of a series of Cu-mineralized intrusions located near Mamainse Point, Ontario, Canada, on the northeastern shoulder of the ~1.1 Ga Midcontinent Rift System (MRS), hosts unique porphyry-style Cu-(Mo) mineralization in an intra-plate, rift-related large igneous province setting. A new high precision 206Pb/238U zircon age of 1090.90 ± 0.35 Ma (CA-TIMS) constrains the formation of the Jogran porphyry to the waning of the main Rift Stage (1102-1090 Ma) and transition to the Late-Rift stage (1090-1083 Ma). Zircon geothermometry and oxybarometry indicate crystallisation conditions of 900-670 °C and a fO2 range of ΔFMQ = -1.3 to +0.6. As temperatures decrease, ΔFMQ values increase along a trend subparallel to the SO_2-H_2S buffer. The presence of sulfide inclusions in zircon, confirms sulfide saturation during crystallization.

The zircon crystals display an increase in [Yb/Gd]n ratios and depletion in Th/U (1.0-0.4) in the rims relative to the cores consistent with single stage fractionation and crystallisation upon emplacement. Melt inclusions (MIs) range in composition from 65-70 wt.% SiO_2 with 5.5-8.3 wt.% K_2O and K_2O/Na_2O ratios of ~1.5-3.5, suggesting the parental melt was alkalic to shoshonitic. Low Cs concentrations, coupled with high Rb, Ba, and Nb, in MIs indicate minimal crystal fractionation of a near-primitive, mantle-derived composition. In contrast, whole-rock data show lower alkali contents (4.0 wt.% K2O) and have a subalkalic affinity, suggesting crustal contamination or alteration obscured the primitive magmatic signature.

Felsic magmatism and porphyry-style mineralization at Jogran occurred during maximum lithospheric weakening/crustal thinning during the shift from extensional tectonics to thermal subsidence in the late stages of the MRS. Although small, Jogran highlights the potential for porphyry-style mineralisation in non-subduction tectonic contexts and underscores the need to better understand metallogenic pathways beyond the traditional subduction models.

Exploration and Genesis Information Revealed from Comparative Copper Isotope Analysis of Sulfides from PCDs, IOCGs, and Mantos

Ryan Mathur[1], Adam Simon[2], Vladimir Lisitsin[3], Christopher Emproto[2], Linda Godfrey[4], Jorge Morales[5], Matthew Towne[6], Hannah Aird[6], Irene de Real[7], Ioan Sanislav[8], Peter Rea[9], Gisella Palma Lira[10]
[1]Juniata College, Huntingdon, United States, [2]University of Michigan, Ann Arbor, United States, [3]Geological Survey of Queensland, Brisbane, Australia, [4]Rutgers University, Piscataway, United States, [5]University of La Serena, La Serena, Chile, [6]California State University, Chico, United States, [7] Pontificia Universidad Católica de Chile, Santiago, Chile, [8]James Cook University, Townsville, Australia, [9]MIM Glencore, Mt Isa, Australia, [10]Universidad Mayor Santiago, Santiago, Chile

This talk shows how copper isotope compositions of chalcopyrite and pyrite can be used as a vector to mineralization as well as identify metal transport mechanisms during deposit genesis. Chalcopyrite and pyrite from PCDs (16 deposits, >150 samples) and IOCGs (8 deposits, >200 samples) display a general pattern of relatively lower values in the cores of the systems to higher values in the periphery. The starting points in the cores of the systems are different and PCDs show a larger range of values within each deposit type. In contrast, Mantos (4, ~40 samples) have significantly lower mean copper isotope values. We will explore the global and regional implications of these observations in the context of exploration and ore deposit genesis.

Geometallurgy of the Olympic Cu-Au Province Deposits, South Australia

Vanessa Liebezeit[1], Kathy Ehrig[1], Yan Li[1], Edeltraud Macmillan[1], Smith Michelle[1], Salomon Slabbert[1], Benjamath Pewkliang[1], Phillippa Ormond[1], Shaun Light[1], Alexandra Roslin[1]
[1]BHP Copper South Australia - Geoscience, Adelaide, Australia

The Olympic Dam Fe-oxide Cu-U-Au-Ag (IOCG) deposit not only contains elevated concentrations of Cu, U, Au, Ag but also a diverse range of currently, non-revenue elements including, but not limited to, REEs, F, C, S, As, Se, Bi, Te, Sb, Co, Zn, Pb, Mo, Sn, W. These elements, along with the typical gangue mineralogy elements such as Na, Mg, Al, Si, P, K, Ca, Ti, Mn, Sr, Ba occur in >125 minerals. This suite of elements/minerals is also characteristic of the other IOCG deposits (Oak Dam, Prominent Hill, Carrapateena) in the Olympic Cu-Au Province, although the absolute concentrations of the elements/minerals do vary at each deposit.

Geometallurgy aims to relate the deposit chemistry and mineralogy to the important metallurgical responses for the operations. At Olympic Dam the processing plant consists of 1) grinding/flotation, Cu-concentrate smelting and then electrorefining to produce Cu-cathode and Au-Ag bullion, 2) flotation tailings leach, solvent extraction, U-calcining and electrowinning (EW) to produce UOC and EW Cu-cathode, and 3) tailings storage. The processing circuits at Prominent Hill and Carrapateena consist 1) grinding/flotation to produce copper sulfide concentrates and 2) tailings storage.

Copper occurs predominately as chalcopyrite, bornite, chalcocite with significantly lesser amounts of covellite and native Cu. Because the sulfides (including pyrite) are all primary, with very minor post mineral oxidation, sulfide flotation recovery is high. The Cu-concentrate grade is controlled by the relative sulfide abundances. Sulfide concentrate impurities (e.g. As, Bi, Sb, Se, Te, Zn, Pb, Co, Mo, F, U) depend on their absolute concentrations, associations with primary sulfides and processing routes. The gangue minerals largely impact materials handling processes such as grinding, leaching (at Olympic Dam only), solid-liquid separation and ultimately tailings disposal. The geometallurgical learnings from the operating mines are part of the technical studies supporting the ongoing development of the Oak Dam deposit.

Improving Remediation Strategies for Legacy Mines: Case Study of Mt. Morgan Gold and Copper Mine, Queensland, Australia

Adithyanti Febriana[1], Samadhi U. Gunathunga[2], Peter Damian Erskine[1], Gordon Southam[2,3]
[1]Centre for Mined Land Rehabilitation, Sustainable Minerals Institute, The University of Queensland, Brisbane, Australia, [2]School of the Environment, The University of Queensland, Brisbane, Australia, [3]W.H. Bryan Mining and Geology Research Centre, Sustainable Minerals Institute, The University of Queensland, Brisbane, Australia

Due to their geological traits, various mine sites worldwide are affected by Acidic Metalliferous Drainage (AMD). AMD is generated when sulfide minerals are exposed to water, air, and iron- and sulfur-oxidizing bacteria. Once initiated, it is challenging to stop AMD generation since it is a spontaneous reaction that continues if all four components are present in an environment. Characteristically, AMD is low in pH and contains elevated concentrations of dissolved metals. These conditions pose serious concerns given their potential detrimental effects on surrounding ecosystem, wildlife, and human health. The concerns are more acute at legacy mines, where financial and human resources for remediation are drastically more limited than at active mine sites.

This study presents an integrated assessment on AMD impacts across multiple environmental systems at the historic Mt. Morgan gold and copper mine in Central Queensland, Australia. Although mining activity that started in 1882 ceased in early 1990s, AMD issues from the site persist until today.

A combination of chemical and mineralogical analyses, and 16S rRNA gene sequencing on field samples revealed potential pathways for metals transportation, transformation, and persistence across water, soil, and plant systems. Chemical assays of water, soil, and plant samples indicated low pH conditions (2.5-3.9) and elevated metal concentrations i.e. arsenic, lead, and copper. Investigation of mineralized biofilms using SEM-EDS showed biologically induced formation of jarosite and gypsum. Further synchrotron-based XFM analysis highlighted the presence of Fe-Mn-oxide associated with the base metals, with Fe concentration reaching up to ~ 20% by mass. The gene sequencing data showed dominant presence of acid-tolerant iron-oxidizing and sulfate-reducing bacteria and algae. Overall, geochemical, mineralogical, and microbial insights from the Mt. Morgan site highlighted the potential for harnessing native microbial communities in developing targeted and sustainable AMD remediation strategies.

Bacteriologically Driven REE Recovery from Mary Kathleen Uranium Mine Tailings: Potential Formation of Biogenic REE-Minerals.

Paul Evans[3], **Samadhi Gunathunga**[1], Alan Levett[2], Gordon Southam[1,4]
[1]School of The Environment, Faculty of Science, The University of Queensland, Brisbane, Australia, [2]WSP, Brisbane, Australia, [3]School of Chemistry and Molecular Biosciences, Faculty of Science, The University fo Queensland, Brisbane, Australia, [4]Sustainable Minerals Institute, The University of Queensland, Brisbane, Australia

The global transition to renewable energy technologies and circular economic principles has fuelled an increasing demand for rare earth elements (REE) due to their unique physicochemical properties. Given their difficulties in processing, securing a stable and sustainable supply of REE for critical technologies to supporting the green energy transition is a significant challenge. Currently, global REE supply relies on a combination of recycling and primary production; however, these sources are unlikely to keep pace with projected future needs. Hence, geochemical waste streams, such as mine tailings, present a promising but underutilised source of REE.

Relinquished in 1986, tailings at the historic Mary Kathleen (MK) uranium mine site in North-Western Queensland have very high (~3 wt%) total REE concentrations, that have generated REE-containing leachates for nearly four decades. Initial geochemical analyses of the mine tailings indicated the presence of ~5 wt% water soluble REE. Based on BSE-SEM analysis of the tailing materials, the REE (Ce: 399 ppm, La: 273 ppm, Nd: 70 ppm, Pr: 31 ppm in solution) are likely leaching from highly concentrated, ultra-fine mineral particles. In a laboratory experiment, the REE-impregnated acidic water (pH=4.30) produced by water washing MK tailings was periodically fed into a cyanobacteria-sulfate reducing bacteria bioreactor (180 days) and samples (water and biofilm) were collected bi-monthly for ICP-OES, SEM-EDS, metagenomics, and proteomics analyses. Preliminary data obtained for ICP-OES analysis of the water collected from the bioreactors demonstrate 90.0-99.9% removal of REE from the solution, indicating bacteriologically mediated intracellular or extracellular organo-metal composites, or secondary REE-mineral precipitation, as REE-CO_3, REE-S, REE-O, and/or REE-PO_4.

Recovering REE from solution via traditional methods, i.e., crystallization ponds, is challenging due to their similar physicochemical properties. With selective biogeochemical processing, biologically mediated REE recovery may offer novel, REE processing pathways providing potential REE-fractionation, transforming a waste into a resource.

From Penalty to Opportunity: Geometallurgical Characterisation of Gold Concentrates and Tailings for Critical Metal Recovery Using Novel Solvents

Gawen Jenkin[1], Helena Moosberg-Bustnes[2], Daniel Smith[1], **Lara Tritton**[1]
[1]University Of Leicester, Leicester, United Kingdom, [2]Mandalay Resources, Bjorkdal Gold Mine, Sweden

Bismuth, tellurium and tungsten are widely classed as critical metals due to their strategic importance and limited supply. It is crucial to diversify the sources of these metals, whilst minimising environmental impact. By-product recovery at operational gold mines could be a promising part of the solution, as gold deposits are often enriched with other rare metals . Recovery of by-products from existing mines would increase supply and reduce the need for new mines, avoiding additional environmental impact and long lead times. Unrecovered by-products may report to tailings, metallurgical wastes or facility emissions, incurring penalty fees and adding to the environmental burden of mines.

This work is a case study with Björkdal Gold Mine in Sweden. It investigates geometallurgical properties of Bi, Te, and W minerals, tracing these through mineral processing, and tests a recovery process involving environmentally-benign Deep Eutectic Solvents (DESs).

Samples from the mineral processing input and outputs underwent detailed mineralogical characterisation and bulk geochemical analysis. Automated SEM-EDS revealed near-total deportment of both bismuth and tellurium into tsumoite (BiTe), and 100% deportment of tungsten into scheelite. The characteristics of both tsumoite and scheelite differed greatly between the different concentrates and the tailings, with particle size, grain liberation, and susceptibility to froth flotation the key parameters for recovery. Two concentrates had promising tsumoite characteristics for a tailored DES-based recovery with an oxidising agent. Dissolution rates were examined through optical profilometry, which, together with the mineralogical data, informed a model for bulk leach behaviour to compare with experiments. Efficient, selective leaching of Bi and Te is possible even at 30°C. Although scheelite does not leach, a multi-step process can produce a tungsten-rich residue.

Selective leaching of bismuth and tellurium from gold concentrates is promising for on-site by-product recovery, and highlights the importance of geometallurgy for understanding by-product behaviour in mineral processing.

Responsible Mining – the Pyrite Challenge

John Thompson[1]
[1]Petrascience Consultants, Vancouver, Canada

The concept of "responsible mining" is used widely both as an aspirational goal and a guide to best practices. In reality, it is poorly defined particularly in the context of ore deposit types and mining methods that are most likely to produce positive outcomes. The concept of environmental models for mineral deposits, linked to geological deposit models, is not new, but the application of these concepts in exploration, resource assessment and responsible mining is limited.

Pyrite is a common mineral in many ore systems, varying from a minor component of ore to a major constituent in millions or billions of tonnes of wall rock, much of which becomes waste. Oxidation and weathering of pyrite contribute significantly to acid generation and release of metals requiring costly management. In the case of copper, increasing demand will be met significantly by mining low grade (<0.2% Cu) porphyry mineralization predominantly in clay-dominant alteration with limited buffering capacity. Volcanic-hosted base and precious metal systems also contain abundant pyrite although the volumes involved in mining are less extreme, while magmatic Cu-Ni sulfide, IOCG, and sediment-hosted Cu deposits can have low to virtually no pyrite.

Reducing the amount of pyrite mined requires an increasing focus on low pyrite systems, more selective mining, and ultimately in-situ recovery. Given that these approaches may not be sufficient to meet demand, new ways to process pyrite concentrates efficiently need to be developed including variations on traditional roasting as well as bio- and enhanced chemical leaching. In all cases, the production of sulfuric acid and other valuable sulfate products must be matched with recovery of all critical metals that occur in pyrite as products that can enter supply chains. The industry can not pretend to be responsible without addressing pyrite and as a result translating a challenge into a major opportunity.

Selective Bioleaching of Critical and Precious Metals from Northern Queensland Mine Waste Using Indigenous Bacteria

Eric O. Ansah[1], Samadhi U. Gunathunga[2], Ali Yadollahi[3], Pamela Curi Rivera[1], Olivia Mejías González[1], Gordon Southam[1,2], Anita Parbhakar-Fox[1]
[1]WH Bryan Mining Geology Research Center, Sustainable Minerals Institute, 40 Isles Rd, Indooroopilly 4068, Queensland, Australia, Brisbane, Australia, [2]School of the Environment, The University of Queensland, St Lucia 4072, Queensland, Australia, Brisbane, Australia, [3]Centre for Water in the Minerals Industry, Sustainable Minerals Institute, St Lucia 4072, Queensland, Australia, Brisbane, Australia

Mine waste in Northern Queensland poses environmental challenges due to reactive metals released via sulfide oxidation, which drives acid and metalliferous drainage (AMD). These wastes also serve as secondary sources of valuable metals, including copper (Cu), zinc (Zn), and gold (Au). Efficient extraction methods that minimize waste and environmental harm are essential. Bioleaching, an eco-friendly metal extraction technique, leverages microbes to recover metals with minimal ecological harm. However, the use of indigenous microbes for selective recovery of valuable metals and bioremediation of toxic elements like arsenic (As) and lead (Pb) in varied mine waste compositions remains underexplored.

This study evaluated indigenous and inoculated iron- and sulfur-oxidizing bacteria for metal extraction from mine waste at Baal Gammon (BGA) and Rishton (RSA) in Northern Queensland, aiming to limit toxic metal mobilization. Shaker flask bioleaching experiments used Fe- and S-oxidizing microbes from the abandoned Mt. Morgan mine. Complementary tests—net acid generation (NAG), acid consumption, and paste pH—evaluated AMD potential to optimize bioleaching conditions.

The results highlighted RSA, with high clay and feldspar content, was non-acid forming (NAG pH 7.5, paste pH 7.8), naturally attenuating AMD. Conversely, BGA, rich in sulfides, was potentially acid forming (NAG pH 2.1, paste pH 3.5), favoring bioleaching. Further, RSA mobilized Au (15%) > Cu (2.1%) > Pb (1.7%) > As (0.5%), while BGA yielded Cu (28.2%) > Zn (22.2%) > Pb (3%) > As (0.4%). Acid consumption tests showed RSA required 21 days and 800 μL of acid per 10 g of material in 100 mL lixiviant to reduce pH from 7.7 to 2.0, necessitating acid pre-treatment for efficient bioleaching.

This study demonstrates that indigenous microbes enable selective metal recovery from mine waste, supporting resource valorization and mine closure. By tailoring bioleaching to site-specific properties (e.g., AMD-potential, mineralogy), this approach maximizes metal extraction and minimizes environmental harm.

Interrogating LA-ICP-TOF-MS Mineral Maps: Methods and Applications to Critical Metal Research

Jeffrey Oalmann[1], Ashley Norris[2], Henry Schaumburg[1], Alice MacDonald[1], Carlos Carrasco[1], Jack Ward[1], Ivan Belousov[1]
[1]University of Tasmania, Codes Analytical Laboratories, Hobart, Australia, [2]Norris Scientific, Oxford, United Kingdom

Efficient and sustainable extraction of critical metals as by-products or from legacy mine waste requires a thorough understanding of the deportment of the metals of interest and penalty elements. LA-ICP-MS mapping is well suited for deportment studies, and time-of-flight (TOF) LA-ICP-MS systems make it possible to measure nearly all elements from single laser pulses at speeds up to 100 times faster than possible with quadrupole systems. Because nearly all elements are measured, each pixel can be grouped into prospective mineralogical classifications. Interpreting these large high-dimensional data sets benefits from novel data science approaches, such as machine learning algorithms.

We present an app (LazAbMapper) written in Python with a Shiny user interface, implementing some of the vast number of algorithms available in the Python ecosystem. Highlights of LazAbMappers's functionality include: 1) displaying and exporting single maps, composite RGB maps, maps calculated from multiple elements, and multi-map panels; 2) classifying minerals using library matching, compositional criteria, or K-means clustering; 4) selecting regions of interest and extracting their compositions; 5) performing statistical analyses of the data using a variety of algorithms.

We demonstrate LazAbMapper's capabilities using a map of a magnetite crystal with silicate inclusions from the Kara deposit in Tasmania. Data were collected at CODES Analytical Laboratories with a Tofwerk icpTOF R and RESOlution-SE laser ablation system using a 12 μm square beam scanning at 360 μm/s, firing at 30 Hz. Spectral baseline subtraction was performed with Tofware software, and data were reduced using LADR's new imaging capabilities. Nearly all pixels were classified as either magnetite or silicates. Sn, V, and Ti are hosted primarily in magnetite, with Sn exceeding 1200 ppm in some zones. Cu and Zn occur primarily in the silicate phases and in some magnetite domains. Small inclusions of calcite were identified.

Sericite and Chlorite Type and Composition within and around Porphyry Deposits: Tools to Vector Towards K-Silicate Alteration at Depth

Farhad Bouzari[1], Shaun Barker[1]
[1]MDRU - Mineral Deposit Research Unit, The University of British Columbia, CANADA, Vancouver, Canada

Porphyry copper deposits have a large (>25 km2) alteration footprint hosting a smaller (<0.5 km2), typically sub-vertical, body of copper mineralization. Sericite and chlorite are the most common alteration minerals that occur in shallow levels of porphyry deposits and can provide vectors for exploration and drilling.

During this study, samples from Camp Creek calc-alkalic Cu-Mo-Au porphyry deposit were selected and compared with samples from Gibraltar calc-alkalic Cu-Mo and Mount Milligan alkalic Cu-Au porphyry deposits. Alteration at the near-surface at Camp Creek is characterized by advanced-argillic assemblage, which transitions laterally to sericite, sericite-chlorite, and distal chlorite. K-silicate alteration occurs at depth, which hosts the mineralization. At Gibraltar, mineralization is hosted by sericite-chlorite alteration, but no K-silicate alteration is recognized at depth. K-silicate alteration hosts mineralization at Mount Milligan, and it is locally overprinted by sericite-chlorite, with intense sericite alteration occurring along structures.

A total of 44 samples were selected from these deposits. Samples were studied using petrography, SWIR, and sericite, chlorite mineral chemistry by EPMA and LA-ICP-MS. Sericite composition varies from muscovite (white sericite) to phengite (green sericite) at depth, with K-deficient sericite occurring at shallow levels with pyropyllite.

Sericite is alkali-deficient, Al- and Na-rich above Camp Creek and becomes alkali-rich, Fe-Mg-rich, and Na-poor at depth. Trace elements Zr, Rb, Tl, Zn, and Ti of sericite provide vectors. Remnants of K-rich muscovite with a higher Rb:Ti ratio suggest K-silicate alteration at depth. Chlorite composition is influenced by the hostrock. Fe-Mg in chlorite provides vectors in mafic host rocks, and Al-in-chlorite provides vectors in intermediate/felsic rocks. Several trace elements in chlorite, such as Ti, As, Co, Li, provide vectors. The sericite and chlorite compositional data can identify mineralization signatures typically not recognized by SWIR and whole rock geochemistry and provide vectors for K-silicate alteration at depth in areas with known shallow-level alteration.

Trace Element Compositions of Micro Inclusions Extracted from Mixed LA-ICP-MS Analyses and Their Use in Exploration

Ivan Belousov[1], Axel Cima[2], Leonid Danyushevsky[3], David Cooke[2], Mitchell Marcelissen[4], Pete Hollings[4]
[1]CODES Analytical Laboratories, University Of Tasmania, Hobart, Australia, [2]CODES, University of Tasmania, Hobart, Australia, [3]Friendly Solutions, Hobart, Australia, [4]Lakehead University, Thunder Bay, Canada

Micro-inclusions in minerals are observed in a range of magmatic as well as hydrothermal minerals. In magmatic environment they reflect co-crystallising assemblages and could be used to better fingerprint melt composition as well as temperature, pressure and redox conditions. In the environment of hydrothermal alteration compositions could be either inherited from the original mineral or reflect conditions of formation of the host minerals and composition of fluids involved. Use of compositions of micro-inclusions in exploration is limited due to them often being too small to obtain confident analyses by LA-ICP-MS. However, they are frequently ablated accidentally when analysing host minerals for mineral chemistry or geochronology. Currently mineral inclusions are mostly excluded from the integration intervals at data reduction stage. However, trace element compositions of micro-inclusions in many cases could be extracted from such mixed analyses.

We present examples of apatite inclusions in zircon from porphyry Cu deposits as an example of magmatic environment and show that LREE contents of apatite inclusions in zircon can be accurately quantified through deconvolution of mixed LA ICPMS signals and potentially can identify fertile intrusions and provide discrimination between different deposit types and stages of magma evolution. When used in conjunction with host zircon compositions they can improve prediction of magma fertility for exploration. We also present compositions of rutile and titanite inclusions in chlorite from propylitic alteration environment around porphyry Cu deposits and show how their compositions could complement use of chlorite compositions for vectoring and fertility studies.

Inclusions are often too small to analyse directly and use of deconvoluted composition of inclusions can provide a cost-effective addition to an exploration when using LA-ICP-MS analyses for mineral chemistry or geochronology, however analytical conditions and downhole fractionation correction are crucial for obtaining accurate data.

Assessing the Precision and Accuracy of In Situ Re-Os Dating of Molybdenite and Considerations for Ore Systems Research

Duane Petts[1], Christopher Lawley[1], Robert Creaser[2], William Davis[1]
[1]Geological Survey of Canada, Natural Resources Canada, Ottawa, Canada, [2]Department of Earth and Atmospheric Sciences, University of Alberta, Edmonton, Canada

Precisely constraining the timing of mineralization remains a key component in the development of robust ore deposit models. Recent developments in laser ablation-based, tandem inductively coupled plasma mass spectrometry (e.g., LA-ICP-MS/MS) allow for the application of in situ beta-decay geochronometry to ore deposits research, including the use of in situ Re-Os isotope analysis of molybdenite to directly date mineralization. Here we assess the precision and accuracy of the in situ Re-Os method and evaluate its applicability to ore deposit research.

In situ (LA-ICP-MS/MS) and conventional (isotope dilution – negative ion – thermal ionization mass spectrometry, ID-N-TIMS) Re-Os measurements were undertaken on three molybdenite samples for potential development as reference materials and for evaluation of the in situ technique, including samples from: 1) a mineralized ca. 2.66 Ga pegmatite in the Moly Hill/La Corne region (Quebec), 2) a Neoarchean Cu-Zn VMS deposit related to the past producing Geco mine (Ontario) and 3) a Mesoproterozoic pegmatite associated with the Grenville orogeny (Gatineau region, Quebec). Additional in situ Re-Os isotope analyses were conducted on previously ID-N-TIMS dated molybdenite samples from the ca. 154-145 Ma Endako and ca. 80 Ma MAX Mo porphyry deposits (British Columbia).

While the in situ Re-Os isotope data yields accurate age interpretations, precisions of 1 to 2% are reported for the calculated ages. The precision of the in situ technique is highly dependent on the age and Re content of molybdenite, with the latter, varying between 2 and 50 ppm in the samples studied here. With these precision and accuracy considerations in mind, the Geco data highlights potential challenges with application of the in situ Re-Os technique to ore deposits research in Archean rocks, including utilizing the in situ data to precisely distinguish between contemporaneous volcanism and mineralization at 2.72 Ga and metamorphism at 2.67 Ga.

New Developments in Geochronology to Place Mineralisation Processes into Better Constrained Tectonic and Geodynamic Context.

Stijn Glorie[1], Jack Mulder[1], Alexander De Vries Van Leeuwen[1,3], **Dr. Sarah Gilbert**[2]
[1]Earth Science, University Of Adelaide, Adelaide, Australia, [2]Adelaide Microscopy, University of Adelaide, Adelaide, Australia, [3]Mineral Exploration Cooperative Research Centre, University of Adelaide, Adelaide, Australia

Mineral deposit formation and modification is one of the most chemically complex processes that occurs in the Earth's crust and presents significant challenges for geochronology methods applied to constrain timing. It is well demonstrated that traditional geochronological methods such as U-Pb and Ar-Ar can provide high fidelity age constraints on the timing of mineralisation. However, these methods can also have significant limitations, for example significant common Pb in U-Pb systems, and time consuming methodology in the case and difficulties with in situ targeting in the of Ar-Ar.

An alternative approach to the geochronologic interrogation of mineral deposits is with beta-decay geochronology analysed via reaction cell laser ablation. The method combines the rapidity and spatial precision of laser-ablation geochronology with a wide range of isotope systems and minerals relevant to ore deposit formation.

At the University of Adelaide, the use of reaction cell technology is routine for beta-decay geochronology applications, with more than 200,000 analyses in the last 4 years. The analytical techniques developed and optimised at the University of Adelaide provide fast, high-volume analysis of Rb-Sr, Lu-Hf and Re-Os radiogenic systems that constrain age in combination with trace element compositions of a range of minerals typically associated with mineralization (molybdenite, apatite, muscovite/sericite, carbonates, fluorite) in their paragenetic context.

Laser ablation reaction cell beta-decay age-dating methods offer a flexible portfolio of approaches to interrogate the timing of mineral deposit formation and modification. The spatial precision of laser-based data acquisition puts the petrographer back into a central position to inform the targeting process. The trace element information that accompanies age acquisition creates a nexus between geochemists, tectonicians and geodynamists that will better illuminate when, how and why mineral deposits form.

In this presentation we demonstrate the utility of reaction cell beta-decay geochronology to augment the understanding of mineral deposit formation.

The Geological Footprint of Alteration and Mineralization at the High-Grade Resolution Porphyry Cu-Mo Deposit, Arizona

Takeshy Coaquira[1], David Cooke, Lejun Zhang
[1]University of Tasmania, Adelaide, Australia

The Laramide Resolution Cu-Mo porphyry deposit is one of the world's largest and highest grade hypogene copper deposits. Four major paragenetic stages contributed to copper mineralization. Stage 1 was related to the emplacement of rhyodacite porphyries and magmatic-hydrothermal associated with chalcopyrite and molybdenite mineralization. Stage 2 was associated with the emplacement of inter-mineralization quartz- porphyries and hydrothermal breccias produced chalcopyrite and bornite mineralization with phyllic alteration (white mica - quartz - pyrite - anhydrite - APS). Stage 3 advanced argillic alteration (topaz - pyrophyllite - dickite - kaolinite - APS) was associated with late-stage hydrothermal breccias. Stage 4 produced a leached cap and supergene chalcocite enrichment zone for up to 300 m beneath the Tertiary volcanic rocks. Resolution has a highly productive and strongly telescoped hydrothermal system with multiple paragenetic events adding significant mineralization–a key to high grade hypogene hybrid porphyry-HS deposits.

Distinctive alteration zoning patterns and changes vein types and abundances define a geological footprint to Resolution that extends up to 6 km laterally from the ore zone. There is a transition from distal chlorite (> 4 km) to medium epidote - albite - pyrite (2-4 km) to proximal K-feldspar - biotite alteration (< 2 km). Stage 2 formed deep-seated phengite alteration overprinting potassic alteration in the core of the deposit and transitioned upwards (400 m) to pyrite-rich muscovite alteration. Lateral transitions from phengite outwards to illite – phengite (> 1.5 km) and muscovite alteration assemblages (> 0.5 km) occur around the main 1 km wide phyllic upflow zone that overprinted the potassic-altered core. Stage 3, advanced argillic alteration is restricted to domain of 500-m wide. It is structurally-controlled and overprinted phyllic alteration. Dickite – topaz – zunyite assemblages transitions outwards to kaolinite at 200 m, and hinsdallite transitions upwards at 200 m to svanbergite and woodhousite.

Adakitic Signatures from Permian Granitoids Around the Georgetown Region – Potential for Porphyry Cu Mineralisation and Implication for Tectonic Setting

Jonghyun Lee[1], Ioan Sanislav[1], Helen McCoy-west[1]
[1]James Cook University, townsville, Australia

The Permian granitoids around the Georgetown region (GT) have distinctive adakitic characteristics such as higher Sr and LREE contents; high Sr/Y and (La/Yb)N ratios; low Y and HREE contents and steep REE patterns with no or weak negative Eu anomalies. These characteristics differentiates them from granitoids in the other provinces like Herberton and Daintree provinces, which have been considered as sources of W or Sn mineralisation. These granitoids were analysed to investigate their petrogenesis, tectonic setting and potential for mineralisation, especially for porphyry Cu deposits. Discrimination diagrams for adakitic sources indicate that these granitoids have compositions similar to continental crust-derived adakites (partial melting of lower continental crust or melting due to delamination). These observations are consistent with the Hf isotopes in zircons, which show negative εHf values, reflecting derivation from continental crust sources and as their ages became younger, the values decrease indicating thickening of the continental crust in GT as opposed to the Herberton area where similar age granitoids have positive εHf values indicating thinning and mantle input. Zircon geochemistry indicates that during Permian, the less felsic granodiorites plot in the porphyry Cu(-Au) productive fields showing high oxidation and hydration degrees (higher (Ce/Nd)/Y and Eu/Eu*).

Adakitic signatures are commonly associated with porphyry Cu-(Au) mineralisation even when they occur in non-arc settings such as the Tibetan examples and this may be the case in the GT region. In addition, two distinct metallic Permian mineralisation corridors (Au, Ag, base metal) occur in close spatial association with the Permian granitoids suggesting a fertile environment. Therefore, we suggest that the less felsic Permian granitoids from the GT region have the potential to produce porphyry Cu mineralization and the newly discovered Cu mineralization associated with the Permian Yataga intrusive complex indicates that that may be the case.

Global perspectives on exploration and ESG

Sarah Mackenzie[1]
[1]University of Queensland, Australia

With the global drive to decarbonize, the extractives industry is facing challenges that go beyond geotechnical complexities of the green energy transition. Across jurisdictions, social and cultural factors are having an increasing impact on project success or failure at the exploration phase. The search for the critical minerals needed to make this transition possible is taking place in locations that face complex environmental, social, and governance challenges, where exploration or mining could have serious consequences for local ecosystems and communities. Explorers can no longer responsibly claim minimal social or environmental footprint as a pathway to “land access” or “social acceptance” for exploration activities. Short-term transactional approaches are widely understood to be detrimental in the medium and long-term for project-affected communities, and the skill sets required from early exploration go well beyond geology and geophysics. Over the past five years, researchers at the Centre for Social Responsibility in Mining have brought evidence to the scale of these challenges. This opening presentation at the ESG session of the SEG Conference will set the scene from a broad global perspective, showcasing research that highlights the extent and scale of the challenges and reflecting on the implications for exploration teams seeking to better understand and address the social aspects of exploration.

Practitioner perspectives on current approaches to Community Relations at Exploration: Highlights from Northern Europe, Latin America, North America, and Central Africa

Sarah Mackenzie[1], **Kathryn Kochan**[1]
[1]University Of Queensland, Brisbane, Australia, [2]Centre for Social Responsibility in Mining, Brisbane, Australia

In response to the growing global demand for metals and closer regulatory scrutiny around sustainability performance, explorers must focus on developing strategies and policies to secure the supply of critical minerals responsibly and efficiently. This is taking explorers into new jurisdictions and increasing their presence in others. While explorers are facing growing community opposition to exploration in certain jurisdictions, expectations of economic growth and benefit are creating immense development pressures in others.

As the social context for exploration evolves, consent and access to land are becoming increasingly determinant factors in project success. To meet societal expectations of a responsible transition, geoscientists must address logistical challenges, such as long lead times between discovery and mining, and engage with communities early on. Geoscientists are now expected to have the knowledge base and capability to address the complexities of community relations across diverse and socially complex regions. In response, companies are implementing social performance and community relations skills, data, and resources ahead of shifts in policy and governance agendas.

The Centre for Social Responsibility in Mining (CSRM) at the University of Queensland has conducted research to understand how exploration stakeholders are building and strengthening community relations capability in the drive to decarbonize. This draws on CSRM's experience training global exploration teams, and interviews with geoscientists, exploration community relations and sustainability personnel, and local community representatives in four resource-rich regions: Northern Europe, Latin America, North America, and Central Africa. The case studies illustrate how company, regulatory, and community approaches and capabilities are changing in response to geopolitical tensions, increasing land pressures, and the social justice implications of critical mineral exploration. Both missteps and successful approaches to integrating social considerations into exploration activities offer insights into how broad uptake of emerging practices by exploration teams will be critical to the success of future projects.

Geospatial Modelling of Environmental, Social and Governance (ESG) Risks for Mineral Exploration and Mining Feasibility Assessments in Finland

Bijal Chudasama[1], Toni Eerola[1], Helena Hulkki[1]
[1]Geological Survey of Finland, Espoo, Finland

Mining-related Environmental, Social and Governance (ESG) risks are typically assessed during the advanced stages of mining permit procurement. Delays and disruptions in acquiring these permits due to ESG risks cause increased costs and threaten timely and cost-effective execution of mineral exploration and mining projects. Hence, early identification and mitigation of these risks prevent potential conflicts and guide policy making.

This study presents a conjunctive a-priori evaluation of geological favorability and ESG criteria for mineral targeting and mining feasibility assessments using geospatial data modeling to minimize ESG risks. The integrated approach is applied to Pyhäsalmi-Vihanti Volcanogenic Massive Sulphide (VMS) belt in Finland. The first step involves development of quantitative metrics for ESG-related risks. Next, these metrics are represented as spatially mappable criteria for spatial-mathematical assessments. The spatial mathematical integration is implemented using knowledge-driven inference networks, and data-driven machine learning; utilizing the available data on exploration permits, mines, and mineral deposits for training the algorithms. The outputs show that the highest ESG risks in the study area are associated with the ecological regions conserved by law. Risks emanating from the social factors are primarily over the second homes along the lakes in the southern part of the belt with known cases of mineral exploration disputes. This separates the belt into two domains regarding social acceptance (northern and southern); the northern domain being characterized by old mine sites indicating acceptance towards mining activity. Because the delays in permit approvals occur over the environmentally and socially sensitive zones, the governance risks spatially correlate with the environmental and social risks; hence are strongly controlled by the ES metrics. This study demonstrates that geospatial modelling enhances ESG assessments by providing spatial context, mathematical consistency and hence objectivity, thereby facilitating early identification of potential conflicts, and guiding informed policy decisions in mining project evaluations.

Project-AIMEX, Business Finland.

Biomass Carbon Emissions from Nickel Mining Have Significant Implications for Climate Action

Evelyn Mervine[1,2], Rick Valenta[1], James Paterson[2], Gavin Mudd[3], Tim Werner[4], Ilyas Nursamsi[1], Laura Sonter[1]
[1]The University of Queensland, St Lucia, Australia, [2]The University of Edinburgh, Edinburgh, United Kingdom, [3]British Geological Survey, Nottingham, United Kingdom, [4]The University of Melbourne, Melbourne, Australia

Global nickel demand is projected to double by 2050 to support low-carbon technologies and renewable energy production. However, biomass carbon emissions from clearing vegetation for nickel mining are rarely included in corporate sustainability reports or considered in mineral sourcing decisions. Here, we compiled data for 481 nickel mines and undeveloped deposits to show that the footprint of nickel mining could be 4 to 500 times greater than previously reported (depending on the mine site), and thus the environmental impacts of nickel products, including batteries, have been underestimated to date. We found large variation in biomass losses among mines, and, in many cases, these unaccounted carbon emissions were significant relative to other Scope 1 and Scope 2 emissions from nickel extraction and processing. Reporting emissions from biomass losses from mining is key for strategic decision making on where to source nickel needed for effective climate action.

Mineral Landscapes: Connecting Geoscience to Stakeholders

Karin Olson Hoal[1], Mark Pearce[2], Vinicius Louro[2], Humera Rind[2], Aditi Sinha Roy[1]
[1]Cornell University, Ithaca, United States, [2]CSIRO, Kensington, Australia

The minerals sector has made some steps in innovation such as dry stack tailings, modular plant design, and integrated sensors. Also, some multidisciplinary initiatives have been innovative in linking steps along the minerals value chain: these include mine-to-mill-to plant approaches, ore body knowledge, and geometallurgy. In many cases, these latter are allocated to specific teams, creating new silos rather than breaking them down. Major, transformative change in our industry is still lacking. We propose that truly transformative innovative is likely to come from outside the sector, and this requires new, transdisciplinary communication and knowledge-share frameworks. Transdisciplinary thinking across mineral resources is necessary as the world moves toward an environmentally impacting and geopolitically sensitive era of expanded minerals extraction. It further presents an opportunity for communication and sustainable development across the sector. The very nature of rocks, ores, minerals, and element deportment we take for granted as geoscientists, particularly with respect to environmental and social impacts throughout the mine life and after closure, is not well communicated to stakeholders. This creates barriers to effective best practice and sets up more silos, distrust and outsidership. Stakeholder input to sustainable practices, approaching ESG frameworks, also has important financial risk benefits to the sector as a whole and to companies in particular.

Evidence-Based Solutions for Improving Equity, Diversity and Accessibility: Informing Policy and Governance in Economic Geology

Alanis Olesch-Byrne[1,2], Erin Martin[1], Romana Dew[1], Neethu Madhukumar[1], Madison Styles[1], Tara Djokic[1], Grace Shepard, **Fernanda Alvarado-Neves**[1], Leah Lynham[1]
[1]Women in Earth and Environmental Science Australasia (WOMEESA), Sydney, Australia,
[2]University of Melbourne, Melbourne, Australia

With increasing global scrutiny of the mineral sector, meeting societal expectations for equitable and inclusive workplaces has become integral to policy and governance within economic geology. Extensive research states that organisations fostering diversity are more innovative, resilient, profitable and adept at managing complex societal and environmental challenges, aligning closely with contemporary governance expectations.

Despite balanced gender representation in tertiary geoscience education and early-career roles globally, a significant gender gap emerges at advanced career stages in both academia and industry, a trend described as the "leaky pipeline." This attrition severely limits sector performance, notably in senior roles, where representation of women remains disproportionately low. For example, women only account for 21.9% of CEO positions nationally and in Australia's mining industry, women only occupy just 8% of CEO positions.

Barriers identified include unconscious bias, inadequate parental and family support policies, limited visible role models, unequal opportunities for professional advancement, and persistent workplace harassment and exclusionary practices. These systemic issues not only deter women from progressing but also negatively impact the sector's reputation, sustainability, and compliance with societal expectations.

This presentation will highlight evidence-based strategies and practical recommendations for addressing these challenges through targeted policy reforms and enhanced governance practices. By implementing inclusive workplace policies, transparent governance structures, and robust anti-harassment frameworks, the economic geology sector can significantly improve diversity and equity, aligning itself more effectively with societal standards and expectations.

Cobalt Potential from IOCG Tailings and Copper-Cobalt Deposits of the Atacama Region, Chile: Bioleach Alternatives for Sustainable Recovery

Brian Townley[1], Patricio Martinez-Bellange[2], Natalia Zúñiga-Prohaska[2], Roberto Collao[3], Pilar Parada[2], Duncan Blount[4]
[1]University of Chile, Santiago, Chile, [2]CBS, University Andres Bello, [3]Independent Consultant, [4]Chilean Cobalt Corporation

Cobalt, a critical element for renewable energy and electric mobility, is primarily sourced from the Democratic Republic of Congo, followed by Indonesia, Russia, Australia, and Canada. Chile, which hasn't produced cobalt since 1944, holds untapped potential through mine tailings from IOCG (Iron Oxide Copper Gold) deposits and primary cobalt ore in copper-cobalt deposits in the Atacama region.

Studies in the Punta del Cobre IOCG district show that cobalt is primarily deported in pyrite, with concentrations in head feed ranging from 50 to 300 ppm, increasing up to 1,800 ppm during copper flotation, due to collateral pyrite concentration, as determined in the cleaner stage scavenger tailing. Determinations of cobalt in pyrite indicate levels between 4,000 and 4,800 ppm. Simple flotation can increase pyrite content in the tailings to over 70%, yielding cobalt concentrations up to 2,800+ ppm.

In the Las Cobalteras–San Juan district, cobalt occurs in vein deposits, often associated with copper, cobaltite, chalcopyrite, and pyrite. Ongoing exploration suggests that these deposits, which are genetically linked to major deformation zones along the Atacama Fault Zone, are similar to copper-cobalt lode deposits found globally. Hydrothermal alteration in the region, characterized by pervasive biotite potassic alteration, suggests formation temperatures well above 350°C, consistent with the proposed geological model.

Cobalt recovery experiments from pyrite concentrates using bioleaching technology have shown over 80% recovery within 120 days. The patented bacterial consortium, Kobald B, was cultured from onsite samples. Bioleaching has proven to be a cost-effective and sustainable option compared to higher-capital methods, especially for low grade cobalt in pyrite concentrates or small-scale deposits like those in Las Cobalteras–San Juan.

In conclusion, Chile has significant, untapped cobalt resources, both in tailings and primary deposits. Innovative recovery methods, such as bioleaching, offer not only economic opportunities but also environmental benefits, aligning with global sustainability goals.

Environmental Impact of Abandoned Mining Activities in the Monte Amiata district (Italy) and Implications for a Sustainable Environmental Management

Hala Chmes[2], Francesco Ciani[2], Pilario Costagliola[1], Pierfranco Lattanzi[1], **Guia Morelli**[1], Alessia Nannoni[2], Valentina Rimondi[2]
[1]National Research Council, CNR, Geoscience and Earth Resources Institute - IGG, Florence, Italy, [2]Department of Earth Science, University of Florence, Florence, Italy

Increasing demand of metal resources should deal with site-specific management to prevent environment degradation, mainly due to the release of potentially toxic trace elements (PTE) to freshwater systems.

In the Mt. Amiata mining district (Italy), the 3rd former Hg producer worldwide, epithermal deposits were historically exploited (Hg, Sb) without management strategies. After 40 years from mine decommissioning, a legacy of contaminated sediments affects river channels/floodplain (Paglia-Tiber and Fiora rivers). In the upper Paglia River floodplain the interplay of mine waste deposition and changes of river morphology in the last 130 years formed a 20 km^2 contaminated area (Hg up to 100 mg/kg); its tributaries still supply Hg (up to 5400 mg/kg) to the main streams. In the Fiora River up to 3300 mg/kg Hg were observed in stream sediments.

Erosion of overbank sediments, intensified by extreme flood events driven by climate change, represent long-term PTE intermittent sources, with a long-distance impact up to the Mediterranean Sea, an important biodiversity hotspot and a food reservoir. The mass loads of Hg to the Sea from the whole district could be as high as 6.6 t/year.

The extent of contamination and low resilience suggest that river self-restoration seldom occurs in extensively exploited mining areas and, clean-up remediation strategies are not feasible. Management solutions should consider the coexistence with pollution, ensuring minimum risk for anthropogenic activities, also considering the opportunity of revising legislations on local PTE thresholds. A sustainable mitigation approach in the Paglia River is the cultivation/harvesting of riparian biomasses, since they store PTE (Hg up to 59.9 mg/kg), limit overbanks erosion and can be used for energy production. This is a pragmatic choice applicable to similar non-remediable areas to minimize the transport of PTE and a cost-effective circular remediation solution.

Software Tool Development for Evaluating the Beneficiation Potential of Mining Waste

Dandara Ataide Salvador[1], Mohammad Jooshaki[4], Rita Kallio[2], Kimmo Kärenlampi[1], Tommi Kauppila[5], Minna Markkanen[3]
[1]Geological Survey of Finland (GTK), Outokumpu, Finland, [2]Geological Survey of Finland (GTK), Rovaniemi, Finland, [3]Geological Survey of Finland (GTK), Espoo, Finland, [4]Geological Survey of Finland (GTK), Kokkola, Finland, [5]Geological Survey of Finland (GTK), Kuopio, Finland

The Geological Survey of Finland (GTK) is actively engaged in research projects focused on tailings management and re-mining, encompassing mineral reserve studies and beneficiation tests. These projects aim to minimize mining waste by extracting valuable resources from tailings and exploring innovative applications. The REPower-CEST project, funded by the EU's Recovery and Resilience Facility, emphasizes the transition to green energy and its demands for a wide range of critical raw materials (CRMs). To support the efficient extraction and diversification of CRM sources, as well as the reduction of mining and processing waste, a software tool is being developed at the Geological Survey of Finland. The tool conducts mining waste analysis primarily through representative automated mineralogy assessments, complemented by essential chemical and mineral analyses. Automated mineralogy data is integrated into the software via MS Access databases, compatible with various SEM-AM systems such as MLA, AMICS, and INCAMineral. Moreover, the software includes an extensive mineral database with information on beneficiation properties, application domains, and value assessments. It combines the produced datasets with an embedded mineral property database and mineral processing information to achieve a repeatable and systematic identification and visualization of the potential utilization of minerals. This comprehensive data integration enables the characterization of materials, assessment of applications and value, determination of processability, and evaluation of health and environmental risks. The software significantly accelerates sample analysis while reducing error-prone manual operations. The results can be used to produce estimates of the potential value of minerals and metals for the clean energy transition and serve as a roadmap for evaluating the potential of tailings beneficiation, aiding the preliminary selection of optimal unit processes and technologies.

Rare Earth Elements' Secondary Prospectivity of Mine Waste in South Australia

Francesco Colombi[1], Adrian Fabris[2], Carmen Krapf[2], Anita Parbhakar-Fox[1], Jane Thorne[3]
[1]The University of Queensland - Sustainable Minerals Institute, Brisbane, Australia, [2]Geological Survey of South Australia - Department of Energy and Mining, Adelaide, Australia, [3]Geoscience Australia, Canberra, Australia

In recent years, the demand for rare earth elements (REEs) has significantly increased due to their essential role in a broad spectrum of modern high-tech applications and geopolitical challenges linked to their supply. As a result, mine waste has recently attracted interest as an unconventional yet promising secondary source of critical metals. In South Australia, various forms of REE mineralisation occur associated with primary commodities of copper, gold, and uranium.

To assess the potential of mine waste in South Australia as a secondary source of REEs, the content and distribution of critical metals were investigated at two key sites: Paratoo, a sediment-hosted copper (Cu), and Port Pirie, the site of a former uranium processing and monazite REE extraction facility. At Paratoo, waste rock samples were collected, while at Port Pirie, the sampling focused on tailings, sludge, and thorium-rich waste. The collected samples were analysed using ICP-MS, XRD, MLA, and LA-ICPMS. At Paratoo, the primary REE-hosting minerals were identified as azurite and malachite, which are enriched in both light rare earth elements (LREEs), such as La (82,000 ppm), and heavy rare earth elements (HREEs), such as Gd (9,600 ppm). In contrast, at Port Pirie, REE enrichment is associated with monazite and REE-carbonates. Monazite is predominantly enriched in LREEs (Ce at 552,000 ppm), while the REE-carbonates show higher concentrations of HREEs (Dy at 8,200 ppm).

Metal deportment of rare earth elements was assessed in host mineral phases at both sites. Gaining a clear understanding of REE deportment is essential for evaluating the economic feasibility of their recovery and for developing effective extraction and processing methods. At Paratoo, bioleaching presents a potential approach for extracting REEs from secondary copper minerals. Meanwhile, at Port Pirie, pug leaching could be investigated as a method for recovering REEs from monazite and REE-carbonate minerals.

Carbon Dioxide Removal Using Serpentine-Rich Mine Residues

Greg Dipple[1]
[1]Arca Climate Technologies Inc., Vancouver, Canada

The greenhouse gas carbon dioxide (CO2) is removed from the atmosphere using alkaline industrial waste, such as ultramafic mine tailings, as a reactive feedstock. The captured CO2 is stored durably in carbonate minerals. Mine tailings derived from serpentinized ultramafic ore passively capture CO2 at slow rates, but the rate of capture can be enhanced through the activation of the feedstock during mineral processing and the manipulation of tailings after deposition in the tailings storage facility.

Arca Climate Technologies Inc. is a start-up company based in Vancouver, Canada that is built on 20 years of research from the Mineral Deposit Research Unit at The University of British Columbia. Here we present results from Arca laboratory and field pilots of activation and Smart Churning technologies. Smart Churning is integrated with continuous monitoring of CO2 capture. The tailings surface is roughened using physical manipulation techniques to increase the rate of CO2 capture from air. Key parameters that impact the CO2 capture rate are the surface area, water content, and pH of the tailings, all of which can be modified by Smart Churning. CO2 removal is directly monitored at a point-scale using commercially available dynamic closed chambers, and at a larger-scale with eddy covariance monitoring. The two independent methods measure similar CO2 removal rates. Results demonstrate that the rate of capture from Smart Churning is significantly increased over baseline passive air capture rates. Mineral trapping is confirmed with electron microscopy imaging of carbonate mineral cements that bind milled tailings grains. The reactivity of serpentine to CO2 increases substantially after thermal, mechanical, and chemical activation. Laboratory air capture tests using activated serpentine demonstrate a 150-fold increase in CO2 capture capacity, and similar enhancement of capture rate. Together, activation and Smart Churning deployed at mine sites globally could remove CO2 from the atmosphere at gigatonne scale.

New Insights for the Porphyry Model from Complex Overprinted Porphyry Systems

Angela Escolme[1], David Cooke[1], Scott Halley[1], Lejun Zhang[1], Rhiannon Jones[1], Takeshy Coaquira[1], Jaime Osario[1], Erik Fabreschi[1], Michael Baker[1]
[1]CODES - University Of Tasmania, Hobart, Australia

Within the porphyry system a transition zones exists in the region between the deep potassic core and shallow epithermal environment. Complex and variable alteration assemblages of overprinting white micas and clays may occur in this region as a result of variations in fluid pH and temperature, fluid to rock ratio, and lithology. Assemblages may be synchronous or temporally distinct, grade additive or grade destructive. This complexity presents challenges to both explorers and miners.

The spatial and temporal distribution of minerals, and their compositional variation, in the transition zone have been explored though five detailed case studies at Northparkes (Australia), Resolution (USA), Tujuh Bukit (Indonesia), Spence and Encuentro (Chile). Geological logging, petrography, whole rock geochemistry, hyperspectral and mineral chemistry (by EMPA, SEM and LA-ICP-MS) datasets were collected at each deposit during the project. Based on these case studies, three new models for weak, moderate and strong alteration overprints are presented.

Our observations have important implications for the porphyry genetic model and provide new insights to aid exploration, including vertical and lateral alteration vectors. Mineral chemistry studies (specifically looking at white mica and pyrite) have enabled a greater understanding of trace element deportment that enhances our interpretations of whole rock geochemistry data and provides a new frame of reference for levels of anomalism. These datasets reveal consistent patterns in chemical zonation between regions and degrees of overprint. Whilst the absolute concentrations may vary, we observe that the same patterns are repeated. Mapping of mineral species and their abundance has also generated new mineral maps for different degrees of porphyry overprint that provide a foundation framework for ore body knowledge studies in complex porphyry systems.

Magmatic Plagioclase as a Porphyry Copper Fertility Indicator in Arc Systems

Daniela Parra-Encalada[1], Teresa Ubide[1], Yamila Cajal[2], Gideon Rosenbaum[1], Catharine Wang[1], Andy Tomkins[3], Ian Campbell[4], David Paterson[5]
[1]The University of Queensland, Brisbane, Australia, [2]University of Tasmania, Hobart, Australia, [3]Monash University, Melbourne, Australia, [4]Australian National University, Canberra, Australia, [5]Australian Synchrotron, ANSTO, Melbourne, Australia

Subduction zones host not only Earth's most hazardous volcanoes but also its largest porphyry copper deposits, which supply ~75% of global copper production. Both the formation of porphyry deposits and volcanic eruptions are linked to magmatic processes within the arc. However, it remains unclear why magma sometimes erupts, sometimes stalls and contributes to crustal growth, or in other cases drives the formation of economic mineralisation.

The Andes hosts some of the world's largest porphyry copper deposits, along with complex volcanic structures. Our study targets three key components of the Andean arc: porphyry Cu-Mo deposits (El Teniente), barren intrusions (within El Teniente district, Central Chile, and global datasets), and a diverse suite of Central Andean volcanoes (stratovolcanoes, cones, and volcanic fields). We analyse magmatic plagioclase—the most common mineral in arc magmas—using novel in-situ microanalytical techniques. We measured major and trace elements, as well as Sr isotopes, using electron microprobe and LA-ICP-MS (quadrupole and multicollector), and assessed textures with petrographic microscopy and X-ray Fluorescence Microscopy.

Our results reveal chemical and textural similarities between plagioclase from mineralised porphyries and volcanoes located in tectonically anomalous segments of the arc. Porphyry-related plagioclase displays dominant oligoclase–andesine compositions, sharp Sr increases with minor Fe variations, and a decoupled relationship between An and Sr. In contrast, crystals from barren intrusions and volcanoes along the main arc show dominant labradorite compositions, typical of arc settings, and Sr–Fe trends indicative of mafic magma recharge. These findings show that the magma sources and processes involved in porphyry formation—likely influenced by subduction geometry and crustal structure—share similarities with those in certain anomalous Andean volcanoes and may also be crucial in metal accumulation. This new approach highlights plagioclase as a key mineral for identifying fertile magmatic systems in active arcs, with direct applications in future mineral exploration.

Insights Into the Petrogenesis of the Taca Taca Bajo Cu-Mo Porphyry Deposit (Argentina) and Its Significance Within the Andean Retro-Arc

Madeleine Ince[1], Steffen Hagemann[1], Nora Rubinstein[2], Marco Fiorentini[1], Tony Kemp[1], Chris Fisher[1], Tim Ireland[3], Santiago Gigola[4]
[1]University Of Western Australia, Crawley, Australia, [2]Universidad de Buenos Aires, Buenos Aires, Argentina, [3]First Quantum Minerals Ltd., Vancouver, Canada, [4]Independent Consultant, Mendoza, Argentina

The magmatic processes that lead to porphyry Cu deposit formation in continental retro-arc settings remain poorly understood, creating uncertainty in exploration within these regions. New case studies are therefore essential for improving targeting success, particularly as copper is considered a critical resource. The Taca Taca Bajo Cu-Mo-Au deposit (11.7 Mt contained Cu), located in the retro-arc of the Central Andes, offers valuable insights as a well-mineralized retro-arc expression of the middle Eocene to early Oligocene metallogenic belt.

Mineralization is spatially and temporally correlated with a series of north to northwest-trending rhyodacitic porphyry dikes, from which six characteristic samples were analyzed for whole-rock geochemistry and zircon petrochronology (U-Pb, O, and Lu-Hf isotopes as well as trace elements). The crystallization ages of these Oligocene rhyodacitic porphyry intrusions range from 30.3 ± 0.5 Ma to 29.1 ± 0.3 Ma (zircon U-Pb SHRIMP). Based on zircon isotopic signatures, the source of the rhyodacitic porphyry dikes is interpreted to be a mantle-derived arc magma (εHf = +5.3 ± 2.4 to +7.6 ± 0.7; δ18O = 5.6 ± 0.5 to 5.8 ± 0.2‰ [2 SD]), which interacted with older, predominantly Permian crustal material (270-250 Ma, i.e., Choiyoi arc magmatism) in the lower- to mid-crust.

Zircon trace element geochemistry data indicates that the Oligocene Taca Taca intrusions are moderately hydrous (mean Eu/Eu* = 0.25 to 0.34) and oxidized (mean ΔFMQ = +0.2 to +1.0), but less than other Cu porphyry deposits within the same middle Eocene to early Oligocene metallogenic belt (e.g., Escondida, Chuquicamata, El Salvador). It is suggested that the Taca Taca Bajo deposit may have failed to reach the supergiant-scale (>24 Mt) potentially due to the isolated, relatively small volume of magma and shorter duration of magmatism (1–1.5 Myr). This is attributed to its position in the retro-arc with thinner crust and less intense compression.

Anatomy of a Moderately-Telescoped Porphyry–High Sulfidation System: Insights of a 2-km Vertical Profile of Alteration at Valeriano Cu-Au Deposit, Chile

Jaime Osorio[1], David R. Cooke[1], Lejun Zhang[1], Ivan Belousov[1], Matt Heizler[2], Robert A. Creaser[3]
[1]University of Tasmania, Hobart, Australia, [2]New Mexico Bureau of Geology and Mineral Resources, New Mexico Tech, Socorro, USA, [3]Department of Earth and Atmospheric Sciences, University of Alberta, Edmonton, Canada

The Valeriano Cu-Au deposit is located within the El Indio metallogenic belt in the Chilean Central Andes. In the last decade, deep drilling beneath the Valeriano lithocap resulted in the discovery of a concealed Cu-Au ± Mo porphyry system at depth, providing a 2-km vertical profile of alteration from the shallow high-sulfidation epithermal to the porphyry environment. Porphyry-style mineralization was temporally and spatially associated with the emplacement of a series of hornblende-biotite-bearing porphyries of granodioritic composition, and related magmatic-hydrothermal breccias between 10.39 ± 0.19 and 10.12 ± 0.18 Ma, based on zircon U-Pb ages. These intrusions at depth produced Cu-Au (>0.7% CuEq) mineralization of chalcopyrite and lesser bornite and molybdenite coupled with K-silicate alteration between 10.55 ± 0.05 and 10.1 ± 0.04, as determined by molybdenite Re-Os ages. K-silicate alteration is overprinted by classic quartz-muscovite ± chlorite (sericitic) alteration containing chalcopyrite and pyrite. This assemblage transitions upward to a thick zone of quartz-muscovite ± pyrophyllite at the base of the lithocap that is intimately associated with low-grade, high-sulfidation state Cu sulfides and Cu-As sulfosalts dominated by bornite, digenite, chalcocite, and minor covellite. The upper part of the lithocap consists of concentrically zoned advanced argillic alteration characterized by different proportions of quartz, alunite, kaolinite, and pyrophyllite, centered on stratiform or structurally controlled zones of massive and vuggy quartz. These zones host discrete Au-bearing enargite epithermal mineralization, which occurs in breccias, veins, as well as open space fillings, which seem to be the remnants of a larger eroded high-sulfidation system. Muscovite and alunite Ar-Ar ages from zones of sericitic and advanced argillic alteration, respectively, yielded ages between 10.151 ± 0.024 and 10.038 ± 0.014, that overlap with the age of porphyry intrusion and associated mineralization, suggesting that porphyry and epithermal mineralization at Valeriano formed over a short period between 690–510 k.y.

Geology, Mineral Alteration and Mineralization of the Santa Cecilia Porphyry Au-Cu Deposit, Maricunga Belt, Northern Chile.

Antonio Arribas[1], Luciano Bocanegra[3], **Jose Franco Moraga**[1], Roman Karczewski[2], Miguel Maldonado[2], Piotr Paleczek[2], Simon Reyes[2]
[1]The University of Texas at El Paso, El Paso, United States, [2]Torq Resources, Vancouver, Canada, [3]Petra Tocall Consulting, Coquimbo, Chile

The Cerro del Medio system, within the Santa Cecilia-Caspiche district of the Maricunga Belt, records magmatic and hydrothermal processes from the late Oligocene to early Miocene (25.2–24.7 Ma). It lies adjacent to the Caspiche deposit (1,091 Mt @ 0.55 g/t Au), separated by a 250 m-wide valley, and formed during the late Oligocene (~25 Ma). This study aims to refine the nature of ore mineralization at the Cerro del Medio target while ongoing research continues on other targets within Santa Cecilia. The objectives include identifying mineral assemblages and the timing of hydrothermal alteration, evaluating Cu-Au grades across alteration stages, and analyzing relationships between alteration and mineralization centers using petrographic, mineralogical, geochemical, and isotopic techniques. The results indicate that Cerro del Medio exhibits a complete porphyry copper-style vertical alteration zonation, consisting of four main zones: (1) biotite-magnetite (potassic), hosting the highest Cu grades; (2) chlorite-white mica (chlorite-sericite); (3) white mica (sericitic); and (4) aluminosilicate (advanced argillic lithocap), with the latter suggesting deeper porphyry-style mineralization potential. Three mineralization stages were identified: (1) magnetite-quartz, with chalcopyrite-rich magnetite veinlets introducing copper; (2) quartz-rich, where disseminated chalcopyrite occurs in quartz veinlets; and (3) sulfide-sulfate, dominated by pyrite-anhydrite veining, which remobilized and diluted copper mineralization. The impact of the sulfide-sulfate stage on copper grade highlights the importance of targeting zones unaffected by this overprint. Comparisons with Caspiche reveal similar alteration and mineralization patterns, suggesting that both systems share a common magma chamber as the source of the porphyry copper systems. This genetic link underscores the regional significance of magmatism in driving hydrothermal activity and highlights the potential for additional high-grade Cu-Au mineralization at Cerro del Medio and the broader Santa Cecilia property. Further studies at the Pircas and Gemelos targets will refine the deposit model and improve the understanding of its magmatic, hydrothermal, and structural evolution.

Discovery of the Ferrobamba Deeps Deposit, Las Bambas, Southern Peru

James Cannell[1], Hugo Rios[1], Paolo Petersen[1], Carl Schnell[1], Rex Berthelsen[2], Alex Santos[1], Julio Cordova[1], Zhaoshan Chang[3]
[1]MMG Las Bambas, Lima, Peru, [2]MMG, Melbourne, Australia, [3]Colorado School of Mines, Denver, US

The Ferrobamba porphyry-skarn deposit (2024 open pit remaining resources: 600 Mt @ 0.57% Cu, 180 ppm Mo, 2.3 g/t Ag, and 0.03 g/t Au) is part of the Las Bambas cluster in the Andahuaylas – Yauri belt. Pre-, intra- and late-mineralisation intrusions span ages from ~38 Ma to 33 Ma, with the skarn and porphyry mineralisation related to two Hornblende Monzonite (MZH) stocks dated from ~35 – 34 Ma. High grade exoskarn mineralisation occurs in the Ferrobamba Formation limestone adjacent to the MZH stocks and in between the other intrusions. Higher tonnage but lower grade stockwork veining ore occurs in all but the latest intrusions. Porphyry alteration and veining zonation studies indicate a single mineralising system centered on a deep, barren, intensely quartz veined core with pervasive K-feldpsar alteration, focused on the southern margin of the main MZH stock. The core is surrounded by a shell of biotite alteration in the intrusions accompanied by mineralised B-type veins, which passes laterally to distal propylitic alteration. The mineralised exoskarn is restricted exclusively within the biotite alteration domain.

Understanding the zoning patterns together with primary and secondary controls on mineralisation in the deposit provided support and aided targeting of the deep extensions to the deposit. Deep drilling campaigns began tentatively in 2020, and rapidly increased in scale culminating in 42,000m drilled in 2024 as the potential of the Ferrobamba Deeps deposit was realised. A Scoping Study in 2024 defined potential underground mining volumes totaling 680 Mt @ 0.37% Cu and 190ppm Mo, 0.02 ppm Au and 1.4 ppm Ag of Measured, Indicated and Inferred Mineral Resource. The resource potential is open at depth and deep drillholes, up to 2000m, indicate mineralisation extends below the scoping mining volumes. Drilling and mining studies are ongoing to advance the potential development of this significant new discovery.

New Discoveries in a Mature Terrain Through Holistic Exploration Practices: A Story of Safford, Arizona

Matthew Wetzel[1], Michaela Young-Mitchell[1], Michael Tharby[1], Matthew Dunlap[1], Simone Runyon[1], William Stavist[1], Paul Albers[1]
[1]Freeport-McMoRan Inc., Tucson, United States

Many mature mineral belts are often written off as being over-explored, and the Laramide porphyry belt of the American Southwest may be considered one of the most "mature" porphyry belts in the world. However, these "mature" terrains still hold significant discovery potential that has been overlooked for decades. These mature terrains have been discounted due to the extensive amounts of near-mine shallow exploration that has occurred because of the Exploration-to-Mining cycle demands on resource evaluation. Once a mine starts production, the emphasis almost always shifts from discovery-focused exploration to resource replacement styles of exploration. This leads to the shelving of quality technical targets and leaves opportunities for the next generation of explorers to uncover large new orebodies that were previously overlooked due to economic forces.

Freeport-McMoRan Inc. has recently focused on re-examining our near-mine opportunities in the Safford porphyry copper district of Arizona, leading to significant new geological and potentially economic discoveries, including one new porphyry center and the discovery of a faulted portion of the Dos Pobres deposit. The use of new technologies, collection of large amounts of detailed geochemistry, improved geophysics, updated detailed geologic mapping, implementation of SWIR and XRD data into 3D space, and rapid targeting updates were critical in the definition of these new porphyry centers. These efforts have also resulted in the identification of three previously unrecognized rock types, stratigraphic controls on volcanic rocks, and much better constraints on structural controls within the Safford district. This integrated holistic evaluation of districts will lead to a new wave of discoveries in mature terrains, which will likely have similar results in Greenfields applications.

Reevaluating the Schultze Granite: A New Temporal Model for the Globe-Miami Magmatic-Hydrothermal System and Its Implications for Porphyry Exploration

Lawrence Carter[1,2], Robert Lee[3], Adam Gorecki[3], Dan Condon[4], David Selby[5], Holly Stein[6,7], Robert Creaser[8], Simon Chenery[4], Matthew Power[9], Benjamin Buse[10], Ester Jolis[11]
[1]LC Geoscience Ltd, United Kingdom, [2]Camborne School of Mines, University of Exeter, United Kingdom, [3]BHP, Tuscon, USA, [4]British Geological Survey, Keyworth, United Kingdom, [5]Durham University, United Kingdom, [6]AIRIE Program, Fort Collins, USA, [7]University of Oslo, Norway, [8]University of Alberta, Canada, [9]Vidence Inc, Burnaby, Canada, [10]University of Bristol, United Kingdom, [11]Geological Survey of Finland, Finland

Porphyry copper deposits form from magmatic-hydrothermal fluids generated by large and long-lived magmatic systems, typically in subduction-related settings. Although extensively studied, our understanding is fragmentary due to a paucity of exposed magmatic root-zones. Key questions remain as to how sufficient volumes of mineralising fluids emanate from parent magmas and are focused into the deposit-forming environment, and over what timescales. To address this, we have integrated the results of field, micro-textural, geochemical and geochronological studies into a self-consistent 4-D framework for the Schultze Granite – previously interpreted to be the exposed root-zone of the Globe-Miami District, Arizona, United States. We show that mineralization within related deposits (Resolution, Pinto Valley, Diamond H and Copper Cities) occurred over a period of ~1.3 Myrs (~65.1-63.8 Ma), which is coincident with the age of a major porphyry dike at Pinto Valley (~64.5 Ma), but significantly pre-dates (by ~1-2 Myrs) emplacement of the Schultze Granite, which occurred rapidly (in <~600 kyrs, from ~62.6 Ma) from an estimated depth of ~20 km. The Schultze Granite did not therefore contribute to the formation of these deposits. Instead, the magma which formed it evolved and upwardly injected aplitic-porphyry dikes which underwent syn-emplacement first-type boiling and fluid exsolution to produce quartz USTs, miarolitic cavities and early DQ and A-type quartz veins. Cu-Mo mineralization occurred where and once these fluids cooled below ~600 °C. The aplitic-porphyry dikes served as fluid conduits for at least 400 kyrs, feeding ~1.1 Myrs (~62.6-61.5 Ma) of hydrothermal mineralization at Miami and the eastward located Ocelot and Old Dominion mineralized system.

Whilst providing significant new constrains for our understanding of hypogene mineralization in porphyry systems worldwide, our new framework demonstrates the value in integrating fundamental field observations with micro-textural study and high-precision geochronology rather than relying on geochemical ratios alone when exploring these multi-faceted systems.

Occurrence of Invisible Au in Carlin-Type Au Deposits

Zhuojun Xie[1]
[1]Institute Of Geochemistry, Chinese Academy Of Sciences, Guiyang, China

A key diagnostic feature of Carlin-type gold deposits is the predominance of invisible gold in arsenian pyrite, contrasting with other gold deposit types. Here, we used the focused ion beam combined with scanning electron microscope (FIB-SEM), spherical aberration-corrected transmission electron microscope (AC-TEM), and atom probe tomography (APT) to examine invisible Au and how it evolved through later geologic events. TEM analyses reveal extensive Au-enriched domains within high-grade arsenian pyrites. These domains exhibit characteristic Au-associated trace element signatures (Sb, Tl, Hg, Cu) in energy-dispersive X-ray spectroscopy (EDS), while showing no detectable metallic nanoparticles. APT data further demonstrate that invisible gold occurs in complex, oscillatory-zoned micro- to nano-scale zones, with Au distributed homogeneously at atomic scales rather than as discrete nanoclusters. Critically, AC-TEM confirms gold's structural incorporation through substitution at iron lattice sites and vacancy positions. The widespread occurrence of stacking faults and nano fluid inclusions in Au-bearing arsenian pyrite indicates rapid crystallization conditions. Curved dislocations pinned by solid inclusions further record subsequent annealing processes.

Combining nanoscale textures with geologic information, we suggest that invisible Au was initially incorporated into the crystal structure of arsenian pyrite during rapid precipitation. Subsequent post-ore thermal events initiated the annealing of the ionic Au-bearing arsenian pyrite, leading to the re-distribution of trace elements and formation of Au-bearing nanoparticles. The dominant occurrence of ionically bound Au in arsenian pyrite indicates that ore-forming fluids were undersaturated with respect to gold during pyrite crystallization. This gold deposition from Au-undersaturated, As-rich ore fluids ultimately produced the giant Carlin-type gold deposits.

Inwards-Younging of the Annular Twihinate Carbonatite-Aillikite Complex (Southern Morocco) and Its Implication for REE Distribution

Lucas Tatnell[1], Marco Fiorentini[1], Mohammed Bouabdellah[2,3], Wissale Boukirou[2], Chris Fisher[1], Andrea Giuliani[4,6], Isra Ezad[1], Laurent Sapor[5]
[1]Centre for Exploration Targeting and ARC Centre in Critical Resources for the Future, School of Earth and Oceans, The University of Western Australia, Perth, Australia, [2]Département de Géologie, Faculté des Sciences, Université Mohammed Premier, Oujda, Morocco, [3]Geology and Sustainable Mining Institute, Mohammed VI Polytechnic University, Benguerir, Morocco, [4]Institute for Geochemistry and Petrology, Department of Earth Sciences, ETH Zürich, Zürich, Switzerland, [5]Groupe Managem, Casablanca, Morocco, [6]Department of Terrestrial Magnetism, Carnegie Institution of Science, Washington, United States of America

Twihinate, situated in the Oulad Dlim Massif on the northwestern margin of the West Africa Craton, is an excellent example of a ring-shaped carbonatite complex with an outer calcite carbonatite ring and a core of dolomite carbonatite and aillikite. Calcite carbonatite tends towards higher proportions of biotite, magnetite, diopside±andradite towards the boundary between calcite carbonatite and granitic gneiss host rock ("antiskarn"). Here, we employ a variety of targeted spatially constrained geochronologic tools (mica Rb-Sr in calcite carbonatite and antiskarn, apatite U-Pb in antiskarn, zircon U-Pb in syenite, and perovskite U-Pb in aillikite) to investigate the sequence of events and timeframe of emplacement of Twihinate. Antiskarn and calcite carbonatite ages at Twihinate range from 89.0 ± 1.3 to 83.5 ± 1.3 Ma, younging towards the centre of the complex. Aillikite from a dike crosscutting the dolomite carbonatite in the centre of Twihinate is dated at 82.0 ± 0.5 Ma. Crosscutting relationships show that dolomite carbonatite dikes crosscut calcite carbonatite, therefore broadly constraining dolomite carbonatite emplacement between aillikite and calcite carbonatite (although there were probably several batches of dolomite carbonatite). The inwards-younging trend of carbonatite and aillikite at Twihinate is interpreted as resulting from repeated pulses of carbonatite and aillikite melt injected into the same lithospheric pathway over ca. 7 million years, and implies that liquids are channelled towards the centre of annular carbonatite structures. Global carbonatite complexes with similar geometry might have been constructed through a similar process of repeated pulses of carbonatite and related melts, younging inwards. This process can foster the re-dissolution, extraction and concentration of critical elements including REE in late-stage melts and fluids hence explaining why REE are distributed in the centre of the complex at Twihinate and in other major carbonatite complexes such as Mt Weld in Australia.

Data Integration of Hyperspectral and Whole-Rock Geochemistry: New Insights into Mt Weld Carbonatite-Hosted REE Deposit

Siyu Hu[1], Carsten Laukamp[1], Ross Chandler[2], Naina Goswami[1], Bobby Pejcic[1], Michael Wawryk[3]
[1]CSIRO Mineral Resources, Perth, Australia, [2]Australian National University, Canberra, Australia, [3]Geological Survey of Western Australia, Perth, Australia

Rare earth elements (REE) are essential for producing current and developing future technologies. Understanding existing REE deposits and identifying new viable REE resources is critical due to fragile supply chains and increasing demand in coming decades. The Mt Weld carbonatite-hosted REE deposit is one of the richest REE deposits in the world, however, remains relatively poorly understood.

This study extrapolates previously characterized major mineralogy and REE mineral distribution of selected samples from the fresh carbonatite, to an entire drill core (MWEX10270), providing unbiased and contiguous distribution of major carbonates and REE-host minerals. This is achieved by translating hyperspectral signatures – collected with HyLogger3™ technology - into relative mineral abundance and chemical composition. Six distinct zones are recognized within the central carbonatite: (I) weathering profile, (II) magnesiocarbonatite, (III) phosphate-siderite-rich magnesio- and ferro-carbonatite (MF carbonatite), (IV) phosphate-rich MF carbonatite, (V) phosphate-poor MF carbonatite, and (VI) calcio-carbonatite. Whole-rock geochemical reassessment, based on the newly defined zones, reveals a gradual variation in λ values (a quantitative descriptor of REE patterns) of MF carbonatite REE geochemistry, reflecting the progressive magmatic fractionation from Mg-rich to Fe-rich carbonatite magma. Meanwhile, the total concentration of REE is initially elevated in Mg-rich carbonatite but subsequently decrease as phosphorus concentrates within Fe-rich carbonatite. REE precipitate as monazite during the hydrothermal evolution of each magmatic stage, transitioning to REE-fluorocarbonates in the late-stage magma when REE and phosphorus become depleted.

The application of Hylogger3™ for Mt Weld carbonatite analysis demonstrates that the HyLogger is an effective tool for rapidly identifying REE-hosting minerals and their distribution precisely in fresh carbonatite drill cores. This study also establishes new knowledge that is important to better understand global carbonatite-hosted REE deposits and future explorations.

Geology and Genesis of the REE-Bearing Ironstones of the Yin Complex, Gascoyne Province, Western Australia

Lachlan Anderson[1], John Mavrogenes[1,2], Marissa Higgins[1], Ross Chandler[1,2], Luke Blais[3]
[1]The Australian National University, Canberra, Australia, [2]ARC Centre in Critical Resources for the Future, Canberra, Australia, [3]Dreadnought Resources, Perth, Australia

Part of the larger Gifford Creek Carbonatite Complex (Capricorn Orogen, Western Australia), the Yin dykes feature over 43km of rare earth element (REE)-rich ferrocarbonatite. Mineralogically, the dykes are dominated by Fe-carbonates (siderite and/or ankerite) at depth, transitioning to goethite-dominated "ironstones" near the surface, interpreted as forming via the supergene alteration of the underlying ferrocarbonatite by oxidised meteoric waters. While superficially similar along strike, there are distinct differences in the REE-bearing mineralogy between different sections of the dyke, with the ferrocarbonatite at Yin Central being bastnäsite-(Ce)-rich, and the Yin North ferrocarbonatite being monazite-(Ce)-rich. This study investigates the Yin Complex ironstones, their REE-bearing mineral assemblages and genesis via near surface weathering.

At Yin Central, weathering fluids caused replacement of Fe-carbonates by goethite and dissolution of other primary carbonates, fluorapatite and barite; with accompanying loss of soluble components (CO32-, Ca, Mg, lesser F, Sr and S). The key primary REE-bearing mineral, bastnäsite-(Ce), was replaced in-situ by rhabdophane-(Ce) following reaction with PO34- released by fluorapatite dissolution. In contrast, at Yin North, primary monazite-(Ce) resisted alteration, and remnant primary monazite is the principal REE phase in the ironstones.

Bulk geochemistry and porosity, as summarised in a weathering isocon, suggest that the light REE (LREEs) are immobile on a deposit-scale during weathering and are enriched only due to mass loss of other elements, in contrast to other styles of supergene mineralisation. Total mass loss during ferrocarbonatite weathering (~20%, as estimated from porosity) was limited by the high concentration and immobility of Fe. LREE enrichment during weathering was correspondingly low (~1.19x) compared to the extreme enrichments observed at supergene calcio- or magnesiocarbonatites elsewhere in Australia and globally (2x to 10x enrichment).

Preservation of Na-REE Mineralisation at Kangankunde Carbonatite Complex, Malawi, with Implications for Magma Evolution

Nicholas Mappin[1], Sam Broom-Fendley[1], Frances Wall[1]
[1]University Of Exeter, Cornwall, United Kingdom

Carbonatites account for the majority of global light rare earth element (LREE) supplies and are vital in the transition towards low-carbon technology. Current models for REE-enrichment in carbonatites are conflicting, with some invoking hydrothermal REE transport, whilst others support extreme magmatic fractionation to an alkali- and REE-rich brine, commonly referred to as 'brine-melts'.

Conceptually, brine-melts are distinct from typical hydrothermal fluids owing to a lack of observable evidence for immiscibility between a melt and hydrothermal fluid. Experimental results suggest evolved carbonatite melts should be alkali enriched, but high alkali contents are rarely preserved in natural carbonatites. Their absence may be due to the water-soluble nature of Na-carbonate phases, which are rapidly lost during weathering and consequently their role in REE-mineralisation is overlooked. Here, we use core samples from the Kangankunde Complex, Malawi, which sample below the carbonatite weathering horizon, enabling the preservation of primary Na-REE-phases. Burbankite (Na-REE-carbonate), which has previously been postulated to act as a precursor to monazite mineralisation at Kangankunde, is described here for the first time. It occurs in all samples, preserved as inclusions. Moreover, other alkali-carbonate phases, such as bradleyite (Na-carbonate-phosphate) and eitelite (Na-carbonate) are also preserved, indicating that late-stage carbonatite at Kangankunde is enriched in Na, Sr, Ba, P, and LREEs. Ephemeral alkali components are subsequently removed, and the primary mineral assemblage is replaced by more stable phases. Our results move towards reconciling the gap between experimental studies and natural examples, confirming the importance of alkalis in an evolved carbonatite magma and highlighting their influence on primary rare-earth mineralisation.

Rare Earth Element Ion Adsorption Clay Deposits: A Mineral Systems Approach for Exploration

Samantha Russo[1,2], Ignacio González-Álvarez[2,3,4], Helen A. Cocker[2], Alex J. McCoy-West[1,2]
[1]IsoTropics Geochemistry Laboratory, Earth and Environmental Science, James Cook University, Townsville, Australia, [2]Economic Geology Research Centre, James Cook University, Townsville, Australia, [3]CSIRO, Mineral Resources, Discovery Program, Perth, Australia, [4]University of Western Australia, Perth, Australia

The transition to a 'green energy' future requires increased production of critical elements, such as rare earth elements (REE). REE Ion Adsorption Deposits (IACD) have sparked international interest as a heavy rare earths (HREE) source, supplying ~80% of global HREE. Therefore, a mineral systems approach is applied herein to constrain REE source, mobility, deposition, and preservation required for IACD formation. REE IACD are formed through intense weathering, where REE are liberated from REE-bearing minerals (e.g., REE-(fluoro)carbonates and REE-phosphates) and mobilised within the weathering profile before adsorbing onto clay minerals (e.g., kaolinite and halloysite). Granitic rocks are common IACD source rocks, however, other igneous, metamorphic, and sedimentary rocks, as well as REE-rich fluids and brines, are potential REE sources for in-situ and/or transported weathering required for IACD formation.

Once liberated, REE are mobilised within the weathering profile as complexes with ligands (e.g., F-, CO_3^{2-}, SO_4^{2-}, PO_4^{3-}) or existing as hydrated REE species. REE complexation and fractionation are strongly controlled by pH, where non-fractionating complexes dominate under acidic to circumneutral conditions (e.g., hydrated REE mineral species and REE-sulfate complexes) and fractionating complexes dominate under circumneutral to alkaline conditions (e.g., REE-carbonate and -phosphate complexes). Therefore, pH and REE complexation control REE adsorption onto clay minerals, and relative light REE (LREE)- and HREE-enrichments throughout the weathering profile. REE adsorption is most favourable at circumneutral pH, where competition with H+ ions is minimal, and hydrolysis is limited. REE adsorption, and therefore IACD, are preserved under low erosional settings. Climates with excessive rainfall (e.g., tropical humid climates) are problematic for preservation, where excessive rainfall promotes clay dissolution. The conceptual model based on a mineral systems approach provided herein presents a framework to be built upon in the coming years as our knowledge and global exploration of IACD continue.

The Cradle and Future of Rare Earth Elements

Ignacio Gonzalez-Alvarez[1, 2, 3]
[1]CSIRO, Mineral Resources, Discovery Program, Perth, Australia, [2]University of Western Australia, Centre for Exploration Targeting, Perth, Australia, [3]James Cook University, Earth and Environmental Science, Queensland, Australia

The world is experiencing rapid geopolitical, financial, and technological shifts, alongside a fast-tracked transition to green technologies. Rare earth elements (REE), especially Nd, Dy, Tb, and Pr, are central to this transition due to their role in Strategic technology production. Permanent magnets, accounting for ~45% of REE use in 2023, are critical for energy, defence, and digital competitiveness.

In the 1980s, the U.S. was the largest REE supplier. In 2024, global Rare Earth Oxide (REO) production reached ~390K t, with reserves estimated at ~90 Mt. The Indo-Pacific dominates the REE sector, producing ~85% of global output and holding ~65% of reserves. However, only two major producers, Northern Rare Earth (China) and Lynas Rare Earths (Australia), were profitable in 2024, with REE prices declining since 2022. This volatility has created uncertainty, prompting the development of new supply chains and advancements in REE science from exploration to processing. The Indo-Pacific's REE backbone includes carbonatite deposits in China and Australia, clay-hosted REE in Myanmar and China, and mineral sands in India and Australia. Vietnam's hydrothermal deposits, with ~3.5 Mt of reserves, produced ~300 t of REE in 2024, showcasing future development potential. Myanmar exported ~45K t of REE to China in 2024, representing ~11.5% of global production.

Carbonatites account for ~80% of global REE production, while heavy mineral sands (~35% in garnet-zircon-monazite content) contribute ~40% of global reserves, mainly mined in India and Australia. Ionadsorption clay deposits in Brazil and South China, including Caldeira (~410 Mt @ 2,625 ppm) and Rocha da Rocha (510 Mt @ 1,513 ppm), offer new opportunities. This presentation contextualises key variables in the REE ecosystem, from exploration to processing, supply chain architecture, and geopolitical dynamics in the new multi-polar financial system, to enable informed, low-risk assessments.

Automating Lithology Prediction Using Python and ioGAS: A Streamlined Workflow for Geochemical Analysis

Ingrid Flemons[1], Adam Eskdale[1], Michael Whitbread[1]
[1]Rio Tinto Exploration, Australia

Geochemical analysis has traditionally relied upon manual methods, such as exploratory data analysis workflows, to interpret and define relationships between chemical data and the feature of interest, be it rock type, alteration, mineralisation or other parameters. These methods have evolved within mineral exploration to incorporate advanced similarity techniques (e.g., UMAP), enhancing the ability for geoscientists to handle large multi-element datasets.

Handling complex and diverse geochemical datasets still brings complications, such as missing or poorly recorded values and complex spatial relationships between variables. Greater controls and standardized workflows are essential to facilitate geoscientific efforts to leverage these datasets successfully.

This project aims to streamline the workflow for predicting lithology by leveraging similarity algorithms within ioGAS software and novel data science techniques powered by Python. Our approach maintains user control within the familiar ioGAS environment, which includes many published geochemical relationships and robust visualization capabilities that effectively communicate data within the geology and exploration communities. This retains the excellent capabilities within ioGAS to link plots and attribute data intuitively, keeping it accessible to geoscientists. Python packages can add ever-expanding algorithms and code bases (both public and in-house) to this environment.

By utilizing freely available Python packages alongside ioGAS tools, we developed a proof-of-concept workflow that integrates data science into geology and geochemistry processes within a large organization with much reduced intermediate product file handling. Key outcomes from this include: (i) a more efficient workflow for lithological prediction; (ii) reduced time-on-task for the responsible geoscientist, increasing capacity for stronger interpretation; (iii) a demonstration of potential future advancements in this field; (iv) evidence for the wider applicability of combined supervised-unsupervised workflows to facilitate successful discovery; (v) making data science techniques in Python accessible to non-Python proficient geoscientists.

From Data to Discovery: Targeting Sedimentary Copper in South Australia Using a Mineral Systems Approach

Genevieve Luketina[1], Tom Czertowicz[1]
[1]Kenex Ltd, Lower Hutt, New Zealand

Sedimentary copper deposits are globally significant but remain relatively underexplored in Australia. The Adelaide Superbasin in South Australia hosts known occurrences but has seen little systematic targeting using modern mineral systems frameworks.

Here, we applied a data-driven mineral systems approach to identify new opportunities for sedimentary copper. Historic drilling, surface geochemistry, geology, and geophysics were compiled into a unified GIS database, and weights of evidence modelling was applied to generate a sedimentary copper prospectivity map.

This process identified several key geological controls, including the Tapley Hill Formation as a critical stratigraphic host unit, tillite as an additional host rock, mafic sediments and volcanics as significant copper sources, diapiric breccias as a proxy for a brine source, major faults and crustal boundaries as transport pathways, and Cu-Co anomalies in drillholes and rock chips as evidence for metal deposition.
A prospectivity map was generated and used to delineate high-priority targets, which were compared against existing tenements, leading to strategic ground acquisition and discussions with tenement holders regarding potential partnerships.

Field validation confirmed high-grade copper and silver mineralisation at a top-ranked target and identified untested historic workings. These results demonstrate the effectiveness of systematically combining geological knowledge with data-driven prospectivity modelling.

Our work highlights that significant near-surface exploration opportunities remain available in South Australia, challenging assumptions that exploration must shift under deep cover. It also underscores the value of revisiting historic datasets with modern, systematic approaches to generate actionable targets.

Technological Innovations in Exploration and Mining – Success Stories from a Copper Miner

Michaela Young-Mitchell[1], Michael Tharby[1], Nicole Arnush[1], Matthew Dunlop[1]
[1]Freeport-McMoRan Exploration Corporation, Tucson, United States

Technological advances applicable to the mining industry have expanded, and with increased demand for data, Freeport-McMoran Exploration looks at advances in analytical geochemistry, mineralogy, geometallurgy, geomechanics, core scanning, software, data analytics, and AI/ML, and has successfully applied them to exploration efforts, ore characterization, and downstream products for mine planning, blasting, and geometallurgy.

We built a centralized core processing facility and mineralogy lab capable of processing up to 10,000 samples/month, collecting trace element and quantitative mineralogy data with XRF, XRD and SWIR on 10-foot/3-meter drillholes samples. The data complements assay and trace element geochemistry from the analytical results for over 50 elements and analytes, also on every sample. Combined with high-resolution photography, core scanning, data analytics and machine learning, application of these technologies enables us to perform auto-RQD measurements and detail logging with more accuracy, consistency, and efficiency, improve the identification and differentiation of lithology and alteration, leading to better targeting/discoveries. We added items in our resource models that help predict downstream processing issues, e.g., calcite for acid consumption or pyrite for acid generation on leach pads, feldspars and quartz affecting hardness for blasting and comminution, clays affecting milling and blending/routing, and deleterious elements that affect flotation.

To handle the significant increase in data, we expanded our GIS capabilities for compiling and viewing geologic mapping, surface geochemistry, geophysics, drillholes, and land status. We are merging data from multiple sites into a unified database to improve consistency and efficiency in logging and oretype characterization. Data accessibility is enhanced by innovative platforms such as Power BI and web-based viewing platforms for our database software. With these developments, we continue to enhance the orebody knowledge of our mineral deposits, which improves targeting, routing, and processing decisions, and positions us well for strategic growth opportunities.

Prospectivity Mapping of Gold Deposits in the Qinling Orogen Belt: A Data-Driven Framework

Haicheng Wang[1], Ruiwen Shen, Liangsheng Ge, Zhili Du
[1]China Geological Survey, Beijing, China

In the context of increasingly scarce resources and growing environmental constraints, there is an urgent need to adopt advanced technological approaches to enhance exploration success rates and efficiency. In recent years, supervised machine learning methods have emerged as powerful tools for mineral prospectivity mapping, which fully leverage the spatial distribution and metallogenic characteristics of known deposits to train predictive models. Data-driven mineral prospectivity mapping offers new ideas and methodologies for delineating target areas in mineral resource exploration. This study focuses on the Qinling Orogenic Belt as a case study to illustrate the strong adaptability and considerable potential of data-driven methods for gold prospectivity mapping, which is a one of China's most significant gold metallogenic zones. Numerous large and medium-sized gold deposits have been discovered in this region, indicating significant potential for further exploration. This study aims to develop a data-driven prediction model that integrates geological, geochemical, and geophysical data to enhance the accuracy and efficiency of delineating prospectivity targets, providing quantitative support for gold exploration and decision-making. A multi-dimensional index system was constructed, including metallogenic geological factors, geochemical indicators, and geophysical anomalies. The known gold deposits within the study area were used as positive training samples, while negative samples were selected from areas far from known mineral deposits and areas with unfavorable metallogenic conditions to form a training dataset.

The results showed that stratigraphy, faults, intermediate-acidic volcanic rocks, and geochemical anomalies exhibit high importance in the trained model, indicating their significant roles in controlling gold mineralization. Based on the predicted results, several prospective gold targets were delineated. Field investigations of these targets revealed distinct indications of mineralization, confirming the applicability and reliability of the proposed model. These results demonstrated strong adaptability and considerable promise of data-driven modeling, injecting new vitality into geological exploration workflows.

Using Open-Access Remotely Sensed Data to Support Greenfield Mineral Exploration: Methods and Lessons from Case Studies

Mana Rahimi[1]
[1]Mineral Deposit Research Unit (MDRU) - University Of British Columbia (UBC), Vancouver , Canada

The availability of open-access remote sensing data is increasingly becoming an important information source for mineral exploration strategy and planning, especially in greenfield areas. These datasets offer the ability to rapidly assess geological features, alteration patterns, and structural controls across large and often inaccessible areas. Greenfield regions often lack detailed ground mapping, making remotely sensed observations a critical first step toward defining geological features linked to understanding subsurface mineralization.

Classic and advanced methodologies for extracting value from open-access remotely sensed data to support mineral exploration are reviewed; traditional techniques such as band ratios, spectral indices, principal component analysis, and structural interpretation are widely used for mapping alteration zones and geological features. Advanced workflows apply machine learning algorithms and fuse satellite imagery with airborne geophysical and elevation datasets to enhance exploration targeting. The best-suited methodologies are selected depending on the exploration scale, the geological context, and the targeted mineralization style. The advantages and challenges associated with the remote data processing methods are discussed, offering practical insights into their application in greenfield settings.

Case studies from mineralized regions demonstrate how using open-access remotely sensed data, combined with integrated exploration methods, can strengthen targeting decisions, improve exploration efficiency, and help reduce geological uncertainty in greenfield environments. These examples highlight the advantages and the practical challenges of applying remotely sensed data in real-world exploration projects.

SMX - Sustainable Mining Explorer Platform: A Geoscience-Based Approach for ESG Risk Management in Mining

Vinicius Louro[1]
[1]CSIRO Mineral Resources, Kensington, Australia

The exploration, extraction, and processing of minerals significantly interact with the environment and nearby communities, with impacts varying by commodity. Each stage of the mining lifecycle—from initial geophysical surveys and drilling to waste management and mine closure—carries distinct environmental and societal implications. This highlights the increasing importance of Sustainability and ESG (Environmental, Social, and Governance) practices in mining. Although Australia maintains one of the world's most comprehensive geoscientific data collections, this resource remains underutilized in addressing ESG-related challenges, particularly in the context of Critical and, to a lesser extent, Strategic Minerals. Mining strategies directly influence a project's environmental footprint and are increasingly evaluated under ESG frameworks, and briefly outlined in the revised Australian JORC (Joint Ore Reserves Committee) Code.

In response, CSIRO is developing the Sustainable Mining Explorer (SMX) platform, which integrates precompetitive geoscientific data and physical models to assess the social and environmental implications of mining activities. SMX enables users to select a geographic region in Australia, specify a mine type, and explore ESG-related topics to obtain preliminary insights into the potential challenges and opportunities associated with mine development in that area. This abstract introduces the SMX platform and presents its initial results, focusing on two ESG topics from the 'Air' category in Attachment 1 of the JORC Code: Dust Production and Noise, the latter of which has been expanded to include seismic noise.

From Ore to PV: A Sustainable Tellurium Value Chain for Net-Zero Goals

Homay Fath[1], Simon Jowitt
[1]Nevada Bureau of Mines and Geology, United States

A global shift to renewable energy depends on securing a sustainable supply of critical metals and minerals, but the mining industry, essential for providing these materials, faces escalating environmental, economic, and social challenges. We highlight the role of critical materials in low and zero-CO2 energy storage and generation technologies using the example of tellurium in solar photovoltaic technologies, emphasizing the importance of building a resilient and agile supply chain in delivering critical materials and achieving decarbonization. Thin-film cadmium-telluride (CdTe) photovoltaics have emerged as a promising alternative to conventional silicon solar panels. However, the cadmium and tellurium required to produce these modules are almost entirely obtained as byproducts from base and precious metal mining operations. This study presents a proxy-based approach to estimate the hidden and unrealized tellurium potential in some active gold and base metals mining operations. We emphasize that improved knowledge from studies of existing ore deposits containing tellurium, followed by coordinated actions by governments, industry players, and other stakeholders, are crucial to ensure a reliable supply of materials vital for decarbonizing mining energy systems. Enhanced domestic recovery efforts can markedly decrease import reliance, strengthen supply-chain resilience, and accelerate the global energy transition. Moreover, the proxy-based methodology introduced here provides a versatile framework for assessing the recovery potential of other critical byproduct metals across diverse mining operations.

Critical Minerals Potential in the Delamerian Orogen Margin: Exploring the Sustainable Prosperity

Yanbo Cheng[1], Adrienne Brotodewo[2], Sebastien Meffre[3]
[1]Geoscience Australia, Canberra, Australia, [2]Mineral Exploration Cooperative Research Centre (MinEx CRC) and Future Industries Institute, University of South Australia, Adelaide, Australia, [3]ARC Research Hub for Transforming the Mining Value Chain & Centre for Ore Deposit and Earth Sciences, University of Tasmania, Hobart, Australia

The collaboration between Geoscience Australia's Exploring for the Future (EFTF) program, Resourcing Australia's Prosperity (RAP) initiative, and MinEx CRC's National Drilling Initiative (NDI) marks a significant advancement in Australia's geoscientific research and resource exploration. The EFTF program (2016–2024) enhanced the understanding of Australia's geology, supporting economic growth and sustainable resource management by providing comprehensive pre-competitive geoscientific data. Building on EFTF's success, the 35-year, $3.4 billion RAP initiative, launched in July 2024, focuses on accelerating the discovery of critical minerals, groundwater, and other resources. It aims to support a resilient and sustainable resource sector through innovative data, research, and analysis. The NDI, managed by MinEx CRC, complements these efforts by conducting extensive drilling campaigns to map regional geology and identify potential mineral systems, leveraging advanced drilling technologies to maximize data collection and improve geological knowledge.

In the Delamerian Orogen, this collaboration is pivotal. The Cambrian-Ordovician-aged Delamerian Orogen, formed during the transition from a passive margin to an active subduction margin along Gondwana's eastern edge, is one of Australia's largest and most under-explored provinces. Despite being mostly covered by younger rocks and sediments, mineralisation has been encountered in drilled areas, particularly along the eastern margin. This collaborative project analyses the in-situ chemistry of sulfides and alteration minerals from key prospects across the Delamerian Orogen aiming to assess critical mineral potential across the metallogenic province. The in-situ analyses highlight distinct differences in sulfides and alteration chemistry between the different deposit styles across the region with showing strong variations in their regional prospectivity of multiple critical minerals which may cause by complex paragenesis and element deportment mechanisms. These efforts, integrating new geophysical data, stratigraphic drilling, and geochemical analysis, aim to provide a new understanding of the Delamerian Orogen's critical mineral potential, aiding mineral exploration and supporting sustainable resource management in the region.

Evaluating the Tectonic and Climatic Controls on Sediment-Hosted Ore Deposits in Deep Time

Sheree Armistead[1], Simon Williams[2], Bruce Eglington[3]
[1]CODES, University of Tasmania, Hobart, Australia, [2]Institute for Marine and Antarctic Studies (IMAS), University of Tasmania, Hobart, Australia, [3]Department of Geological Sciences, University of Saskatchewan, Saskatoon, Canada

Sediment-hosted ore deposits are the world's largest source of the critical mineral cobalt, the second largest source of copper, and an important source of base metals such as lead and zinc. These deposits are essential to supporting the global energy transition, yet the processes that control their formation remain poorly understood.

This study integrates a newly updated global database of sediment-hosted ore deposits with a full-plate tectonic reconstruction spanning the last billion years. By linking the timing and location of known deposits to evolving tectonic settings and paleoclimatic conditions, we aim to quantitatively assess the key environmental and geodynamic criteria that favour ore formation. Using plate reconstruction software and global paleoclimate models, we evaluate paleolatitudes, tectonic regimes, and climate proxies through time to better understand the spatial and temporal patterns of deposit formation. In doing so, we test long-standing assumptions about the relationship between ore formation and subtropical evaporite belts and explore the potential influence of major climate events such as Neoproterozoic glaciations. Preliminary findings suggest that a range of climatic and tectonic factors may influence the formation and preservation of these systems.

We also present new results from numerical analyses that show statistically significant spatial correlations between sediment-hosted deposits and fossil rift margins. These findings support the idea that long-lived extensional settings, and their subsequent reactivation, play a key role in creating favourable conditions for Cu-Co mineralisation.

Together, these results contribute to a new quantitative global framework for understanding where sediment-hosted deposits are most likely to occur. This work has direct implications for identifying underexplored but highly prospective sedimentary basins in Australia and globally, with relevance for critical mineral exploration strategies.

Insights into the Delamerian Orogen Mineral Systems through Pyrite Chemistry.

Adrienne Brotodewo[1,2], Yanbo Cheng[3], Sebastien Meffre[4]
[1]Future Industries Institute, University of South Australia, Australia, [2]Mineral Exploration Cooperative Research Centre (MinEx CRC), Australia, [3]Mineral System Branch, Geoscience Australia, Australia, [4]ARC Research Hub for Transforming the Mining Value Chain & Centre for Ore Deposit and Earth Sciences, University of Tasmania, Australia

MinEx CRC is the world's largest collaborative mineral exploration research initiative, uniting industry, government, and research organisations. As part of its mission, the National Drilling Initiative (NDI) brings together geological surveys, researchers, and industry to drill in under-explored regions with high mineral potential, while also testing new exploration technologies.

The Cambrian–Ordovician Delamerian Orogen in central and eastern Australia is one such region, spanning five states but largely concealed beneath younger sediments. Limited surface exposure has led to minimal exploration, despite known mineral occurrences. To address this, two NDI drilling campaigns have been conducted in partnership with MinEx CRC member organisations. These include case studies in eastern South Australia in collaboration with the Geological Survey of South Australia and in the Loch Lilly–Kars area of New South Wales, in partnership with Geoscience Australia through the Exploring for the Future (EFTF) program.

The Delamerian Orogen formed during a significant tectonic shift being the transition from passive to active margin along eastern Gondwana. Known magmatic-hydrothermal and other mineral prospects occur along the eastern Delamerian margin therefore a larger collaborative project looking into these systems to assess critical mineral potential across the metallogenic province was established. As part of this research effort, samples from porphyry Cu-Au, volcanic-hosted massive sulfide (VHMS), orogenic gold, and orthomagmatic Ni-Cu-PGE systems from South Australia, Victoria, and New South Wales are assessed. In-situ analyses of pyrite show variations in texture and chemistry across these different deposit styles suggesting a complex paragenesis. Additionally, pyrite from mineralised systems not yet fully classified are being studied, offering new insights into their genesis. Together, these findings are shedding light on the broader mineral potential of the Delamerian Orogen and are key to guiding future exploration across this under-explored yet prospective province.

Solubility of Alkali Elements and Chlorine in Sulfide Melts Under Magmatic Conditions: Implications for Ore Genesis

Maria Cherdantseva[1], Michael Anenburg[2], John Mavrogenes[2], Marco Fiorentini[3]
[1]The Centre For Exploration Targeting, The University of Western Australia, Crawley, Australia, [2]Research School of Earth Sciences, Australian National University, Canberra, Australia, [3]ARC Centre in Critical Resources for the Future, The University of Western Australia, Crawley, Australia

A notable characteristic of magmatic sulfide occurrences within mafic intrusions is their frequent spatial association with hydrous and alkali- and chlorine-bearing minerals. The reason for this consistent proximity has been a subject of debate. Although prior interpretations have often invoked post-magmatic fluid activity or late alteration processes, the specific affinity of these mineral types for sulfides has persisted as an open question, suggesting a possible fundamental genetic relationship underlying these occurrences. To explore this potential link, we conducted experiments focused on quantifying the capacity of sulfide melts to incorporate sodium, potassium, and chlorine under conditions mimicking those found within the Earth. The outcomes of our investigations demonstrated a substantial degree of dissolution for alkali metals (reaching concentrations up to 3.5 wt %) and chloride (up to 0.3 wt %) at elevated temperatures (1100–1250 °C) and high pressures (0.5–1 GPa), under oxygen fugacities typical of the Earth's mantle (ΔFMQ = –2 to 0). From these experimental findings, we infer that potassium, sodium, and chlorine may play a significant role in the mobilization of metals from the mantle source. This could be accomplished through their influence on reducing the melting temperatures of sulfides and by facilitating their upward movement via density reduction. Subsequently, during the cooling and solidification of these sulfide melts, the incorporated elements would be released. This expulsion could lead to the enrichment of adjacent silicate minerals or contribute to the formation of a later fluid phase. Consequently, this mechanism offers a potential explanation for the observed mineral assemblages and could also enhance the capacity of magmatic-derived hydrothermal fluids to form ore deposits.

The Queensland Mineral Deposit Atlas Series

Tom Evans[1], Rick Valenta[1], Paul Gow[1], Nick Cook[1], Sally Allan[1], Jaime Poblete[1], Katerina Savinova[1], Nathan Fox[1], Jenna McGovern[1], David Esser[1], Juliana Vilhena[1], Mark Hinman[1], Sasha Aivazpourporgou[1,2], Vladimir Lisitsin[2], Al-Tamini Tapu[2], Courteney Dnharam[2]
[1]Sustainable Minerals Institute, University Of Queensland, Indooroopilly, Australia,
[2]Geological Survey of Queensland, Department of Resources, Brisbane, Australia

The Queensland Mineral Deposit Atlas Series has been created as part of the Department of Resources' Strategic Resources Exploration program and New Economy Minerals Initiative. The overall aim of this project is to assemble all relevant geoscientific studies of the key mineral deposits across Queensland, and to summarise the important features of each deposit.

This work details 69 mineral deposits in 52 chapters, across 2 volumes. Each chapter comprises a document summarising the deposit, a collection of open file data for the deposit and a 3D model of that data that can be opened in the free version of the Geoscience Analyst software. Data is compiled from the Geological Survey of Queensland's Open Data Portal, published academic literature, post graduate theses and company publications.

This compilation is intended to provide a technical reference and conceptual framework that exploration and mining companies can use to select ground and design exploration programs. Access to organised, curated datasets of all the significant mineral deposits in the state helps streamline the work of exploration geologists, allowing more time and resources to be spent completing exploration programs.

Alkaline Magmatism in the Adang Volcanic Complex, Mamuju District (Western Sulawesi, Indonesia): A Preliminary Study on Rare Earth Elements Enrichment

Fadlin Fadlin[1], Renaldi Suhendral[2], Godang Shaban[3], Muhammad Rayyan[1], FX Anjar Laksono[4]
[1]Geological Engineering Department, Jenderal Soedirman University, Purwokerto, Indonesia, [2]Research Center for Geological Resources, National Research and Innovation Agency, Bandung, Indonesia, [3]Independent Geochemist, Geochemistry Tectonic Modelling and REE Researcher, Jakarta, Indonesia, [4]Doctoral School of Earth Sciences, University of Pécs, Pecs, Hungary

Alkaline igneous rocks are one of the major sources of rare earth elements (REEs). The largest REE deposits are primarily discovered in alkaline rocks and carbonatites (e.g., Bayan Obo, China; Mountpass, California; Mountweld, Australia), which typically form in post-collisional and continental rift-related tectonic environments. The Adang Volcanics is one of the uncommon volcanic systems that resulted from continental extensional settings, composed of mafic alkaline rocks, which have alkalinity ranging from ultrapotassic to sodic and have not been explored for their relationship to the enrichment of REEs. This study investigates the processes behind the formation of the Adang volcanic rocks and their role in REE concentration. Multiple analytical methods were used in this study, including Petrography, XRD, SEM-EDS, XRF, and ICP-MS. The results of petrographic observation show that the samples are mostly composed of leucite-bearing minerals characterized by textures such as trapezohedral and skeletal forms. The whole-rock geochemistry data indicate that the Adang Volcanic developed in within-plate continental extension, which magmatism is derived from decompression of the enriched SCLM [characterized by $TiO_2 > (-1.1610 + 0.1935 \times Al_2O_3)$] and in certain zones appears to have involved the post-collisional environment. The metasomatism processes have substituted the leucite ($KAlSi_2O_6$) with analcime (($NaAlSi_2O_6H_2$), resulting in a reduction of potassium and other mobile elements, including REEs. The magmatism resulted from the enriched SCLM, which has auto-generated the higher Th/Nb, Th/La, and Th/Ce ratios. The alkaline rocks of Adang Volcanic have a high REE content, with total REE (TRE+Y) ranging from 1069.2 to 1454.7 ppm, compared to the concentration of the REEs in the Earth's crust, which ranges from around 130 to 240 ppm. The enrichment of REEs in this formation is driven by metasomatism processes and secondary upgrading through hydrothermal alterations, which are embedded in REEs-bearing minerals (e.g., Florencite (La-Ce-Th), Loparite (La-Ce-Th), and Johnsenite (Ce-Th)).

National-Scale Critical Mineral and Strategic Material Potential Assessments to Support Decision-Making in the Net Zero Transition

Arianne Ford[1], Jonathan Cloutier[1]
[1]Geoscience Australia, Canberra, Australia

With the projected increase in demand for critical minerals and strategic materials required to meet net zero objectives by 2050, a series of national-scale mineral potential assessments are being delivered as part of the Resourcing Australia's Prosperity initiative at Geoscience Australia. In addition to providing an evidence-base for supporting government decision-making, these precompetitive assessments aim to support the discovery of deposits that host critical minerals and strategic materials by reducing the exploration search space and promoting increased exploration investment. The mineral potential assessments integrate both new and legacy geoscience data and aim to delineate prospective belts or districts where favourable geological, geophysical, and geochemical criteria that represent evidence of relevant mineral systems processes are present. To date, assessments for nickel-copper-platinum group element, sediment-hosted base metal, carbonatite-related rare earth element, and iron oxide-copper-gold mineral systems have been undertaken at a national-scale by Geoscience Australia as part of previous programs, with additional mineral systems currently being assessed. The methodology used in each assessment is reviewed on a case-by-case basis due to the widely varying availability of known deposits and occurrences that can be used to train data-driven models using statistics or machine learning for each mineral system. Using a mineral systems-based approach as a basis, the Resourcing Australia's Prosperity initiative aims to deliver a series of commodity potential maps for all of Australia's critical minerals and strategic materials over the 35-year duration of the initiative. An example of how this may be achieved for copper will be presented, with an interim map highlighting Australia's copper potential based on the mineral systems modelled to date.

Phreatomagmatic Ash as a Natural Drill Core: Unraveling Subvolcanic Hydrothermal Systems in Active Arc Volcanoes

Muhammad Andriansyah Gurusinga[1,2,4], Tsukasa Ohba[2], Takashi Hoshide[2], Wildan Nur Hamzah[2], Gabriela Nogo Retnaningtyas Bunga Naen[1], Indranova Suhendro[5,6], Dini Nurfiani[3], Aoi Hirata[2]
[1]Center for Regional Revitalization in Research and Education, Akita University, Japan, [2]Department of Earth Resource Science, Akita University, Japan, [3]Research Center for Geological Disaster, National Research and Innovation Agency of the Republic of Indonesia (BRIN), Indonesia, [4]Department of Geological Engineering, Faculty of Mineral Technology, Universitas Pembangunan Nasional "Veteran" Yogyakarta, Indonesia, [5]Department of Environmental Geography, Faculty of Geography, Universitas Gadjah Mada, Indonesia, [6]Center for Disaster Studies (PSBA), Universitas Gadjah Mada, Indonesia

Active arc volcanoes host dynamic hydrothermal systems that exhibit alteration processes similar to those in porphyry-epithermal ore environments. Phreatomagmatic eruptions—driven by magma-water interaction—can excavate material from depths reaching hundreds of meters to ~2 km, in contrast to the shallower origins of phreatic eruptions. These events eject lithic fragments from active alteration zones, providing a unique opportunity to investigate subvolcanic hydrothermal processes and fluid-rock interactions. This study analyzes altered lithic fragments from Bromo and Ciremai volcanoes (Indonesia), supplemented by comparative data from the Akita Yakeyama and Chokai volcanoes (Japan). The samples are characterized using petrography, componentry, SEM-EDS, and X-ray diffraction (XRD) to identify and characterize the alteration textures and assemblages.

The results reveal three distinct alteration regimes: [1] Ciremai samples are dominated by alunite (SO_3 ~45 wt%), silicate-rich (SiO2 ~ 97.7 wt%) assemblages, kaolinite-group minerals, irregular to laminated pyrite, and rare copper oxide grains observed (30.85% in isolated particles). The Akita Yakeyama 1997 samples exhibit similar quartz-alunite-kaolinite associations. Both indicate high-sulfidation conditions under acidic, steam-heated regimes. [2] Chokai samples are characterized by smectite and pyrophyllite dominance, reflecting intermediate argillic alteration at near-neutral pH conditions. [3] Bromo samples feature Mg-Fe phyllosilicates (chlorite with MgO ~23.4 wt%, illite with K_2O ~5.45 wt%), Fe-Ti oxides (magnetite, ilmenite with TiO_2 ~70.5 wt%), and Zinc enrichment (ZnO ~ 2.8 wt%). This assemblage suggests higher-temperature alteration compared to the acidic systems in Ciremai and Akita Yakeyama.

The distinct alteration assemblages on Ciremai - Akita Yakeyama, Chokai, and Bromo indicate different hydrothermal conditions beneath these arc volcanoes. Ongoing investigations on additional Indonesian (Marapi, Lewotobi, and Ruang) and Japanese (Aso) volcanoes will integrate altered/juvenile particle ratios, mineralogical fingerprints, and host magma characteristics to advance our understanding of how modern hydrothermal systems develop in arc settings, with implications for both volcanic hazards and the fundamental ore-forming systems.

Empowering Exploration with National Scale Maps

Nichole Knepprath[1], Eloise Beyer[1], Guillaume Sanchez[1], Colleen Bryant[1], Madeline Buddee[1], Murray Woods[1], Lindsay Highet[1], Cathy Brown[1], Marie-Aude Bonnardot[1], Jackie Hope[1]
[1]Geoscience Australia, Symonston, Australia

Geoscience Australia (GA) is committed to enabling a strong resources sector through its landmark Resourcing Australia's Prosperity initiative. A focus for this initiative is integrated and standardised geoscientific mapping to improve understanding of what lies on and beneath the land surface. National-scale surface and subsurface geological maps and derivative thematic maps that highlight geology of interest provide the framework needed to map Australia's onshore resource potential, including for critical minerals and strategic materials, thereby reducing exploration risk and accelerating new discoveries. Digital flagship products such as the national-scale Layered Geology of Australia and Alkaline Rocks Atlas of Australia are compiled from published datasets at a variety of scales and with geological information harmonised across state and territory borders. The 1:1 million scale Layered Geology dataset extrapolates Australia's surface geology beneath cover to produce seamless geology for five time slices. Geological units captured in these layers are attributed with rich information sourced from the Australian Stratigraphic Units Database. Augmenting these layers with depth points using GA's Estimates of Geological and Geophysical Surfaces database will add that critical third dimension and fundamentally change the exploration landscape, opening up that depth component and broadening the resource search space. Higher resolution products such as the Alkaline Rocks Atlas capture smaller scale units of interest, like alkaline igneous rocks, which are volumetrically minor but a significant source of critical minerals such as rare earth elements. Thematic maps like this can provide vectors to regions of potential mineralisation and are important inputs for national resource prospectivity assessments.

Chromite and Sulphide Mineralization of the Uitloop Ultramafic Bodies in the Northern Limb of the Bushveld Complex, South Africa

'Mabatho Mapiloko[1], Marina Yudovskaya[1,2]
[1]University Of Witwatersrand, Johannesburg, South Africa, [2]Institute of Geology of Ore Deposits, Mineralogy, Petrography and Geochemistry, Russian Academy of Sciences, Moscow, Russia

The Uitloop farm, located in the northern limb of the Bushveld Complex, represents a geologically complex and understudied region. Historically, the area was described as hosting two Lower Zone (LZ) offshoots; Uitloop I and II, intruded into the Archaean basement and Transvaal Supergroup. However, recent deep drilling has revealed presence of a Platreef offshoot as well as Marginal Zone sills.

Differentiating the LZ from the Platreef Unit is challenging due to overlapping mafic/ultramafic lithologies, chromite, and abundant sulphide mineralization. However, distinct mineral chemical signatures (olivine, chromite, and pyroxene) and whole-rock geochemical trends enabled stratigraphic discrimination. Correlations with adjacent Turfspruit and Macalacaskop regions confirmed alignment of the Uitloop Platreef offshoot with the main Platreef Unit, even though it lacks the Main Zone hanging wall and is instead commonly enveloped by Duitschland Formation metasediments. Mineralization in the Uitloop Platreef offshoot exhibits classic 'Platreef-style' patterns; (i) contact-style mineralization, marked by elevated Ni-Cu and PGE concentrations in sulphide-rich magmatic sequences and footwall rocks, and (ii) reef-style stratabound mineralization associated with chromitite seams in the upper magmatic sequences.

Sulphur isotope analyses of Uitloop sulphides revealed assimilation of ca. 2.4 Ga marine sulphate evaporites from the Duitschland Formation. Thus, we suggest that during and after the Great Oxygenation Episode, sulphate evaporites became an important S source to the mafic/ultramafic magmatic systems, facilitating formation of world-class Ni-Cu-PGE mineralization.

The global demand for Ni and Cu surged, driven by the rapid adoption of renewable energy technologies because of stricter emissions regulations aimed at decarbonizing economies. While demand for PGMs, traditionally dominated by the automotive industry, has declined amid the shift toward battery electric vehicles, emerging technologies such as hydrogen fuel cells and green hydrogen production are poised to revitalize PGE markets. Detailed investigations of poorly understood magmatic sulphide intrusions are essential to secure future metal supplies.

Archaean BIFs from Karnataka, India: genetic insights from the Dharwar craton, India, and implications for post-depositional hematitic iron deposits

Perumala Venkata Sunder Raju[1]
[1]National Geophysical Research Institute, HYDERABAD, India

The inferred genesis of the Dharwar craton encompasses initial ocean plateau formation leading to an early oceanic crust upon which widespread BIF sedimentary rocks were deposited in the Western Dharwar Craton (WDC); later deformation and metamorphism during cratonisation led to distinct greenstone belts and embracing granitisation, but where the once sedimentary successions now extensively altered are correlatable and coeval. The BIFs themselves include both oxidised and enriched deposits of iron as well as extensive sulphidic facies BIF.Here we provide a brief overview of the BIF deposits of the Dharwar craton of India, and their genesis, and then shift to a brief discussion of the implication of their unique geodynamic history for a later hydrothermal iron deposit enrichment, as postulated for the Pilbara craton. Pilbara (Western Australia) and the Kaapvaal (Southern Africa) cratons encompass relatively minor BIF lithologies within crustal greenstone belts, which were deformed, metamorphosed as the proto-cratons grew through island arc complex accretion and late stage and voluminous granites enveloped these to form the initial cratonic basement. These early crusts are overlain by widespread sedimentary successions of chemical and clastic affinity, largely laid down in shallow marine environments under initially low continental freeboard conditions, including widespread sheetlike and very large BIF deposits such as those of the Hammersley (Pilbara) and Transvaal (Kaapvaal) basins. These BIF represent enormous oxide facies iron ore sources, initially believed to have formed at c. 2.7-2.6 Ga, and enriched and oxidised to hematitic ore bodies during the ‘Great Oxidation Event’ at c. 2.2-2.0 Ga. However, recent work on Pilbara (with some minor comparisons to Kaapvaal), applying direct dating of platy hematite in the ore deposits clearly demonstrates that the oxide mineralisation dates to a much later, c. 1.4-1.1 Ga geodynamically driven hydrothermal enrichment event.

Deciphering the Role of Remobilisation in the Formation of High-Grade Gold Deposits

Sumail[1,2], Nicolas Thebaud[2], Quentin Masurel[2], Julien Perret[2], Denis Fougerouse[3]
[1]Model Earth Pty Ltd, Perth, Australia, [2]Centre for Exploration Targeting, The University of Western Australia, Perth, Australia, [3]School of Earth and Planetary Sciences, Curtin University, Perth, Australia

High-grade native gold mineralisation in gold deposits is often attributed to secondary enrichment processes, such as gold remobilisation. Evidence for such processes is typically indirect and often restricted in elucidating their spatial extent and their relative contribution to the overall gold endowment. In this study, we investigate the trace element composition of native gold coupled with the characterisation of pyrite textures and chemistry to gain insights into the formation of native gold at the high-grade Jundee gold deposit. This innovative approach reveals gold mineralisation was polyphased within a single veining phase. The main mineralisation event is hosted in quartz-carbonate shear vein (VJ2B), characterised by the early formation of invisible gold associated with oscillatory zoned pyrite (PyVJ2B) mantles that formed from an As-rich fluid. Subsequent PyVJ2B growth was characterised by the formation of As-poor rims surrounding the PyVJ2B mantles and the coetaneous deposition of extensive native gold during the same veining event (VJ2B). Native gold occurs in distinct habits, including native gold inclusions in PyVJ2B along with extensive native gold grains external to PyVJ2B. Differences between the trace element composition of native gold inclusions within PyVJ2B and native gold external to PyVJ2B indicate that not all native gold associated with the main mineralisation stage at Jundee formed via secondary enrichment processes and therefore manifest a primary introduction. Mass balance calculations further demonstrate that the latter introduction of native gold accounts for the high-grade nature of the Jundee deposit with limited contribution from remobilisation processes.

Geochemical Signature of Au-Cu Mineralization in Soil Profiles of the Wadi Al Jaww Prospect and Applications for Exploration, Saudi Arabia

Hassan Abbas[1]
[1]Ma'aden, Saudi Arabia

A geochemical soil survey was carried out over the Wadi Al Jaww prospect located in the western part of the Arabian Shield between the Hijaz and Jeddah terranes. The area is approximately 150 km southwest of Maaden's Mahd Ahd Dahab mine and 150 km northeast of Jeddah city, and 200 km south of Madinah city. No historical work was reported at Wadi Al Jaww. The Maaden discovery was made from exploration geochemistry methodologies including silt, soil and rock samples. The gold and copper mineralization are hosted by the Shiwan complex of tonalite to granodiorite composition belonging to Kamil suite. The soil samples were collected from the A horizons of a soil profile with an 800- x 200-m grid spacing and have been statistically examined using descriptive statistics, bivariate correlations, cluster and principal component analysis to determine elemental associations. The first PC explains 47% of data variation and outlines the association of Au-Cu-Te. Gold is the main indicator of mineralization in soil profiles. Its concentrations reach up to 2.39 ppm, indicating the position of close proximity to mineralization. While gold may be mobile in arid regions where the dissolution of elemental Au occurs by oxic, acidic, chloride-rich waters, the close spatial association of soil anomalies and hypogene gold mineralization indicates limited mobility of gold over the Wadi Al Jaww prospect. Copper is believed to be highly mobile in arid environments and usually incorporated into Fe-, Mn oxides and hydroxides. The composition of soil profiles and the arid environment at project area explains the distribution of metals in soils. The soil geochemistry survey was the main contributor for discovery of project Au-Cu and outlined association of elements could be used as a guide for future discoveries of similar types of mineralization associated with intrusive complexes along suture zones of Arabian Shield.

Applications of Mineral Chemistry to Petrogenesis and Exploration in the Borralan Carbonatite Complex and Associated Alkaline Silicate Rocks

Abdulkadir Mohamed Abdulkadir[1], Hilary Downes[2], Sudeshna Basu[1], Dan Barfod[3], Ian Millar[4]
[1]Department of Earth Sciences, University College London, United Kingdom, [2]School of Natural Sciences, Birkbeck University of London, United Kingdom, [3]NERC Argon Isotope Facility, East Kilbride, United Kingdom, [4]Geochronology and Tracers Facility, British Geological Survey, United Kingdom

The Loch Borralan Carbonatite Complex (LBC) and associated alkaline silicate rocks in NW Scotland provide insights into the interplay between mineral chemistry, petrogenesis, and rare earth element (REE) enrichment in carbonatite systems. Mineral chemistry (LA-ICP-MS, EMPA), stable isotopes (C, O, N and Ar), radiogenic isotopic tracers (Sr-Nd-Hf-Pb) and geochronology (U-Pb, Ar-Ar) were used to investigate the LBC rocks. Mineral chemistry revealed that multistage REE mobilisation and deposition was driven by fluorine- and carbonate-rich magmatic-hydrothermal fluids. Early magmatic phases (apatite, perovskite, chevkinite) sequestered REEs via carbonate complexation, while later oxidising-hydrothermal events triggered remobilisation into secondary assemblages (bastnäsite, monazite, ancylite). Redox contrasts are revealed by coexisting sulphides (pyrite, pyrrhotite) and sulphates (barite, celestine) that localised REE deposition along structural conduits (e.g., fenite veins with high REE). Phlogopite zoning and replacement textures (perovskite→chevkinite→ fluorocarbonates; perovskite→lmenite, calcite and ancylite; monazite→britholite) chronicle episodic fluid influx and evolving REE availability. Geochronology links LBC emplacement (~424–438 Ma) to Caledonian subduction with inherited zircons (480–553 Ma) and enriched mantle isotopic signatures (εNd_{430}: −1 to −2), suggesting that subduction of oceanic crust modified mantle sources and controlled formation of the carbonatite-associated REE system. Radiogenic isotope data and analysis of abundance and isotopic composition of C, O, N and Ar revealed that formation of carbonatite and initial enrichment was linked to fluids from recycled marine sediments, although primary mantle signatures are preserved. Mineral chemistry defines key vectors to guide extraction strategies: (1) fluorite-rich fenites mark fluid pathways, (2) sulphide-sulphate redox interfaces with high REE-rich zones, and (3) refractory vs. leachable REE phases (zirconolite vs. fluorocarbonates). The complex mineralogy necessitates advances in beneficiation tailored to mineral speciation. This work demonstrates that interpreting mineral-scale processes can bridge the gap between petrogenetic models and orebody knowledge, emphasizing the role of ligand-driven fluid evolution and tectonic fluid histories in targeting carbonatite-hosted REE deposits.

Exploring for Cu-Au Alkalic Porphyry Deposits from the Fringes

Fabrizzio Abello[1], Shaun Barker[1], Farhad Bouzari[1], Devin Wade[2]
[1]MDRU, Vancouver, Canada, [2]New Gold Inc, Toronto, Canada

The late Triassic - early Jurassic Iron Mask batholith (IMB) in southern British Columbia (BC) hosts several alkalic porphyry Cu-Au mineralized centers such as New Afton, Ajax, Copper King, and Python. This research identifies mineral and chemical footprints of the alkalic porphyry deposits in the IMB through fieldwork, petrography, geochronology, and geochemistry. Results were compared with the footprint of the nearby calc-alkalic Nora and other BC porphyry deposits.

Pathfinder elements in the studied alkalic porphyry deposits extend horizontally from a central core of Pd, Pt, Cu, and Au (~100m wide) to Te, Re, and Se (~700m wide) and further distal V, As, and Sb (~1 km wide). Vertically, at New Afton, a higher concentration of Mo occurs at depth (~750m deep) along with depleted Zn and Mg. This is overlain by a zone of high concentrations of Cu, Au, Pd, and Pt. Near the surface, Te, V, As, and Sb have higher concentrations. Consistent with element zonation, the mineral chemistry of pyrites, chlorites, and epidotes shows distinctive anomalous elements (e.g., Te, Se, Co, V) distribution with proximity to mineralized centers. Similarly, mineral chemistry inferred from short-wave infrared spectroscopy (SWIR) shows zonation in chlorites and white micas, with Mg-Fe-rich chlorites and phengitic white micas proximal to ore and Fe-Mg-rich chlorites and structurally controlled muscovitic white micas distal.

Results corroborate previous studies that hydrothermal alteration in alkalic porphyry deposits is more cryptic than calc-alkalic deposits. Porphyry emplacement and alteration are typically asymmetric and structurally controlled. Similarly, anomalous pathfinder element values are generally less pronounced than those in the calc-alkalic porphyry deposits, except for Te, Pd, As, Ag, and V, which have higher concentrations. These results help to identify the signature of the alkalic porphyry mineralization early during exploration and vector toward potential mineralization.

Mineralization and Ore Genesis of the Neoproterozoic Pan-African Belt, Eastern Ghana: Constraints on Lithology and Mineralogy.

Rauda Addae[1,2], Frank Awua[2], Huayong Chen[1], David Kumah[2]
[1]Guangzhou Institute Of Geochemistry, Tianhe, China, [2]Ghana Geological Survey Authority, Accra, Ghana

From recent field exploration programs by the Ghana Geological Survey Authority, seven distinctive mountains suspected to be mineralized have been established. Of interest to this study is the Akokrowa range which spans the areas, Akokrowa, Asukawkaw and Okanease in the eastern part of the Ghanaian Neo-Proterozoic Pan-African belt. Evidence from petrology and mineralogy makes room for three broad rock classifications: clastic sedimentary rocks, chemical sedimentary rocks and mafic rocks. Two phases of alteration are observed, a pervasive low-grade greenschist metamorphism and an overprinting pumpellyite alteration, both of which are associated with propylitic, phyllic and intermediate argillic alterations. Serpentinization and albitization are worth mentioning. The minerals of interest in these rocks are iron oxides and oxyhydroxides: magnetite, hematite, ilmenite, chamosite and goethite, which are mainly hosted in BIF and ferruginous clastic and chemical sedimentary rocks, and serpentine (including cronstedtite) with associated Cr-spinel (Mg-Al-Cr-Fe and Mg-Si-Cr-Fe), greenalite-Ni, birnessite, and minnesotaite that are mainly hosted by the mafic rocks. Associated sulphides include pyrite, chalcopyrite and pentlandite. Apatite, florencite, barite, talc, manganese oxides and manganese carbonates are common accessory minerals. Mineralization is mainly structurally controlled and disseminated. The studied BIF intercalated with iron rich shales, sandstones and cherts, and mafic intrusions may have formed in restricted basins associated with widespread volcanic and hydrothermal activities of cyclic pulses in spreading centers and/or developing continental rifts, where Fe-rich oxyhydroxide minerals are precipitated from hydrothermal fluids which are later diagenetically transformed to hematite and magnetite whereas, the interbedded chert bands may have precipitated from sea water, hydrothermal vent fluids, or as a result of secondary replacement. The deposition of these precipitates may have occurred distally from the hydrothermal vent source, with its associated ferruginous clastic rocks forming from the reworking of BIF in shallow waters.

Sulfide Dendrites in Silica Scales at the Salton Sea Geothermal Field, California

Aaron Adsit[1], Thomas Monecke[1], Santiago Rocha[2]
[1]Center to Advance the Science of Exploration to Reclamation in Mining, Department of Geology and Geological Engineering, Colorado School of Mines, Golden, United States, [2]CyrQ Energy, Salt Lake City, United States

The Salton Sea geothermal field is located within an active pull-apart basin developed by the San Andreas fault - Imperial fault transtensional step-over in in southern California. The field is hosted by a ~10-km-thick succession of metamorphosed sedimentary rocks and an overlying <5-km-thick succession of Holocene fluvial deposits that were intruded by bimodal volcanic rocks. The geothermal reservoir associated with the volcanic activity has been exploited since the 1960s. The production fluid is unusually saline with a high total dissolved solid content of up to 32 wt. % and base metal concentrations of up to 8 ppm Cu, 790 ppm Zn, and 84 ppm Pb occurring in solution.

Silica scales collected from a production well at the Hudson Ranch 1 geothermal power station contain abundant sulfide minerals in a light brown, smooth, finely banded, non-crystalline silica matrix. The sulfides form multi-branching dendrites that are up to one millimeter in size. Bulk samples of the scales contain up to 7.72 wt.% Zn, 4.09 wt.% Pb, and 5819 ppm Cu. The scales are layered with some layers containing primarily sphalerite and galena dendrites while others are dominated by chalcopyrite. Textural evidence suggests that dendrite growth originally occurred in a gel-like matrix, allowing diffusion of solutes from the fluids flowing over the top of the scale to the site of dendrite growth. At the location of scale formation, two-phase flow conditions occurred at temperatures of ~260°C, with vapor forming ~8 – 12 wt. % of the production fluid. The wellhead sampled was rimmed only 12 months prior to sampling, suggesting that silica deposition and ore mineral dendrite growth occurred rapidly. These observations provide important constraints on the ore-forming processes in epithermal deposits.

What Do the Pyrites Tell Us About the Young-Davidson Orogenic Gold Deposit, Matachewan, Ontario, Canada?

Oya Ak[1], Daniel Gregory[1]
[1]University Of Toronto, Toronto, Canada

The Archean Superior Province within the Canadian Shield plays a key role in Canada's gold production, accounting for approximately 88% of the country's total output. A particularly important region within this province is the Cadillac-Larder Lake Deformation Zone, located in the southern Abitibi greenstone belt of Northern Ontario. The Young-Davidson (YD) orogenic gold deposit in Matachewan, Ontario, Canada is located at the western end of this zone. This deposit primarily hosts gold within syenite, which is surrounded by a widespread hydrothermal alteration halo. The main zones of mineralization include quartz-carbonate and syenite veins with sulfide mineralization, often accompanied by intense hematite alteration within the syenite.

This research focuses on refining the genetic model of the YD deposit and identifying key mechanisms responsible for gold precipitation, using a combination of macro- to micro-scale techniques. We utilize a combination of core logging, reflected and transmitted light microscopy, geochemical analysis, and sulfur isotope analyses from various lithologies, alteration styles, and mineralization types within and distal to the ore zone. The results highlight pyrite as the principal gold-hosting mineral, with high-grade gold commonly found as inclusions in disseminated pyrite, particularly in quartz-rich zones. Whole-rock geochemical data show a strong positive correlation between gold content, pyrite abundance, and total sulfur.

To better understand the deposit's evolution, pyrite textures were examined using reflected light microscopy, and their geochemical characteristics were compared with whole-rock and sulfur isotope analyses. LA-ICP-MS trace element and sulfur isotope (S32, S33, S34; SIMS) analyses were conducted to determine the origin of the gold and to link specific generations of pyrite to gold mineralization. These integrated approaches aim to advance the understanding of hydrothermal gold deposition processes and contribute to the development of predictive tools for estimating gold grades ahead of mining activities.

Geological Features and Diagnostic Characteristics of Gunung Akmil HSE Prospect in Central Java, Indonesia: Vectoring to Concealed Porphyry System

Andreas Rama Alfario[1], Tesalonika Novelita Kate Pongtuluran[1], Azzaini Ghoni[1], Muhammad Farel Diputra[1], Arifudin Idrus[1]
[1]Universitas Gadjah Mada, Sleman, Indonesia

The Gunung Akmil high sulphidation epithermal (HSE) prospect in Central Java's Menoreh range, hosted by andesite and diorite porphyries, displays concentric alteration zones and mineralization patterns indicative of a concealed porphyry system. Structurally controlled by ENE- and NNW-trending faults, the prospect is characterized by four alteration zones progressing from peripheral propylitic (chlorite-epidote±pyrite) to argillic (kaolinite±dickite), advanced argillic (quartz-pyrophyllite-white mica-dickite), and a central silicified zone (quartz-goethite). Silicification intensity and stockwork vein density systematically increase toward the center, paralleled by rising gold grades from 0.03 g/t (distal) to 0.44 g/t (central). Diagnostic minerals like pyrophyllite and dickite in the advanced argillic zone reflect acid sulfate alteration, typical of HSE-porphyry transitions, while the scarcity of alunite suggests deep erosion, potentially exposing deeper system levels. Ground magnetic surveys reveal a high-susceptibility anomaly, interpreted as magnetite-enriched altered rock beneath the HSE zone-a signature consistent with porphyry copper-gold environments. These findings align with established exploration vectors: zonal alteration patterns, increasing metal grades toward structural centers, stockwork density correlating with fluid proximity, and high-temperature mineral assemblages. The integration of lithocap features (HSE mineralization) with geochemical and geophysical data underscores the prospect's potential for concealed porphyry-style mineralization, particularly in the central zone. This multi-method approach highlights the efficacy of combining alteration mineralogy, geochemistry, and geophysics in targeting buried mineral systems as a priority for deeper exploration to confirm porphyry linkages.

Keywords: Gunung Akmil HSE prospect, concealed porphyry, vectoring to ore, Menoreh range

Initial Investigation of Geology and Litho-Geochemistry in North Halmahera, Indonesia: Implications for Porphyry Copper and Epithermal Deposit Potential

Achmad Zulfan Almahdy[1], Fadlin[1], Raden Muhammad Asfaro Fajri[1]
[1]Jenderal Soedirman University, Purwokerto, Indonesia

The Halmahera Volcanic Belt on Halmahera Island may have valuable mineral deposits, similar to the famous Gosowong gold deposit. Our research focuses on Galela in North Halmahera, which has not been previously explored for gold. We aim to identify its potential through geological mapping and geochemistry methods. In this study, we use several methods: petrography, ore microscopy, X-ray diffraction, fire assay-atomic absorption spectrometry (FA-AAS), and X-ray fluorescence (XRF). Our mapping lithologies include pyroclastic breccia, andesite lava, diorite porphyry, diatreme breccia, and limestone. Diorite porphyry is a key host rock for mineralization, and we classify it as hornblende diorite, showing a variety in grain size textures: fine, medium, and coarse textures, associated with calc-alkaline magma. We found two prospect areas of interest: the North and the South. In the North Prospect, we suggest a possible porphyry copper deposit where the hydrothermal fluids are associated with multiple intrusive diorite hornblende with textural variation from fine- to coarse-grained, associated with diatreme breccia. The significance of hydrothermal alteration of this prospect can be classified into two types: chlorite-epidote-magnetite and quartz-calcite±sericite±biotite embedded in both diorite hornblende and diatreme breccia along with the diverse mineralization of sulfide minerals such as pyrite, sphalerite, covellite, chalcocite, and chalcopyrite. Furthermore, according to FA-AAS analysis, data from rock samples shows a copper anomaly level of 228 ppm. The South prospect has different characteristics, with hydrothermal activity influenced by faults. Here, we see pyroclastic breccia intruded by fine-grain diorite hornblende associated with hydrothermal alterations like quartz-alunite alteration and dickite-kaolinite-illite alteration spread following the pyroclastic breccia distribution, which is a unique feature. This prospect area is commonly associated with a quartz vein brecciated system that contains sulfide minerals such as pyrite and galena, with a gold level of 21.6 ppm, indicating potential for an epithermal deposit.

Nature of the Parental Source Material(s) for the Placer Gold-Bearing Banket Conglomerate, Ghana: Constraints from δ18Owater in the Quartz Pebbles

Edwin Animah Obeng[1], Kofi Adomako-Ansah[1]
[1]University Of Mines and Technology, Ghana., Tarkwa, Ghana

Many hydrothermal sources can be invoked for the parent quartz vein materials that were eventually eroded to supply the quartz pebbles for the Banket conglomerate. These fluid sources could have signatures that suggest magmatic-hydrothermal (+6 to +10‰ δ18Owater), seawater-hydrothermal (~0±2‰ δ18Owater), meteoric-hydrothermal (<0‰ δ18Owater), metamorphic-hydrothermal (+4 to +25‰ δ18Owater). These fluids are associated with particular environments or deposit-type such as epithermal gold deposits (meteoric with or without contributions from magmatic fluids), Mesothermal and hypothermal gold deposits (metamorphic and/or magmatic), gold-rich Volcanogenic Massive Sulphide deposits (seawater, with or without contributions from magmatic fluids), porphyry copper-gold deposits (magmatic).

Klemd et al. (2015) showed that the quartz pebbles from the Banket conglomerates mostly contain high density CO2-N2 fluid inclusions, with rare or no aqueous H2O-NaCl at temperatures > 300 to 450 ºC considering immiscibility in the CO2-N2-H2O system and/or ductile environment within the deep crust and a temperature of < 350ºC for dominant H2O-CO2-NaCl composition in a brittle environment. Therefore, based on this deduction, a temperature of 350-450ºC was considered for evaluating the δ18Owater in the Banket conglomerates in this study. Result for fluid preserved in quartz-pebbles (+5.9 to +10.1‰) shows that the calculated δ18Owater values fall within the range for magmatic and/or metamorphic water. This suggest the source(s) of the quartz-pebbles were originally hydrothermal quartz veins from Mesothermal lode-gold and/or porphyry copper-gold deposits. However, the lack of copper in the Banket conglomerate eliminates porphyry copper-gold deposits as source material. Therefore, ancient Mesothermal lode-gold deposits are the most likely sources for the gold-bearing reefs of the Banket conglomerates.

Although Mesothermal lode-gold deposits are genetically similar, deduction structural analysis across the Birimian-Tarkwaian contact by Perrouty et al. (2015), suggest that these eroded ancient Mesothermal lode-gold deposits are older than the currently known Mesothermal lode-gold deposits along the left flank of the Ashanti metavolcanic belt.

Geology and Geochemistry of Nikolauz Mafic-Ultramafic Intrusions, Peru: A Gondwanide Ni-Cu Sulfide Mineralized Conduit System in an Andean Orogenic Environment

Abraham Arana[1,2], Cesar Fonseca Ferreira Filho[2], Jonas Mota e Silva[1]
[1]Rio Tinto, Lima, Peru, [2]Universidade de Brasilia, Brasilia, Brazil

The demand of Ni and Co as critical metals renewed the interest in the exploration of deposits associated with mafic-ultramafic magmas. Orogenic environments underexplored for Ni-Cu-PGE deposits provide important opportunities for exploration. The Nikolauz intrusion, located in the Eastern Cordillera of Peru, is the first Ni-Cu sulfide mineralized conduit system reported within an Andean orogenic setting. It occurs as elongated bodies up to 4 km long, consisting of gabbroic rocks crosscut by ultramafic pulses emplaced into metamorphic schists of the Marañon Complex. Despite pervasive alteration of igneous olivine-pyroxene-plagioclase, magmatic textures are largely preserved, revealing tubular bodies with well-developed zoning consisting of adcumulus dunites at the core, which grade into peridotites, pyroxenites and gabbros toward the margins. Ni-Cu mineralization hosted within the ultramafic rocks consists of sulfide disseminations, blebby textures, and net-textured aggregates. Whole-rock geochemistry and petrography indicate that the Nikolauz intrusion consists mainly of variably altered olivine, pyroxene and plagioclase cumulates, with dunite as the principal ultramafic rock, evidenced by high MgO (>35wt.%) and Cr (>2500ppm) contents. Ultramafic rocks enriched in chalcophile elements (Ni: 2500–6000 ppm; Pd+Pt: 20–150 ppb) consistently exhibit elevated Cu/Zr (>100) and Ni/MgO (>50) ratios. Trace element systematics reveal high LREE/MREE ratios in the ultramafic rocks (La/Sm > 5) and a calc-alkaline affinity in the mafic-ultramafic suite (MgO/FeOt = 0.5–0.8; La > Nb; Th/Yb = 0.1–10; [Nb/Th]PM < 1). U-Pb zircon geochronology yields crystallization ages between 312 and 323 Ma, linking the Nikolauz magmatism to post-compressional processes related to the late Gondwanide event. Geological and geochemical evidence suggest an origin by partial melting of a pyroxenitic mantle source during lithospheric decompression triggered by slab break-off from Gondwanide orogenic. The recognition of a Ni-Cu sulfide mineralized dynamic conduit system in the orogenic environments of the Peruvian Andes highlight the potential for Ni-Cu deposits along the Eastern Cordillera.

REE and HFSE Fractionation and Mobility in the Tapira and Salitre Carbonatite Complexes Southeast, Brazil

Jo Hannah Asetre[1,2], Ignacio Gonzalez-Alvarez[1,3], Siyu Hu[1], Claudio Gerheim Porto[4]
[1]CSIRO, Mineral Resources, Kensington, Australia, [2]Macquarie University, Macquarie Park, Australia, [3]University of Western Australia, Perth, Australia, [4]Federal University of Rio de Janeiro, Rio de Janeiro, Brazil

Brazil produces ~90% of the global niobium (Nb), which is primarily used in high-strength steel and superconductor production. Globally, most Nb and rare earth elements (REE; ~80%) deposits are associated with carbonatites.

The Alto Paranaíba Igneous Province (APIP) in southeastern Brazil hosts several alkaline-silicate and carbonatite complexes known for their exceptional concentrations of niobium (Nb), phosphorus (P), titanium (Ti), and REE. Among these, the Araxá complex is the world's primary Nb source. Other key localities, such as Tapira and Salitre, contribute significantly to P extraction and remain relatively underexplored, despite their recognised potential for Ti, Nb, and REE.

This study is a preliminary assessment of REE and HFSE mobility within the bedrock and weathering profiles of the Tapira and Salitre complexes. Through integrated petrological and geochemical analyses, key elemental gains and losses across the weathering gradient are identified and linked to the mineralogical transformations that shape these geochemical signatures.

Our results reveal REE enrichment in the Cover from ΣREE ~0.03 wt% to 1.1 wt% in fresh rocks vs. 0.37 to 2.4 wt% in the saprolite. Reprecipitation of secondary phosphates (e.g., crandallite, monazite) in the weathered profile led to LREE enrichment of up to 2.0 wt% (relative to 0.13-1.4 wt% In the primary apatites). Titanium oxide contents significantly increased from 0.2-7 wt% in the fresh rocks to 19-28 wt% in the saprolite associated with anatase formation. Whole rock Nb concentrations can reach up to 0.4 wt% in the lower saprolite, which is consistent with the preservation of trace pyrochlore and perovskite (up to 1 vol% and 5 vol%, respectively).

Tapira and Salitre provide a compelling case for understanding REE and HFSE (particularly Nb) mobility and fractionation under intense, deep weathering. These deposits serve as references for similar geological-weathering contexts associated with carbonatites in West Africa, India, and Australia.

Unlocking the Hidden Potential of the Cargo Deposit: New Geochronological and Geochemical Insights from the Macquarie Arc

Jessica Askew[1], Yamila C. Contreras, Sebastien Meffre, David Cooke
[1]University of Tasmania, Hobart, Australia

The Macquarie Arc within the Lachlan Orogen, central New South Wales, is recognised as Australia's most significant porphyry copper-gold (Cu-Au) province, hosting a total metal endowment exceeding 83 Moz of gold and 17.2 Mt of copper. The arc has been the focus of major exploration efforts and contains world-class deposits such as Cadia, Cowal, Northparkes, Marsden and Copper Hill.

Although historically considered a minor prospect, the Cargo deposit may represent a critical link between these major centres. Situated in a structurally complex region, Cargo exhibits features consistent with porphyry-style mineral systems, yet it has a subdued surface footprint. This may be attributed to structural overprinting and the presence of late-stage mineralising events that have complicated its recognition and exploration history.

This study presents new and ongoing petrological, geochronological and geochemical investigations of the Cargo system. Integrated U-Pb zircon geochronology, whole-rock geochemistry and petrography are being used to constrain the timing and evolution of multiple intrusive and mineralising events within this multiphase magmatic complex.

Preliminary results suggest that Cargo may represent a significant, underexplored porphyry system. Further analytical work, including U-Pb dating of monazite and apatite, as well as trace element analysis of sulphides and hydrothermal minerals, will refine interpretations of fluid evolution, magmatic oxidation states, and provide comparisons to better characterised systems such as Cadia and Cowal.

This integrated approach offers new insights into porphyry system development within arc terranes and carries broader implications for exploration targeting in regions with complex magmatic and structural histories.

The Geology and Paragenesis of the Akyem Orogenic Gold Deposit, Ghana

Fuseini Atanga[1], Michael Baker[1], Sheree Armistead[1], Lejun Zhang[1]
[1]Centre for Ore Deposit and Earth Sciences (CODES), University of Tasmania, Hobart, Australia

Akyem is hosted in the Birimian metasedimentary and meta-volcaniclastic rocks of the northeastern Ashanti greenstone belt, Ghana. The Ashanti belt hosts the largest orogenic gold deposit in West Africa, the Obuasi deposit. Most world-class gold deposits are situated in the central and southern parts of the Ashanti belt. As such, these areas have been the focus of numerous studies and most exploration activities, neglecting other prospective areas. Consequently, the northern part of the Ashanti belt is relatively underexplored and has the potential to shed light on the regional geological and tectonic processes that control gold deposition and may host additional gold resources.

Our recent sampling campaign, petrographic analysis and Pb—Pb age dating have revealed that mineralization at Akyem occurred during the Eburnean D2 regional deformation, ca. 2,142—2109 Ma. We identified two sets of quartz – carbonate veins with a mutual cross-cutting relationship, associated with mineralization: crack-seal veins (V2) and extension veins (V3) which overprint V1 boudinaged quartz-carbonate veins. Hydrothermal alteration associated with mineralization is typically rock-buffered, with a mineral assemblage of quartz – pyrite – albite ± white mica and a carbonate alteration overprint of mainly siderite and ankerite. We identified pyrite as the main ore mineral, occurring as disseminations in the host rock or along vein walls. Backscatter SEM imaging revealed two generations of pyrite: (1) earlier fine-grained and anhedral pyrite, and (2) later coarse-grained and euhedral pyrite. Gold occurs as visible native gold and as inclusions in the lattice of both generation of pyrites. We interpret that gold precipitation likely resulted from sulfidation of the wall rock during a low arsenic fluid-rock interaction. This is evident in the lack of arsenopyrite at Akyem relative to nearby gold deposits in the Ashanti belt such as Obuasi, Takwa, Iduampriem and Bogoso/Prestea.

µ-XRF Atlas of Epithermal and Porphyry Deposit Textures

Aaron Atkins[1], Antonio Arribas
[1]University Of Texas At El Paso, EL PASO, United States

Porphyry copper deposits are large hydrothermal ore systems associated with an intrusive igneous body and the porphyry copper and epithermal mineralization/alteration systems that lie above it. The magma body is usually andesitic to dacitic in composition, and it is rich in fluids and volatiles that fracture the host rock and precipitate mineralization as it rises to the surface. The chemical composition of these fluids influences the alteration, textures, and mineralogy that are observed along structural controls like fractures, from which we can identify three types of epithermal deposits in addition to porphyry: High Sulfidation (HS), Intermediate Sulfidation (IS), and Low Sulfidation (LS). This classification correlates with variations in sulfur content, sulfur oxidation state, and sulfidation state within the ore mineralization site. The main objective of this project is to characterize the mineralogy and textures of High, Intermediate, and Low Sulfidation Epithermal and Porphyry Deposits, using micro energy-dispersive X-ray fluorescence (µ-XRF) technology. Samples have been collected from several well-known and representative deposits around the world and scanned using a Bruker M4 Tornado to create high-resolution maps of the distribution of elements contained in each and color-coded for interpretation. We found that there are many chemical and distributive similarities among deposits occurring around the world. To illustrate this point, the samples studied have been superimposed onto existing models of porphyry and epithermal systems proposed by previous researchers (e.g., Sillitoe, 2010; Hedenquist and Arribas, 2022).

Mineralogical and Geochemical Characteristics of the Epithermal Au–Ag Mineralization at the Felicia Orebody, Lepanto, Mankayan District, Philippines

Yasuhiro Awano[1], Pearlyn Manalo[1], Ryohei Takahashi[1], Andrea Agangi[1], Hinako Sato[1], Shion Nohara[1], Akira Imai[2], Jillian Gabo-ratio[3], Leo Subang[4], Japhet Matula[4]
[1]Akita University, Akita, Japan, [2]Kyushu University, Fukuoka, Japan, [3]University of the Philippines, Quezon, Philippines, [4]Lepanto Consolidated Minig Company, Makati, Philippines

Recent exploration activities in the Mankayan District, Philippines delineated new sets of veins of the Felicia deposit. The veins exhibit brecciated and crustiform textures and are classified into four mineralization stages based on cross-cutting relationships.

Stage 1 veins consist of gray quartz and pyrite. Quartz occurs as elongated crystals, while pyrite is subhedral. Silver-bearing galena occurs as inclusions within pyrite. Stage 2 veins are characterized by abundant quartz, anhydrite, and sulfide minerals. Electrum occurs interlocking with sphalerite, pyrite, galena, and chalcopyrite. Petzite and Ag-bearing galena occur as inclusions in pyrite. Sphalerite, chalcopyrite, and galena commonly occur as fracture-fills and replacement of earlier pyrite. Stage 3 veins are dominated by rhodochrosite with very small amounts of sulfide. Stage 4 veins consist of coarse- to fine-grained anhydrite associated with electrum.

Bulk vein geochemical analysis shows that stage 1 veins contain up to 1.0 ppm Au and 17 ppm Ag. Higher gold contents were observed from the stage 2 and stage 4 samples, reaching up to 5.5 ppm and 6.2 ppm, respectively. Stage 2 veins contain a higher amount of Ag (194 ppm) than stage 4 veins (12 ppm). Electron probe microanalysis of sphalerite from stage 2 reveals FeS content ranging from 0.29 to 2.7 mol%. Sulfur isotope analysis shows that δ34S values of pyrite and sphalerite from stage 2 veins range from 0.0 to +1.3 ‰ and -0.4 to +3.0 ‰, respectively. Anhydrite from the stage 2 veins has δ34S values ranging from +15.7 to +25.4 ‰. Calculated sulfur isotope equilibrium temperatures from anhydrite and sphalerite pairs range from 300 to 450 °C. These results indicate that the gold and base metal sulfides of the Felicia veins were deposited from an intermediate sulfidation state fluid at relatively high temperatures.

Textures and Geochemistry of Pyrite and Arsenopyrite from the Leon Sediment-Hosted Gold Deposit, Ratatotok District, Indonesia: Implications for Ore-Forming Conditions

Muhammad Arba Azzaman[1,4], Ryohei Takahashi[1], Andrea Agangi[1], Pearlyn Manalo[1], Hinako Sato[1], Manuel Nopeia[2], Arifudin Idrus[3], Dolly R. Simbolon[5]
[1]Graduate School of International Resource Sciences, Akita University, Akita, Japan, [2]Center for Regional Revitalization in Research and Education, Akita University, Akita, Japan, [3]Department of Geological Engineering, Universitas Gadjah Mada, Yogyakarta, Indonesia, [4]Department of Geological Engineering, Halu Oleo University, Kendari, Indonesia, [5]PT Sumber Energi Jaya, West Jakarta, Indonesia

The Leon gold deposit, situated in Ratatotok district, North Sulawesi, Indonesia, is hosted by Miocene silty and fossiliferous limestone, which is overlain by Pliocene andesitic volcanic rocks and intruded by porphyritic andesite. Ore mineral paragenesis in the limestone and porphyritic andesite is divided into three stages: pre-ore, ore (early, main, and late), and post-ore. The pre-ore stage is characterized by framboidal pyrite disseminated and filling fractures in limestone. The limestone-hosted early-ore substage consists of anhedral to subhedral pyrite and marcasite, which occur as overgrowths of pre-ore pyrite and replacement of foraminifera. Early-ore substage in porphyritic andesite has subhedral-euhedral marcasite and pyrite with rutile and apatite inclusions. The main-ore substage in both rock types is characterized by arsenopyrite and arsenian pyrite, occurring as euhedral-subhedral, elongated, and skeletal grains or surrounding the early-ore pyrite. The late-ore substage in the limestone includes arsenian pyrite overgrowing the main-ore substage pyrite, with minor stibnite, realgar, orpiment, and trace chalcopyrite, sphalerite, and cinnabar in vugs of limestone. The late-ore substage in the porphyritic andesite includes arsenopyrite that overgrows the earlier arsenopyrite. The post-ore stage includes hematite and goethite replacing pyrite and realgar. $\delta^{34}S$ values of pyrite in the pre-ore stage and early- and main-ore substages range from +5.0 to +8.9‰, -1.6 to +1.2‰, and -3.1 to +2.3‰, respectively, while those of realgar and orpiment in the late-ore substage are -3.1‰ and -0.1‰, respectively. Arsenopyrite of the main- and late-ore substages contains 31–33 at.% As and 30–31 at.% As, respectively. This suggests formation temperatures and sulfur fugacity of ~427–475 °C (log fS_2 = –6.3 to –5.3) for the main-ore substage and ~363–395°C (log fS_2 = –7.7 to –6.9) for the late-ore substage. Gold occurs mainly as solid solution in arsenian pyrite of the main-ore (average 150 ppm Au) and late-ore substages (average 139 ppm Au).

Iron-Oxide-Copper-Gold (IOCG) Type Hydrothermal-Alteration Recorded in the Lawa Gold Deposit: North Singhbhum Mobile Belt, Eastern India

Paromita Banerjee[1], Rajarshi Chakravarti[1], Crystal LaFlamme[2]
[1]Indian Institute of Technology, Roorkee, Roorkee, India, [2]Université Laval, Quebéc City, Canada

The Lawa gold deposit in the Proterozoic North Singhbhum Mobile Belt (NSMB) is the only active gold mine in eastern India. There, mineralization is confined to a ~1000 m-long, NW-SE-trending shear zone along the contact between ferruginous quartzite and phyllite of the ~1600 Ma Chandil Formation. A pervasive Fe-K-(Na-Ca) alteration assemblage, rather than discrete quart-veins, exerts primary control on mineralization. Stage-I alteration comprises albite + actinolite + pyrite-I + chalcopyrite ± magnetite and reflects infiltration of oxidized and acidic magmatic-hydrothermal fluids (>350°C). This is supported by Py-I geochemistry (Co/Ni = 1–1.5; mean Se ~300 ppm; below detection As) and in-situ $\delta^{34}S$ values of Py-I ranging from +1.0 ‰ to +6.6 ‰ (LA-QQQ-ICP-MS). These fluids induced extensive brecciation of the host lithology. Stilpnomelane + magnetite + chalcopyrite + carbonates ± native gold (Stage-II alteration) overprints the Stage-I assemblage and occupies pre-existing breccias. This Fe-K dominated assemblage and a lack of gold-sulfide association suggest transport of gold as $AuCl_2^-$ complexes. Fluid decompression along the preexisting breccias and H_2O-CO_2 unmixing raised fluid alkalinity and triggered native gold precipitation. Dolomite associated with native gold along brittle fractures and coeval stilpnomelane + magnetite (pH >7) supports this inference.

The Stage-III alteration comprises abundant pyrite-II + chlorite + hematite ± chalcopyrite indicative of increasing $a\sum S$. Gold, exclusively occurring as Pb-Bi-Te-Au polymetallic inclusions along pyrite-II rims reflects remobilization of the pre-existing native gold as $AuHS^-$ complexes. Subsequent mixing of this fluid with meteoric water led to an increase in fO_2 and consequent deposition of Pb-Bi-Te-Au polymetallic inclusions. This contention is supported by the variable Co/Ni ratios, mean Se ~550 ppm, mean As ~100 ppm in Py-II and progressively lighter $\delta^{34}S$ values (-4.0 ‰ to +6.5 ‰) relative to Py-I. We infer that the Lawa deposit likely represents a hitherto unknown IOCG-type mineralizing system in the NSMB.

Late Au-Bearing Quartz - Carbonate Veins System Overprinted the Base Metal Skarn Mineralization in the E44 Deposit, Northparkes District, NSW

Billy Beas Caceres[1], David Cooke[1], Lejun Zhang[1]
[1]CODES / University of Tasmania, Hobart, Australia

The E44 Au-Cu-Zn-Pb deposit is located in the Junee-Narromine Belt of the Macquarie arc, 26 km NW of Parkes, NSW. E44 consists of mineralized skarn and Au-bearing IS epithermal veins hosted by the Goonumbla Volcanics. E44 is one of numerous mineralized occurrences that have been discovered in the Northparkes district, which formed during the Late Ordovician to early Silurian (Wells et al., 2021; this study). Alkaline Cu-Au porphyry deposits, epithermal Au systems, skarn prospects at E6, E7, and E44, and unmineralized lithocaps at Nash's Hill occur in the district (Coke et al., 2007).

E44 skarn includes precious and base metals in andraditic skarn lenses hosted in calcareous and andesitic volcaniclastic units (Jones, 1991). CMOC's exploration drilling identified Au-bearing carbonate and Cu-bearing veins that cut the skarn. The skarn system is ~15 m thick and ~300 m long and comprises patches of galena–sphalerite overprinting andraditic–magnetite skarn. Proximal skarn grades to medial andradite–calcite–epidote and distal epidote–chlorite skarn zone, with minor andradite–calcite patches developed in the overlying mafic to intermediate Goonumbla Volcanics. An unmineralized recrystallized marble zone, up to 20 m thick, occurs below the garnet skarn. Quartz–crustiform calcite–chalcopyrite–pyrite–galena–sphalerite and quartz–calcite–pyrite–chalcopyrite epithermal veins with Au and Te crosscut the skarn deposit and contain significant precious metal assays. Post-mineralized calcite–fluorite vein stage overprinted both systems. A widespread white mica alteration zone has been identified at depth in the E44 system, affecting the volcanic and intrusive rocks containing copper mineralization in veins. At considerable depths (~350m) to the NW of E44, deep drilling intersected the Cu-Au-Mo-K-feldspar-quartz cemented Two-Thirty magmatic hydrothermal breccia complex (Wells et al., 2021), implying proximity to a porphyry Cu-Au centre.

Geochemical and Mineralogical Signatures of the Rocklands Cu-Au-(Co) Deposit, Northwest Queensland, Australia

Alkis Kontonikas-Charos[1], Courteney Dhnaram[1], Vladimir Lisitsin[1], Lisa Kearney[2], **Elena Belousova**, Daniel Killen[1], Al-Tamini Tapu[1]
[1]Geological Survey of Queensland, Brisbane, Australia, [2]Queensland University of Technology, Brisbane, Australia

The Rocklands Cu-Au-(Co) deposit is one of numerous mineralogically and geochemically diverse IOCG deposits in the Mount Isa Province of Northwest Queensland, Australia. Rocklands consists of multiple structurally controlled ore bodies, including Las Minerale and Rocklands South, hosted within Proterozoic metasedimentary and volcanic rocks. The deposit contains a supergene enrichment blanket and hypogene mineralisation with dominantly sodic and calcic-ferric alteration.

The Geological Survey of Queensland (GSQ) characterised seven drillholes from Rocklands using a range of hyperspectral, geochemical, and microanalytical techniques. Continuous hyperspectral and XRF core scanning with systematic whole rock multi-element geochemistry data provided an effective method to classify alteration and lithology variations throughout the deposit. Selective samples were analysed using Bruker Tornado µXRF imaging and SEM-TIMA to resolve detailed paragenetic relationships and fine-scale mineral associations. Based on Tornado and SEM-TIMA imaging, multiple sulfides and Fe oxides were selected for Fe and Cu isotope analyses, as part of a systematic GSQ program of geochemical and isotopic fingerprinting of IOCG deposits in the Cloncurry district.

Hyperspectral and microanalytical analyses revealed the occurrence of distinct alteration zones of dominant albite, actinolite, carbonate and, locally, K-feldspar. Comparison between whole rock multi-element geochemistry data and microanalytical analysis of selective samples highlighted a decoupling between the association of Cu and Co. Detailed SEM-TIMA analysis enabled us to investigate the distribution of Cu and Co and to recognise the occurrences of Cu and Co-bearing mineral phases. Cobalt-bearing phases included carrollite, linnaeite, and pyrite (in the lattice and within Co mineral inclusions), which were typically associated with chalcocite, bornite, and chalcopyrite.

The detailed characterisation of the Rocklands Cu-Au-(Co) deposit provides new insights into the nature of Co and Cu mineralisation and offers a framework for enhancing the understanding of mineralogical and geochemical signatures and the critical element deportment within IOCG deposits of the Mount Isa Province.

Tectonic Controls on Long-Lived Sn Mineralization in the Martinamor Extensional Dome (Salamanca, Spain)

Daniel Bermejo López[1], Lorena Ortega[1], Pedro Castiñeiras[1], Lorenzo Tavazzani[3], Santos Barrios[2], Elena Crespo[1], Cyril Chelle-Michou[3], Juan Gómez Barreiro[2]
[1]Universidad Complutense De Madrid, Spain, [2]Universidad de Salamanca, Spain, [3]ETH Zürich, Switzerland

Regional tectonothermal evolution plays a crucial role in the formation of mineral deposits. One of the key processes in this context is the gravitational collapse of orogens, widely documented in belts such as the Variscan. The Martinamor gneissic dome, located in the Central Iberian Zone of the Iberian Massif, is a well-constrained structure related to the late-Variscan extensional collapse. Its formation is associated with crustal delamination and exhumation along a major subhorizontal extensional detachment located at the top of the dome. The dome hosts a significant tin-tungsten (Sn-W) mineralization, including cassiterite-bearing quartz veins and pegmatites, as well as stratabound scheelite skarns.

This study focuses on cassiterite-bearing mineralization. Cassiterite U-Pb dating from different deposits reveal a prolonged mineralizing event lasting ~25 million years (324.1 ± 5.9 to 300.7 ± 5.4 Ma). The mineralization began with early pegmatitic Sn deposits and evolved into cassiterite-bearing quartz veins that formed during syn- to late-extensional deformation. These deposits occur across various levels of the Martinamor shear zone and show structural patterns consistent with progressive deformation within the extensional regime.

Our results suggest that the E2 extensional phase, traditionally constrained to 315–300 Ma, began earlier, around 325 Ma, marking the onset of local syn-collisional extension. This evolved into dominant extensional tectonics from 315 to 300 Ma. The downward propagation of cassiterite mineralization over time suggests a structural control linked to progressive deepening of the brittle-ductile transition zone and evolving detachment architecture.

Comparable mineralization ages across the Iberian Massif and in other European regions point to a regionally significant metallogenic episode. The Martinamor dome thus offers a key record of prolonged tectono-magmatic-hydrothermal activity, highlighting the importance of sustained heat flow, crustal melting, and major extensional structures in the genesis of post-orogenic Sn-W deposits.

Clay-Hosted REE Deposits Associated to Eucla Basin Paleovalleys in Southern Australia

Gabriel Valentim Berni[1], Mario Iglesias-Martinez[1], Nathan Reid[1], Robert Thorne[1], Mark Pierce[1]
[1]CSIRO, Kensington, Australia

The growing demand for rare-earth elements (REE), driven by the transition to renewable energy technologies and geopolitical supply uncertainties, has intensified the search for alternative REE sources. Current economic REE deposits are in regions of tropical weathering, where REE are absorbed onto the surfaces of clay minerals. During the late Cretaceous to Eocene, much of Western Australia experienced tropical climates that likely generated ionically bound REE deposits. These weathering profiles have been partially eroded and covered by sediments from the Eucla Basin and its associated paleovalleys. Since the Miocene, arid weathering, influenced by seasonal drainages, shaped the current landscape. Reworking of the tropical weathered profile formed: (1) preserved lower saprolite REE enriched zones located at the base of the paleovalleys (Esperance-Western Australia); (2) confined layers within the paleovalley sediments with REE that could be extracted by in-situ leaching (Wudinna-South Australia). We have studied three REE prospects associated with paleovalleys from Western and Eastern sides of the Eucla basin. Our preliminary results show that the majority of the REE are hosted by secondary phosphate and minor carbonate minerals at the lower saprolite zone. The REE mineralisation within paleovalley sediments at eastern Eucla is likely hosted by ionically bound REE in clay minerals, as no REE mineral phases were identified to account for the reported grades. This presence of ionically bound clay hosted REE could relate to more recent climate differences between the two regions, where the more wet climate of Esperance could be responsible for leaching ionically bound clays, reprecipitating it as secondary phosphate minerals at the lower saprolite , while the eastern side of the Eucla retains un-leached ionic clays. These observations highlight the importance of understanding the geometry, and sedimentation regimes of the Eucla basin paleovalleys, and climate history for generating clay-hosted REE exploration targets within southern Australia.

A Prolonged Magmatic History to the Dismembered Mid Miocene Arc, Manus Island, Papua New Guinea

Joel Blake[1], Matthew Campbell[1], Scarlett Blewett[2], Rob Duncan[2], Mike Erceg[3]
[1]Rio Tinto Exploration, Bowen Hills, Australia, [2]Rio Tinto Exploration, Bundoora, Australia, [3]Canterbury Resources, Pyrmont, Australia

Papua New Guinea's Manus Island is today an isolated segment of a previously dismembered volcanic arc on the Pacific plate. Remnants of arc magmatism are preserved within a series of middle Eocene to early Miocene volcanic and volcaniclastic packages (Tinniwi Volcanics) and a northwest-southeast trending belt of middle Miocene age intrusive rocks (Yirri Intrusive Complex). Exploration has revealed several porphyry Cu-Au and epithermal Au occurrences spanning the width of the island. Given the isolation of Manus Island today from other parts of the arc, linking magmatism to other regional tectonic events is difficult, and the evolution of the arc remains unclear. We present new U-Pb geochronology of magmatic zircon and high precision Re-Os geochronology of molybdenite to (1) examine the age of magmatic rocks across the Yirri Intrusive Complex, and (2) constrain the age of Cu-Au porphyry mineralisation from known prospects. Results demonstrate a prolonged continuation of magmatism and associated Cu-Au porphyry mineralization that migrated over time from the northwest to the southeast within the Yirri Intrusive Complex. Younger ages from the southwestern segment of the Yirri Intrusive Complex extend the magmatic history on Manus Island from that previously known and potentially provide new age constraints on the cessation of southward dipping subduction and the docking of the Ontong Java Plateau. Results improve the constraints on the tectonic reaction to the docking of the Ontong Java Plateau, propagating outwards along the arc from the proximal trench zone, westward to the location of Manus Island at the time.

Exploring a Sleeping Giant: Chlorite Chemistry as a Vectoring Tool in the World-Class Tennant Creek Au-Bi-Cu Province

Damian Braize[1], Jeffrey A. Steadman[1], Sebastien Meffre[1]
[1]CODES, University of Tasmania, Hobart, Australia

The Tennant Creek mineral field in the Northern Territory is one of Australia's largest historical mining provinces, with significant past production of gold, copper and bismuth. Tennant Creek deposits are tentatively classified as iron-oxide copper-gold (IOCG) type and are typically hosted in iron oxide-chlorite-quartz bodies known locally as ironstones. Unlike many other IOCG systems, these deposits are characterized by spatially restricted chlorite-rich hydrothermal alteration haloes, making exploration challenging, especially undercover. This study investigates the potential of chlorite chemistry as a vectoring tool to mineralization in the province. Using laser ablation inductively coupled plasma mass spectrometry (LA-ICP-MS), chlorite grains were analyzed from mineralized, proximal and distal barren zones across several high-grade deposits.

Results show systematic variations in key trace elements (such as, Fe, Co, V, Ba) with proximity to mineralization. These patterns extend well beyond visible alteration, delimiting the geochemical fingerprint of mineralized ironstones and effectively distinguishing barren from mineralized zones. Elemental gradients are observed at the deposit scale and provide insights into fluid flow pathways around the deposits. Chlorite chemistry also reflects differences in metal endowment between deposits, potentially delineating cobalt-, uranium-, and bismuth-rich mineralization. Additionally, multivariate statistical analyses show compositional similarities between Tennant Creek chlorite and that from other IOCG systems, confirming the classification of Tennant Creek deposits in the IOCG class.

Our findings highlight the usefulness of chlorite chemistry as a powerful mineral exploration tool in the Tennant Creek province. Further application in greenfield or undercover terrains could help in the search for the next generation of world-class discoveries at Tennant Creek.

Orebody Knowledge from the Sinclair Deposit: A New Template for Cesium Exploration

Nigel Brand[1,2], **David Crook**[3], Stuart Kerr[4], Haydn Daxter[5]
[1]Geochemical Services Pty Ltd, West Perth, Australia, [2]CET, University of Western Australia, Crawley, Australia, [3]Ore Source Pty Ltd, West Perth, Australia, [4]OMNIGEOX Pty Ltd, North Fremantle, Australia, [5]Power Metals Corp, Vancouver, Canada

Globally significant discoveries of Cesium-rich pollucite deposits are exceptionally rare, with only three commercial operations reported to date: Bikita (Zimbabwe), Bernic Lake (Canada), and Sinclair (Western Australia). Pollucite mineralisation forms exclusively within extremely differentiated lithium–Cesium–tantalum (LCT) pegmatites, making Cesium deposits challenging to explore for, and highly sensitive to the specific conditions of pegmatite emplacement and fractionation.

Despite a surge in global exploration for LCT pegmatites during the Lithium Demand Rally (2015–2025), driven by renewable energy and electric vehicle markets, only one new pollucite mine (Sinclair, discovered in 2016) and one new resource (West Joe, Ontario, release in 2025) were added to the global Cesium inventory. Additional encouraging drill intercepts were reported from two other Canadian properties and one Angolan project, highlighting the extreme rarity of economic pollucite mineralization even under intense exploration focus.

The discovery and development of the Sinclair Cesium Mine has provided a rare and invaluable opportunity to investigate the geological and geochemical controls on high-grade Cesium mineralization. Detailed study of the Sinclair orebody has improved understanding of LCT pegmatite evolution at the extreme end of fractionation, providing new insights into the timing, setting, and mineralogical signatures critical for pollucite development and Cesium enrichment.

This new orebody knowledge offers a robust template for future exploration targeting Cesium-rich pegmatites. Furthermore, it assists in the re-evaluation of historical LCT pegmatite fields, including the interpretation of emerging discoveries such as West Joe. The Sinclair model illustrates the importance of detailed deposit studies to guide the identification of rare critical metal resources, supporting future discoveries essential for high-technology and energy transition industries.

The Prominent Hill Deposit: A Modified, High-Grade IOCG System Hosted in Hematite-Altered Sedimentary Breccias

Riccardo Brigante[2,1], Hamish Freeman[1,2]
[1]BHP Exploration, Adelaide, Australia, [2]BHP Copper South Australia, Adelaide, Australia

Prominent Hill is an iron-oxide copper-gold (IOCG) deposit located on the southern margin of the Mount Woods Domain, approximately halfway between Olympic Dam and Coober Pedy in the Gawler Craton of South Australia. It was discovered by Minotaur Resources in 2001 by drilling of a high frequency gravity anomaly. The deposit occurs in overturned stratigraphy near an unconformity that separates two different basins.

The orebody is primarily hosted by an older sequence of sedimentary rock that is unmetamorphosed and has undergone intense hematite, sericite, chlorite, and carbonate alteration. The copper-gold mineralised hematite breccias with various sedimentary and igneous clasts are believed to have formed from turbiditic and debris flow protoliths. Studies have shown that these rocks experienced a quartz-stable porosity-forming event before hematisation. Carbonate rocks are found near the Hanging Wall Fault, which marks the boundary between unmetamorphosed sedimentary rocks (south) and greenschist facies metasedimentary (north). Detrital zircon studies of the host sequence indicate no input from the Gawler Range Volcanics (GRV), and provenance ages of 1.79-1.69 Ga.

Stratigraphically overlying the mineralised sequence are strongly altered mafic to intermediate correlatives of the ca. 1.59 Ga GRV, suggesting that mineralisation occurred after 1.59 Ga. South of the GRV package, an overturned succession of fluvial or alluvial oxidized clastic rocks may be correlatives of the Pandurra Formation.

Sulfide mineralisation at Prominent Hill is predominantly comprised of chalcocite, bornite, and chalcopyrite. Replacement of early formed pyrite by copper sulfides is common. Some zones feature supergene leaching textures, characterised by abundant earthy hematite and/or hydrated iron oxides. Like the Olympic Dam deposit, there is geochronological evidence for significant hydrothermal activity during the “boring billion”.

As Prominent Hill transitions from a truck haulage operation to a shaft, and greater depth and lateral extents are explored, geological knowledge will continue to develop.

Statistical Analysis of Geochemical Data from In-Bearing Mineralization in the Marta Centro Vein, Pingüino Deposit, Patagonia, Argentina

Juan Matías Buratti[1], Luciano López[2], Pablo Martín Demetrio[3]
[1]Facultad De Ciencias Naturales Y Museo, La Plata, Argentina, [2]Instituto de Recursos Minerales (INREMI), Facultad de Ciencias Naturales y Museo, La Plata, Argentina, [3]Centro de Investigaciones del Medio Ambiente (CIM, CONICET-UNLP), La Plata, Argentina

Indium is a critical metal due to its growing demand in modern technologies such as photovoltaic cells and high-resolution displays, contrasted by its limited global production. Although typically scarce in epithermal environments, indium has been found in notable concentrations within base metal-rich polymetallic and intermediate sulfidation epithermal systems. The Pingüino deposit, situated in the low sulfidation epithermal metallogenic province of the Deseado Massif, Patagonia, Argentina, represents a unique case in the region due to its atypical indium enrichment. Mineralization at Pingüino occurred in two distinct events: an early polymetallic event enriched in In, Zn, Pb, Ag, Cd, Au, As, Cu, Sn, W, and Bi characterized by a complex sulfide assemblage; and a later Ag–Au intermediate sulfidation event. The early event, which is the focus of this study, has been classified as an In-rich polymetallic epithermal vein system.

This work presents a statistical analysis of geochemical data from In-bearing mineralization in the Marta Centro vein, where indium displays its highest concentrations within the Pingüino deposit. Exploratory analysis revealed several relevant patterns and features. Significant anomalies were observed in elements such as In, Ag, Pb, Zn, Cd, and Sb. Correlation coefficients and binary diagrams indicate strong relationships between In–Sn, Au–Ag–Cu–Sb–Sn, Ag–Cu–Pb, and Zn–Cd.

These results help to provide valuable insights into the processes controlling indium enrichment and contribute to the development of exploration strategies for critical metals in polymetallic epithermal systems.

From Regional Insight to Discovery: Integrated Targeting for Sapphire Mineralization in Southeastern Madagascar

Evelyn Caiza[1]
[1]Central University of Ecuador, Quito, Ecuador

Sapphire deposits in southeastern Madagascar remain underexplored despite the region's known gem potential and favorable geological setting. This study applies an integrated exploration workflow—from regional targeting to project-scale delineation—aimed at advancing new discoveries within the Anosy region, across the NANTIN and SIAM licenses (~11,910 ha).

The study area encompasses four distinct geological domains characterized by northward structural trends, predominantly comprising foliated marbles and pyroxenites intruded by granitic bodies, shaped by high-temperature metamorphism and deformation events linked to shear zones hosting sapphire. Zircon U-Pb ages of 523–510 Ma confirms the metamorphic framework conducive to mineralization.

We combined conventional methods—regional geology, known sapphire occurrences, remote sensing (ASTER and Sentinel), and airborne geophysics (magnetometry and radiometry)—with non-conventional approaches. Machine learning techniques, including Fuzzy logic and Self-Organizing Maps (SOM), were applied to generate mineral potential maps. This led to the identification of five new prospective zones. Among them, Ankaramavo, Antirimena, and Satrokala emerged as high-priority targets, while Ankazoavo and Andranondambo were ranked medium-priority.

Ankaramavo was elevated to project-scale focus due to its strong mineralogical and spectral indicators, including calcite-phlogopite associations and AlOH spectral anomalies. Ground-truthing through petrography, mineralogy, and geochemistry (notably enrichments in Al, Si, K, Rb, Y, Nb) supports a robust geological-geochemical model for sapphire mineralization.

These results highlight the effectiveness of integrating remote sensing, geophysics, geochemistry and machine learning for gemstone exploration in complex metamorphic terrains, providing a scalable framework for future discoveries across the Central-East African belt.

Exploration Vectoring Tools in Complex Magmatic-Hydrothermal Systems: Application to the Potrerillos District in Northern Chile

Yamila Cajal[1], David Cooke[1], Michael Baker[1], Ivan Belousov[1], Matthew Cracknell[1], Carolina Rodriguez[2], Felipe Jimenez[3], Sergio Pichott[2]
[1]CODES, University of Tasmania, Hobart, Tasmania, Australia, [2]Gerencia Corporativa de Exploraciones, CODELCO, Santiago, Chile, [3]Exploraciones Mineras Andinas, Santiago, Chile

The Potrerillos Mining District is located in the southern part of the world-class Eocene-Oligocene metallogenic belt of Northern Chile, central Andes. The district hosts several ore systems, including porphyry, epithermal and skarn occurrences, which contain important resources of Cu, Au, Mo, Ag and Mn. The geology of this region consists of Jurassic to Palaeocene sedimentary (marine and continental) and volcanic sequences that were intruded by porphyries and hydrothermal breccias between 40 and 31 Ma. The diverse host rocks produced a complex arrangement of hydrothermal alteration and mineralization, which makes this district an ideal case to test simultaneously different vectoring tools.

Previous studies have demonstrated that the composition of hydrothermal minerals, such as epidote and chlorite (green rocks), and quartz and calcite (white rocks) provide powerful tools to track the pathways of hydrothermal fluids and vector towards the centre of magmatic-hydrothermal systems. This study aims to test and improve these tools by the application of a combination of them in the complex Potrerillos District, and to enhance orebody knowledge in the district.

We have collected eighty-five outcrop samples, covering several sedimentary, igneous and volcanoclastic lithologies and alteration assemblages. Field observations, along with short-wave infrared data, show that the carbonate rocks and propylitic alteration have a wide distribution within the district. The carbonates exhibit a range of colours and intensities of short-wave ultra-violet fluorescence (SW-UVF), which indicate different trace element compositions and fluid sources. Preliminary results of mineral chemistry using LA-ICP-MS show significant variations in Ti and Mg in chlorite, As and Sb in epidote, Ti/Sb in quartz and Mn/Mg in carbonates, as well as a SW-UVF response, towards the ore deposit centres, especially closer to porphyries. These results are promising for the use of the combination of these tools in complex magmatic-hydrothermal districts.

Assessing the Timing of the Magmatic to Hydrothermal Transition and Onset of Porphyry Mineralization Across the Yerington Porphyry System, USA

Lawrence Carter[1], Simon Tapster[3], David Selby[4], Ben Williamson[2,5], Gavyn Rollinson[2]
[1]LC Geoscience Ltd, United Kingdom, [2]Camborne School of Mines, University of Exeter, United Kingdom, [3]British Geological Survey, Keyworth, United Kingdom, [4]Durham University, United Kingdom, [5]University of Bristol, United Kingdom

Porphyry Cu(-Au-Mo) deposits form in the upper ~2-8 km regions of large, long-lived magmatic-hydrothermal systems. The timescales over which sufficient volumes of mineralizing fluids are extracted and focused from magma source regions into the narrow, shallow zones of mineralization remains unclear. Here, we provide fundamental constraints on the relative timing and duration of porphyry-style mineralization from detailed field, mineralogical, geochemical and geochronological studies of the classic Yerington porphyry-district of Nevada, where Cenozoic extension and tilting has exposed an exceptional top to bottom ca. 8 km cross-section through the porphyry system. From high precision zircon U-Pb dating, the exposed Luhr Hill granite cupola of the Yerington batholith, and multiple generations of aplitic- and porphyry dikes, on which the porphyry-type deposits are centred, were emplaced within only a few 100 kyrs. Cu mineralization within miarolitic cavities and USTs in the dikes, and associated quartz veins, was paragenetically late in the magmatic to hydrothermal transition, not occurring until the fluids responsible had cooled below ~600 °C. From molybdenite Re-Os age determinations, hydrothermal mineralization in the Bear and Ann Mason deposits was coeval with and much longer lived (>1 Myrs) than magma emplacement. We suggest that the early, high temperature fluids travelled up into the epithermal environment and it was fluids from deeper, which had cooled over time, which moved up through permeable, partially crystallised 'mush' dikes which produced hypogene mineralization. We suggest that this process is considered in all new genetic, exploration and numerical models for porphyry and similar types of magmatic-hydrothermal systems.

High-Grade Gold Controls and Spatial Chemical Zonation at The Snip North Prospect of The Iskut Project, Canada

Antonio Arribas[1], **Daniel Castano Madrigal**[1]
[1]University Of Texas At El Paso, El Paso, United States

The Iskut project, in northwestern British Columbia, Canada, lies within the Intermountane Belt, formed by the convergence of the Pacific and North American plates. This subduction-related setting has led to intense magmatic activity, faulting, and folding. The area comprises a Triassic-Jurassic volcano-sedimentary sequence of siltstones, sandstones, and basalts, intruded by Cretaceous monzodiorites and younger intrusive bodies associated with hydrothermal processes. At the Snip North prospect, multiple vein systems cut across all lithologies, defining at least three stages of mineralization. The first stage consists of pyrite, quartz, chalcopyrite, and carbonate veins. The second and most economically significant stage hosts the highest concentrations of gold and copper, along with sphalerite, galena, quartz, and carbonates in veins with white mica halos. The final stage comprises late pyrite and quartz veins with minor chalcopyrite. At greater depths, veins with secondary biotite halos are linked to pervasive biotite alteration in the host rock, and quartz-monzodiorite intrusions at depths exceeding 1 km are associated with high molybdenum concentrations. Geochemical analysis by u-XRF reveals that copper is not solely associated with sulfides but also appears disseminated in the host rock, requiring further study to determine whether it represents an additional mineralization stage. Additionally, high concentrations of magnetite, with localized hematite zones, indicate variations in oxidation conditions during hydrothermal events. The overprinting of secondary biotite alteration by white mica alteration suggests a shift in hydrothermal conditions, influencing the development of higher-grade mineralized veins. These findings highlight the complexity of the deposit and the need for further research, including Re-Os and Ar-Ar dating, to refine the mineralization timeline. A better understanding of non-sulfide copper mineralization will enhance exploration strategies and resource evaluation at Snip North.

Cobalt Enrichment in Sulfides from Iron Oxide Copper-Gold Mount Colin Deposit, NW Queensland

Gabriel Cellier[1], Avish Kumar[1], Elizabeth Mariño[1], Ioan Sanislav[1], Alex McCoy-West[1,2], Alkis Kontonikas-Charos[3], Rhiannon Jones[3]
[1]Economic Geology Research Centre (EGRU), James Cook University, Townsville, Australia,
[2]IsoTropics Geochemistry Laboratory, James Cook University, Townsville, Australia,
[3]Geological Survey of Queensland (GSQ), Brisbane, Australia

Iron Oxide Copper-Gold (IOCG) deposits have traditionally been associated with production of copper, gold and iron; however, it is known that they can have a wide range of endowments, including potential for many critical metals. In this context, cobalt has recently attracted considerable interest among researchers, as understanding its occurrence is crucial for both scientific and economic reasons. While the number of publications studying cobalt endowment has grown, most studies have focused on other types of hydrothermal deposits. In this study, we analyzed samples from the Mount Colin IOCG deposit located in the Mount Isa Inlier, Cloncurry province, NW Queensland, where pyrite represents the main cobalt-bearing phase. A combination of petrography, EPMA imaging, LA-ICP-MS and EPMA spot analyses were employed to conduct a detailed mineralogical investigation of the sulfides. Three main forms of Co-rich distribution have been identified including: (1) pyrite presenting core-mantle-rim textures with variation in concentrations, (2) pyrite with a more irregular zoning, and (3) nano-inclusions within the sulfides. Results from the EPMA analysis show cobalt contents ranging from 0.01 to 3.6 wt%. LA-ICP-MS trace element data illustrates that Co-rich pyrites can reach concentrations up to 32056 ppm, with the highest averages associated with the first type of distribution. EPMA elemental maps reveal that the Co-rich inclusions also show higher concentrations of Ni, a behavior not observed in other cobalt enrichments, which typically vary between Co-rich/Ni-poor and Co-poor/Ni-rich compositions. EPMA data confirmed the high concentrations of Ni on some of the inclusions, possibly indicating the presence of cobalt-nickel sulfides. This would indicate more than one form of cobalt occurrence in this deposit. Understanding the Co-Ni enrichment patterns provides insights into the ore-forming processes, while the characterization of the textures and inclusions provides insights into the hydrothermal activity and possible overprinting from multiple mineralizing events at Mount Colin deposit.

Structural Controls on Epithermal Ag-Pb-Zn Vein Systems in the Eastern Erzgebirge, Germany

Jan Cerny[1], Sam Thiele[1], Marie Guilcher[1], Uwe Lehmann[2], Henrik Kaufmann[2], Jens Gutzmer[1]
[1]Helmholtz-Zentrum Dresden-Rossendorf, Helmholtz Institute Freiberg For Resource Technology, Freiberg, Germany, [2]Landesamt für Umwelt, Landwirtschaft und Geologie, Freiberg, Germany

The Freiberg District in Saxony, SE Germany, is home to one of Europe's most historically significant Ag-Pb-Zn vein systems, yet its structural controls and tectonic setting have long remained poorly described. In the present study an integrated structural geology approach was applied to characterize the tectonic controls on this epithermal system and the distribution of ore within it. The analysis reveals that NW-SE trending second and third-order dextral fault systems parallel to the Elbe Shear Zone, a major fault zone located to the East of the Freiberg District, likely bound the main mineralized areas. NE-SW trending dilation (normal) faults linking these dextral systems created persistent dilational zones during transtension related to late- to post-orogenic stages of the Paleozoic Variscan Orogeny. These structural corridors localized fluid flow and possibly felsic magmatism, serving as key fluid pathways and trap sites. Paleostress data support regional transtension and localized extension consistent with syn-mineral fault dilation and vertical fluid migration.

Moreover, structural reconstructions suggest that repeated fault reactivation during Permian to Mesozoic, mostly extensional events, contributed to local remobilization and preservation along significant lithospheric structures (faults and shear zones). Importantly, our structural reconstructions highlight the influence of fault linkage, rheological contrasts, and post-Variscan reactivation on mineral system evolution, providing insights for targeting still-covered extensions of the Freiberg system. These results also show how reinterpreting classical districts through a modern structural geology and mineral systems lens can unravel the tectonic architecture behind ore-forming processes and open new exploration frontiers in Europe and beyond.

Link Between Deformation and Pegmatite Emplacement in the Mataki Greenstone Belt, North East Zimbabwe Craton

Pharisie Chibaya[1], Brian Mapingere[1], Maideyi L. Meck[1], Fadzanai B. Mupaya[1]
[1]University Of Zimbabwe, Harare, Zimbabwe

The Mataki greenstone belt (MGB) in the north-eastern part of the Zimbabwe Craton hosts several pegmatite occurrences, including the Neoarchaean lithium-bearing Benson mine pegmatites. While the emplacement of structurally controlled pegmatites is fairly documented globally, the link between deformation structures and pegmatite emplacement remain poorly understood in the MGB. This study integrates structural mapping and petrographic analyses to assess the deformation history of the MGB and its influence on pegmatite emplacement. Three deformation events and pegmatite generations are recorded in the MGB. The earliest recognisable deformation event (D1) formed bedding-parallel schistosity S1 in thinly-laminated schists in the biotite zone of greenschist facies metamorphism. D2 folded bedding-parallel schistosity S1 into cm-m-scale tight to isoclinal folds F2 associated with moderately E-dipping fold axial planes and schistosity under upper greenschist to lower amphibolite facies metamorphism. Lithium-bearing pegmatites, including the Benson mine pegmatites were folded during D2 suggesting that they likely emplaced along S1 planes during the late stages of D1. The second generation of lithium-barren pegmatites crosscuts D2 structures. Lastly, D3 operated under medium amphibolite to granulite facies conditions in the greenstone belt and the structurally overlying migmatitic gneiss terrain. D3 formed moderately NNE-dipping schistosity and mylonitic foliation S3 carrying moderately ESE-plunging mineral and stretching lineations associated with top-up to the west sense of shear. The last generation of pegmatites emplaced along S3 planes and preserve foliation marked by preferred alignment of muscovite which is concordant to S3 in the host rocks. These pegmatites do not carry lithium in their system. Our findings suggest that the most prospective pegmatites for lithium in the MGB are the ones folded during D2.

Multi-Stage Fluid Alteration of the Agua Santa Pegmatite (Brazil): Implications for Mineral Zonation and Exploration

Sofia Chinaglia[1], Andrew Miles[1], Tania Martins, Kathryn Goodenough, Lee Groat
[1]University Of Leicester, Leicester, United Kingdom

Zoning is commonly observed in lithium bearing pegmatites, however overprinting by metasomatic and/or hydrothermal alteration obscures and complicates the identification and interpretation of zoning. The 65 m wide Agua Santa (AS) pegmatite in Minas Gerais, Brazil, is a spodumene-bearing pegmatite within a NE/SW trending shear zone that hosts significant Li deposits (such as the Cachoeira -CBL- and the Xuxa -Sigma- mines). Four distinct mineralogical zones have been identified in the field:

- Zone 1 (top of the pegmatite): Quartz + albite + tourmaline, overprinted by albitization.
- Zone 2: Quartz + K-feldspar + minor tourmaline/mica, with graphic textures.
- Zone 3: Quartz + K-feldspar + tourmaline.
- Zone 4 (innermost): Spodumene clusters (variably altered) + quartz K-feldspar,+ marking lithium saturation.

The AS pegmatite is considered to be a good example of a fractionated pegmatite as evidenced by white micas geochemistry results with elevated Cs (up to 2500 ppm) and Rb (up to 6000 ppm), and low K/Rb ratios (e.g. 18) suggest this pegmatite formed from a highly fractionated granitic melt. Furthermore, high aluminium (e.g. 40 wt.%) contents in these micas suggest that these melts were sourced from Al-rich metasedimentary protoliths, or strongly interacted with albitizing or other late-stage fluids.

Multi-stage alteration of magmatic minerals is evident from: (1) pervasive albitization, with albite primarily replacing quartz, K-feldspar and most importantly spodumene; (2) secondary quartz-mica veining; and (3) late clay overprinting. Albitization of spodumene diminishes the economic potential of pegmatites, but the origin of such fluids remains uncertain. Further stable isotope analysis of quartz will help identify the source of these metasomatizing fluids and whether they are meteoric or hydrothermal in origin.

A New Process-Based Mineral System Classification Linked to Commodities

Jonathan Cloutier[1], David Huston[1,2], Yanbo Cheng[1], Arianne Ford[1]
[1]Geoscience Australia, Canberra, Australia, [2]Australian National University, Canberra, Australia

Several deposit and mineral system classifications are available in the literature, with most focused on detailed analysis of a few deposit types using the mineral systems approach. However, these classifications are often not comprehensive and/or are too complex for non-expert users. This study presents a new mineral system classification which aims to address these problems. The new classification is exhaustive and incorporates new deposit types to account for current and potential future critical mineral resources (e.g., high purity silica). In addition, we review and classify the presence of economically extractable elements for each deposit type according to their global and local significance, with the aim of creating a simplified mineral system classification where mineral systems and deposit groups are linked to potential economic commodities.

For the classification, deposit types are grouped into broad mineral systems using similarities in critical processes that relate to metal, fluid and ligand sources, energy sources, fluid-flow drivers and pathways, architecture and depositional gradients. The mineral systems are then associated with broad geological features instead of geodynamic and tectonic settings, as commonly used in other classifications, to make the classification easier to apply, particularly given that geodynamic and tectonic settings are often debated in the geological record, especially for the Archean and Proterozoic. Furthermore, some mineral systems can develop in more than one geodynamic or tectonic setting, making these less reliable criteria to apply in a classification due to the many-to-many relationships.

The new classification comprises 202 deposit types categorized into 86 deposit groups that are associated with 13 mineral systems and six mineral system classes. This approach to mineral system classification may help explorers in regions where the geology is complex or relatively unknown by linking geological features to possible elemental enrichment and may assist industry, academia and government in conducting commodity-based mineral potential assessments.

Oxidized Mantle Sources in Cretaceous Komatiites from Gorgona Island: Insights into Mantle Redox Structure and Impacts on Ore-Forming Systems

Camilo Andrés Conde Carvajal[1], Iván Mateo Espinel Pachón[1], Ana Elena Concha Perdomo[2]
[1]University of Geneva, Geneva, Switzerland, [2]Universidad Nacional de Colombia, Bogotá, Colombia

The Gorgona Island, located 50 km off the Pacific coast of Colombia, is a well-known locality for the only known Phanerozoic examples of komatiites. These ultramafic volcanic rocks are suggested to be formed as part of the Caribbean Large Igneous Province (CLIP) during the Late Cretaceous (~88 Ma) within a complex, compositionally and thermally heterogeneous mantle plume (Kerr, 2005). This study aims to analyse the spinifex-textured komatiites on Gorgona Island to determine the mantle conditions during the Cretaceous, evaluating whether there are any differences between these and the Archean komatiites.

In this study melt oxygen fugacity has been estimated at ΔQFM + 3 ± 1 by spinel-olivine pairs (Ballhaus et al., 1991) using adquired data, and ΔQFM + 2.1 ± 0.1 by the partitioning of V between olivine and melt (Shishkina et al., 2018) using published data from Kamenetsky et al (2010). This discovery extends the evolutionary record proposed by Nicklas et al. (2019) of a relative reduced mantle in the Archean with ΔQFM values near 0 at ~3.6 Ga, that transitioned to ΔQFM ~ +1.3 by ~1.7 Ga, at least until the Cretaceous.

This determination by two independent methods challenges the conventional model that place komatiites origin from reduced mantle sources, raising questions about the heterogeneity of the mantle and the redox evolution of magma sources. These insights are critical for the genesis of ore models in which redox state influences metal solubility, and transport. To understand if there is an oxidised mantle domain that contribute to magmatic fertility may be critical to define new exploration criteria and expand the range of considered metallogenic settings previously underestimated.

Chemical and Isotopic (Sr, Nd) Analysis of the Nurra Basin Bentonites (NW Sardinia, Italy)

Sebastiano Coticelli[1], Giuseppina Balassone[1], Piergiulio Cappelletti[1], Massimo D'Antonio[1], Valeria Di Renzo[1], Concetta Rispoli[1], Fabio Granitzio[2], Andrea Testa[3], Nicola Mondillo[1]
[1]Dipartimento Di Scienze Della Terra, Dell'ambiente E Delle Risorse, Università Federico II, Naples, Italy, [2]Società Sarda di Bentonite, Cagliari, Italy, [3]Argillitti S.r.l., Cagliari, Italy

The Nurra bentonite basin, located in northwestern Sardinia, represents one of the most significant bentonite mining districts in Europe. Among its deposits, the S'Aliderru site stands out for its production, exceeding 300,000 tons/year of material. The bentonite mineralization in this basin is genetically linked to the alteration of calc-alkaline pyroclastic rocks of Oligo-Miocene age. This study investigates the magmatic provenance and alteration processes responsible for bentonite formation by analyzing samples from the two main deposits, S'Aliderru and Sa Pigada Bianca, and from the Mandra Ebbas exploration permit area. A total of 30 samples, including tuffs and associated bentonites, were characterized through geochemical analysis of major and trace elements, as well as Sr-Nd isotopic analysis. The geochemical data reveal a systematic depletion in SiO_2, K_2O, and Na_2O, coupled with an enrichment in Al_2O_3, MgO, and Fe_2O_3, consistent with argillic alteration. Trace element patterns are enriched in Ba (with the highest variability), Li and V, and are depleted in Sr and Rb. Isotopic compositions show $^{87}Sr/^{86}Sr$ values ranging from 0.706752 to 0.710448, and $^{143}Nd/^{144}Nd$ values ranging from 0.512279 to 0.512492. Geochemical and isotopic variations are observed at both deposit and basin scales. The poorly variable Nd isotopic signatures are coherent with the Oligo-Miocene volcanic rocks of the region. Furthermore, the broader variability in Sr isotopic ratios implies the involvement of multiple alteration processes, potentially including interaction with seawater and/or hydrothermal fluids. Interestingly, the highest quality facies have extremely high $^{87}Sr/^{86}Sr$ values. These findings underscore the geochemical complexity of the Nurra basin and point to the need for further multidisciplinary investigations to construct a comprehensive genetic model at the basin scale.

Structural Characterisation of the Ribeirão Grande Shear Zone, Southern Brazil

Graciany da Silva[1,2]
[1]SRK Consulting, Perth, Australia, [2]Universidade Federal do Paraná, Curitiba, Brazil

The Ribeirão Grande Shear Zone (RGSZ) is a major tectonic structure within the Apiaí Domain, southern Brazil, separating the Votuverava Formation and Perau Formation of the Açungui Group. The Votuverava Formation consists of sericite schists, phyllites, slates, metarenites, rare carbonate rocks, and metabasalts, reflecting a marine to transitional depositional environment. The Perau Formation features slates, phyllites, metarenites, metaconglomerates, carbonate rocks, and metavolcanic rocks, and is associated with lead and zinc mineralisation due to its carbonate-rich lithology and tectonic setting. This study examines the structural and tectonic significance of the RGSZ and its role in controlling Pb, Zn, and Cu mineralisation. A multidisciplinary approach integrates Digital Elevation Model (DEM) interpretation, field mapping, and microtectonic analysis. DEM interpretation reveals anastomosing lineaments with stepped geometries, indicating shear-related patterns. Field mapping identifies ductile deformation near the shear zone core and transitional ductile-brittle structures in distal areas. Microtectonic analysis highlights multiple deformation phases, with alternating dextral and sinistral shear movements, confirming the polyphase nature of the RGSZ. Stress inversion analysis defines principal axes: σ_1 trending NE-SW (compressional), σ_2 vertical, and σ_3 trending WNW-ESE (extensional). The RGSZ exhibits ductile deformation followed by a ductile-brittle transition, displacing three blocks: western and eastern blocks moving northeastward, and the central block shifting southwestward. These findings enhance understanding of tectonic controls on mineralisation within the Perau Formation, particularly in carbonate-rich lithologies. By analysing fault network geometry, strike-slip systems, and kinematic segmentation, the study provides a framework for predictive exploration models. Structural controls and fluid flow pathways identified in the RGSZ guide exploration strategies, optimising the discovery of economically viable Pb, Zn, and Cu deposits across the Apiaí Domain.

Geochemical and Structural Controls of the Gold-Bearing Carbonaceous Veins Hosted Malenggang Project, West Kalimantan, Indonesia

Arif Zardi Dahlius[2], Alison Purba[1], Nayarudin Nofiya Rahmat[1], Kelik Prasetyo Purnomo[1]
[1]PT Rekabuana Penjuru Mineral, Jakarta, Indonesia, [2]Institute of Technology Bandung, Bandung, Indonesia

The Malenggang project in Sanggau is hosted by Triassic metasedimentary rocks and conglomerate which are overlain by alluvium and paleo-placer sedimentary rocks. The mineralization is associated with a northwest-striking, regional-scale shear zone that dips gently and contains parallel vein systems. A recent and systematic exploration program—including detailed geological and alteration mapping, trenching, lithogeochemical surveys, diamond drilling, geophysical surveys, and mineralogical studies—has been undertaken to target the main veins. Quartz-albite veins at Malenggang shows pinch-and-swell texture both along the strike and the deepening, indicating syn-tectonic formation. Hydrothermal alteration is dominated by albite and silica, with minor chlorite and calcite. Gold occurs in several forms, including: (i) native gold within quartz-albite veins, (ii) along carbonaceous-rich metasediments, and (iii) within the mylonitic shear zone. Gold shows a strong positive correlation with arsenic content (median of 1.4 ppm Au; 870 ppm As, n=2535). High gold grades belong to fractured highly deformed mylonitic carbonaceous mudstone and sulfide-bearing veinlets. The gold in metasediment-hosted orogenic systems is proposed to be sourced from metamorphic devolatilization during orogenic events, whereby metamorphic fluids transport gold and other metals into the shallower crust.

Resolving the Syn-Genetic Age of the King Metamorphosed VHMS Deposit, Yilgarn Craton, Western Australia

Cendi Dana[1], Steven Hollis[1], Kazuhide Mimura[2,3], Moei Yano[2,4], Junichiro Ohta[2,4], Yusuke Kuwahara[2,4], Lorenzo Tavazzani[5], Cyril Chelle-Michou[5], Stijn Glorie[6], David Selby[7], Yasuhiro Kato[2,4], Vanessa Pashley[8], Megan James[9], Darryl Podmore[9]
[1]School of GeoSciences, The University of Edinburgh, Edinburgh EH9 3FE, United Kingdom, [2]Ocean Resources Research Center for Next Generation, Chiba Institute of Technology, Chiba 275-0016, Japan, [3]National Institute of Advanced Industrial Science and Technology, Ibaraki 305-8567, Japan, [4]Frontier Research Center for Energy and Resources, The University of Tokyo, Tokyo 113-8656, Japan, [5]Department of Earth Sciences, ETH Zurich, Zurich 8092, Switzerland, [6]Department of Earth Sciences, University of Adelaide, Adelaide, SA 5005, Australia, [7]Department of Earth Sciences, University of Durham, Durham DH1 3LE, United Kingdom, [8]Geochronology and Tracers Facility, British Geological Survey, Nottingham NG12 5GG, United Kingdom, [9]Black Raven Mining, Perth, WA 6872, Australia

The exploration of VMS deposits in high-grade metamorphic terranes can be quite challenging due to intensive deformation and metal remobilization. Metamorphism can also cause isotopic redistribution, obscuring the syn-genetic age of VMS mineralization and leading to misinterpretations of regional metallogenic models. In this study, we employed multiple geochronological approaches to constrain the temporal evolution of the King metamorphosed VMS deposit in the Eastern Goldfields Superterrane, Yilgarn Craton.

The peak metamorphism age, constrained by Lu-Hf garnet geochronology, yielded an isochron age of ca. 2680 ± 27 Ma. This age is consistent with the M2 regional-contact metamorphism associated with the emplacement of Ca-rich granitoid intrusions. On the other hand, the footwall felsic unit, which is the host rock of massive sulfide mineralization, yielded a weighted mean age of 2725 ± 10 Ma.

Detailed petrographic observations revealed that pyrrhotite was formed by the desulfidation reaction of pyrite during metamorphism. Consequently, pyrrhotite is expected to record the metamorphism age, whereas pyrite would preserve the syn-genetic mineralization age. Systematic Re-Os geochronology revealed that recrystallized pyrite grains yielded a Re-Os isochron age of 2729 ± 34 Ma, consistent with the felsic footwall. Meanwhile, pyrrhotite yielded an isochron age of 2654 ± 39 Ma, consistent with the metamorphism age. Additionally, we analyzed molybdenite hosted by late granitoid intrusions, which yielded a weighted mean age of 2648 ± 16 Ma. Molybdenite occurring along the foliation of footwall biotite-quartz schist also consistently yielded a weighted mean age of 2658 ± 23 Ma.

In the Eastern Goldfields Superterrane, significant VMS mineralization is typically restricted to the Kurnalpi Terrane between approximately 2705 and 2680 Ma. However, this study demonstrates that an older period of VMS mineralization also occurred in the Eastern Goldfields Superterrane, coeval with the youngest period of VMS mineralization in the Youanmi Terrane, Western Yilgarn (ca. 2725 Ma).

Garnet Geochemistry as a Tracer of Timing and Fluid Pathways in Pb-Zn Systems

Joëlle D'Andres[1], John A. Mavrogenes[1,2], Nick Dyriw[3]
[1]Research School of Earth Sciences, Australian National University, Canberra, Australia, [2]ARC Training Centre for Critical Resources for the Future, Australia, [3]MMG Ltd, Australia

Garnet can be a powerful tracer of metamorphic and hydrothermal processes in ore systems, preserving chemical and isotopic signatures that reflect its formation conditions and evolution. We present a multi-proxy geochemical study of garnet from two contrasting Pb-Zn deposits in the Eastern Succession of the Mount Isa Inlier: A Broken Hill-type (BHT) system hosted by migmatitic gneisses and a clastic dominated Pb-Zn deposit hosted in shales.

Garnets were collected across ore zones, alteration halos, and distal host rocks, and analysed for major and trace elements (EPMA, LA-ICP-MS), oxygen isotopes (SHRIMP), and Lu-Hf geochronology on selected samples (LA-ICP-MS/MS). In both deposits, garnets near ore show elevated Mn concentrations and $\delta^{18}O$ values up to ~3‰ higher than distal garnets, consistent with metasomatic overprint by external fluids—such as metal-rich evolved basinal brines, Mn-enriched metamorphic fluids derived from dehydrating sediments, or rock-buffered fluids that equilibrated with high-$\delta^{18}O$ sediments. In addition, Zn enrichment in near-ore garnets and, in the shale-hosted system, Lu-Hf ages overlapping with the timing of mineralisation, further support a genetic link between garnet growth and ore-forming processes.

Despite these shared trends, garnets from the two systems show distinct signatures. Notably, in the BHT system, near-ore garnets are also enriched in Ca, display positive Eu anomalies and lower $\delta^{18}O$ values (up to ~11.5‰ vs. ~13.5‰ in the shale-hosted system) possibly reflecting higher-temperature conditions, interaction with Ca-rich lithologies and/or more reducing fluid conditions.

These results demonstrate that garnet Mn contents and elevated $\delta^{18}O$ can serve as reliable path finders to mineralisation. The integration of stable isotope geochemistry, trace element signatures, and in-situ geochronology of garnet provides a robust framework for reconstructing fluid evolution and timing in sediment-hosted ore systems, with garnet offering a unique, time-integrated archive of metal-bearing fluid pathways.

Genesis of IOA-IOCG deposits, Coastal Cordillera, Peru: Insights from Iron ($\delta^{56}Fe$) and Triple Oxygen ($\delta^{18}O$, $\Delta^{17}O$) Isotopes in Magnetite

Jose Carlos Davila[1], Fredrik Sahlström[1], Fernando Tornos[2,4], Valentin Troll[3], John Hanchar[4], Ilya Bindeman[5], Xin-Yuan Zheng[6]
[1]Pontificia Universidad Católica Del Perú, Lima, Peru, [2]Consejo Superior de Investigaciones Científicas, Madrid, Spain, [3]Uppsala University, Uppsala, Sweden, [4]Memorial University of Newfoundland, St. John's, Canada, [5]University of Oregon, Eugene, USA, [6]University of Minnesota – Twin Cities, Minneapolis, USA

Iron-oxide – apatite (IOA) and iron-oxide – copper-gold (IOCG) deposits represent significant metal resources needed for the green energy transition. However, their genesis, and possible relationships between these deposit types, remain poorly understood and are a matter of intense debate. The Cordillera de la Costa of southern and central Peru hosts several IOA and IOCG deposits that have been comparatively little studied in a global context, and they may thus be able to provide new insights into the genesis of these enigmatic but economically important ores. We report new iron ($\delta^{56}Fe$) and triple oxygen ($\delta^{18}O$, $\Delta^{17}O$) isotope data for 23 magnetite samples collected from two IOA (Marcona, Hierro Acari) and two IOCG (Mina Justa, Raul-Condestable) deposits in Peru, as well as from the nearby Pampa del Pongo iron skarn deposit. Petrographic characterization of magnetite-bearing samples from Mina Justa and Raul Condestable was also done to provide a textural context for the isotope data. The measured magnetite isotopes ranges are $\delta^{18}O$ = 0.17 – 5.30 ‰ and $\delta^{56}Fe$ = - 0.06 – 0.49 ‰. The values from all five deposits essentially overlap with the ranges determined for ortho-magmatic magnetite, thus indicating a predominantly magmatic origin. Nevertheless, the magnetite $\Delta^{17}O$ values (- 0.097 – - 0.016 ‰) are in part heavier and in part lighter than typical magmatic values. This suggests local alteration by low-temperature fluids and input from evaporites, respectively. Further analysis of our new isotope data will provide new insights into the genesis of and relationships between IOA, IOCG, and iron skarn deposits in Peru. Our insights will assist in the development of novel exploration strategies for these important metal resources in Peru and likely also on a global level.

Relative Timing of Au Mineralization at the Epithermal Calm Before the Storm Zone, Treaty Creek Area, Golden Triangle, British Columbia

Nikola Denisova[1], Kate Rubingh[1], Foteini Drakou[1], Shaun Barker[1]
[1]Mineral Deposit Research Unit, University of British Columbia, Vancouver, Canada

The Golden Triangle area in northwest British Columbia is part of the North American Cordillera and comprises Triassic and Jurassic age rocks of the Stikine island-arc terrane. Porphyry Cu-Au, epithermal Au-Ag, and base metal volcanogenic massive sulfide deposits and occurrences are hosted along prospective horizons, commonly unconformities, or pre-existing lineaments.

The Treaty Creek area occurs at the north termination of the Sulphurets trend, a series of world-class Cu-Au porphyry deposits along a northerly structure. In Treaty Creek, the Goldstorm Au-Cu porphyry deposit (Indicated Mineral Resource of 730.2 Mt @ 1.19 g/t AuEQ) is linked with variably Au-mineralized epithermal-style systems, such as the Calm Before the Storm (CBS) zone.

The high sulfidation CBS zone comprises systematically zoned advanced argillic assemblages that formed at or within 1000 m of the paleosurface. Gold occurs in the alunite- and pyrophyllite- dominant zones; however, Au deportment is unclear. Petrographic and mineral chemistry studies suggest that Au is associated with pyrite, which is ubiquitous across the alteration assemblages. Pyrite grains rarely exceed 200 μm. Anhedral pyrite (<50 μm) occurs dispersed in the host rock. Crosscutting veins contain pyrite as anhedral-subhedral grains with locally abundant inclusions, and as euhedral-subhedral clean grains, locally zoned with inclusion-rich rims or cores. Element maps acquired using laser ablation inductively coupled mass spectrometry are used to determine the deportment of Au and other elements hosted in pyrite.

Constraining the Au deportment will help resolve the relative timing of Au mineralization in relation to the advanced argillic alteration assemblages at CBS, the other epithermal Au-mineralized zones, and the paleosurfaces they represent. This will assist integration of the Goldstorm deposit with Treaty Creek's mineralized zones into a comprehensive model of the magmatic-hydrothermal system in the Treaty Creek area and aid exploration in the Golden Triangle and similar camps worldwide.

Contribution of Auger and Reverse Circulation to the Definition of Waraba Mineralization (Loulo Permit)

Zoumana Diallo[1]
[1]National School of Engineers ABT, Bamako, Mali

This end-of-studies project focuses on the contribution of auger and reverse circulation (RC) drilling in the definition of mineralization at Waraba, a target located at the northernmost point of the Loulo permit. The mapping of the regoliths shows that Waraba is an area covered by transported laterite, i.e. dominated by the depositional regime. In 2021, an auger drilling program was conducted to test geochemical anomalies over an area over of 800mx1500m and utilising a grid of 200mX50m. Ideally, 2 samples of 1 meter are taken for analysis in each vertical hole: the first at the base of the laterite or pedolith and the second level of full saprolite. At Waraba, the profile is defined by the transported cover, the mottled clay zone and the saprolite. This program confirmed that the geochemical anomalies were "in situ" and RC drilling program was then executed in order to define the continuity of the mineralization at depth and also its geometry. The lithologies intercepted include sandstones, breccias, mafic intrusions and gossans. The alterations observed are silicification, albitization, hematization, limonitization, chloritization, tourmalinization and carbonation. Mineralization is primarily related to disseminated fine pyrite in silicified and albitized medium to coarse sandstones, as well as silica carbonate veins. The mineralized zone strikes NNE (028/80E) to N (360/63E) with the main shoot having an average width of 4 m and plunging northwards. Drilling confirmed a vertical depth of 100m for the orebody.

Polyphase Cobalt Mineralisation in the Rajapalot Au-Co Deposit, Finland

Lily Dickson[1], Taija Torvela[1], Rob Chapman[1], Jukka-Pekka Ranta[2]
[1]University Of Leeds, Leeds, United Kingdom, [2]University of Oulu, Oulu, Finland

The Paleoproterozoic greenstone belts of northern Finland host orogenic gold deposits with 'atypical metal associations' [1]. This group includes Rajapalot, situated in the Peräpohja Belt, which is enriched in cobalt [2]. The timing and processes of cobalt mineralisation, and potential genetic link to the Au mineralisation, at Rajapalot (and other Au-Co deposits in northern Finland) has been explored [3] but remains only partially understood.

In this study, detailed optical and scanning electron microscopy, has formed the basis of a paragenetic interpretation [4], elucidating multiple stages of cobalt mineralisation associated with different p-T conditions, deformation and alteration phases. Major and trace element geochemistry for Co and Fe sulphides and sulpharsenides from EPMA and LA-ICP-MS respectively further characterises the different stages of cobalt mineralisation.

Cobaltite (Cbt; CoAsS) solid solution (ss; where Co substitutes for Ni and Fe) minerals form across four paragenetic phases at Rajapalot, identified by distinct textures, associations and chemistries. Glaucodot, a Fe-rich cobaltite ss mineral, forms early in the paragenesis and is associated with pyrite, which shows fluctuation in cobalt and arsenic activity through its growth zoning. Cbt1 and Cbt2 record compositions between alloclasite and cobaltite ss members. They are both fine grained and inclusionless but have different nickel concentrations. These stages have experienced ductile deformation, resulting in an alignment with the foliation of the host rock, and dynamic recrystallisation, evidencing prolonged mineralisation over changing p-T conditions. The fourth cobaltite ss phase, Cbt3, is coarse grained, Co-rich, and commonly hosts gold inclusions, indicating coeval mineralisation. Finally, the mobilisation of trace elements and their signatures in different sulphide/sulpharsenide generations from LA-ICP-MS point and map analyses are also discussed.

[1] Hector et al. (2023). OGR, 154.
[2] Rantala et al. (2021) NI43-101 Report, AFRY.
[3] Sayab et al. (2025) MinDep, 60, 213-231.
[4] Webb et al. (2024) GSL, 541, 273-297.

Morphochemical Analysis of Gold in the Pinalejo Placer: The First Quantitative Analysis of Gold Wear Rate During Fluvial Transport

Kelvin Dos Santos Alves[1], Santos Barrios Sánchez[1], Daniel Lopez Bermejo[2], Lorena Ortega Menor[2], Juan Gómez Barreiro[2]
[1]Universidad De Salamanca, Salamanca, Spain, [2]Universidad Complutense de Madrid, Madrid, Spain

Understanding the morphochemical evolution of gold during its transport is essential for developing effective prospecting models for new gold deposits. Although numerous authors have used morphological indices to describe the shape and flattening of gold particles, to date, no studies are known to quantify gold mass loss due to mechanical wear during transport.

In this study, the morphochemical evolution of gold was analyzed at 11 sampling points along a 6 km stretch of the Pinalejo Placer, from its release at the source area. The main objective was to understand and quantify gold particle mass loss caused by mechanical abrasion during fluvial transport.

Two populations of gold particles, A and B, were identified, distinguished by their average size and morphological maturity. Population A ranged in size from 0.02 to 0.68 mm, while Population B ranged from 0.07 to 1.16 mm. The morphological contrast between these populations, together with differences in the area's geomorphology, directly influenced the particle abrasion rate.

In this regard, the wear coefficient (k) values for Population B were higher than those for Population A. Furthermore, the steeper section of the placer (P3–P6) exhibited higher k values for both populations compared to the smoother zones (P7–P11). The k values for sections P3–P6 and P7–P11 were 3.21 km^{-1} and 1.18 km^{-1} for A, and 3.84 km^{-1} and 2.77 km^{-1} for B, respectively.

Compositional analysis of gold from three source zones revealed the following Au–Ag ratios: $Au_{72-96}Ag_{28-4}$ (first source), $Au_{91-97}Ag_{9-3}$ (second source), and $Au_{91-97}Ag_{9-3}$ (third source), which are typical of orogenic gold.

Is the Zabrnjica Prospect on Ober License in Southwestern Serbia a Previously Unrecognized Reduced Intrusion Related Gold Systems (RIRGS)?

Dragan Dragic[1]
[1]Konstantin Resources, Belgrade, Serbia

Exploration work conducted by Konstantin Resources on its 100% owned Ober exploration licenses within the Priboj Project area in southwestern Serbia, revealed to date unrecognized Au mineralization in this part of the Serbia (Janković & Petrascheck, 1987, Janković, 1990, Janković et al., 1992, Janković et al., 2003, Jelenković, 1999). Ober exploration licenses occupy central eastern limb of the Javorje anticline comprising undifferentiated Devonian/Carboniferous schists and meta sandstones in the core and undifferentiated Permian/Triassic meta sandstones and conglomerates occurring on the flanks of the anticline in contact with Jurassic Ophiolite complex. Field work to date including geochemical work of stream sediments, soil and rock chip sampling indicated presence of gold in the undifferentiated Perm-Triassic clastic sediments. This gold mineralization is associated with silica-sericite altered terrigenous sediments, predominantly sandstones and conglomerates with quartz veining that are partially leached. Zabrnjica prospect is most prominent prospect at the Ober license and main focus on exploration activity on Ober license to date.

Reduced Intrusion Related Gold Systems (RIRGS) are associated with the ilmenite intrusive series plutons forming in weak post-collisional extension behind a thickened continental margin tectonic settings and emplacement of depth between 3 and 8 km. These systems are characterized by arrays of sheeted quartz veins where gold bearing fluids are Au-Te-Bi complex. (Thompson et al., 1999; Lang et al., 2000; Lang and Baker, 2001; Hart et al., 2002; Baker et al., 2005, 2006; Hart., 2007; Hart., 2024).

Reviewed field observation along with the geochemical signature of the soil and rock samples raised working hypothesis that Zabrnjica prospect can be previously unrecognized RIRGS and therefore attempt was made to compere Zabrnjica characteristic with the genetic characteristic of the RIRGS and known deposits of the group.

Geological Controls on Critical Metal Endowment in Sedimentary-Hosted Zn-Pb Deposits: Insights from the Canadian Cordillera

Foteini Drakou[1], Zoe Lynn[1], Darius Kamal[1], Jinglin Xu[1], Shaun Barker[1], Kenneth Hickey[1]
[1]MDRU - Univeristy of British Columbia, Vancouver, Canada

Sediment-hosted massive sulfide (SHMS) Zn-Pb deposits are increasingly recognized as potential secondary sources of critical metals, including germanium (Ge), gallium (Ga), indium (In), and tin (Sn). These elements are commonly incorporated into sphalerite and can be recovered as by-products during zinc smelting. The Canadian Cordillera hosts several major SHMS districts - including Howard's Pass and Macmillan Pass (Selwyn basin), the Gataga district (Kechika Trough), and the Purcell district - formed within Mesoproterozoic to Devonian-aged sedimentary sequences.

Trace element analysis of sphalerite from several ore bodies across the four districts reveal significant variability in critical metal concentrations, reflecting a combination of the original hydrothermal fluid compositions and the effect of post-depositional chemical modifications. The Macmillan Pass district exhibits the highest Ge contents of the region, with sphalerite from the Boundary Zone and Tom deposits reaching up to 320 ppm Ge. At the Tom deposit, Ge shows a strong correlation with Cu that is consistent with the coupled substitution mechanism (2 $Zn^{2+} \leftrightarrow Cu^{+} + Ge^{4+}$). In contrast, elevated Ge at the Boundary Zone appears to reside in micro-inclusions, likely linked to localized post-depositional processes. In the Purcell basin, sphalerite from the amphibolite-facies metamorphosed Sullivan deposit is chemically homogeneous and depleted in most critical metals except In. Elevated whole-rock Sn concentrations reported for the Sullivan deposit are hosted entirely in cassiterite rather than sphalerite. Across the four districts, Ga concentrations in sphalerite are consistently low (<15 ppm), probably reflecting a Ga-poor signature of the mineralizing fluids.

These findings highlight the important role of post-depositional processes in controlling the redistribution and retention of trace metals within the SHMS deposits. Advancing our understanding of trace metal mobility in these settings is essential to unlocking the largely untapped critical metal potential of sedimentary-hosted Zn-Pb systems.

Oak Dam West IOCG deposit, South Australia – Evolving Geological Knowledge as Resource Definition, Advanced Exploration and Mining Studies Continue

Michael Fonti[1], Marcel Menicheli[1], Lauren Brown[1], Camla Rochat[1], Kathy Ehrig[1]
[1]BHP Copper South Australia, Adelaide, Australia

Oak Dam West is a breccia hosted iron-oxide copper-gold (IOCG) deposit located approximately 65 km southeast of the Olympic Dam IOCG deposit, along the eastern margin of the Gawler Craton. Even though the first drill hole into the deposit was completed in 1981 to a depth of 1256.2m, it intersected hematite-rich, Cu-barren breccias from 727m to end-of-hole. The overlying 727m of undeformed and unaltered Meso- to Neoproterozoic sediments and ~530m of barren breccias were not considered of interest because the potential mineralized rocks occurred well below the economic mining depth of the 1980s. However, two further holes were completed within a few years, with one reaching basement of weakly altered and unmineralized granitoids at ~875m depth.

The prospectivity of the area was re-evaluated by the BHP Exploration team in 2017. In mid-2018, AD22 intersected chalcopyrite-pyrite mineralized breccias. AD23 intersected high copper-grade chalcocite-rich hematite breccias, along the contact with the central Cu-barren breccias, with 425.7m at 3.04 % Cu and 0.59 ppm Au, becoming the discovery hole. Since then, resource definition, advanced exploration activities and mining studies continue, with the first Mineral Resource (1340 Mt, 0.66 % Cu, 0.33 ppm Au) reported in August 2024.

The challenges, opportunities and ore body knowledge evolution by the Oak Dam geoscience team have been significant. Because the deposit occurs below ~750-900m of cover and mineralization is steeply dipping, it is not unusual for drill holes to exceed 2km in length, necessitating the use of navigation drilling. Assaying of all basement drill core samples for an extended suite of elements, supported by measured mineralogy has improved the classification of alteration zonation patterns as well as assessing AMD risk. Seismic surveys, completed to assist with rock mass characterization of the cover sequence, were instrumental in understanding the underlying deposit structural architecture and revealing new drilling targets.

Quantitative Analysis of Mineral Orientation in Sulphide Ore Deposits

Tom England[1], Nathan Daczko[1], Nicholas Dyriw[2], Siyu Hu[3], Elena Belousova[4]
[1]Macquarie University, Sydney, Australia, [2]MMG Dugald River, Mount Isa, Australia, [3]CSIRO, Kensington, Australia, [4]Geological Survey of Queensland, Brisbane, Australia

Numerous ore deposits within the globally significant base metal province of the North West Queensland Mineral province are complicated by multiple deformation events. Interpretations for timing of mineralisation vary from during, and after tectonic activity.

Detailed surficial and underground mapping and observations of ore bodies aim to identify key deformation structures such as regional foliations, shear zones, and fault zones to assist in the structural paragenesis for each deposit. Sometimes due to numerous over printing relationships interpretation is not always clear.

Traditionally, structural investigations of these deposits have relied on methods such as regional mapping, hand sample analysis, thin-section petrography, and scanning electron microscopy (SEM). However, Electron Backscatter Diffraction (EBSD) has recently emerged as a powerful tool in structural and metamorphic geology, enabling high-resolution, quantitative analysis of crystallographic orientation, microstructure, and deformation processes down to the nanometre scale.

Despite the complex deformation histories of many sulphide deposits, EBSD has not yet been widely applied to examine the microstructural evolution of sulphide ore minerals in relation to well-characterised regional structures.

This study investigates the potential of EBSD as a microstructural characterisation technique for sulphide ore deposits. We start with a case study of the Dugald River Zn-Pb-Ag deposit in the Eastern Succession of the Mt Isa Inlier. This deposit is subject to at least three deformational-metamorphic events. Previous research conclusions differ in the assessment of how deformation has affected sulphide precipitation. Using EBSD we will assess deformational features across the ore lode including curious massive sulphide textures. We aim to better constrain the microstructure, microchemistry, and mineral assemblages of the deposit, elucidate the timing and nature of deformation events, and refine the structural paragenesis and mineralisation history through the dating and characterisation of key mineral assemblages.

Geochemical and Mineral Inclusion Analysis of Alluvial Gold as Exploration Tools: Experiences from the Central Iberian Zone, Spain

Vanessa Leonor Escobar Duche[1], Santos Barrios, Kelvin dos Santos, Juan Gómez Barreiro
[1]Universidad De Salamanca, SALAMANCA, Spain

Gold occurrences in the Central Iberian Zone have been known since pre-Roman times, principally in placers, where primary sources remain uncertain. This study applies a methodology that includes microgeochemical analyses established by Electron Microprobe Analysis (EMPA) and the mineralogical study of mineral inclusions using Scanning Electron Microscopy with Energy Dispersive X-ray Spectroscopy (SEM-EDS) on two Holocene placer deposits — the Los Ángeles River placer and the Pino del Oro placer — providing a distinctive geochemical 'signature' that aids in tracing their primary sources. In the Río Los Ángeles placer, three gold types were identified: Au_1 (Au-Ag alloy, mean 86.1 wt% Au), Au_2 (high-purity rims, mean 97 wt% Au), and Au_3 (fine-grained gold associated with Fe-oxides and clays). Mineral inclusions, present in 23% of particles, mainly consisted of sulfides (e.g., galena, pyrite, arsenopyrite) and sulfosalts (e.g., boulangerite). The geochemical data indicating a mixture of gold populations, possibly related to vertical zoning in an orogenic source and/or to different hydrothermal mineralization pulses. In the Pino del Oro placer, a similar methodology revealed two gold types: Au_1 (electrum with 16–24 wt% Ag) and Au_2 (high-purity rims). Mineral inclusions, found in 11.8% of particles, included arsenopyrite, galena, sphalerite, and members of the cobaltite group. The compositional data suggest a hydrothermal origin at lower temperatures (<200°C) and indicate complex zonation in the original lode system. The consistent application of combined geochemical and mineral inclusion analyses across both sites demonstrates that both placer deposits originated from distinct stages of orogenic (amagmatic) mineralization. Furthermore, secondary supergene processes, notably silver leaching, significantly altered the gold particle rims during sedimentary transport. These results underline the effectiveness of integrated compositional and mineral inclusion analyses of gold particles in reconstructing the provenance and post-depositional history of alluvial gold.

Constraining Magma Evolution Associated with Alkalic Porphyry Mineralization in British Columbia, Canada: A Melt Inclusion Study

Maria Paula Estrella[1], Pilar Lecumberri-Sanchez[1], Xinyue Xu[1], Chris Rees[2]
[1]University Of Alberta, Edmonton, Canada, [2]Imperial Metals Corporation, Vancouver, Canada

As global copper demand rises, there is growing interest in relatively scarce but high-grade Cu-Au systems such as alkalic porphyries. This study aims to determine the source of alkalinity in magmas related with porphyry-style mineralization through the study of mineral chemistry and melt inclusions.

Alkalic porphyries are localized in two main metallogenic provinces worldwide: British Columbia, Canada and New South Wales, Australia. Although these share similarities with their calc-alkalic counterparts, a large question in alkalic porphyries is what is the source and/or processes that drive magma alkalinity, as this influences where metals are acquired and how they are enriched. Here we study mineralization-related volcanic and intrusive units from the Late Triassic – Early Jurassic Mount Polley alkalic deposit to determine such processes.

All magmatic pulses in Mount Polley show significant carbonate and sulfate components in melts responsible for clinopyroxene crystallization. Melt inclusions hosted in unaltered clinopyroxene crystals from four chronologically constrained units at Mount Polley (pre-, syn- and post-mineralization) contain up to 50% volume contents of sulfate and/or carbonate components. Additionally, one sample analyzed by Focused Ion Beam – Secondary Electron Microscope (FIB-SEM), revealed ~5%vol. of chloride, suggesting high chlorine content.

Calc-alkalic porphyry-related melts are known to become sulfate-saturated during crustal evolution. The presence of sulfate, carbonate, and possibly halides in Mount Polley melt inclusions suggests involvement of evaporitic material over a prolonged magmatic history. However, whether this reflects subducted evaporites or crustal assimilation at this point remains uncertain.

Melt sources, assimilation, and immiscibility determine whether porphyries are fertile or barren. The distinct metal and geochemical signatures of alkalic porphyries may reflect their unique source and/or magmatic evolution. In these deposits it is likely that magma interaction with evaporites could impart both alkalinity and a redox signature, which ultimately will influence metal partitioning and consequent metal endowment.

Global Controls on the Formation of Porphyry Systems: Insights from Machine Learning and Plate Tectonic Reconstructions

Ehsan Farahbakhsh[1], Sabin Zahirovic[1], Brent McInnes[2], Fabian Kohlmann[3], Maria Seton[1], R. Dietmar Müller[1]
[1]EarthByte Group, School of Geosciences, The University of Sydney, Sydney, Australia, [2]John de Laeter Centre, Faculty of Science and Engineering, Curtin University, Perth, Australia, [3]Lithodat Pty. Ltd., Melbourne, Australia

Porphyry systems account for the majority of global copper production and host significant gold and molybdenum resources. As demand for copper grows, driven by the expansion of renewable energy technologies, understanding the controls on porphyry formation has become increasingly important. These deposits are typically associated with subduction-related volcanoplutonic arcs, but a limited understanding of the processes within the subducting and overriding plates, as well as the mantle wedge, constrains our ability to improve prospectivity mapping and exploration targeting. In this study, we develop a global prospectivity model for porphyry mineralisation using a machine learning approach trained on a global database of porphyry occurrences. This model incorporates spatiotemporal features from a global plate tectonic reconstruction, including ocean basin evolution from 540 Ma to the present. Our results highlight three primary factors influencing porphyry formation: total carbon density of the subducting plate, distance to the trench, and pelagic sediment thickness on the downgoing slab. Porphyry deposits are most commonly found 200–400 km from the trench, typically associated with subducting slabs that have carbon densities exceeding 10 t/m^2 and carry pelagic sediments thicker than 200 m. We postulate that high carbon densities may be related to volatile release into the mantle wedge, promoting metasomatism, magmatism, and mineralisation. Trench distance likely reflects slab dip, with the degree of mechanical coupling between the subducting and overriding plates as a secondary factor influencing crustal deformation and volatile pathways. Thick pelagic sediments may indicate higher water content subducted with the slab, which is released as fluids that metasomatize the overlying asthenospheric mantle wedge—conditions favourable for the redistribution and enrichment of soluble metals (e.g., Cu, Au, Pd) in metasomatised mantle regions. These findings offer new quantitative insights into the tectonic and geochemical controls on porphyry formation, establishing a robust framework for global-scale exploration targeting.

Decoding Structural Influences on Vein Formation in the Japudali Prospect, Indonesia

Dian Yesy Fatimah[1], Abu Shidiq[2], Muhammad Sujaka[2]
[1]Universitas Pertamina, South Jakarta, Indonesia, [2]Minevesting Resources Indonesia, West Jakarta, Indonesia

Indonesia, located within the tectonically active Pacific Ring of Fire, hosts extensive volcanic and magmatic arcs formed by the complex convergence of the Pacific, Indo-Australian, and Eurasian plates. The Sunda-Banda magmatic arc in the southwestern archipelago (Wulandari, 2015) exemplifies a long-lived subduction-related magmatic system. The interplay of arc magmatism, crustal deformation, and hydrothermal fluid evolution has positioned Indonesia as a major metallogenic province, hosting widespread occurrences of metallic mineralization across varied lithospheric domains. This study investigates the relationship between structural geology and the spatial distribution of hydrothermal alteration and mineralization within a 6×10 km² area in Sukabumi Regency, Southwestern Java. The research was guided by a multidisciplinary approach that integrated literature review, remote sensing, field mapping, and laboratory analyses. The geological structure is characterized by three major fault systems: the Japudali thrust left slip fault (NW–SE), normal right slip fault (NNW–SSE), and the Pasawahan right slip normal fault (WSW–ENE). And there are three subordinates NE-SW trending faults: the Cihaur left thrust slip fault, Cihaur normal left slip fault, and Cihaur lag right slip fault. These structures are observable from the outcrop to micro-scale features such as mineral grain microfractures. Vein distribution near minor faults indicates localized hydrothermal fluid flow, while extensive alteration zones (propylitic, argillic, advanced argillic, and phyllic) are widespread. Findings suggest that the fault networks, active during both pre-mineralization and syn-mineralization stages, acted as principal conduits for hydrothermal fluid migration and subsequent mineral deposition. This paper presents all details about structural geology information within the hydrothermal alteration, with the outcomes of the distribution of hydrothermal alteration occurrences in Sukabumi Regency, Southwestern Java, to serve as a recommendation for detailed exploration in the future.

Keywords: structural control, hydrothermal alteration, Sukabumi regency

Pyrite Chemistry as a Tracer for Metallogenic Evolution in the Au-Ag-Te Kochbulak Deposit (Chatkal-Kurama Region, Uzbekistan)

Danis Ionut Filimon, Rustam A. Khalmatov[2], Paolo Fulignati[1], Anna Gioncada[1]
[1]Earth Sciences Department, University of Pisa, Pisa, Italy, [2]Institute of Geology and Geophysics, University of Geological Sciences, Tashkent, Uzbekistan

Pyrite trace element signature may contribute to untangle hydrothermal processes, particularly in cases with complex geochemistry and mineralogy such as gold-tellurides deposits. The Chatkal-Kurama region (Uzbekistan) is located in the Central Asian Orogenic Belt, specifically in the Middle Tien Shan Metallogenic Belt. It hosts the Valerianov-Beltau-Kurama magmatic arc, originated by the subduction of the Turkestan Ocean. From Middle-Late Ordovician to Early Permian, the geodynamic setting favoured magma ascent and storage in the crust, leading to the genesis of porphyry copper, epithermal and skarn ore deposits. One of these is the Au-Ag-Te Kochbulak deposit, which includes three different orebodies consisting of high- and low-angle veins and steep lenticular to tubular breccias. The geochemical signature, with positive anomalies of Te, Se, Sb, Cu, Zn, Pb and Bi, manifests in a complex mineral assemblage of sulfides, sulfosalts and tellurides associated with gold and electrum. A multi-stage ore-forming process has been identified, with cyclic input of hydrothermal fluids, including pre-ore, main-ore (subdivided into I, II, IIIa and IIIb phases) and post-ore stages. This complex process involves physico-chemical changes of the fluids, such as pH changes, decreasing T and changing fS_2 and fTe_2, in addition to reopening events of the mineralized veins.

In this work we determined by LA-ICP-MS the trace element composition of pyrite, which is present in the mineral assemblage at all stages. The variable Cu-Zn-As-Se-Ag-Te-Au-Bi-Pb contents in pyrite, joined to the identification of polymetallic inclusion, track the hydrothermal fluid evolution and suggest a role for low-melting-temperature Te-rich melts in scavenging Au from the fluid, providing insights into the genesis and evolution of this high-grade epithermal deposit. Defining the genesis of the Au-Ag-Te Kochbulak deposit will contribute to understanding the metallogeny of the Chatkal-Kurama epithermal deposits, in addition to providing constraints useful for the comprehension of epithermal gold-telluride deposits elsewhere.

Integrated Exploration Approach Unlocks a Deep-Seated High-Grade Copper Mineralization in the Jervois Range, Northern Territory, Australia

Jalu Bias Firdausi[1], Atiqullah Amiri[1], Salma Difa Masti[1]
[1]KGL Resources Limited, Brisbane, Australia

The Jervois region in the eastern Arunta Orogen, Northern Territory, is a significant Proterozoic base metal province, representing a high-priority target for copper-lead-zinc exploration. However, exploration remains challenging due to complex faulting and folding and lack of surface anomalies. This study integrates geological mapping, geophysical methods, drillhole data, and geochemical analysis to evaluate the economic potential of the deep-seated ore bodies.

Geophysics has played a critical role in exploration. Regionally, aeromagnetic and gravity surveys have defined major J-fold tectonic structures. At the deposit scale, induced polarization (IP) and resistivity have delineated mineralized zones, while Down-Hole Electromagnetic (DHEM) surveys have successfully identified mineralization at Rockface, Reward and Cox's Find.

The principal deposits—Rockface, Reward, and Bellbird—are subvertical stratabound, predominantly hosted in garnet-chlorite-magnetite schists, with lateral facies variations into quartz-tourmaline, banded epidote, and calc-silicate rocks. At Rockface, mineralization consists of massive chalcopyrite-pyrite in brecciated magnetite, while Reward and Bellbird feature galena-sphalerite boudins within skarn-like calc-silicates. Subsurface geochemical data suggest a metamorphosed SEDEX-style system overprinted by oxidized hydrothermal fluids of which fluid is essential in enhancing metal mobility and enrichment.

Drilling conducted in 2024 led to a 17.4% increase in resources and confirmed mineralization continuity along the key shear zones. Notably, hole KJD627D1 at Rockface intersected a brecciated sulphide-magnetite vein with 5.08 m at 6.74% Cu, 5.36 g/t Au, 18.41% Zn, and 8.42% Pb from 1013.05 m. Integrated analysis of geological, geophysical, and geochemical datasets have discovered the deep-seated high-grade copper mineralization, substantially increased confidence in ongoing target delineation, further establishing Jervois as a highly prospective area within the global copper supply framework.

Integration of Petrographic, Structural, and Geochemical Data in Weathered and Challenging Prospective Areas: Case Study of the Descoberto Prospect, Brazil

Beatriz Fontes Andrade[1], Carolina P. N. Moreto[2], Ricardo O. A. Mabub[1]
[1]AngloGold Ashanti, Nova Lima, Brazil, [2]University of Campinas (UNICAMP), Campinas, Brazil

The Iron Quadrangle (IQ) is a world-renowned mineral district with a complex tectonic and structural history. Located in the São Francisco Craton in Minas Gerais, Brazil, the IQ is related to historical gold extraction in the country since the 18th-century, during the Portuguese colonization. In this context, scientific studies of gold prospects and responsible mining are encouraged. This study focuses on a better understanding of petrography and the structural architecture that hosts the mineralization at the deposit scale. The area is within the geological context of the metavolcanosedimentary sequence of the Archean Rio das Velhas greenstone belt, with the mineralization hosted in quartz veins. The Descoberto prospect is intensely weathered and located in a dense vegetation cover area, presenting a scarcity of outcrops, and well-developed soils due to the tropical climate. Thus, the integration of petrographic, structural and geochemical data with field mapping greatly assists in geological understanding at local and regional scales. Soil sample data were analyzed and diamond drillhole samples were selected for thin sections covering all the lithotypes in the area. The recent interpretation suggests that the prospect is not located within the north flank of a recumbent fold, as thought, but in a NWW-SEE shear zone. The existence of NEE-SWW diabase dyke cutting the structure is also new evidence and it's marked by high concentrations of Nb-Ta-Ti immobile elements in soil chemical analysis. Petrographic data of a Nb-Ta-Ti-rich drillhole intercept and partition coefficients for minerals such as titanite and ilmenite, in basaltic to andesitic magmas, support the dyke interpretation. The study underscores the value of well-collected and reliable data, as well as the importance of periodically reevaluating such data in light of new reinterpretation opportunities. It also highlights the potential of tools such as geochemistry to improve geological understanding in highly weathered and challenging prospective areas.

Structural Controls, 3D Geometry and the Evolution of Vein-Hosted Copper Mineralization at Frontier Mine, Democratic Republic of Congo

Heike Fourie[1,2,3], Alexander Kisters[2], Ian Basson[2,3]
[1]SRK Consulting, Perth, Australia, [2]Stellenbosch University, Stellenbosch, South Africa, [3]TECT Geological Consulting, Cape Town, South Africa

The Central African Copperbelt is world-renowned for its sediment-hosted stratiform copper (SSC) deposits. Throughout the Copperbelt, a series of deposits occur which 1) do not conform to the general SSC style of mineralization and/or 2) are hosted within stratigraphic units other than the Lower Roan Subgroup, the typical host for SSC deposits. One example is the understudied Frontier Mine, a structurally-controlled, vein-hosted copper deposit which is distinct from the majority of the deposits within the Zambian Copperbelt sub-basin.

The successive emplacement of chalcopyrite-quartz-carbonate veins record the interaction between progressive structural evolution of the deposit, documenting northeast-southwest shortening, and associated hydrothermal fluid flow and sodic alteration (albitization). These veins are hosted by greenschist-facies metasedimentary rocks of the Mwashya Subgroup, located in the hinge zone of a first-order, regional-scale, refolded recumbent fold (F1/F2). Veins can be discretised into at least three sets, wherein early bedding/S1-parallel veins (Vein Set 1) formed during layer-parallel shearing and the commencement of recumbent folding (F1). Although relatively poorly mineralized, this vein generation was associated with near-pervasive, bedding-parallel albitization, leading to replacement of the predominantly metapelitic wallrocks. This resulted in reaction-hardening of the wallrock, in turn leading to brittle deformation during subsequent upright folding (F2) and the formation of a second generation of highly mineralized, saddle-reef type veins (Vein Set 2) in the hinges of the folds (F1/F2). The controls and collective geometry of Vein Set 2 defines the overall shoot-like geometry of the orebody, parallel to SE-plunging folds. Vein Set 3 is defined by brecciated textures and a subhorizontal orientation, underlining the formation of these veins under continued northeast-southwest shortening of the brittle wallrocks during the lock-up of the first-order refolded fold. The close relationship between vein sets, alteration and regional folding indicates long-lived, syn-tectonic, multi-phase mineralization during the Lufilian Orogeny.

The Epithermal to Skarn-Porphyry Transition at La Colorada Deposit (Zacatecas, Mexico)

Pedro Francisco[1], Joshua Ebner[1], Zhaoshan Chang[1], Chris Emerson[2], Dante Juarez[2], Patricio Perez[2]
[1]Colorado School of Mines, Golden, USA, [2]Pan American Silver Corp., Vancouver, Canada

The Ag-Zn-Pb La Colorada deposit in Mexico is a large system featuring intermediate sulfidation (IS) epithermal, carbonate replacement (CRD), skarn, and porphyry styles of mineralization. With resources of 195 Moz Ag (254g/t), 0.6 Mt Zn (2.40%), and 0.3 Mt Pb (1.38%) in IS veins, and 367 Moz Ag (35g/t), 9.1 Mt Zn (2.79%), and 4.2 Mt Pb (1.29%; 06/30/2024) in skarn-CRDs, La Colorada is currently the world's second-largest Zn-Pb skarn in our database with >1700 skarns.

The volcanic-hosted IS veins transition downward into CRD veins in carbonate-rich sequences, with the same mineral assemblage, but more calcite. The veins transition into massive CRD, and deeper skarn with the appearance of pyroxene and/or garnet. In the skarn, garnet:pyroxene ratio increases, pyroxene color changes from brown to green, and garnet color darkens towards causative intrusions with red garnet endoskarn or porphyry-style veins. In addition, microprobe analysis show Mn decreasing from 30.6 to 0.1wt% in pyroxene and 1.9 to 0.1wt% in garnet. Skarn zoning patterns and Mn trends reveal two syn-skarn intrusive centers.

Illite crystallinity, a temperature proxy, of hydrothermally altered volcanic rocks next to IS veins range from 0.6 (~245℃) to 1.9 (~300℃). Fluid inclusion homogenization temperatures in sphalerite and quartz from IS and CRD veins range from 234℃ to 325℃. The spatial trend of both data sets indicate two heat and fluid sources that coincide with the two syn-skarn intrusive centers.

Metal zonation features highest Mo grades inside the porphyry, Cu in the porphyry and proximal skarn, Zn-Pb in distal skarn-CRD, and Ag (>100g/t) in distal CRD and IS veins. Lower Ag grades (>25g/t) show plumes rising from the deep skarn into the epithermal environment.

The findings at La Colorada reveal connections between different styles of mineralization in a skarn system and can be used as an exploration model in similar systems.

Geochemistry Modeling and Mineralogy of the Banded Iron Formation, West Moeda Syncline, Quadrilátero Ferrifero Region

Carolina Frésca Alves[1], Letícia Santos Barbosa[1,2], Guilherme Prosdocimi[1], Marcos Vinicius Monteiro Carvalho[2], Rodrigo Sérgio De Paula[1], Jorge Roncato[1]
[1]Federal University of Minas Gerais, Institute of Geosciences - Graduate Program in Geology, Manoel Teixeira da Costa Research Center, Belo Horizonte, Brazil, [2]Vallourec Mineração, Belo Horizonte, Brasil

At the western flank of Moeda Syncline, a BIF-hosted with variable grades lithotypes: Hematite (>=64 wt.% Fe), Rich Itabirite (58 to 64 wt.% Fe), Intermediáte Itabirite (45 to 58 wt.% Fe), Poor Itabirite (20 to 45 wt.% Fe), Poor Limonitic Itabirite (20 to 45 wt.% Fe), Intermediate limonitic Itabirite (45 to 58 wt.% Fe) and Manganese-rich Itabirite (20 to 45 wt.% Fe). The metasedimentary sequence is correlatable with phyllites and quartzites of the Caraça Group at the base of the Minas Supergroup. The sedimentary/diagenetic bedding of the metamorphosed BIF (itabirites of the Cauê Formation) is generally transposed by a planar schistosity. The 76 drill holes carried out between 2005 and 2014, and nine additional samples taken from the deposit were classified into different facies of itabirite based on the degree of enrichment in iron, aluminum, and manganese in the chemical analyses, and the degree of compactness of the material, resulting in eleven different lithotypes (with their respective average iron values). Lithotypes with an Al2O3 content above 2% are called Limonitic, and those above 1% of Mn are Manganese. Both XRD and optical microscopy studies indicate a mineralogical composition of quartz, granular hematite, goethite, and a very fine-grained mixture of iron oxides and hydroxides (limonitic material). The diverse and extensive lithotypes pose considerable difficulties for exploration - as well as opportunities regarding different sample media and the formation of resources within them. The development of appropriate geochemical understanding is exceedingly important for better exploratory planning. This data aims to provide a concise introduction to the ore geology of the part of Moeda Syncline iron ore deposit.

Keywords: Iron ore, Quadrilátero Ferrífero Region, Geochemistry Modeling, Mineralogy, XRD.

Chile's First Defined Intrusion-Related Gold System: Ternera's Hydrothermal Footprint in the El Zorro Gold District

Eduardo Fritis[1], Nicholas H. S. Oliver[2,3], Julie V. Rowland[1], Michael C. Rowe[1], Ruben Cahuana[4], Zeffron C. Reeves[4], Kevin Faure[5], Laure A. J. Martin, Shane Askew
[1]School of Environment, The University of Auckland, Private Bag 92019, Auckland 1142, New Zealand, [2]HCOVGlobal consultants, 6 Hancock Rd, Alligator Creek, Australia, [3]Economic Geology Research Unit, Division of Tropical Environments and Societies, James Cook University, Townsville, Australia, [4]Tesoro Gold Limited, Level 48, 152-158, St Georges Terrace, Perth, Australia, [5]GNS Science, 1 Fairway Drive, Lower Hutt, New Zealand, [6]Centre for Microscopy, Characterisation & Analysis, University of Western Australia, 35 Stirling Highway, Perth, Australia

The El Zorro gold district is the most recent gold discovery in northern Chile and the first intrusion-related gold (IRG) system identified in the country. Ternera is the most significant deposit, with resources of 1.282 Moz, and is surrounded by over eight active IRG prospects within a 30 km gold corridor. Mineralization occurs in Upper Triassic to Lower Jurassic felsic to intermediate, ilmenite-series intrusions of the Relincho and Cuevitas plutons (~205–185 Ma) and in Devonian to Carboniferous metasedimentary rocks of the Las Tórtolas Formation. Hydrothermal alteration is confined to mm- to dm-scale selvages adjacent to mineralized veins, without large-scale halos typical of porphyry or IOCG systems. The mineralization consists of Bi-Te-Ag-W-Sn-Sb ± As-S assemblages, closely linked to high-temperature fluid circulation along structurally controlled vein arrays.

Carbon, oxygen, and sulfur isotope data support a dominantly magmatic source with minor crustal contributions. Carbon and oxygen isotope compositions from calcite veins and whole-rock samples indicate magmatic-derived fluids with CO_2 from degassing during fractional crystallization of mafic magmas and assimilation of high-$\delta^{18}O$ metasediments in the magma source, with limited fluid–rock interaction at the emplacement site. Sulfur isotope values ($\delta^{34}S$ = –8 to –2‰) reflect a reduced magmatic-hydrothermal system some metasedimentary input, mostly in the melt source region and no significant contribution from evaporitic or seawater-derived sulfur.

Spatial and temporal features suggest a magmatic-to-hydrothermal transition, marked by aplitic dikes evolving into quartz–sulfide veins, miarolitic cavities, and evidence of magma mingling. These features imply that mafic–felsic magma interaction may have triggered fluid exsolution and ore formation. Combined geological, geochemical, and structural evidence supports classification of El Zorro as a reduced IRG system, similar to the Tintina Belt IRG intrusions in Yukon/Alaska. This expands the metallogenic framework of northern Chile, highlighting the potential for IRG mineralization beyond traditional Andean porphyry and IOCG models.

Unveiling Diamond Potential in the Meya Kimberlite Dyke: Petrographic and Thermobarometric Insights

Tenoch Galeana Cornejo[1]
[1]University Of British Columbia, Vancouver, Canada

This MSc research project aims to characterize distinct kimberlite units within the Meya dyke system in Sierra Leone and assess their diamond potential. Located adjacent to the Koidu kimberlite (West Africa's largest diamond mine) the Meya dyke is near vertical, 1–3 meters thick, and extend approximately 2.5 kilometers. Accurate geological models are essential for optimizing diamond exploration, and this study integrates mineralogical, petrographic, and geochemical data to define the origin, relationships, and economic relevance of the kimberlite units.

A central component of the project involves applying Ni-in-garnet thermometry to estimate the sampling depths of the kimberlite melts, using the local geotherm of 38 mW/m². Understanding these depths aids in evaluating the likelihood of diamond preservation and concentration. This research is supported by Meya Mining and SRK Consulting, with direct economic significance highlighted by the discovery of two high-value Type IIa diamonds from the Meya deposit, each valued at over USD $15 million.

The methodology includes detailed petrographic description of 45 thin sections from three kimberlite units (KIMB1, KIMB2, KIMB3), combined with scanning electron microscopy (SEM) for mineral identification and electron microprobe (EMPA) analysis of spinel, perovskite, and phlogopite to determine major-element compositions. Additionally, 111 garnet macrocrysts have been analyzed using LA-ICP-MS to quantify Ni concentrations, supporting thermobarometric modeling used to evaluate diamond potential. Laboratory work was conducted at the University of British Columbia between 2024 and 2025, with data interpretation and thesis completion expected by November 2025.

Long-Lived High-Temperature Metamorphism in the Late-Archean Granulite-Hosted Katanning Gold Deposit, Southwest Yilgarn Craton

Jonathan Garcia[1], Noami Tucker[1], Margaux Le Vaillant[2], Helen McFarlane[2], Marco Fiorentini[1], Laure Martin[1], Matthew Greentree[3]
[1]University Of Western Australia, Perth, Australia, [2]Commonwealth Scientific and Industrial Research Organisation, Mineral Resources, Kensington, Australia, [3]Ausgold Ltd, Perth, Australia

P-T-t constraints were used to investigate part of the poly-metamorphic history of the granulite-facies greenstone belt hosting the 3 Moz (resources) Katanning gold deposit, in the Corrigin Tectonic Zone, Southwest Yilgarn Craton. We studied five samples of barren mafic granulite and felsic/metasedimentary migmatitic gneisses, using automated mineralogy (TIMA) and whole-rock geochemistry to define peak assemblages and estimate P–T conditions via phase equilibria forward modelling, further refined using EPMA mineral chemistry. Additionally, five garnet-bearing samples from barren and mineralized zones were analysed using Micro-XRF mapping and LA-ICP-MS to characterize garnet composition and obtain in-situ Lu–Hf garnet ages.

The peak metamorphic minerals exhibit from core to rim predominantly homogeneous, Fe-rich compositions (XFe ≥ 0.5) consistent with diffusional homogenization by prolonged high-temperature conditions. P–T estimates indicate high-temperature, suprasolidus conditions (780–890 °C), characteristic of granulite-facies metamorphism. However, the mafic granulites hosting the bulk of the mineralization display lower pressures (4.4–4.8 kbar) and higher thermal gradients (~196 °C/km) compared to felsic (7.5 kbar; ~113 °C/km) and metasedimentary units (5.5–6.5 kbar; ~140 °C/km). Such regimes suggest high to ultra-high-temperature metamorphism under low-P conditions, in line with prior studies in the Corrigin Tectonic Zone, and likely associated with crustal radiogenic heating and/or magmatism-driven high heat flow. Lu–Hf garnet ages from felsic and metasedimentary samples record peak and prograde garnet growth, overlapping with the (zircon U–Pb) regional metamorphic peak at c. 2650 Ma (Heseldell, 2023). In contrast, anomalous silica-rich mafic granulites, up to 70% quartz, yielded older prograde garnet growth ages c. 2670 Ma, within error of the depositional age of the Katanning greenstone belt at c. 2710 Ma (Heseldell, 2023). These results hint at a rapid and protracted high-temperature metamorphic history, and possible hydrothermal processes linked to gold the mineralization. Further micro-scale work on gold associations is underway.

Geochemical and Structural Comparison of the Golden Pike and Dunlops Crossing Stockwork Orebodies, Golden Mile, Kalgoorlie, WA

Ana Sofia Garcia Saltzmann[1]
[1]Northern Star Resources, Kalgoorlie, Australia

The Golden Mile is situated in the Kalgoorlie Terrane, within the Eastern Goldfields of the Archaean Yilgarn Craton in Western Australia. It is one of the world's richest gold deposits, having produced more than 65 million ounces of gold since its discovery in 1893. In 1989, the entire area was combined into Kalgoorlie Consolidated Gold Mines (KCGM), which is now operated by Northern Star Resources. As of March 2025, KCGM holds 13.2 Moz Au in reserves and 31.6 Moz Au in resources.

Gold mineralisation in the Golden Mile is hosted by two main mineralisation styles: Fimiston-style shear veins and Mt Charlotte-style stockwork veining. This study focuses on two Mt Charlotte-style orebodies within the Golden Mile: Golden Pike and Dunlops Crossing. Both are adjacent to the Golden Pike Fault, a NNE-striking, west-dipping, structure marked by a wide mylonitic deformation zone. Although the fault is barren, it juxtaposes two key host units: the granophyric unit of the Williamstown Dolerite and the granophyric Unit 8 of the fractionated Golden Mile Dolerite (GMD).

The Golden Pike stockwork is hosted in the granophyric GMD and features quartz veins with intense hematite alteration halos extending up to 30 cm, often associated with visible gold. In contrast, the Dunlops Crossing stockwork, hosted in the Williamstown Dolerite, is characterised by quartz veins with sericite-carbonate alteration halos up to 10 cm wide. Despite similar vein orientations, the two systems exhibit distinct geochemical signatures and alteration styles.

Geochemical and structural data have been integrated to improve orebody definition, highlighting the critical role of host stratigraphy in controlling alteration and grade. This comparative study offers insights into the structural and lithological controls on mineralisation within the Golden Mile, with important implications for future exploration targeting and resource definition.

Carboniferous Epithermal Gold Deposits in the Khundii Metallogenic Province, Southwest Mongolia

Bat-erdene Gendenjamts[1], Mike MacDonald[2], Erdenebaatar Otgonbaatar[1], Badrakh Bolor-Erdene[1], Khatanbaatar Narantsetseg[1], Erdenebileg Pagva[1], Michael Gillis[3], Peter Dalton[3], Jeffrey Hedenquist[4]
[1]Erdene Mongol LLC, Ulaanbaatar, Mongolia, [2]Independent, Dartmouth, Canada, [3]Erdene Resource Development, Dartmouth, Canada, [4]Independent, Ottawa, Canada

Systematic exploration by Erdene Resource Development Corp. since 2005 in an underexplored region of southwestern Mongolia resulted in the discovery of three deposits with defined resources, including two epithermal gold deposits and one porphyry Mo-Cu deposit (Zuun Mod), plus several prospects and advanced projects that collectively form the newly defined Khundii ('Valley') metallogenic province. The first gold production from the Bayan Khundii gold deposit is planned for Q3 2025.

The province lies within a Carboniferous Trans-Altai island arc terrane, part of the Central Asian Orogenic Belt. The erosion level preserves the tops of epithermal deposits, and in some areas, the tops of porphyry deposits.

The Altan Nar Au-polymetallic deposit was discovered in 2011 and is characterized by veins of coarse crystalline quartz-adularia (309.7 ± 0.5 Ma) and Ca-, Mg-, Mn-, and Fe-carbonate gangue hosting coarse galena, low-Fe sphalerite and minor zones with chalcopyrite.

The Bayan Khundii gold deposit was discovered in 2015 while prospecting ~16 km SE of Altan Nar. Results from an initial drilling program included the intersection of a 7-m interval averaging 27.5 g/t Au commencing at 14 m depth. Discovery of the Khar Mori Au prospect was announced in early 2021, ~3.5 km north of Bayan Khundii, with one drill hole reporting 45 m of 5.97 g/t Au from 10 m depth. Later in 2021, drilling discovered wide zones of disseminated gold at Ulaan Southeast (up to 258 m at 0.98 g/t Au from 92 m depth), 300 m west of Bayan Khundii. The epithermal quartz-adularia-Au veins (336.8 ± 0.5 Ma) of Bayan Khundii have colloform bands with minor pyrite and occur within an envelope of illite alteration at Bayan Khundii. The illite overprints earlier, partly eroded porphyry-related lithocap alteration, including residual quartz and pyrophyllite ± dickite ± diaspore-kaolinite.

Critical Metal Distribution in the Renison Bell Tin Deposit, Western Tasmania

Javier Gil-Rodriguez[1], David R. Cooke[1], Owen P. Missen[1], Wei Hong[1], Ivan Belousov[1], Colin Carter[2], Esther Little[2]
[1]University of Tasmania, Hobart, Australia, [2]Bluestone Mines Tasmania JV PTY LTD, Zeehan, Australia

Critical metals are vital and scarce resources required for modern technology, infrastructure, and the transition to a renewable- and green-energy future. Tasmania's ore deposits have the potential to contribute significantly to growth in both the Australian and international critical metals inventory. Renison Bell is a world-class tin deposit in western Tasmania and is Australia's largest tin producer. It is a carbonate-replacement tin deposit related to a magmatic-hydrothermal skarn system produced by the Devonian Pine Hill Granite. The ore is hosted by Neoproterozoic dolostones that were mineralized with pyrrhotite, chalcopyrite, and cassiterite. Structures associated with the Federal Bassett Fault host high-grade tin mineralization (2% average). Renison Bell has copper as a by-product, which mostly occurs in chalcopyrite. We are assessing Renison's potential for other critical metals in sulfides and oxides within a new cross-section by detailed geological logging and analytical work to determine geological and spatial controls on critical metal deportment.

Petrographic observations and scanning electron microscope (SEM) analyses on ore and wallrock samples from Renison Bell have revealed the presence of minor mineral phases that contain critical metals as major elements, including native bismuth and tungstates (scheelite, ferberite and wolframite). The major sulfide phases (pyrrhotite, pyrite, chalcopyrite, and arsenopyrite), as well as cassiterite and native bismuth, are being analyzed to assess their trace critical metal content. Laser ablation inductively coupled plasma mass spectrometry (LA-ICP-MS) analyses have revealed relative high concentrations of cobalt (<3,800 ppm) and nickel (<1,450 ppm) in pyrite, indium (<1,620 ppm) in chalcopyrite, cobalt (<82,300 ppm), nickel (<18,400 ppm), antimony (<5,500 ppm) and tellurium (<380 ppm) in arsenopyrite, and vanadium (<930 ppm) and scandium (<210 ppm) in cassiterite. Some of these metals are related to the main mineralized structures and dolostone layers, but other metals may have been remobilized from the wallrock.

Geology and Age of the Granite Related Metasomatic Overprinting of the Rosebery VHMS Deposit

Vinicius Godoi Pereira Da Cruz[1], David R. Cooke[1], Rebecca Carey[1], Lejun Zhang[1], Corey Jago[2,3]
[1]Centre for Ore Deposit and Earth Sciences (CODES), University of Tasmania, Hobart, Australia, [2]MMG Limited Rosebery mine, Rosebery, Australia, [3]Sustainable Minerals Institutes, University of Queensland, Brisbane, Australia

The Rosebery deposit in western Tasmania is the oldest continuously mined volcanic-hosted massive sulfide deposit in the Mount Read Volcanics VHMS province. Syn-volcanic mineralisation occurred during the Cambrian in a submarine setting. Extensive replacement of the southern baritic and massive sulfide lenses by magnetite and pyrrhotite bodies has been linked either to Middle Devonian regional metamorphism or to Upper Devonian granite-related hydrothermal event. Here, we describe the metasomatic overprint of the U and V ore lenses, at the southern end of the deposit, and provide the first U-Pb ages for the event based on in-situ LA-ICP-MS dating of hydrothermal monazites and xenotime. Baritic orebodies at V lens were partially replaced by hematite and magnetite bodies whereas the Zn-Pb rich massive sulfides in U and V lenses were partially replaced by pyrrhotite-pyrite-chalcopyrite. Devonian metasomatism results in depletion of Zn-Pb from the ore lenses but enrichment in Cu and Au in the pyrrhotite bodies. The metasomatic overprint was accompanied by local biotite + chlorite alteration overprint. At U lens, above the massive sulfide lens, occurrences of tourmaline-quartz breccias host significant gold, which is interpreted to have been remobilised from the massive sulfides. Hydrothermal monazite and xenotime give U-Pb ages ranging from 369 ± 2 Ma to 356 ± 4 Ma with a distinct age concentration at ca. 360 Ma. Our ages overlap crystallization ages for Sn granites and cassiterite ages in western Tasmania, confirming the genetic link between these events. Together with previous geochronological results the history of major alteration events at Rosebery have been constrained from 500 Ma to 350 Ma. Significant amounts of Zn-Pb were remobilised in the Devonian and Cu may have been added during the Devonian metasomatic event, which has important implications for exploration at Rosebery and elsewhere in western Tasmania.

Trace Elements of Pyrite, Chalcopyrite and Sphalerite Across Distinct Ore Types and Lenses at the Rosebery VHMS Deposit, Western Tasmania

Vinicius Godoi Pereira Da Cruz[1], David R. Cooke[1], Rebecca Carey[1], Ivan Belousov[1], Lejun Zhang[1], Corey Jago[2,3]
[1]University Of Tasmania, Hobart, Australia, [2]MMG Limited Rosebery mine, Rosebery, Australia, [3]Sustainable Minerals Institutes, University of Queensland, Brisbane, Australia

Volcanic-hosted massive sulfides (VHMS) are a major class of polymetallic, base and precious metals deposits. Understanding how trace elements are distributed in sulfides across these deposits can lead to efficient mining and unlock possibilities for critical metals by-product production. The Rosebery Zn-Pb-Cu-Au-Ag deposit in western Tasmania contains over 20 ore lenses with several distinct types of mineralisation. Three main episodes of mineralisation events have been recognized: (1) Cambrian syn-volcanic VHMS mineralisation; (2) Middle Devonian deformation related quartz-carbonate-sulfide veins and (3) Late Devonian metasomatic pyrrhotite and magnetite rich bodies associated with the emplacement of regional Sn granitoids. In this study we obtained and compiled LA-ICP-MS trace element data for pyrite, chalcopyrite and sphalerite from six ore lenses from Rosebery across a range of mineralisation styles. The objective was to identify systematic changes in trace elements patterns across the deposit. Cambrian massive sulfides, Cambrian baritic lenses and Devonian deformation related quartz-carbonate veins show minimal trace elements variations. Pyrite has a fairly constant trace element pattern and shows no statistically significant distinction across the studied mineralisation types. Sphalerite shows the most significant variation between mineralisation styles. When associated with Cambrian mineralisation sphalerite is significantly depleted in Fe and slightly enriched in Ga, whereas sphalerite associated with Devonian metasomatism is enriched in Sn, Fe, Cu, Bi, Ag and In. Cambrian chalcopyrite is characterised by comparatively higher Co, Ge and As whereas Devonian chalcopyrite has higher In, Bi and Au. Our results are concordant with geological observations elsewhere that sulfides from Sn systems are enriched in indium. The results highlight that Cu-Bi-Au have been added to Rosebery during Devonian metasomatism.

Advanced Argillic Alteration and Geochemistry of Alunite in the Iriki Gold Exploration Area, Hokusatsu, Japan

Yuji Gonoi[1], Jun Ito[2], Kotaro Yonezu[2], Akira Imai[2], Akizumi Ishida[3], Tatsuo Nozaki[1,4]
[1]Waseda Univ., Tokyo, Japan, [2]Kyushu Univ., Fukuoka, Japan, [3]Chiba Univ., Chiba, Japan, [4]Univ. of Tokyo, Tokyo, Japan

The Hokusatsu district hosts the largest low-sulfidation epithermal gold-silver system in Japan, with the world-class Hishikari gold deposit included. The Iriki area, located 30 km southwest of the Hishikari deposit, is considered as a prospective gold exploration area, supported by several geological evidence such as presence of the large kaolin deposit exploited by open pit, record of small-scale gold-silver underground mining and drilling results by the government exploration campaign in 1990s. Due to its potential, an Australian-based company has been conducting exploration since 2020. This study reports the results of field survey and laboratory analyses, elucidating the rock textures, alteration mineral assemblages and whole-rock chemical compositions on the surface of the Iriki area. The surveyed area was classified into five zones: unaltered zone, smectite zone, cristobalite zone, alunite + quartz zone and kaolinite zone. Cristobalite zone is distributed separately in the western and eastern parts of the area, with the eastern side exhibiting more extensive and thicker distribution. The most noteworthy feature is an occurrence of the advanced argillic zone dominated by alunite + quartz and kaolinite zones, which distribute 1.5 km east from the kaolin open pit. The common presence of crystalline, acicular alunite with diameters of several hundred μm probably indicates the formation by the ascent of acidic hydrothermal fluids derived from magmatic-hydrothermal systems, rather than supergene oxidation or steam-heated acid-sulfate waters. $\delta^{34}S$ data of the alunite will also be reported for the assessment of the origin of sulfur and the responsible fluid. Although the acid alteration zone lacks residual quartz, mineralization of the Iriki area likely includes a high-sulfidation environment due to the occurrence of the hypogene crystalline alunite. Combined with the previous exploration results, the area might possess both high- and low-sulfidation environments, potentially capable of gold mineralization underground.

Rodinia Breakup and a New Undercover Neoproterozoic Magmatic Cu-Ni-PGE Province at Cape York, Australia

David Begley[1], **Peter Gregory**[1,2], Angus Hornabrook[1]
[1]Rhumb Resources, Brisbane, Australia, [2]GeoDiscovery Group, Brisbane, Australia

Breakup of Laurentia from the North Australian Craton (NAC) in the early Neoproterozoic produced focused mafic-ultramafic complexes, dykes, chonoliths, and one large layered complex in an S-shaped faulted zone near the northeastern margin of the NAC at Cape York, Australia. Evidence is based on interpretation of aeromagnetic and gravity imagery, other regional occurrences of deformed mafic-ultramafic rocks, and anomalous soil geochemistry (ALS ionic leach). Since most of the complexes are under 50-300 m of Laura Basin cover (Middle Jurassic-Early Cretaceous), alluvial Pt, Pd, and Au are recorded in drainages and appear to be reworked from basal conglomerates of the Laura Basin.There are two possible primary sources: the first related to the Neoproterozoic mafic-ultramafic intrusives and feeders, and the second to small mafic-ultramafic intrusive bodies, possibly related to Permian flood basaltic andesites in an aulacogen formed between the major Palmerville and Yintjingga faults (Norilsk-type Ni-Cu-PGE deposit scenario).

Rhumb Resources has defined eight target areas with significant potential for Cu-Ni-PGE-Au-Ag-Co-REE in their nine tenures. Limited partial-leach soil data over magnetic target zones have defined several significant anomalous areas, with the St George-Fairlight North zone defined over 6.4 km x 1 km within an intrusive complex.A Voisey's Bay-style model best fits the data, with an interpreted chonolithic feeder and a large intrusion hosting the Cu-Ni-Pd-Co-Au-Ag-REE anomalies. To the northwest at Dead Horse, several feeder dykes/sills are evident, where 2 km of one dyke also has similar anomalism. A Jinchuan-type mineralisation scenario is suggested here. Reconstructions suggest the major Chinese Jinchuan Ni-Cu deposit (830 Ma) may have formed in reasonable proximity to the NAC. Ground geophysics and drilling are planned.

The interpreted large layered complex has dimensions of 36 km x 14 km, being similar in size to the Stillwater Complex in the US. It has potential for PGE reefs and contact Ni-Cu-PGE deposits.

Quantifying the Financial Implications of Orebody Knowledge Deficiencies in Mining Projects

Haruna M. Grema[1], Shaun L. L. Barker[1], John Steen[2], Cassady Harraden[1], Dave Lawie[3], Ahmad Saleem[3], Andrea Rutley[1]
[1]MDRU - Mineral Deposit Research Unit, University of British Columbia, Vancouver, Canada, [2]BRIMM - Bradshaw Research Institute for Minerals and Mining, University of British Columbia, Vancouver, Canada, [3]IMDEX Limited, Perth, Australia

Technical errors, particularly those related to poor geological interpretation or integration, are among the leading causes of mine project write-downs, often accounting for a substantial portion of overall impairments. Many of these failures stem from insufficient orebody knowledge (OBK) during mine planning, development, and operations. This underscores the critical importance of enhancing OBK across the mining value chain to reduce financial risk, enhance shareholder returns, and ensure long-term project viability across increasingly complex exploration, mining, and mineral processing environments.

In this study, we analyze historical and contemporary exploration and mine project data derived from regulatory filings, including Canada's National Instrument 43-101, Australia's Joint Ore Reserves Committee (JORC) Code, and the United States Securities and Exchange Commission's Form 10-K. Using Natural Language Processing (NLP), we extract key technical metrics—including reserve downgrades, recovery shortfalls, and capital cost overruns—and integrate them with associated financial performance indicators reported in annual reports and filings across stock exchanges (TSX, ASX, NYSE, and LSE).

The resulting integrated dataset allows us to assess patterns and the implications of technical errors on project impairments and financial underperformance from 2001 to 2025. The overall aim is to quantify the financial impact of poor OBK and demonstrate how linking technical and financial data can proactively flag high-risk projects. This integrated approach supports data-driven decision-making in mine planning, improves capital efficiency, and reduces the likelihood of future technical error-related write-downs.

The Archean Troilus Au-Cu-Ag Deposit: An Atypical Syn-Volcanic World-Class Deposit Reworked During Amphibolite-Grade Metamorphism and Deformation (Quebec, Canada)

Pierre-Arthur Groulier[1], **Nicolas Guest**[2], Remy Klick[2], Kyle Frank[2]
[1]Troilus Gold Corp., Montréal, Canada, [2]Troilus Gold Corp., Toronto, Canada

The world-class Troilus Au-Cu-Ag deposit (13.01 Moz Au equivalent) is located in the underexplored Frotet-Evans greenstone belt (Quebec, Canada). Though it was historically mined as an open pit operation from 1996 to 2010, producing more than 2Moz Au and 70kt Cu, the nature of the deposit remains a controversy. It was originally described as a volcanic-hosted Archean porphyry deposit, primarily due to its base metal content, the bulk nature of mineralization and an association with felsic porphyritic dykes. Subsequent work described the deposit as the result of overprinting orogenic events. The dispute over Troilus' origins is the consequence of it displaying features characteristic of a wide range of deposit types (VMS, porphyry/intrusion-related, skarn/contact metasomatism and orogenic). After extensive exploration efforts and ~350,000m of drilling since 2018, a deposit reclassification seems warranted; a polyphased atypical Archean Au-Cu-Ag deposit is proposed.

New field and core observations, combined with comprehensive deposit-scale 3D modelling, point to an early mineralizing event (~2.79 Ga) during the emplacement of the polyphased gabbro-diorite-tonalite syn-volcanic Troilus intrusion into a submarine caldera containing a complex mafic to felsic tholeiitic-transitional-calc-alkalic volcanic sequence intruded by a swarm of felsic dykes. Au-Cu-Ag mineralized zones are dominantly disseminated in nature or as a stockwork and hosted within specific stratigraphic horizons/units, including brecciated volcanic rocks and tuffs, high Fe-Ti basalts, near felsic dykes and within a tonalitic phase of the Troilus intrusion. The deposit has been deformed and metamorphosed under amphibolite facies conditions, resulting in a high-degree of remobilization and a strong structural control within a dextral transpressional shear system. Ore is associated with the presence of biotite, sericite, albite, quartz, carbonate, magnetite, chlorite and locally with alumino-silicates and fuchsite. A retrograde greenschist event has also been recognized, which is interpreted as an orogenic Au overprint and presents as discordant quartz-tourmaline-chlorite veins.

Alteration Zoning and Exploration Implications in the Freiberg Pb-Zn-Ag Epithermal Vein System, Germany

Marie Guilcher[1], Jan Černý[1], Samuel Thiele[1], António Albertino José[1], Jörg A. Pfänder[2], Sandra Birtel[1], Uwe Lehmann[3], Henrik Kaufmann[3], Jens Gutzmer[1]
[1]Helmholtz-Zentrum Dresden-Rossendorf, Helmholtz Institute Freiberg for Resource Technology (HiF), Freiberg, Germany, [2]Technische Universität-Bergakademie Freiberg, Institute of Geology, Freiberg, Germany, [3]Landesamt für Umwelt, Landwirtschaft und Geologie (LfULG), Freiberg, Germany

The Freiberg district is a prime example of an epithermal precious and base metal deposit of Late Paleozoic age in Europe. The origin of magmatic-hydrothermal vein-hosted mineralization is related to felsic magmatism in the aftermath of the Variscan Orogeny. Whilst the infill of the ore-bearing veins has been the focus of various extensive studies, there is still very little known about the effects and the footprint of hydrothermal alteration on the wall rocks. This study offers detailed petrographic and mineral chemical data of hydrothermally altered host rock from vein selvages in different parts of the Freiberg district. It is combined with Ar-Ar age data of muscovite (sericite) that occur in both altered host rock and as vein infill. Results suggest the presence of a complex interplay of sericitic, chloritic and silicic alteration of the host rock in relation to the epithermal mineralizing event. Coarse crystalline muscovite from within mineralized veins, and fine crystalline sericite from the alteration halo, yielded Ar-Ar ages ranging from 316.7±3.2 to 306.8±3.1 Ma. These ages are older than the previously reported age of 276±16 Ma (Rb-Sr dating on sphalerite), yet are in excellent agreement with reported cross-cutting relationships and recent ages suggesting at least two rhyolite magma pulses in the district (between 326 and 297 Ma). Other experiments yielded anomalously old ages attributed to excess argon. This study provides further evidence for the complex relationship between geodynamic evolution, felsic magmatism and the formation of hydrothermal ore deposits in the Erzgebirge.

New Calibration of the Al-in-Hornblende Geobarometer: Evidence for Lower-Crustal Fluid Saturation in Porphyry Copper Magmas

Gonzalo J. Henríquez[1], Robert R. Loucks[1], Marco L. Fiorentini[1]
[1]Centre for Exploration Targeting, The University of Western Australia, Crawley, Australia

Al-in-hornblende geobarometry provides a useful tool to understand emplacement depth and magmatic evolution of granitoids. It was empirically calibrated by Hammarstrom and Zen (1986) and re-calibrated by other investigators over the years. Currently, the calibration by Mutch et al (2016) appears to be the most reliable in the 0.8-5 kbar range, suitable to retrieve pressures equivalent to upper-crustal levels, but it tends to yield large under-estimates in the 8-15 kbar range, which relates to lower crust – upper lithospheric mantle depths. Experiments that report hornblende compositions synthesised on calc-alkalic, metaluminous granitoid melts at 650 to 950°C, 5 to 15 kbar, provide the basis for our re-calibration. Using the total Al (a.p.f.u) content in hornblende synthesised in 91 experiments from 10 published studies, linear regression yields the equation log P(kbar) = –0.2351 + 0.5082 × Al(tot); R^2 = 0.90, with a standard error ±16% (e.g., ±1.6 kbar at 10 kbar). We have applied this calibration to calculate crystallisation pressures of hornblende phenocrysts in ore-forming intrusions parental to major porphyry-type and high-sulphidation epithermal Cu(-Au-Mo) ore deposits worldwide. Plots of P by our calibration versus wt % H_2O dissolved in the melt (using Ridolfi et al, 2010), indicate onset of fluid exsolution at 8-12 kbar in many cases, as very hydrous granitoid melts begin ascent from storage chambers near the base of the crust in orogenic segments of magmatic arcs. Published experiments indicate the fluid/melt partition coefficient of CuCl is ~10 times higher at 10 kbar than at 1 kbar. Therefore, high-pressure fluid segregation from exceptionally hydrous melts as magmas begin their trans-crustal ascent from Moho depths yields efficient Cu extraction and resolves the Cu supply problem. This new Al-in-hornblende geobarometer can be used as an exploration tool by relating high-Al content in hornblende to high pressure volatile exsolution and Cu-metallogenic fertility.

Clumped Isotope Thermometry in MVT Carbonates: Insights into Ore-Forming Temperatures and Fluid Evolution

Hashindra Herath[1], Bryce Barney[2], Zeyang Sun[2], Phillip Hays[1], Ethan Grossman[2], Adriana Potra[1]
[1]Department of Geoscience, University of Arkansas, Fayetteville, United States, [2]Department of Geology & Geophysics, Texas A&M University, College Station, United States

Carbonate clumped isotope (Δ47) thermometry is a novel technique that allows for the determination of carbonate mineral precipitation temperatures independent of the oxygen isotopic (δ18O) composition of the precipitating fluid. The Δ47 temperatures coupled with the measured δ18O of the carbonate allow for the calculation of the δ18O composition of the precipitating fluid. The application of this method in ore geology, particularly in the context of Mississippi Valley-type (MVT) Pb-Zn deposits, is an area of research that could answer several long-standing questions on basin evolution and these deposits.

Twenty-two carbonate samples from six MVT districts in the US midcontinent (Northern Arkansas, Tri-State, Southeast Missouri, East Tennessee, Central Tennessee, and Upper Mississippi Valley) have been analyzed to constrain the ore fluid composition and the ore-formation temperatures. This study is the first application of the novel clumped isotope thermometer to these deposits. Typically hosted by platform carbonates, these hydrothermal deposits form from hot, saline basinal fluids. Secondary carbonates form at different stages throughout the paragenetic sequence and can vary between ore districts. All the dolomite (11 samples from Northern Arkansas and East Tennessee) and two calcite samples (Central Tennessee) recorded temperatures comparable with published fluid inclusion data, suggesting ore-stage formation. Cooler Δ47 temperatures measured in the remaining calcites suggest precipitation during the post-ore stage (32-80 °C). Northern Arkansas samples representing multiple mines in the district recorded precipitation temperatures between 118 °C and 141 °C. East Tennessee samples collected from a single mine recorded temperatures ranging from 155 to 173 °C, suggesting a cooling trend in mineralizing fluids migrating from southeast to northwest. Our results provide insights beyond fluid inclusions or 18O thermometry, enhancing our understanding of ore formation in carbonate-hosted hydrothermal systems.

Mineralogical, Geochemical, and Sulfur Isotopes Studies of the Partolang Volcanogenic Massive Sulfide Deposit, Wetar Island, Indonesia

Tomy Herawan[1], Ryohei Takahashi[1], Andrea Agangi[1], Pearlyn Manalo[1], Hinako Sato[1], Manuel Nopeia[2], Arifudin Idrus[3], Andi Pahlevi[4]
[1]Graduate School of International Resource Sciences, Akita University, Akita City, Japan, [2]Center for Regional Revitalization in Research and Education, Akita University, Akita City, Japan, [3]Universitas Gadjah Mada, Sleman, Indonesia, [4]PT Batutua Kharisma Permai, Merdeka Copper Gold, Indonesia

Wetar Island is part of the Sunda Banda arc. Its stratigraphy consists of Neogene volcanic and marine sedimentary rocks hosting polymetallic volcanogenic massive sulfide (VMS) deposits. The Partolang VMS deposit is located ~5 km east of the Kali Kuning deposit, and its characteristics and process of ore formation are remains less studied compared to nearby deposits. This study reports mineralogy, geochemistry, and sulfur isotopes characteristics of the Partolang deposit.

An ore horizon of the Partolang deposit is bounded by footwall and hanging-wall rocks. The footwall consists of rhyodacite lava, while the hanging-wall consists of andesitic pyroclastic breccia and dacite lava. The rhyodacite and dacite lavas exhibit calc-alkaline affinity, while the andesitic pyroclastic breccia is transitional between tholeiitic and calc-alkaline affinities. The rhyodacite and dacite lavas sourced from a continental arc setting, while andesitic pyroclastic breccia from a transitional oceanic-continental arc setting. Alteration of the host rocks includes silicification, advanced argillic, argillic, and propylitic alteration.

Mineralization occurred in at least four stages: (1) early stage pyrite with minor chalcopyrite-bornite and trace sphalerite-galena, (2) main stage pyrite-enargite-tennantite-tetrahedrite with minor chalcopyrite-marcasite and trace sphalerite-galena, (3) late stage pyrite-marcasite with minor covellite-chalcocite-enargite-tennantite-tetrahedrite, and (4) supergene chalcocite-covellite-jarosite. Copper, lead, and zinc concentrations of ores suggest that the deposit is a bimodal-mafic type VMS deposit. Pyrite-I is collomorphic, vuggy, and zoned (avg. 0.9 wt.% As, 0.4 wt.% Cu), while Pyrite-II is euhedral and massive (0.1 wt.% As, 0.1 wt.% Cu). δ34S values of sulfides in siliceous ores, brecciated sulfide ores, and massive sulfide ores narrowly range from 9.3 to 9.7 ‰, 7.9 to 8.0 ‰, and 7.9 to 8.1 ‰, respectively. A sample of barite from a brecciated sulfide ore yielded a δ34S value of 21.2 ‰.

Developing Soil Sampling Workflows for Lithium Exploration in SW England

Edward Hill[1], Benedikt Steiner[1]
[1]University Of Exeter, Penryn, United Kingdom

The Cornubian granite batholith of SW England hosts one of the largest undeveloped Li resources in Europe. Li has been historically overlooked by regional and local scale exploration programmes due to a lack of demand, with current exploration activity largely focussed on brownfield sites, leaving the greenfield Li exploration potential of SW England under-investigated.

Soil sampling, a well-established greenfield exploration tool, is typically applied without significant consideration of variability between deposit types or natural variations across a sampling area. This study focusses on two known and contrasting Li occurrences in SW England: the Tregonning Granite, a medium-grained and variably altered Li-mica rich granite pluton; the Meldon Aplite, an aphanitic Li-mica and petalite bearing aplite dyke. Trial orientation pits were dug at both sites to collect bulk soil samples from different soil horizons, with sufficient material to separate each into five size fractions for fully quantitative (ICP-MS) geochemistry. Additional material was prepared as resin-mounted blocks for advanced mineralogical analysis using SEM-EDS, QEMSCAN and RAMAN spectroscopy.

Li distribution in soil size fractions shows soils overlying the Tregonning Granite have elevated Li in the 500-2000 μm fraction, and those overlying the Meldon Aplite have Li in the 75-150 μm fraction. Multielement geochemical data analysis highlights the application of U/Th ratios, distinguishing granitic (U/Th > 0.43) from metasedimentary-derived material. SEM-EDS and QEMSCAN have identified mineralogical sources of the U and Th, as well as providing bulk mineralogical data about the sample material, with RAMAN spectroscopy being used for the rapid identification of Li mineral speciation.

Initial results from these samples reinforce the need to tailor soil sampling programmes for specific deposit types, with geochemistry and mineralogy showing a strong correlation to the source material properties.

Geochemical Characterisation of the Orion Massive Sulfide Deposit and Surrounding Prospects, Wunaamin Miliwundi Orogen, Western Australia

Sam Hill[1], John Mavrogenes[1], Antony Burnham[1,2]
[1]Australian National University, Canberra, Australia, [2]Geoscience Australia, Canberra, Australia

Massive sulfides have been recently intercepted undercover at the Orion prospect in the Paleoproterozoic Wunaamin Miliwundi Orogen, Western Australia. The deposit style has so far been unclear with limited exploration drilling.

The mineralogy of the Orion prospect consists primarily of pyrite–pyrrhotite–chalcopyrite–sphalerite ± galena ± sulfosalts hosted by the Paleoproterozoic Ruins Dolerite and Marboo Formation metasediments. δ34S data from sulfides at Orion measured in situ by SHRIMP-SI had a range of 2.4 – 8.3 ‰, consistent with a Volcanogenic Massive Sulfide (VMS) origin. Sulfur is modelled to have been derived from mixing leached volcanic sulfur and reduced seawater sulfate. δ18OH2O values of 1.0 – 4.5 ‰ were calculated based on fluid temperatures of 250–350 °C, comparable to VMS deposits.

Assemblages of Se-enriched galena, Bi-Sb alloys and various Bi-Ag-Pb sulfosalts are present throughout the ore. These are low melting point chalcophile elements (LMCEs) which lower the melting point of sulfides. Based on their chemical and textural evidence, the sulfide assemblages are interpreted to be the result of sulfide partial melting during greenschist facies metamorphism. Precursor sulfosalts likely acted as nucleation sites for melting which facilitated diffusion of trace LMCEs from the surrounding sulfide into the melt. Immiscible melt fractionation resulted in assemblages of early sulfides followed by sulfosalts and alloys.

Orion is interpreted to be a metamorphosed VMS deposit that has undergone partial sulfide melting. Remoblisation may have occurred on a larger scale, potentially invalidating the standard VMS exploration model at this deposit. Cassiterites in the ore were dated to a 208Pb-corrected 206Pb/238U age of 457 ± 12 Ma using SHRIMP-RG. This age likely represents a resetting of the cassiterite U-Pb system during a Phanerozoic alteration event postdating the youngest recognised c. 590-500 Ma Wunaamin Miliwundi Orogeny, potentially extending the duration of this orogeny.

REE Mineralization at Koppamurra, South Australia: Insights into the Transport, Vertical and Lateral Distribution of REEs in a Low-Temperature Environment

Jasmin Hiller[1], Stefan Löhr[2], Laura Morrissey[1], Andre Baldermann[3], Justin Payne[1], Casey Doolette[1], Carmen Krapf[4], Carl Spandler[2]
[1]University of South Australia, Adelaide, Australia, [2]University of Adelaide, Adelaide, Australia, [3]Graz University of Technology, Graz, Austria, [4]Geological Survey of South Australia, Adelaide, Australia

The demand for rare earth elements (REE) is expected to increase in the near future as new technologies develop and supply chain issues force countries to identify their own resources. Ion-adsorption REE mineralisation is known from tropical weathering profiles developed on igneous rocks mainly in South China and Myanmar. Recent research by Löhr et al. (2024) on the unusual ion-adsorption REE deposit at Koppamurra, South Australia, identified REE mineralization hosted in authigenic clays, secondary formed carbonates, and cerianite, which acts as a Ce sink. This deposit has formed above a marine limestone unit and thus lacks an underlying REE-rich protolith, and it has not experienced tropical weathering conditions, which contrast common ion-adsorption REE deposits. Instead, the mineralization is proposed to have formed through rapid (< 2 Ma), pedogenic weathering-induced redistribution of REEs within lacustrine sedimentary material.

Our research provides new insights into the spatial and lateral distribution of REE in the Koppamurra deposit and their controlling factors. In addition, we show that REE-rich horizons are predominantly associated with neo-formed smectitic clays (Fe-smectite or nontronite-like) and secondary carbonates. Steep pH gradients within the weathering profile exert a significant control on the immobilisation of REE. Our new data provides additional constraints on the availability and origin of REEs in Koppamurra.

Multiple Mineralisation Events Demonstrated in Complex Li-Cs-Ta Pegmatites at Buchanan Creek, Queensland

Benjamin Hines[1], Lisa Kearney[1,2], Daniel Killen[1], Andrew Christy[2,3], Mitchell de Bruyn[3], Renjie Zhou[4]
[1]Geological Survey of Queensland, Brisbane, Australia, [2]School of Earth and Atmospheric Sciences, Queensland University of Technology, Brisbane, Australia, [3]Central Analytical Research Facility, Queensland University of Technology, Brisbane, Australia, [4]School of the Environment, University of Queensland, Brisbane, Australia

Characterisation of pegmatite dykes from Buchanan Creek, located in the Georgetown Inlier, Northeast Queensland, has produced a detailed story of polyphase deformation within the sampled Li-Cs-Ta (LCT) pegmatite dykes. The primary lithium ore mineral spodumene has recently been identified in this system, in addition to other Li-bearing phases, including amblygonite, lepidolite, and rare elbaite. Caesium and rubidium mineralisation is strongly associated with micas, including the first reported instance of sokolovaite (Cs-polylithionite) in Australia. A multidisciplinary approach encompassing regional mapping, bulk rock geochemical sampling, visual and hyperspectral logging of drill core, and microanalysis (micro-XRF, LIBS, TIMA, Raman, LA-ICPMS) have been applied to characterising the Buchanan Creek LCT pegmatite system. The ca. 1550 Ma Buchanan Creek pegmatites are hosted within tightly-folded, lower greenschist facies Paleoproterozoic dolerite and metasedimentary rocks. Complex pegmatites of the Buchanan Creek area are lenticular in shape, and are typically oriented northwest, following the strong regional S1 foliation. Dykes display strong internal zonation, commonly having a quartz core, with a lepidolite dominated (± quartz, albite, amblygonite) intermediate zone, and an albite-muscovite dominated wall zone. Where present, spodumene and amblygonite are most abundant in the core to intermediate zones of the studied dykes. Cassiterite, tantalite, and microlite are late in the mineralisation sequence, and are commonly associated with albite, and late crosscutting aplitic dykes. Spodumene appears to have formed early in the paragenetic sequence, and subsequently has been largely replaced by amblygonite. Further deformation and remineralisation of incompatible elements (Li, Cs, Rb) is indicated by common brecciation of amblygonite, infilled by early trilithionite, and later polylithionite. Collectively, these observations demonstrate multiple fluid-flow and deformation events affecting the distribution of critical metals and the mineralogy of this system, providing key insights for the future exploration of this system.

The Genesis of Podiform Chromitites From the Baryulgil Serpentinite, NSW, Australia

Tim Huxley, Carlos Spier
[1]WSP, Brisbane, Australia

The Baryulgil Serpentinite in New South Wales, Australia, hosts small podiform chromitite deposits that have been largely understudied, despite their petrogenic value and potential to host platinum group elements (PGEs) mineralisation. The major, minor, and trace element geochemistry of 13 chromite samples from six chromitite deposits reveals that the chromitites are of the high-Al type, with Cr# ranging from 11 to 57 (mean = 48ppm) and show a notable depletion in Ti content, ranging from 211 to 1,391 ppm (mean = 630 ppm). Parental melt composition estimates indicate Al_2O_3 content between 15.4 and 17.3 wt.% (mean = 16.2 wt.%) and TiO_2 content between 0.16 and 0.61 wt.% (mean = 0.33 wt.%). These results suggest a mid-ocean ridge basalt (MORB) affinity for the chromitite parent melts; however, the Ti content is more depleted than typical MORB values, indicating a more complex melt evolution. Batch and fractional melting models suggests that the Baryulgil chromitites formed through a two-stage melting process. Initially, melt extraction from a fertile mantle source produced high-Al, high-Ti magmas, leading to Ti depletion in the mantle. A subsequent, more limited phase of partial melting generated low-Ti magmas, whilst still rich in Al. This process likely occurred during subduction initiation in a nascent fore-arc setting, where such melt evolution is feasible. The extent of partial melting does not suggest significant PGE extraction into the parental magmas, and petrographic studies show no magmatic sulphides within the chromitites, indicating limited potential for PGE mineralisation.

Zircon and Apatite Geochemical Insights into Porphyry Cu Fertility in the Artvin District During Subduction–Collision Transition

Ali Imer[1], **Akin Cil**[1], Sinem Koç Oğuz[2], Massimo Chiaradia[3], Alexey Ulyanov[4], Robert Moritz[3]
[1]Middle East Technical University, Ankara 06800, Türkiye, [2]Galata Mineral Madencilik San. ve Tic. A.Ş., Ankara, Türkiye, [3]University of Geneva Department of Earth Sciences, 1205 Genève, Switzerland, [4]University of Lausanne Institute of Earth Sciences, 1015 Lausanne, Switzerland

The Artvin-Bolnisi transect between NE Türkiye and Georgia hosts clustered VHMS, porphyry, and epithermal systems formed during Late Cretaceous–Eocene subduction- to collision-related magmatism. Volcano-plutonic centers in the Artvin district, the belt's western segment, contain several notable epithermal Au-Ag systems including Hod Maden, Derinköy, and Hızarlı and porphyry Cu-Mo(-Au) systems such as Ardala, Balcılı, and Dereiçi.

This study evaluates zircon and apatite geochemical characteristics of intrusions from the Artvin district to assess the variability in porphyry fertility from Late Cretaceous arc-related to Eocene collisional magmatism. Late Cretaceous samples were collected from host intrusions of the Balcılı porphyry Cu-Mo-(Au) prospect (~78 Ma and ~83 Ma) and the barren Pırnallı Pluton (~77.50 Ma), whereas Eocene samples represent pre-, syn-, and post-mineralization intrusions of the Ardala Cu-Mo-Au system (49.63 ± 0.41 Ma to 48.23 ± 0.47 Ma).

Zircons from the barren Pırnallı Pluton differ slightly from Balcılı host rocks, with lower Eu/Eu* (0.11 vs. 0.23), lower ΔFMQ values (–0.41 vs. +0.07–0.44), and higher zircon crystallization temperatures (821°C vs. 777–693 °C). Eocene Ardala zircons are characterized by higher Eu/Eu* (0.59–0.70), ΔFMQ (+0.33–1.26), and lower zircon crystallization temperatures (682–759 °C). Pre-mineralization zircons have notably higher Th/U, Dy/Yb, and crystallization temperatures (>750 °C), and lower Hf/Y and Ce/Nd, demonstrating their less evolved nature compared to syn- and post-mineralization intrusions. Ardala apatites also have elevated Cl/F and SO_3 concentrations than those from Late Cretaceous intrusions.

Our results, combined with published whole-rock lithogeochemical data, suggest an increasing trend in magma water contents and oxidation states along the Eastern Pontide orogen. Ardala parental magmas underwent amphibole-dominated fractionation within a thickened crust following collision. In contrast, less oxidized and relatively H_2O-poor magmas of Late Cretaceous intrusions of the Artvin district were emplaced within a thinner crust during arc extension and experienced plagioclase-dominated fractionation, thereby limiting Cu fertility.

Characterisation of Clay-Hosted Rare-Earth Element Deposits in Australia

Yongjun Lu[1], Manuel Knorsch[1], Michael Gazley[1], Muhammet Kartal[1], Agnieszka Piechocka[1], Eliza Trunfull[1], Kat Lilly[1], **Madeleine Ince**
[1]RSC, West Perth, Australia

Rare earth elements (REEs) are critical for renewable technologies, electromobility, and the defence industry. While research has focused on ionic clay-hosted REE deposits in South China and Madagascar, little is known about the mineralogy and economic viability of clay-hosted REE projects in Australia. Australian clay-hosted REE deposits have seen 90 new discoveries in the past four years due to strong global demand. Our research focuses on the deportment of REEs in regolith, crucial for developing effective processing strategies. Results indicated an abundance of relict primary monazite in the plasmic zone, with minor relict bastnaesite, allanite, xenotime and fergusonite. Secondary REE minerals in the regolith include rhabdophane, secondary bastnaesite, cerianite, aluminium phosphate sulphate (APS) minerals, and churchite. These relict primary minerals and newly formed secondary REE minerals dominate the deportment of REEs in the plasmic zone. The clay mineralogy is primarily kaolinite and halloysite. The study found overall low ionic REE proportions with an average of ~2%, highlighting the need for further research on processing pathways for REEs locked in minerals. The prevalence of REE phosphates and fluorocarbonates suggests that current exploration often uncovers non-ionic REE mineralisation, necessitating strategies to recover REEs from minerals like REE phosphates.

Geochemical Characteristics and Zircon U-Pb Geochronology of the Late Triassic Granite in the Pilok Tin-Tungsten Deposit, Kanchanaburi, Thailand

Mallika Intachai[1], Ryohei Takahashi[1], Andrea Agangi[1], Marlina Elburg[2], Punya Charusiri[3], Pearlyn Manalo[1]
[1]Akita University, Akita, Japan, [2]University of Johannesburg, Johannesburg, South Africa, [3]Department of Mineral Resources, Bangkok, Thailand

The Pilok Sn-W deposit, situated in the Southeast Asian tin belt, is a significant Sn-W deposit in Thailand. The main Sn-W mineralization occurs as veins, greisen, and veinlets. This study aims to investigate whole-rock chemical compositions, U-Pb ages, and trace element characteristics of zircon within granites in the deposit to determine the petrogenesis and tectonic setting of the granitic magmatism.

The granites are divided into biotite-muscovite granite and tourmaline-biotite-muscovite granite. The biotite-muscovite granite is composed of quartz, plagioclase, K-feldspar, and muscovite, with accessory zircon, biotite, apatite, and opaque minerals. Whole-rock geochemistry indicates that the granites are peraluminous (ASI = 1.2 - 1.5) and calc-alkaline in composition, classifying them as S-type granites.

Zircons are classified into four types based on morphology and cathodoluminescence (CL) textures. Type I zircon shows oscillatory zoning while Type II zircon shows weak oscillatory zoning. Both types show REE patterns characterized by gradually increasing concentrations from La to Lu with positive Ce and negative Eu anomalies, typical for magmatic zircon. In contrast, Type III zircons show dark CL, whereas Type IV zircons display a porous texture, irregular zoning, and elevated total REE contents. Type III and Type IV exhibit less negative Eu anomalies and subdued Ce anomalies compared to Type I and Type II. These patterns are typical for hydrothermal and metamict zircon.

U-Pb dating of magmatic zircons yielded 211.7 Ma and 211.1 Ma for the biotite-muscovite granite and the tourmaline-biotite-muscovite granite, respectively, indicating Late Triassic magmatism. These ages are consistent with a post-collisional tectonic setting following the Sibumasu–Indochina collision. A combination of zircon geochronology, internal textures, and whole-rock chemistry suggests that the granites in the Pilok Sn-W deposit were likely formed from crustal sources under reducing conditions, with subsequent fluid interaction that favored Sn-W mineralization.

Origin of the Aït Abdellah Copper Deposit (Bou Azzer, Anti-Atlas, Morocco): Perspectives from Mineralogical, Stable Isotope, and Fluid Inclusion Analysis

Mohamed Ait Addi[8], Ismail Bouskri[4,5], Ali El-masoudy[5,6], Moha Ikenne[1], Said Ilmen[2], **Marieme Jabbour**[1], Lhou Maacha[5], Mohamed Oulhaj[5], Ilya Prokopyev[7], Mustapha Souhassou[4], Basem Ahmed Zoheir[3]

[1]Laboratory Of Applied Geology And Geo-environment, Ibn Zohr University, Agadir, Morocco, [2] Polydisciplinary Faculty of Ouarzazate, Ibnou Zohr University, Agadir, Morocco, [3]Department of Geosciences, King Fahd, University of Petroleum and Minerals (KFUPM), Dhahran, Saudi Arabia, [4]Laboratory, Polydisciplinary Faculty of Taroudant, Ibnou Zohr University , Agadir, Morocco, [5]MANAGEM Group, Twin Center, Casablanca, Morocco, [6]Applied Geology Laboratory, Moulay Ismail University, Er-Rachidia, Morocco, [7]Sobolev Institute of Geology and Mineralogy, Siberian Branch of the Russian Academy of Sciences, Novosibirsk , Russia, [8]Department of Geology, Moulay Ismail University, Meknes, Morocco

The Aït Abdellah copper deposit, situated within the Bou Azzer-El Graara inlier in the Moroccan Anti-Atlas, represents a significant case of mineralization controlled by structural and lithological factors. This study integrates structural, petrographic, geochemical, isotopic, and fluid inclusion analyses to better understand the origin of copper mineralization. The mineralization primarily occurs in feldspathic sandstones of the Tiddiline Group, which lie above the Bou Azzer ophiolite. It is confined to a NE-SW trending shear zone featuring mylonitic textures, calcite veins, and fracture networks, formed by an initial ductile deformation followed by a brittle phase with faults, brecciated veins, and stockwork zones. These tectonic features facilitated fluid migration and copper sulfide precipitation, with the mineralization consisting mainly of chalcopyrite, bornite, pyrite, chalcocite, digenite, covellite, as well with Cu-bearing carbonates and silicates such as malachite, azurite, and chrysocolla. Sulfur isotope values range from 5.9‰ to 22.8‰, indicating sulfur contributions from both ophiolitic and volcano-sedimentary sources. Carbon and oxygen isotopes suggest fluid interactions with marine carbonates and meteoric waters, possibly linked to the Snowball Earth glaciation during the late Neoproterozoic. Fluid inclusion data reveal a wide range of homogenization temperatures and salinities (195°C to 310°C, 5.7 to 23.2 wt.% NaCl equiv.). Fluid mixing between magmatic-hydrothermal and volcano-sedimentary fluids is proposed as the main mechanism for ore formation. Overall, this study enhances our understanding of copper deposit formation in the Bou Azzer inlier, highlighting the importance of the Precambrian basement and tectonic-volcanic processes in driving mineralization.

Classification of Au Systems in the Mazaruni Greenstone Belt, Guyana, Using Pyrite Trace Element Chemistry

Tramaine James[1], Stephane Perrouty[1], Daniel Kontak[1]
[1]Laurentian University, Greater Sudbury, Canada

"Orogenic gold" is a widely applied generalized term that typically encompasses a variety of Au deposit types within greenstone belts, with the term (s.s.) representing the more common mesozonal lode-gold systems. Other syn-orogenic Au-deposit types, such as intrusion-related and/or porphyry-like Au systems (i.e., magmatic-hydrothermal) are becoming more widely recognized. However, classification of Au deposits into these types remains challenging due to some apparent similarities. Here, the geochemical signature of pyrite grains analysed via laser ablation- inductively coupled plasma-mass spectrometry (LA ICP-MS) is used to discriminate Au deposit types in the Mazaruni greenstone belt, Guyana. Plots of the pyrite LA data using the time-slice domain (TSD) protocol in Au vs Ag binary diagrams show distinct patterns: Type 1 deposits exhibit randomly dispersed Au:Ag ratios, whereas for Type 2 deposits data clusters at Au:Ag = 0.1, 1 or 100. Principal component analysis (PCA) on selected trace elements identified four significant components (PC1 - PC4). Notably, PC1 showed that elements with the highest loadings have similar charge and size, indicating probable substitution into the pyrite lattice based on these properties. K-means clustering on PC1 - PC4 defined four clusters. One cluster showed that Type 1 deposits are enriched in Cu, As, Se, Ag, Sb, Au, Pb, and Bi, whereas three clusters showed Type 2 deposits are depleted in these elements but enriched in Co, Ni, and Te. This methodology suggests that As, Se, Ag, Sb, Au, Pb, and Bi are plausible discriminators for differentiating Type 1 and Type 2 Au systems in the Mazaruni greenstone belt. Literature data and petrographic observations suggest Type 1 settings represents magmatic-hydrothermal Au systems, whereas Type 2 represents mesozonal lode-gold systems. Such data analytics approach on pyrite mineral chemistry could therefore assist in distinguishing Au deposit types and be broadly applicable to greenstone belts worldwide.

Bantug Lithocap, Negros Island, Philippines: Mineralogy, Textures, and Chemistry

Carlos Jimenez[1,2], David Cooke[2], Noel White[2,3], Lejun Zhang[2], Michael Baker[2]
[1]Caldera Geoservices, Brisbane, Australia, [2]CODES-ARC University of Tasmania, Hobart, Australia, [3]Ore Deposits and Exploration Centre, Hefei University of Technology, Hefei, China

Negros is a volcanic island located in Central Visayas, Philippines. The geology of Negros comprises Cretaceous to Quaternary volcanic and sedimentary sequences. Active volcanism occurs in the north and south of the island, with active geothermal activity associated with volcanism in the southern area. The Bantug area is located in the central-east portion of the island, ~17 km west of Tayasan. Advanced argillic mineral assemblages and silicic alteration crop out over an area of ~7 km by ~8 km and have overprinted rocks from the Pleistocene-Holocene Canlaon Volcanic Complex (CVC; 1.82 ± 0.11 Ma to 1.71 ± 0.16 Ma, LA-ICP-MS U-Pb on zircon). Advanced argillic alteration at Bantug, constrained using SWIR spectroscopy and SEM, consists of alunite, natroalunite, diaspore, pyrophyllite, APS minerals, and zunyite. Advanced argillic-altered rocks are characterised by patchy-wormy or gusano texture, and silicic-altered rocks are characterised by massive and vuggy textures. A natroalunite-cemented hydrothermal breccia body cuts the advanced-argillic altered rocks. The lithocap is truncated by the Pleistocene Bantug Intrusive Complex (BIC; 1.37 ± 0.10 Ma to 1.24 ± 0.05 Ma, LA-ICP-MS U-Pb on zircon). The BIC comprises at least three phyllic and propylitic-altered porphyritic diorites, some of which display miarolitic cavities filled with epidote and quartz. The oldest intrusion, a diorite porphyry, displays incipient, millimetre-thick, wavy, quartz-pyrite veins. Base and precious metal results from whole rock geochemical analyses along and across the lithocap are below detection limit. However, REE, Sr, and TiO2 define anomalies which coincide with the occurrence of high temperature minerals (i.e., pyrophyllite, diaspore, zunyite) identified using SWIR spectroscopy and SEM. This information, coupled with LA-ICP-MS mineral chemistry from alunite, pyrite, epidote, chlorite, and magnetite, have helped to locate heat sources and potential zones of magmatic-hydrothermal mineralisation in the lithocap at Bantug.

Patchy-Wormy Textures in Lithocaps: Examples from Bantug, Philippines and La Zanja, Peru

Carlos Jimenez[1], Lejun Zhang[2], David Cooke[2], Noel White[2,3]
[1]Caldera Geoservices, Brisbane, Australia, [2]CODES-ARC University of Tasmania, Hobart, Australia, [3]Ore Deposits and Exploration Centre Hefei University of Technology, Hefei, China

Patchy-wormy textures, also known as "gusano" textures (worm in Spanish), occur at the base of several lithocaps around the world. These textures were first recognised in the Yanacocha district, Peru, during the 1990s. Further research has highlighted that these textures are a common feature rather than a rarity across lithocaps around the globe. Patchy-wormy textures have been reported at Yanacocha, La Zanja, Caspiche, Oyu Tolgoi, Bantug, and Tampakan. The lithocap at Bantug in Negros, Philippines, extends over ~7 km by ~8 km. As yet, no ore-grade mineralisation has been discovered at Bantug. The lithocap at La Zanja in the Yanacocha district extends over ~20 km by ~33 km; it hosts several high-sulfidation orebodies and porphyry targets. Patchy textures in both lithocaps are characterised by amoeboid-like clots, 1 to ~5 cm diameter, of fine-grained, white to pink coloured clay and sulfate minerals. They occur in a grey, dull, quartz groundmass. Pyrite is ubiquitous in patchy features at La Zanja and Bantug, and enargite is common at La Zanja. Wormy textures at both lithocaps are characterised by ~1 cm thick, wavy veins of grey-dull quartz. SWIR spectroscopy, SEM-EDS, and XRD analyses of patchy features revealed the presence of alunite, pyrophyllite, APS minerals, diaspore, zunyite, quartz, pyrite, and enargite. APS minerals commonly occur at the core of alunite crystals of <50 μm. Alunite, APS minerals, along with diaspore, zunyite, pyrite and/or enargite crystals, occur in a fine-grained pyrophyllite groundmass. SEM-CL analyses on quartz groundmass from patchy features revealed they are made of sub-rounded quartz crystals <50 μm in diameter, and show complex concentric zonation patterns within individual grains. SEM-CL analyses of wormy quartz veins revealed zoned euhedral features distinctive from the groundmass quartz. Mineralogical and geochemical variations in patchy-wormy textures in lithocaps may indicate vectors towards heat sources and potential mineralised centres.

Gold-Sulphide Mineralization within Sheared, Subvertical Banded Iron Formation, Jaspilite and Dolomite, Pickstone-Peerless Mine, Chegutu Greenstone Belt, Zimbabwe

Caitlin Jones[1], Corné Koegelenberg[1,2], Jonathan Gloyn-Jones[1], Steven Duma[3]
[1]Tect Geological Consulting, Somerset West, South Africa, [2]Department of Earth Sciences, Stellenbosch University, Stellenbosch, South Africa, [3]Dallaglio Investments, 121 Borrowdale Road, Gunhill, Harare, Zimbabwe

Pickstone-Peerless Mine is located 120 km southwest of Harare within the Chegutu Greenstone Belt which forms part of the broader Midlands Greenstone Belt of Zimbabwe. Detailed surface and underground litho-structural mapping and downstream 3D geomodelling of the mineralized BIF package (BIF, jaspilite and dolomite) and surrounding wallrocks (variable chlorite-carbonate-talc schists and lesser black shale) allowed for a robust structural interpretation of gold-quartz-sulphide mineralization and shearing within the east-west trending, steeply-dipping tabular deposit. The mine exploits three discrete open pits, namely the Pickstone, Peerless and Duchess orebodies, with Pickstone being the largest, exhibiting a maximum known down plunge and strike extent of 700 m and 1800 m, respectively. The rheologically competent Pickstone BIF package is laterally continuous but exhibits a locally pronounced pinch-and-swell boudin geometry. Swells (boudins) contain chaotic stockworks and boxwork breccia textures, indicative of relatively higher fluid throughput, alteration (silicification) and auriferous mineralization (pyrite-chalcopyrite±arsenopyrite massive sulphide loads and sheeted stringers). Where the BIF package pinches and shearing increases, mineralization is only present within isolated sulphide stringers, along with a noticeable increase in talc content within the enveloping schists. Additionally, underground mapping revealed that the lithological packages have undergone both sinistral strike-slip and top-to-the-north dip-slip kinematics. This, along with the pinch-and-swell geometry of the units, can be attributed to forming under subhorizontal north-south directed shortening as well as vertical and east-west horizontal extension. Generally, ductile strain localization is ubiquitous within the relatively less competent enveloping schists while more brittle deformation and, in turn, increased permeability and subsequent mineralization, are localized within the adjacent more competent BIF package, as strain is localized as a function of contrasting rheological properties. Overall, gold mineralization at Pickstone-Peerless Mine highlights the importance of (1) lithologically driven strain heterogeneity (structural control) and hydrothermal fluid throughput in an (2) overall BIF stratigraphy (litho-chemical control).

Eocene Intrusion-Related Gold in the Golden Triangle of British Columbia

Randall Karcher[1], Simon Kocher[1], Manuele Lazzarato[3], Thomas Monecke[1], Simone Pujatti[3], Jim Reynolds[1], Roger Rosmus[3], Holly Stein[2]
[1]Colorado School of Mines, Golden, United States, [2]AIRIE Innosphere Ventures, Fort Collins, United States, [3]Goliath Resources, Toronto, Canada

The Golddigger property of British Columbia is a recently discovered gold prospect hosting a series of base and precious metal-rich quartz veins. It is in the Golden Triangle of British Columbia, a region home to several noteworthy gold and silver producers, past producers, and prospects. Most Au-Ag systems in the Golden Triangle are believed to be related to the Mesozoic Stikine island arc, and early workers interpreted Golddigger to be a Mesozoic epithermal deposit. Extensive field research was conducted at Golddigger during the 2022-2024 drilling seasons and complemented by laboratory work to unravel the nature and timing of mineralization. Petrographic investigations, cathodoluminescence imaging, and fluid inclusions petrography reveal two distinct generations of quartz. Early quartz formed under lithostatic conditions whereas later quartz was deposited under hydrostatic conditions. The elevated CO2 content in primary fluid inclusions entrapped at hydrostatic pressures suggests that the paleodepth of vein formation was >5-6 km below the paleosurface. Paragenetic relationships indicate that the sulfides and precious metals post-date both generations of quartz. Enrichment of Au-Bi is most pronounced in quartz veins overprinting ilmenite-series granitoid dikes and volcanic host rocks whereas enrichment of Ag-Sb occurs further away from the inferred causative intrusion within sedimentary host rocks. There is textural evidence that the CO2-rich hydrothermal fluids responsible for mineralization underwent phase separation, which was accompanied by exsolution of Au-Bi-Te and Ag-Sb-S melt droplets – a process that has not previously reported to occur in hydrothermal environments. Geochronological investigations yielded a zircon U-Pb age of 52.0 ± 1.5 Ma for the granitoid dikes crosscut by the mineralized veins and a molybdenite Re-Os age of 50.7 ± 1.0 Ma for the sulfides. The study demonstrates that Golddigger is an Eocene intrusion-related deposit, which has significant exploration implications at the Golddigger property and the greater Golden Triangle.

Sulfide Sponges — Sulfide Associations with Spinels and Other Minerals

Kathryn Keane[1], Maria Cherdantseva[2], Michael Anenburg[1]
[1]Australian National University, Canberra, Australia, [2]The University of Western Australia, Perth, Australia

The spatial association between sulfides and spinels (such as chromite) is well documented in nature, as is the association of sulfide with orthopyroxene and olivine, particularly in layered mafic intrusions. In previous unrelated experiments, we observed an enigmatic and unintended spatial association between sulfide and chromite, resembling natural observations. We therefore devised experiments to test the sulfide-chromite spatial association in which we placed alternating layers of sulfide, chromite, olivine, and silicate melt. Experiments ranged from simple to more complex geometries and contained varying base metal sulfides. Both silica tubes and alumina crucibles were used. Chromite was obtained from crushed natural samples. Silicate melt was made from crushed and combined natural samples ranging from intermediate to mafic composition. SEM-EDS and reflected light microscopy were used to analyse experiments. Regardless of geometry and degree of initial separation, we found sulfide liquids sticking to, yet paradoxically not wetting, chromite grains. In instances where orthopyroxene and olivine were present, sulfide liquid exhibited similar 'sticking' behaviour. Sulfides were observed to have several textural forms, including mushroom-like, sub-spherical, and spherical, appearing to 'splash' on certain grains, indicating a mechanical rather than thermodynamic relationship. Sulfides are known to float via attachment to vapour bubbles, a behaviour that was routinely observed in our experiments. We therefore present a hypothesis explaining the observed sticking phenomena. We suggest that volatiles carry sulfides upwards, then the sulfide 'splashes' against certain crystals (spinel, orthopyroxene, and olivine) and initially 'sticks' to the grain through wetting. The varying sulfide morphologies observed are attributed to the low silicate viscosity, with the silicate melt creeping along grain boundaries to separate mineral grains from sulfides but still retaining their spatial association.

The State and Fate of Noble Metals in Bornite and Digenite

Samuel A King[1], Cristiana L Ciobanu[1], Nigel J Cook[1], J Richard Kyle[2], Kathy Ehrig[1,3]
[1]School of Chemical Engineering, The University of Adelaide, Adelaide, Australia, [2]Department of Earth & Planetary Sciences and Bureau of Economic Geology, Jackson School of Geosciences, University of Texas at Austin, Austin, USA, [3]BHP Olympic Dam, Adelaide, Australia

Bornite (Cu_5FeS_4; Bn) and digenite (Cu_9-xFexS_5, x~0.2; Dg) are common components of a broad range of Cu-Ag-Au deposits, including skarns. At high temperature, Bn-Dg forms a solid solution (ss) with cubic symmetry (a~5.5 Å). Superstructures, defined by multiples of a, or other structural derivatives, e.g., anilite (Cu_7S_4), rhombohedral (R) Dg (Cu_9S_5), can form during cooling of Bn-Dgss. The potential of bornite and digenite as noble metal carriers is assessed using complementary micron- to nanoscale petrographic/mineralogical analysis in a case study from the world-class Ertsberg East Skarn System (EESS), Papua, Indonesia. The study aims to answer the question of whether noble metals (Au, Ag, Pd) occur in solid solution in Cu-(Fe)-sulfides, or as nanoparticles (NPs), and how this distribution changes during cooling. Results are significant for understanding precious metal enrichment mechanisms in Cu-Au skarns and analogous deposits in which bornite-digenite intergrowths are exsolution products of high-temperature solid solutions in the system Cu-Fe-S. Gold-rich bornite-digenite assemblages are analyzed from EESS magnesian skarns.

Bornite and digenite display lamellar intergrowths and form dense basket-weave textures at the nanoscale comprising host Bn2a, and exsolved Dg1a+anilite+Dg-R. (Sub)-micron inclusions of Au-Ag-Pb-Te-(Se) phases and merenskyite ($PdTe_2$) occur within lamellar bornite–digenite intergrowths, highlighting their role as important carriers of noble metals. Bornite from prograde skarn contains tens of ppm Au; nanoscale imaging confirms the generation of Au-Ag-Te NPs during phase transformation of parent digenite into lower temperature anilite. Lattice-bound Au is estimated at <10 ppm. In retrograde assemblages, a comparable redistribution of lattice-bound Au and Ag into NPs and micron-sized gold is facilitated through reworking of lamellar intergrowths. An enigmatic Cu-telluride phase identified within all Au-Ag-Te NPs may act as a catalyst for noble metal enrichment. Merenskyite inclusions in bornite contain minor Au, emphasizing the role of high-temperature bornite-digenite solid solutions in scavenging noble metals from fluids.

New Insights into Ore Genesis and Exploration Models from Multi-Element Geochemistry

Harry Brookes[1], Simon Griffiths[1], **Aidan Kitchener**[1]
[1]Third Planet Exploration Services Ltd., Bristol, United Kingdom

Geochemical data processing has resulted in enhancements in ore genesis and exploration models for massive sulphide, carbonate replacement and other sediment hosted deposits. These models are here applied to exploration for copper, zinc, gold and silver mineralisation.

Pyrite mineral chemistry reflects the process of formation, which might be diagenetic, hydrothermal, magmatic or metamorphic. Bulk chemical changes through the stratigraphic column are characteristic of individual sedimentary facies.

We use four-acid, multi-element exploration data from drill core to ascertain which samples contain pyrite and other sulphides. Sulphide bearing sample groups are then assessed using multivariate analysis, specifically Principal Component Analysis. Additionally, four-acid, multi-element exploration data were used to identify variations in RedOx Sensitive Trace Elements (ROSTEs) enrichment and covariance. Modelling and interpretation of these features has identified important controls on mineralisation.

We show that diagenetic and hydrothermal pyrite stages are distinguishable using trace element vectors from industry-standard drill core sampling data. This has enhanced alteration modelling, geochemical targeting and geometallurgical characterisation workflows that have been applied to exploration projects and mining operations at several pipeline stages.

Using a multivariate approach to assess how ROSTEs covary with one another and sulphidation, we generate a robust interpretation of oxidation state during deposition of sedimentary facies. This effective tool for identifying reduced sediment (reactive) horizons has been applied to several marine sediment hosted ore deposits. Often these units do not display clear textural or mineralogical differences and remain undifferentiated during core logging.

Our work demonstrates the adaptation of improvements in orebody knowledge to routinely collected exploration data. As such, we add value through more detailed geometallurgical models for both ore and waste material, shortening the time to discovery and development of mineral deposits.

Structural Evolution and 3D Modelling of the Giant Vasilkovskoye Gold Deposit, Northern Kazakhstan

Corne Koegelenberg[1,2], Anatoliy Franchuck[3], Ian Basson[1,2], Michael McCall[1], Caitlin Jones[1], Jonathan Gloyn-Jones[1]
[1]Tect Geological Consulting, Somerset West, South Africa, [2]Department of Earth Sciences, University of Stellenbosch, Stellenbosch, South Africa, [3]Glencore Canada Corporation, Toronto, Canada

Kazzinc's Altyntau Koshetau, Kazakhstan's largest gold mine, is centred on the giant and structurally complex Vasilkovskoye Gold Deposit (VGD). The VGD is spatially associated with the Zerendinsky Intrusive Complex (ca. 442 Ma) which is hosted within proto-continental basement and volcano-sedimentary and/or accretionary sequences that record protracted deformation from 635 Ma to 252 Ma as part of the Kazakhstan Collage System and greater Altaid Orogenic Belt. Structurally, the VGD is located at the intersection of two major fault systems, viz. the NE-trending Vasilkovskoye (VFS) and NW-trending Dongulagashsky (DFS) faults, which are both associated with Au mineralization. On a deposit-scale, studies of kinematics, cross-cutting relationships and controls on Au mineralization for the FVS and DFS have been challenging; whilst on a regional-scale, the temporality and development thereof are ambiguous, largely due to a lack of outcrop, high-resolution mapping or detailed geophysical data interpretation. This study follows a fully-constrained structural analysis and 3D geomodel, based on an all-encompassing pit mapping and drillhole database, combined with local- to licence-scale geophysical data interpretation. The 3D model depicts the initial sinistral VFS, which comprises multiple, well-developed anastomosing brittle-ductile shearing, in response to strain localization along regional sub-Zerendinsky intrusive suite contacts. This was followed by the cross-cutting, dextral DFS, comprising a combination of incipient to well-developed, en-echelon tensional fractures and hybrid-tensional shears, related to progressive regional belt contraction. Cross-cutting relationships and Au-grade discontinuities suggest that high-grade Au mineralization was introduced by the DFS, whilst local re-activation and partial transposition of the locally-overlapping pre-existing VFS-associated structures resulted in discordant (re-) mobilization of Au along VFS trends. Overall, the VGD represents a complicated, but 'text-book' example of how the intersection of major faults system can locally enhance fluid-flow and Au-mineralization. This rationale accentuates exploration opportunities, including continuation of VGD at depth, as well as regionally beyond the mine tenements.

Structural Evolution and Constraints on Gold Mineralisation of the North-East Section of the Kibi-Winneba Gold Belt

Kofi Adomako-Ansah[1], Yves Pierre Clement[2], **Samuel Asare Konadu**[1,2]
[1]University Of Mines and Technology, Tarkwa, Ghana, [2]Xtra-Gold Mining Limited, Kwabeng, Ghana

The structural evolution of the northeastern section of the Kibi-Winneba Gold Belt significantly influences gold mineralisation. This study examines geological structures through detailed field mapping, drill-core logging, and geophysical data analysis to understand their role in controlling gold deposition. The region consists of deformed metasedimentary and metavolcanic sequences affected by multiple deformation phases, which have shaped its structural architecture. Analysis reveals at least two major deformation events that produced distinct foliations, folds, and shear zones, which acted as primary pathways for mineralising fluids. Gold-bearing quartz veins occur in various orientations, including extensional, shear, and stockwork systems, with auriferous veins predominantly associated with altered diorite intrusions. The presence of sulphide minerals and deformation features such as brecciation and sigmoidal textures further suggests a complex mineralisation history. Comparative structural studies with the Ashanti and Sefwi belts indicate that the Kibi-Winneba Belt exhibits intense deformation and overprinted bedding, distinguishing it from these more extensively studied gold belts. Variations in structural styles across deposits highlight the role of localised deformation in controlling gold distribution. This research enhances the understanding of structural controls on gold mineralisation within the belt, providing a framework for more targeted exploration strategies. The findings underscore the need for further investigations to refine the structural model and assess the economic significance of gold mineralisation in the region.

The Formation of Granite-Type Uranium Deposits in South China: Insights from Hydrodynamic and Chemical Modelling

Zenghua Li[1], Teng Deng[1], Deru Xu[1]
[1]East China University Of Technology, Nanchang, China

A large number of granite-hosted hydrothermal uranium deposits formed in South China between the Cretaceous and Paleogene periods. These deposits are primarily controlled by fracture zones within the granites and show a spatial association with nearby red bed rift basins and contemporaneous mafic dykes. However, it remains unclear how such deposits are formed under a consistent hydrodynamic framework. Two-dimensional simulations of fluid flow, coupled with rock deformation and heat transfer, were carried out to investigate the mechanisms driving fluid movement related to uranium mineralization under different thermo-mechanical conditions. The modelling results reveal that extensional deformation greatly promotes the deep penetration of oxidizing fluids, and the presence of a deep magmatic heat source encourages the formation of convection cells within the granites despite their relatively low permeability. Modelling results also indicate that the pre-existing NE- and NNE-trending faults provided favorable mineralization space for the dilational zones developed during extensional tectonic movement. Mineralogical, isotopic and chemical modelling indicates that pyrite, hematite and uranium minerals can be co-precipitated from the same ore-forming fluid, and the driving factor of uranium mineralization is due to the increase in pH value caused by water-rock reaction, the decrease in oxygen fugacity. The extensive granite-hosted uranium mineralization in South China likely stems from multiple forms of fluid flow governed by coupled shallow and deep geological processes within an overall extensional tectonic setting.

Carrapateena IOCG Deposit: Discovery to Operating Sub Level Cave Mine and Beyond

Shaun Light[1], Mitchell Neumann[2], Kathy Ehrig[1]
[1]BHP Copper South Australia, Adelaide, Australia, [2]BHP Exploration, Adelaide, Australia

Carrapateena is a breccia hosted iron oxide Cu-Au deposit (IOCG) which is located along the eastern margins of the Gawler Craton in South Australia approximately 100km SSE of the iconic Olympic Dam IOCG deposit. Disseminated Cu-Fe sulfide (bornite, chalcopyrite, pyrite) mineralisation is hosted within the Carrapateena Breccia Complex (CBC); a polymictic hematite-granite breccia with variable hematite-sericite-chlorite-carbonate alteration. The deposit is surrounded laterally by altered Paleoproterozoic Donington Suite granitoid and is overlain by Neoproterozoic Stuart Shelf sediments. Felsic and multiple generations of mafic dykes crosscut the surrounding granite and to some extent crosscut the outer parts of the CBC.

The Carrapateena deposit was discovered in 2005 by RMG Services, a small private Adelaide firm. A subsequent joint venture followed with Canadian major Teck Cominco (now Teck) conducting delineation drilling to further understand the deposit. Carrapateena was wholly purchased in 2011 by OZ Minerals Limited (OZL), who completed further delineation drilling and scoping studies before commencing a box cut and decline from surface in 2016 to access the planned Sub Level Cave mine (SLC). Production from the SLC commenced in 2019, producing first concentrate in December 2019. OZ Minerals was taken over by BHP in May 2023 and became the 100% owner of Carrapateena.

The geoscience knowledge (e.g. grade distribution, structure, mineralogy, geometallurgy, rock mass characteristics and hydrology) required to support the journey from deposit discovery through various study phases and into production has evolved significantly and will continue to grow as the mine transitions into a caving operation. The teams involved throughout the 20 years have built on predecessors work and continually work to extract the most value out of all the rocks available and the information they hold to unlock and support the future of this unique deposit.

Intrusion-Related and Epithermal Gold Systems in North-East Queensland, Australia: Multiple Metallogenic Districts and Epochs, Diverse Geochemical and Mineralogical Signatures

Vladimir Lisitsin[1], Courteney Dhnaram[1], Al-Tamini Tapu[1], Malcolm Stallman[1], Friedrich von Gnielinski[1], Benjamin Hines[1]
[1]Geological Survey of Queensland, Brisbane, Australia

North-East Queensland hosts over 150 diverse gold deposits of multiple types, with highly variable mineralisation styles and geochemical and mineralogical signatures. They were formed by several genetically distinct regional-scale mineral systems which were intermittently active from middle Carboniferous to Early Permian, producing multiple partially overlapping metallogenic districts and camps. Most major gold deposits in the region have genetic links to felsic to intermediate porphyry to sub-volcanic intrusive complexes, thereby representing Intrusion-Related Gold Systems (IRGS). The remainder are orogenic and epithermal gold deposits.

Over the past decade, the Geological Survey of Queensland (GSQ), in collaboration with multiple research and industry partners, have undertaken a large-scale data acquisition and research program of multi-scale characterisation of gold mineral systems in the region. The program involved systematic comprehensive characterisation of numerous deposits and camps for each mineral system, focusing on the mineralogical and geochemical signatures of mineralisation and alteration (>1,000 samples) and geochronology of metallogenically significant geological events (tens of new U-Pb, Re-Os and Ar-Ar dates). The deposits and camps investigated in this study include Ravenswood, Kidston, Mt Leyshon, Mt Wright, Red Dome, Mungana, Mt Carlton, Agate Creek, Alice River and multiple smaller deposits.

Intrusion-related gold deposits (IRGD) exhibit the greatest variability of mineralisation styles and geochemical signatures, reflecting a complex interplay of factors including the composition, fractionation and emplacement depth of and proximity to causative intrusions, and chemical composition of host rocks. Most IRGD formed at the porphyry level and occur as breccias, veins, stockworks and rare skarns, with geochemical signatures characterised by enriched Au-Ag-Te-Bi-Cu-Zn-Pb-As-Sb(±Mo-W-Sn) and distinct zonation within deposits and camps. In contrast, epithermal vein gold deposits have predominant Au-Ag-As-Sb-Te geochemistry – similar to distal epizonal IRGD.

IRGD formed at ~335 Ma, 325-315 Ma and 290-305 Ma (synchronous with most Sn and W-Mo deposits), while most epithermal deposits – at 285-275 Ma.

Evolution of Eocene Magmatism at Swales Mountain, North-Central Nevada, and the Relationship to Carlin-Type Gold Deposits

Shasta Longo Florez[1], Philipp Ruprecht[1]
[1]University Of Nevada-Reno, Reno, United States

Swales Mountain, located in north-central Nevada, USA, features a shallow to mid-crustal Eocene magmatic complex, dated between 39.5 and 37.1 Ma, consisting of various intrusive and subvolcanic bodies tilted and exposed by Basin and Range extension. Northern Nevada hosts multiple Carlin-type gold deposits (CTGDs); some of the largest gold deposits in the world, and their origin is still under debate. One model proposes that Cenozoic magmatism associated with slab delamination and rollback of the Farallon plate swept NE to SW across Nevada from ~45 to 17 Ma, with Eocene magmatism in northern Nevada possibly driving mineralization of CTGDs. Located <20 km northeast of the Carlin trend, the Swales Mountain intrusions could have served as a heat source for CTGD mineralization.

Here, we present a detailed study that explores the spatial, temporal, and geochemical evolution of the Swales Mountain system using new $^{40}Ar/^{39}Ar$ and U-Pb zircon ages, hornblende barometry, and Sr-Nd-Pb isotopes. The peraluminous, calc-alkaline suite ranges from basaltic andesite to rhyolite and consists of three distinct unmineralized plutons, as well as various dikes and sills interacting with weakly mineralized Paleozoic sedimentary rocks. Geochronological data suggest that magmatic activity began around 39.5 Ma, with the Eastern Swales quartz monzodiorite forming at shallow depths. The mid-crustal Central Swales quartz monzonite was emplaced between 38.5 and 37.1 Ma at depths of 5 to 11 kilometers, supporting models that link magmatic activity to the Carlin trend (37 to 41 Ma). We conclude that Swales Mountain may have provided a magmatic heat source contributing to the mineralization of CTGDs.

Weathering at Depth - Key Process Forming REE-Rich Regolith at the Mt Weld Carbonatite (Western Australia)

Anderson M. Santos, Artur P. Deditius[1], Hans C. Oskierski[1], Arthur V. Oliveira[2], Denis Fougerouse[2], Ganesh Bhat[3]
[1]Murdoch University, Perth, Australia, [2]Curtin University, Perth, Australia, [3]Lynas Rare Earths Ltd, Perth, Australia

Rising demand for green technologies drives interest in rare earth elements (REE), with carbonatite-derived deposits as key exploration targets. However, the genesis of high-grade REE deposits in weathered carbonatites remains unclear due to variable bedrock composition and complex pedogenetic processes. These challenges hinder the development of predictive exploration models and limit the potential for resource extraction. A prime example is the Mt Weld deposit (WA), a world-class REE resource, which consists of a regolith profile formed by the supergene alteration of a 2.06 Ga ferroan dolomite carbonatite (FDC). This study integrates detailed mineralogical investigations and stable and radiogenic isotopic analyses to resolve how magmatic-hydrothermal and weathering processes synergize to concentrate REE.

Between the unweathered FDC and the overlying regolith lies a 60–100m thick transitional zone (TZ), exhibiting partial oxidation of sulphides into hematite/goethite, formation of porosity due to limited matrix dolomite dissolution, and precipitation of Mn-oxides and secondary REE-fluorocarbonates (e.g., bastnäsite-synchysite-parisite). Whole-rock chemistry revealed a twofold increase in REE concentrations from the FDC to the TZ, accompanied by elevated REE content in apatite (avg.=1.07%) and HREE in monazite (avg.=1136ppm). Isotopic values of $\delta^{18}O$ (<21.8‰) and $\delta^{13}C$ (<−3.5‰), higher than mantle signatures, are driven by low-temperature fluid-rock interaction involving oxidized groundwater. This meteoric water overprint can also be seen by the shift in 87Sr/86Sr from ~0.70197 in FDC to 0.70228 in weathered rocks and 208Pb/204Pb from ~34.74 to 35.99. This fluid flows through the newly formed micro-to-centimetre scale porosity, promoting the migration of the alteration front and formation of the TZ.

The transitional zone represents a critical part of the carbonatite-hosted REE ore deposits, where supergene mineral-replacement reactions upgrade the REE-carriers of magmatic-hydrothermal origin. This work highlights the importance of transitional zones during carbonatite weathering, which is an intermediate stage during the transformation of carbonatite source rock into a super resource.

Mechanisms of Nb and Ta Enrichment in Ferricrete at Mt Weld Carbonatite, Western Australia

Anderson M. Santos[1], Artur P. Deditius[1], Hans Oskierski[1], Arthur V. Oliveira[2], Denis Fougerouse[2], Ganesh Bhat[3]
[1]Murdoch University, Perth, Australia, [2]Curtin University, Perth, Australia, [3]Lynas Rare Earths Ltd, Perth, Australia

Ferrigenous cap rocks (ferricrete) are important vectors for exploration in weathered terranes, incorporating trace elements (such as REE, Nb, and Ta) released during weathering of the ore deposits at depth. Yet, the processes and mechanisms leading to the accumulation of Nb and Ta are poorly constrained. Here we report mineralogy and geochemistry of a 10 m thick layer of ferricrete formed on the top of Mt Weld carbonatite, a renowned REE deposit in Western Australia. Apart from REE resources, Nb and Ta ore resources were reported at Mt Weld carbonatite. The bedrock comprises a calcite carbonatite outer rim (~4 km) and a ferroan dolomite carbonatite core (~1.5 km). The ferricrete consists of a mixture of clays and ironstone clasts with fine-grained Nb-bearing goethite, Nb-bearing ilmenite, hematite, Nb-bearing rutile, goyazite, columbite, florencite-Ce, quartz, baddeleyite, and zircon.

Whole-rock analysis revealed 1.6% Nb and 613 ppm Ta in ferricrete, indicating an order of magnitude enrichment in Nb and Ta compared to the underlying regolith. The concentrations of homogeneously distributed Nb and Ta in the core of goethite are 4.57wt.% and 0.12wt.%, respectively. The rounded grains of ilmenite contain up to 19.2wt.% Nb and 1.0wt.% Ta in the 500 μm thick rim. The cores contain (average) 12.5wt.% Nb and 0.52wt.% Ta. The Fe-enriched rutile (7.65wt.% Fe) contains 7.17wt.% Nb and 0.09wt.% Ta. TEM observations and EDS analyses of Fe-(hydro)oxides nodules revealed a sharp contact between a polycrystalline core of Nb-rich goethite surrounded by porous Nb-depleted goethite. A nanoporous box-like structure formed in the rims of altered ilmenite during replacement of ilmenite by Nb-depleted rutile and reprecipitation of Nb-enriched nanoparticulate ilmenite. The high-resolution observations document the key role of Fe-(hydro)oxides in the short-range migration of Nb and Ta during the lateritization of carbonatite, making ferricrete a promising tool for Nb and Ta exploration.

Trace Element Variations of Pyrite and Chlorite Related to Cu Mineralization in the Osarizawa Epithermal Deposit, Akita, Japan

Jonathan Macuroy[1,2], Nozomi Aoki[2], Ryohei Takahashi[2], Pearlyn Manalo[2]
[1]Center for Regional Revitalization in Research and Education, Akita University, Akita City, Japan, [2]Graduate School of International Resource Sciences, Akita University, Akita City, Japan

The Osarizawa polymetallic vein-type deposit in Akita is one of the largest of copper deposits in Japan. The deposit was discovered in 708 and was mined mainly for Cu and Au until 1978. In this study, we aim to reveal the morphological and geochemical characteristics of chlorite and pyrite associated with Cu mineralization in the Utori vein group of the deposit. We employed optical and electron microscopy, electron probe microanalysis, and laser ablation inductively coupled plasma mass spectrometry.

The main gangue minerals in the Utori vein group at +300 masl are quartz and chlorite, and the main ore minerals are chalcopyrite and pyrite with minor sphalerite, galena, hematite, and covellite. Fine-grained euhedral pyrite inclusions in chalcopyrite contain relatively elevated Cd, Sn, Zn, and Sb contents, whereas graphic pyrite contains elevated Au and Mo contents. Atoll pyrite has elevated Co, Ni, and Ag contents, and relatively higher Te content was found in annealed pyrite. Structural formula of chlorite in veins calculated from microprobe data reveals Mg-rich (clinochlore) and Fe-rich (chamosite) phases, whereas chlorite in altered rocks are mainly Mg-rich clinochlore. Concentric banding pattern in radial aggregates of acicular chlorite parallels the variations in Mg, Mn, and Fe contents. Fe/(Fe + Mg) ratio of chlorite ranges from 0.6 to 0.8, with an average of 0.67. Calculated formation temperatures based on AlIV content of chlorite range from 177 to 330 °C, with an average of 274 °C. Chlorite contains minor amounts of Mn, Ti, Na, Ca, K, V, and Zn and trace amounts of Li, Sc, Co, Ni, Ga, Ge, and Sr.

In-situ U-Pb Dating of Monazite and Rutile: Implications to Depositional Age of QPC and U-Au Mineralization Potential, Singhbhum Craton, India

Jyoti Chandra[1], **Smruti Prakash Mallick**[1], Amit Mondal[1], Ashim Kumar Patel[1], Kamal Lochan Pruseth[1], Dewashish Upadhaya[1]
[1]Department of Geology and Geophysics, Indian Institute of Technology,Kharagpur, Kharagpur, India, [2]Department of Earth Sciences, Indian Institute of Technology, Bombay, Bombay, India

Archean quartz pebble conglomerates (QPCs) may host significant uranium and gold mineralisation, depending on whether they were deposited before or after the Great Oxidation Event (GOE) at ~2.4 Ga. Determining the timing of deposition is challenging, as detrital mineral ages only provide maximum age limits, and resetting of these ages in the source rocks can result in dates that postdate actual deposition. In this study, we constrain the depositional ages of four QPCs from the Singhbhum Craton, eastern India, using in-situ U-Pb dating of monazite and rutile, as well as EPMA chemical dating of smaller monazite grains. The upper age limits for deposition vary due to differences in the emplacement ages of the source rocks: 3115 ± 52 Ma (Gorumahisiani–Badampahar), 3390 ± 123 Ma (Tomka–Daitari), 3322 ± 18 Ma (Bagiyabahal–Bartangra), and 2798 ± 48 Ma (Mankarchua). Hydrothermal monazite ages provide lower depositional age constraints: 2908 ± 83 Ma for the first three basins (EPMA) and 2484 ± 47 Ma for the Mankarchua Basin (LA-ICP-MS). The relatively younger age for the Mankarchua QPC may result from resetting of detrital monazite during a tectono-thermal event in the adjacent Rengali Province. Even if these QPCs were originally deposited before the GOE, subsequent geological events have overprinted them, as shown by the younger ages of both detrital and authigenic monazite. These findings suggest that despite post-depositional modifications, the QPCs of the Singhbhum Craton retain potential for modified placer-type uranium and gold mineralisation.

Genesis of the Hybridized Awakmas Gold System, Indonesia

Adi Sulaksono[1], Gyaista Sampurno[1], David Iswanto[1], **Ikrar Mandiri**
[1]PT Geo Fix Indonesia, Jakarta, Indonesia, [2]PT Masmindo Dwi Area, Jakarta, Indonesia

Awakmas is the largest metamorphic rock-hosted gold deposit in Indonesia, but its genetic framework remains poorly understood. This study examined a comprehensive dataset comprising 12,500 meters of diamond drill core logging, 216 thin and polished sections, and 9,456 assay results.

The mineralized zones within the deposit define composite orebodies that record three distinct hydrothermal events: (1) a pre-collisional phase, (2) a syn-collisional phase, and (3) a post-collisional phase. Genetic considerations suggest that the orebodies are hybrid in nature, with each phase contributing to the overall gold endowment.

During the collision between the Sula Spur and the West Arm of Sulawesi, pre-collisional quartz veins underwent metamorphism–evidenced by interlocking quartz grain textures, pervasive undulatory extinction, and parallel-oriented muscovite–and the pre-collisional vein zones simultaneously acted as planar structural anisotropies; gold grades of up to ~5 g/t (avg. ~0.5 g/t) occur exclusively in the metamorphosed pre-collisional quartz veins (Vein1). These structural anisotropies controlled the development of shear zones and localized the flow of hydrothermal fluids, leading to the formation of syn-collisional veins (Vein2). These veins are typically folded and/or discontinuous, mainly consist of quartz-albite-ankerite, lack alteration selvages, and contain low gold grades (avg. 0.25 g/t). Both Vein1 and Vein2 are crosscut by post-collisional quartz-albite veins (Vein3), which are typically accompanied by albite-ankerite selvages. Hydrothermal breccias containing clasts of metamorphic wall rocks, Vein1, and Vein2, cemented by quartz-albite, are common and likely related to the same post-collisional hydrothermal activity, which is interpreted to be associated with post-collisional extension. This phase is considered a major contributor to the district's gold endowment, with Vein3 or hydrothermal breccias containing up to ~10 g/t gold (avg. ~2 g/t).

This hybrid deposit, which contains pre-, syn-, and post-collisional mineralization, helps explain why multiple styles and ages of gold mineralization spatially overlap in several gold deposits worldwide.

Prolonged Mineralization for the Formation of the Southeast Asian Tin Province

Wei Mao, Jiehua Yang[1], Liang Liu[1], Yazhou Fu[1], Xingchun Zhang[1]
[1]Chinese Academy of Sciences, Guiyang, China

The Southeast Asian tin province is a 3,500-km-long belt that accounts for 40-45 % of the world historic tin production. It extends from western Yunnan of China through Myanmar and Thailand to Peninsular Malaysia and the Indonesian Tin Islands. This tin province formed through multiple tectonic events linked to Tethyan ocean subduction and closure. In situ U-Pb dating of cassiterite and wolframite from tin-tungsten deposits in the Southeast Asian Tin province have been conducted to unravel its prolonged mineralization history. Compilation of all available geochronological data identified five distinct episodes of tin-tungsten mineralization: (1) The 285–265 Ma mineralization in eastern Malaysia during Paleo-Tethys subduction; (2) The 239–210 Ma mineralization in Indonesia, western Malaysia, and Thailand after Paleo-Tethys closure; (3) The 124–107 Ma mineralization in the Tengchong block of China during Meso-Tethyan subduction; (4) The 90–42 Ma mineralization in western Yunnan, Myanmar, and Thailand during Neo-Tethyan subduction-linked deposits; and (5) The 31 and 20 Ma mineralization in the Baoshan block in western Yunnan during lithospheric delamination. The repeated episodes of mineralization achieved a cumulated duration >110 Ma, notably longer than in the other major tin provinces (e.g., Cornwall: ~17 Ma; Central Andes: ~47 Ma; South China: ~50 Ma), correlating with Southeast Asia's exceptional endowment. This work highlights the role of recurrent tectonic processes in sustaining metallogenic fertility over geologic timescales, positioning Southeast Asia as the world's largest tin province. Future exploration should focus on structural intersections and zones with repeated magmatism.

The Distribution of Co and Ni in Sulfide and Arsenide Minerals at the Eloise ISCG Deposit,

Queensland

Elizabeth de las Mercedes Marino Morejon[1], Ioan Sanislav[1], Avish Kumar[1], Melanie Finch[2], Alkis Kontonikas-Charos[3], Rhiannon Jones[3]
[1]James Cook University, Townsville, Australia, [2]University of Melbourne, Melbourne, Australia, [3]Geological Survey of Queensland, Brisbane, Australia

Iron-Sulfide Copper Gold (ISCG) deposits host economic concentrations of Cu, Au and Fe and may also contain significant amounts critical metals such as Ni, Co and REE's. This project investigates the mineralogy and geochemical distribution of Co and Ni in sulfide and arsenide minerals at the Eloise deposit, Queensland, Australia. While previous studies have reported the presence of Co and Ni minerals at Eloise, a comprehensive review of the Co-Ni mineralogy and distribution patterns is missing. Therefore, we aim to enhance our understanding of critical metal distribution within ores and provide insights into ore-forming processes and deposit genesis.

We combined textural and geochemical techniques to track Co and Ni distribution at Eloise. Chemical compositions were determined using electron probe microanalysis (EPMA) and laser ablation inductively coupled plasma mass spectrometry (LA-ICP-MS). The EPMA was also used for mapping element distribution patterns and to identify micro-inclusions rich in Co and Ni. The EPMA results show that Co and Ni have an affinity for pyrrhotite over chalcopyrite when only these two phases are present in the ore assemblage. However, in pyrite-rich ore assemblage Co and Ni are preferentially enriched in pyrite followed by pyrrhotite. The EPMA mapping highlights that Co and Ni are concentrated as micro-particles (<10 μm) along grain boundaries and fractures in samples collected from ductile zones of the deposit. The mineral formulas of these micro-particles were estimated by stoichiometry applied to the EPMA data and which revealed compositions similar to glaucodot (Co-rich) and mackinawite (Fe-Ni-rich). Additionally, pyrite grains display compositional variation with cobalt enriched towards grain margin while nickel towards the core. Generally, at the Eloise deposit, pyrite grains from the brittle deformation exhibit low Co and Ni concentrations compared to samples from ductile zones.

Fingerprinting Intrusion-Related Versus Orogenic Gold, Western Lachlan Orogen, Tasmanides, Australia

Helen McFarlane[1,2], Fariba Kohan Pour[1,2], Cameron Cairns[3], Helen McFarlane[1,2], Ross Cayley[3], Zsanett Pinter[4], Michael Verrall[1], Colin MacRae[4], Jacob Walmsley[1], Steven Boger[4], James Lewis[1]
[1]CSIRO Mineral Resources, Kensington, Australia, [2]MinEx CRC, Bentley, Australia, [3]Geological Survey of Victoria, Melbourne, Australia, [4]CSIRO Mineral Resources, Clayton, Australia

The Western Lachlan Orogen (WLO) of southeastern Australia hosts one of the world's largest gold provinces, with multi-million-ounce Ordovician orogenic deposits in the Bendigo and Stawell zones. According to the Lachlan Orocline model, the under-explored Tabberabbera Zone, a potential along-strike extension of the Bendigo Zone, comprises potential for both orogenic-style and intrusion-related gold deposits. However, efforts to constrain the mineralisation style and implementing effective exploration strategies are hindered by the close spatial relationship of Devonian plutons and orogenic-style deposits. To address this challenge, we investigate eight gold deposits located in the WLO, using a macro- to micro-scale analytical approach, integrating micro-XRF MAIA mapping, automated mineralogy, EPMA-CL mapping, and apatite chronochemistry.

Gold mineralisation near the Devonian I-type Beechworth Granite, in the northern Tabberabbera Zone, occurs as free gold in quartz veins, accompanied by native bismuth, Bi-tellurides, lollingite, and hydrothermal apatite coeval with Devonian intrusions, suggesting an intrusion-related style. High Ti-in-quartz and U-Pb thermal reset of detrital apatite further indicate temperatures above 375°C. Conversely, gold at Haunted Stream is locked within sulphides and lacks Bi and Te, indicating a classic orogenic style. Apatite compositions differ between intrusion-related and orogenic-style mineralisation with the latter exhibiting higher Sr/Y, Th/U, HREE/MREE, positive Eu anomaly and lower Mn/Fe inferring crystallisation from oxidising fluids. Hydrothermal apatite ages suggest: (1) gold remobilisation during the emplacement of the Upper Devonian Harcourt Granodiorite overprinting orogenic-style mineralisation at Gill Reef, Bendigo; (2) Triassic mineralisation at Happy Valley driven by distal magmatism; and (3) Jurassic reactivation of the Haunted Stream Fault during the opening of the Otway Basin.

Our findings demonstrate that apatite is a potential indicator for fingerprinting ore-forming fluids and constraining gold mineralisation styles. This study provides new insights for refining mineral exploration models and improving deposit classification and mineral exploration targeting in regions of overlapping intrusion-related and orogenic signatures.

Critical Minerals in Laramide Porphyry Copper and Paleogene Molybdenum-Tungsten Deposits in Southwestern New Mexico, USA

Virginia McLemore[1], Nels Iverson[1], Isabella Sanchez[1], Sarah Mosses[1], Kyle Stafford, Anita Appah[1], Evan Owen[1]
[1]New Mexico Tech/NMBGMR, Socorro, United States

New Mexico lies at the eastern edge of one of the world's great metal-bearing provinces hosting numerous Laramide porphyry copper and Paleogene molybdenum-tungsten deposits. Porphyry copper deposits are large, low-grade (<0.8% Cu) deposits that contain disseminated, breccias and stockwork veinlets of copper and molybdenum sulfides associated with porphyritic intrusions. Paleogene molybdenum-tungsten deposits are found in porphyry molybdenum and skarn deposits. Recent dating has constrained these deposits into three periods. Porphyry copper deposits occurred during two pluses: ~78-71 Ma, including Piños Altos (78.55±1.75 Ma), Copper Flat (Hillsboro district; 75.9±0.66 Ma), Oro (Eureka district; 71.4±0.19 Ma), and ~59-50 Ma, including Lordsburg (59.25-57 Ma), Santa Rita (59.05±0.36 Ma), Hanover-Hermosa Mountain (Fierro-Hanover district, 58.3±0.7 Ma), McGhee Peak (Peloncillo Mountains, 57.28±0.65 Ma), Tyrone (55.2±0.6 Ma), and Lone Mountain (50.6±1.9 Ma) districts. The younger molybdenum-tungsten deposits (~40-30 Ma) includes Camel Mountain, Victorio Mountains, and Tres Hermanas districts. These deposits have potential for various critical minerals, including Cu, Zn, Bi, Co, Ni, rare earth elements (REE), Te, and W that are found as separate minerals, micro-inclusions and/or in solid solution in sulfide minerals. New and published geochemical analyses of ores, mineralized rocks, and mine wastes provide insights into critical minerals in these deposits. High Co concentrations are found in Piños Altos samples (up to 1026 ppm), whereas the Ground Hog mine in the Central district has up to 64 ppm Te. Some of these critical minerals are found in the skarns adjacent to the porphyry copper deposits. Although most samples are low in total REE, a few deposits contain elevated total REE, including Piños Altos (409 ppm), Santa Rita (717 ppm), Fierro-Hanover (937 ppm), and Oro (Eureka, 2272 ppm). Although, these values are not normally economic, the large volumes of porphyry copper deposits mined enhance the economics of REE and critical minerals in these deposits.

Alteration, Mineralization and Geochemical Characteristics of the Arakompa Intermediate Sulfidation Epithermal Gold Deposit, Kainantu District, Papua New Guinea

Olive Lucas Ponyalou[1], David Cooke[1], Lejun Zhang[1], **Owen Missen**[1], Paula Montoya Lopera[1]
[1]Centre for Ore Deposits and Earth Sciences, University Of Tasmania, Hobart, Australia

The Arakompa intermediate sulfidation epithermal gold deposit is located in the Kainantu Cu-Au district, Eastern Highlands Province, Papua New Guinea. It is associated with the NNE-trending Arakompa Fault which crosscuts dioritic stocks of the Akuna Intrusive Complex (~15 Ma) formed during the mid-Miocene Maramuni Arc event. This project aims to increase ore body knowledge and develop a genetic model for Arakompa through detailed characterization of its mineralogy, geochemistry, textures and paragenesis. Hypogene mineralization at Arakompa is characterised by massive sulfide-quartz lodes, sulfide-carbonate-quartz veins, hydrothermal quartz-sulfide breccias and clay-sericite fault breccias. Alteration halos include intense phyllic (i.e. sericite-quartz), sericite-carbonate and sericite-chlorite-epidote assemblages, which overprint earlier propylitic and sodic-calcic alteration. Whole rock geochemistry shows strong depletions in Sr, Na and Ca which correspond to the feldspar-destructive phyllic alteration and high-grade gold mineralization. These Au-rich, Sr-Na-Ca-depleted zones also exhibit sporadic enrichments in K, Mg and Ca relating to muscovite, chlorite, calcite, and epidote alteration. Epithermal quartz displays colloform, crustiform, lattice, moss, cockade and drusy textures indicative of variable hydrothermal fluid conditions and syn-mineral deformation. Multiple generations of quartz, pyrite and chalcopyrite have been recognised based on grainsize, texture, and mineral inclusions. Inclusions in late-stage chalcopyrite include gold (100-200 μm) intergrown with bornite (6 μm), Te-Bi sulfides (70 μm), and pyrite (30 μm). Inclusions in late-stage pyrite include microscopic gold (10-20 μm), chalcopyrite, Cu-Bi sulfide (wittichenite), and Te-Bi sulfide. Bulk rock gold assays also show positive correlations with Te, Bi, Cu, Ag, Fe and As at a factor of 0.7 to 0.5 respectively for drillholes intersecting the epithermal veins. Sulfur isotope (δ34S) analysis of pyrite and chalcopyrite at -0.77 to -3.26 per mill are indicative of oxidized (sulfate-predominant) magmatic-hydrothermal fluids typical of porphyry-epithermal systems in the Maramuni Arc. Detailed characterisation of Arakompa will improve our genetic models for the deposit and district.

Geochemical Characteristics of Copper Mineralization and Exploration History at the Otavi Mountainland, Namibia

Takao Miura[1], Ryohei Takahashi[1], Andrea Agangi[1], Pearlyn Manalo[1], Henri Brunette[1]
[1]Japan Organization for Metals and Energy Security, Minato-ku, Japan, [2]Akita University, Akita, Japan

Japan Organization for Metals and Energy Security (JOGMEC) and Nexa Resources S. A. have conducted exploration of Cu-Pb-Zn resources since 2015 in the Otavi Mountainland, the Damara Belt, the Republic of Namibia. The project is located in the Northern Platform of the Damara Belt, the Neoproterozoic Pan-African mobile belt, in between the Congo and the Kalahari Cratons of Archaean and Proterozoic age. The Damara Belt comprises the Damara orogen together with the Kaoko and the Gariep Belts (Hoffman, 1976; Coward, 1981; Miller, 1983). The Otavi Mountainland is known for the endowment of base metal deposits in association with silver. Ore minerals are hosted in carbonate rocks at the Tsumeb and Kombat deposits, or siliciclastic rocks at the Tschudi deposit (Weatherly Mining Namibia, 2016), and those mineralization styles have been historically classified as Mississippi Valley-type or more generally, stratiform type deposits (Kamona and Günzel, 2007).

Ten years of exploration activity, including satellite image analysis, airborne geophysical surveys, soil geochemical survey, and drilling, revealed the existence of copper mineralization in the style of structurally-controlled calcite-quartz veins. In the early campaigns of exploration, soil geochemical survey worked effectively by extracting copper and lead anomalies, while zinc, silver, and arsenic are also good indicators of mineralization. In a high-grade zone of drill cores, chalcocite, bornite, native copper, and chalcopyrite are the major copper minerals, and they are hosted within diamictite layer or at a boundary of diamictite and dolomite in the Target 13 and 14 areas. Weathering zone after pyrite and chalcopyrite is observed in the surroundings of high-grade mineralized zone.

Magnetite Trace Elements as Petrogenetic Indicators in Exploring Savage River Deposit, Northwest Tasmania

Pratichee Mondal[1,2], Owen P. Missen[1,2], Lejun Zhang[1,2], Julie Hunt[1,2], Roger A. Hill[3], Alexey Lygin[3]
[1]Centre for Ore Deposit and Earth Sciences (CODES), University of Tasmania, Sandy Bay, Australia, [2]Regional Research Collaboration Program: Environmentally Sustainable Production of Critical Metals, University of Tasmania, Sandy Bay, Australia, [3]Grange Resources Tasmania, Burnie, Australia

The Savage River magnetite deposit occurs in the Proterozoic Arthur Metamorphic Complex, in northwestern Tasmania. The deposit type and the ore forming processes of the deposit remain poorly understood. This investigation focuses on the composition of different generations of magnetite to reconstruct the petrogenetic history of Savage River, while also evaluating the critical mineral potential. Testing was carried out on 40 samples collected from two drill holes that intersect the main ore zone in the North Pit of the Savage River deposit. Six different magnetite types were identified based on mineralogy, magnetic susceptibility and associated gangue minerals. The magnetite types were studied in detail using ASD TerraSpec, portable X-ray fluorescence (pXRF), HyLogger, scanning electron microscopy (SEM) and SEM-based automated mineralogy, plus laser ablation inductively coupled plasma mass spectrometry (LA-ICP-MS). Results suggest an iron oxide-apatite (IOA) type mineralisation based on the discriminant diagrams of Ga vs V, Ti vs V, Ni vs V, and V vs Cr, whereas characteristics of both IOA and Fe-Ti-V deposits are indicated by Ti+V vs Ca+Al+Mn, Ti+V vs Ni/(Cr+Mn) and Ti+V vs Al+Mn. Additionally, Sn vs Ga, V/Ti vs Al/Ti, Ni/Cr vs Ti, and Fe vs V/Ti showed all the magnetite types are likely hydrothermal origin. Three of six Savage River magnetite types have a high vanadium content based on LA-ICP-MS data, with grade up to 0.71% V2O5, suggesting the magnetite precipitated from high temperature hydrothermal fluids. Based on the above results and a comparison of textural and compositional variations of Savage River magnetite with magnetite from worldwide IOA deposits, we suggest that Savage River has characteristics of an IOA style deposit hosted in a Fe-Ti-V oxide system with a vanadium critical metal potential.

Miocene Low-Sulfidation Epithermal Deposits of the Colorado Extensional Corridor, USA

Thomas Monecke[1], Lauren R. Terry[1], Greg Seitter[1], T. James Reynolds[2], Erik R. Tharalson[3], Robert Johansing[4], Frank Powell[5]
[1]Colorado School of Mines, Golden, United States, [2]FLUID INC., Denver, United States, [3]Denver Federal Center, Denver, United States, [4]West Point Gold, Vancouver, Canada, [5]Mako Mining Corp., Vancouver, Canada

The Colorado River Extensional Corridor is a 70- to 100-km-wide zone of crustal extension that extends ~200 km from southern Nevada into Arizona. Extension broadly coincided with the shift of the western margin of North America from a convergent to a transform plate boundary. Volcanism commenced at ~22 Ma at Oatman and subsequently swept north, reaching the Lake Mead area at ~12 Ma.

Castle Mountain is the largest low-sulfidation epithermal deposit within the Colorado River Extensional Corridor. The lower section of the host stratigraphy of this low-grade, large-tonnage deposit is composed of trachybasalt and trachyandesite flows and related autobreccia. The bulk of the resource is hosted in volcanic rocks that formed during a period of intense felsic volcanism, which resulted in the deposition of thick aprons of massive to poorly sorted tuff breccia surrounding rhyolite lava flows and domes. Vertical breccia bodies, interpreted to represent phreatic eruption deposits, are spatially associated with these rhyolite centers. In contrast, other deposits in the Colorado River Extensional Corridor are characterized by high-grade veins. The Searchlight district is a relatively small but historically significant producer. Ongoing exploration in the historic Katherine district has resulted in high-grade intercepts at the historic Tyro mine and further south at the former Arabian mine. Present-day mining in the Oatman district focuses on the Moss vein that is hosted by both the Peach Spring Tuff and a monzonite intrusion that postdates deposition of the densely welded trachyte ignimbrite.

The nature of low-sulfidation epithermal deposits in the Colorado River Extensional Corridor is strongly influenced by both volcanological and rheological factors. High-grade precious metal ores primarily formed in competent rocks, where rapid flash vaporization of hydrothermal liquids led to the deposition of silica and ore minerals under far-from-equilibrium conditions.

Lithium Brine Exploration in La Unión, Ahumada, Chihuahua, México

Héctor Montes Carreón[1], Perla Rodríguez Contreras[1], Luis Ángel Jiménez López[2], Rurik Farías[1]
[1]Universidad Autónoma de Ciudad Juárez, Ciudad Juárez, México, [2]Centro Interdisciplinario de Investigaciones y Estudios sobre Medio Ambiente y Desarrollo CIIEMAD-IPN, Ciudad de México, México

La Unión Salar is located 100 kilometers southwest of Ciudad Juárez and 40 kilometers northwest of Villa Ahumada, within the El Barreal Playa Lake, an endorheic basin of approximately 1,200 km² that exhibits key characteristics associated with lithium brine deposits, including an arid climate, tectonic subsidence, potential lithium-bearing source rocks, and a highly saline aquifer. Felsic volcanic rocks, such as rhyolites, rhyolitic tuffs, and ignimbrites, have been identified in the Sierra de la Unión and Sierra de la Candelaria, which delineate the southeastern extension of the basin. The exploration focused on this sector, which encompasses approximately 200 km². These lithologies are interpreted as products of Oligocene ignimbritic pulses and are considered potential lithium sources, as observed in other brine deposits. Vertical Electrical Soundings indicate the presence of a brackish aquifer that likely transitions into a more saline system at depths beyond the reach of the geophysical survey. This aquifer appears to be locally confined or semi-confined by clay-rich sedimentary units. A systematic sampling grid was established, with cells measuring 1,000 by 1,000 meters. A total of 46 sediment samples were collected from the El Barreal basin, covering approximately 40 km². Lithium concentrations in the sediments ranged from 20 to 165 ppm, with an average of 62.6 ppm. In the northern portion of the study area, corresponding to La Unión Salar, the average lithium concentration was 101 ppm. These values are considered anomalous, with peak concentrations reaching up to 165 ppm, comparable to those reported in other lithium exploration projects in geologically similar settings, such as the Animas Valley Project in New Mexico.

The Role of Oligocene Volcanism to Develop Ag-Au-Pb-Zn-Cu Mineralization at San Marcial Area, Western Sierra Madre Occidental, Sinaloa, Mexico

Paula Montoya Lopera[1], Gilles Levresse[2], Marcio Fonseca[3], Luis Coto[3], Marlen Salgado[3], Javier Villegas[3], Miguel Diaz[3], Daniel Schrader[3]
[1]University of Tasmania, Hobart, Australia, [2]Universidad Nacional Autonoma de Mexico, Queretaro, Mexico, [3]GR Silver Mining Ltd, Rosario, Mexico

The Plomosas District and San Marcial Area (PDSMA) host multiple Ag-Au-Pb-Zn-Cu mineralization events, from an early Cretaceous Cu porphyry setting to Eocene high-sulphidation Ag epithermal veins, and Oligocene high-grade Ag-Cu-Pb-Zn-Au tectono-hydrothermal breccias, located at the southern end of the Sierra Madre Occidental (SMOc) in Sinaloa, Mexico. Recent field observations, detailed petrography, U/Pb, and Ar-Ar age dating of the entire stratigraphic sequence were used to characterize the evolution of multiple mineralization styles and obtain new insights into the evolution of the southern extension of the SMOc. The PDSMA does not have a typical SMOc stratigraphic column. The local basement is composed of non-deformed Late Jurassic volcanoclastic units and coeval felsic intrusions, crosscut by the root of Laramide intrusive. Early to late Oligocene silicic ignimbrites, intermediate lava flows, and occasional basalt flows directly overlie the Jurassic basement and Laramide intrusive, separated by an angular discordance. Several Oligocene coarse intermediate porphyry bodies and a few Miocene rhyolitic domes intrude the PDSMA stratigraphy. Within the Oligocene volcanic sequence an angular discordance separates an early and a late period. The recent studies presented here highlight the importance of overprinting explosive volcanism in the genesis of the multiple mineralization episodes in the PDSMA and in the southern SMOc.

The State of Lithium Resources in 2024: A Deposit-Type Analysis for Supply Potential

Brock Moody[1,2], Simon Jowitt[1,2]
[1]Ralph J. Roberts Center for Research in Economic Geology, Department of Geological Science and Engineering, University of Nevada, Reno, Reno, United States, [2]Nevada Bureau of Mines and Geology, Reno, United States

Demand for lithium (Li) has surged over the past decade due to its role in Li-ion batteries for electric vehicles, raising concerns over long-term supply. A 4,000 metric ton shortfall in 2022 drove Li carbonate prices up over 500% in 2023, sparking debate on alternative battery technologies. While mineral resources are often seen as finite, global Li production has steadily grown, with significant increases in reserves and resources from brownfield exploration and discoveries of new deposits. This study compiled internationally reported Li resources by deposit type and project for 2024. At least 143.9 million metric tons (Mt) of Li are reported globally across 146 brine, hard-rock, volcano-sedimentary (VS), and tailings projects, with an additional 6.8 Mt Li in China, yielding a global minimum of 150.7 Mt Li. While hard-rock deposits comprise most projects, they account for only 17.6% of total Li resources. In contrast, brine and VS deposits host more Li per project, representing 56.7% and 25.5% of total resources, respectively. Substantial resources are concentrated in the brines of South America's "lithium triangle" (Bolivia, Chile, Argentina), lithium-cesium-tantalum pegmatites of Western Australia, and VS deposits in the U.S. Basin and Range. Case studies across deposit types highlight ore grade, market prices, and extraction technologies (e.g., direct lithium extraction for geothermal brines) as key constraints on mine development. Brine deposits exhibit a wide variation in cutoff grades, partly due to estimates made during the 2023 price surge, with some lacking defined cutoff thresholds, contributing to inflated resource estimates. Overall, Li resources are sufficient to meet projected demand for decades to come. Future supply limitations will stem from economic, technological, or environmental factors rather than resource scarcity.

Determination of Migration Events of Present Hidrocarbons in the Eastern Llanos Basin, Through the Characterization of Fluid Inclusions

Gabriel David Moreno Aguilar[1,2], Juan Carlos Molano mendoza[1,2], Ariel Oswaldo Cadena[1], Santiago Jimenez[1,2], Angie Cardona[1,2], Nathalie Burbano[2]
[1]Universidad Nacional de Colombia, Bogota, Colombia, [2]Microfluid Spectral Laboratory, Bogota, Colombia

Within the Mesetas sector in the Lejanías area of the Llanos Orientales Basin, defined as part of the ANH's Caracterización de Arenas Bituminosas (CAB) project, six hydrocarbon-impregnated sandstone samples were studied. These samples were collected from two stratigraphic columns (Río Cafre and San Antonio columns) within the Guadalupe Group in the project. Various techniques were used, including petrography, cathodoluminescence, SEM images, and UV epifluorescence, which helped differentiate the diagenetic events of the rock and temporally locate the formation of fluid inclusions (FI) in these processes, followed by FI petrography. Among the fluid inclusion assemblages (FIA's), three main types of FI were studied: Type 1 inclusions consisting of light to dark brown liquid hydrocarbon, gaseous hydrocarbon, and occasionally black irregular solid (insoluble organic material); Type 2 inclusions composed of liquid and gaseous hydrocarbon along with water and sometimes solids (insoluble organic material); and Type 3 inclusions consisting of water and gas. Data on homogenization temperature (Th), final ice melting temperature (Tm), salinity, pressure, Raman spectroscopy, and UV epifluorescence were obtained from these inclusions. The various data allowed for the differentiation of at least two hydrocarbon migration events in the area. One event shows a temperature of around 62 – 86 °C, predominantly greenish-blue fluorescence and salinities around 0.18% wt. NaCl equiv. The other event shows a temperature of around 88 – 100 °C, predominantly strong blue-green fluorescence and salinities around 2.41% wt. NaCl equiv. Both hydrocarbons must have API gravities between 20 and 50 degrees, with the predominantly blue fluorescence indicating lighter hydrocarbons. Finally, a qualitative comparison was made between the fluorescence of hydrocarbons present in fluid inclusions and those in the rock pores, showing significant degradation of the latter, evidenced both by the loss of fluorescence and the characteristics observed in hand samples.

Understanding the Importance of Heavy Mineral Sands as a Source of Rare Earth Elements, Using a Mineral Systems Approach

Paul Duuring, Christopher Gonzalez, Ignacio González-Álvarez, **Sidy Morin-Ka**[1,2]
[1]University of Western Australia, Perth, Australia, [2]Geological Survey of Western Australia, Perth, Australia

Heavy Mineral Sands (HMS) deposits have historically played a crucial role as a source of Rare Earth Elements (REE), though their significance has been overshadowed by the emphasis on carbonatite complexes. While carbonatite complexes are currently the top REE producers, HMS deposits remain a valuable and complementary resource within the REE landscape.

Globally, over one thousand REE-HMS deposits have been reported, characterised by diverse heavy mineral compositions, including ilmenite, rutile, zircon, monazite, xenotime, and garnet. More than 90% of the reported deposits formed within the past 66 million years. The distribution of REE-HMS deposits is shaped by geological, climatic, and surface process factors, with additional considerations such as land accessibility, politics, and exploration maturity also playing crucial roles. Spatially, REE-HMS deposits show distinct patterns, including concentration along modern coastlines, clustering within a 100 km radius, and prevalence in tropical humid areas within 25° N and S of the Equator. Coastal dynamics and ocean currents influence the distribution of coastal REE-HMS clusters, with protected coastal areas exhibiting a higher prevalence compared to regions exposed to major ocean currents. The ore minerals are primarily sourced from regions dominated by cratons, orogenic belts, and metamorphic rock suites. The transportation and sorting of these heavy minerals are mainly shaped by fluvial drainage, winds, and coastal dynamics. Geomorphology, tectonic activity, and climate evolution play a crucial role in HMS preservation.

Thus, understanding evolving environmental conditions and paleoclimates is imperative to fully understand REE-HMS potential discovery beyond modern coastlines. Using a mineral system framework provides a comprehensive approach to characterising REE-HMS deposits and optimising exploration methodologies. These findings enhance our understanding of REE-HMS deposits, emphasising that developing a conceptual model and understanding global REE-HMS distribution will drive more efficient exploration and support a sustainable, diverse supply of REE.

When Dates Lie: The Impact of Mineral Porosity on Isotopic Dating of Base Metal Deposits

Michelle Moxey[1], Nathan Daczko[1], Elena Belousova[2], Jacob Mulder[3], Nicholas Dyriw[4], Thomas England[1]
[1]Macquarie University, Sydney, Australia, [2]Geological Survey of Queensland, Brisbane, Australia, [3]University of Adelaide, Adelaide, Australia, [4]MMG Dugald River Pty Ltd, Cloncurry, Australia

Dating economic mineral systems, such as base metal deposits, poses significant challenges due to the limited availability of conventional geochronometers in ore mineral packages (e.g. zircon, monazite) and the susceptibility of minerals to fluid-induced alteration. Therefore, integrating petrographic observations with in-situ dating is essential for accurately interpreting geochronological data.

Recent advances in laser ablation Lu-Hf and Rb-Sr geochronology offer new opportunities, but this study highlights the limitations of applying these methods to porous minerals. High porosity—a key indicator of coupled dissolution–precipitation replacement—can cause partial isotopic resetting and enhance element mobility below diffusion closure temperatures. Consequently, ages from porous minerals may reflect times between primary crystallisation and later replacement or appear younger due to post-replacement radiogenic loss.

At the Dugald River Zn-Pb-Ag deposit in the Mt Isa Inlier, northwest Queensland, the timing of mineralisation remains poorly constrained. This structurally complex, poly-deformed system has undergone extensive alteration, rendering traditional mineral separate dating ineffective. This study combines ultra-high-resolution BSE imaging with in-situ U-Pb (apatite), Rb-Sr (K-feldspar–muscovite), and Lu-Hf (calcite ± fluorite) geochronology.

Our results identify an albitisation and calcite veining event at ~1535 Ma, and a younger event (~1500–1480 Ma) possibly linked to Zn-Pb mineralisation—providing the first direct age constraints for the Dugald River deposit. These intervals coincide with a regional mineralisation phase at the end of the Isan Orogeny, underscoring its economic significance.

Although U-Pb apatite dating in ore samples was compromised by common Pb, our results suggest that mineral porosity poses an even greater challenge for reliable geochronology. The unclear significance of younger populations (~1430 Ma and ~1140 Ma) in moderately to highly porous minerals highlights this issue. Overall, this study demonstrates the utility of Lu-Hf and Rb-Sr geochronology in complex systems and emphasises the importance of microstructural screening before analysis.

Lithium Oxide in a Weathered Pegmatite in Nyanza District, Rwanda: A Preliminary Data

Theophile Mugerwa[1]
[1]University Of Rwanda, School of Mining and Geology, KIGALI, Rwanda

The global depletion of fossil fuel resources, along with their potential environmental concerns, has prompted the search for alternative clean energy resources for future industrial development. The use of electric cars is growing and creates major concerns for future lithium demand and lithium. In Rwanda, Li-bearing pegmatites are distributed in part of the Nyanza District. Although Li-bearing pegmatite is distributed in that location, the quality of lithium for industrial applications is still undiscovered. As a result, a preliminary assessment of lithium deposits was conducted along the core samples collected in 2024. The results reveal 520 ppm to 730 ppm, 470 ppm to 900 ppm, and 190 ppm to 580 ppm for boreholes 1,2, and 3, respectively. The concentration of Li oxide is not evenly distributed along the core and the results highlight that the moderately weathered pegmatite has a higher concentration of Li oxide compared to the highly weathered pegmatite. Further research will develop a model to evaluate the quantity and quality of lithium for future mining development decision-making.

Structural Architecture of the Kirkland Lake Mining District

Taylor Mugford[1], Bruno Lafrance[1], Ross Sherlock[1], Györgyi Tuba[1], Daniel Ruiz Arriaga[1], Chris Evans[2], Monica Oosterman[2]
[1]MERC, Harquail School of Earth Science, Laurentian University, Sudbury, Canada, [2]Agnico Eagle Corporation, Kirkland Lake, Canada

The Kirkland Lake mining district in the Abitibi greenstone belt, northeastern Ontario hosts Macassa mine, one of the world's highest-grade gold mines (18.4 g/t Au). Gold is primarily hosted along semi-brittle faults and quartz veins that cross-cut volcanic and sedimentary rocks of the Timiskaming assemblage (2676 Ma– 2670 Ma), as well as co-genetic intrusions. These are separated from older volcanic rocks of the Larder Lake Group (2705 Ma) by the Larder Lake-Cadillac deformation zone (LLCDZ), a 200-800 m wide ductile shear zone extending over 250 km across Ontario and Quebec, hosting numerous world-class gold deposits.

The South Mine Complex (SMC) is Macassa's most active zone, with gold localized along narrow, east-northeast trending, shallowly-dipping (0-30°) faults filled with quartz veins, hydrothermal breccia, and sericite-pyrite alteration. These faults lie between the reverse, south-side-up Main Break and the reverse, north-side-up Amalgamated Break. This study focuses on surface mapping of a 2.3 km-wide Timiskaming panel between the Amalgamated Break and the LLCDZ, plus underground structural mapping in the SMC. The panel's structure is dominated by east-northeast-trending, F3 folds with a regional axial planar S3 cleavage, which intensifies in iron-carbonatized LS tectonites along the LLCDZ.

The Timiskaming rocks young towards the LLCDZ and Larder Lake Group, suggesting that the LLCDZ originated as a north-directed thrust fault prior to S3 formation. This thrusting may coincide with the formation of the Amalgamated Break, which evolved from a brittle fault to a ductile shear zone through clay alteration during fluid flow along the fault. Gold within the mineralized AK zone, located in the study area, is controlled by brittle fractures predating regional foliation and the LLCDZ, indicating early gold deposition during a brittle deformation event. This project aims to investigate the relative timing of the semi-brittle faults and ductile LLCDZ and constrain the structural controls on gold mineralization.

On the Nickel-Sulfide Potential of Late-Archean Komatiites from the Dharwar Craton: A Case Study from the Shankaraghatta Complex (Southern India)

Ria Mukherjee[1], Marco Fiorentini, Laure Martin
[1]University Of New England, Armidale, Australia

The Shankaraghatta Complex (12 km × 0.3 km) is located within a Late Archean greenstone belt in the Western Dharwar Craton (WDC) and hosts komatiites that are now entirely serpentinized. These ultramafic rocks represent adcumulate to mesocumulate flow facies and are surrounded by komatiitic basalts, now altered to talc-tremolite-chlorite schists. The komatiitic rocks contain currently uneconomic disseminated sulfide blebs (3–8 modal %), comprising millerite, pyrite, violarite, and minor chalcopyrite.

For the first time, these ultramafic rocks have been dated in this study, yielding a Sm-Nd errorchron age of 2718 ± 107 Ma (MSWD = 2.4), which demonstrates contemporaneity with late-Archean komatiites from the Kalgoorlie Terrane of the Yilgarn Craton (Australia), which are notably mineralized. However, the Dharwar Craton lacks significant endowment. This study therefore further explores the reasons behind this disparity.

Multiple sulfur isotope analyses ($\delta^{34}S$ = -0.8 to +2.5‰; $\Delta^{33}S$ = -0.6 to +0.2‰) from this study reveal a predominantly mantle-like signature with no evidence for mass-independent fractionation, suggesting the absence of any significant crustal sulfur contribution, which is an essential component for economic mineralization, as seen in the Kalgoorlie Terrane. In addition, several factors appear to have inhibited the development of large-scale, komatiite-hosted nickel sulfide mineralization in the Dharwar Craton. Chief among them are the absence of a suitable crustal substrate, the lack of contemporaneous or pre-existing crustal-scale structures (e.g., faults or shear zones), and, most critically, the tectonic setting.

We put forward the hypothesis that in the Late Archean the WDC had already evolved into a mature and stabilized continental crust supporting convergent tectonics, in contrast to the intraoceanic rift environment prevailing in the Kalgoorlie Terrane. It is argued that this fundamental difference in tectonic setting likely played the decisive role in controlling the mineralization potential of the two terranes.

Geology and Polymetallic Mineralization of the Tsagaan Tolgoi Skarn Deposit, South Mongolia

Otgonbayar Munkhtsol[1,3], Jamsran Erdenebayar[1,2], Purevnaym Lkhagvadorj[1,4], Bolorchimeg Nanzad[1]
[1]Mongolian University of Science and Technology, Ulaanbaatar, Mongolia, [2]MiRes Lab, Akita, Japan, [3]Galaxy Geo Mongolia LLC, Ulaanbaatar, Mongolia, [4]Geo-Oron LLC, Ulaanbaatar, Mongolia

The Tsagaan Tolgoi skarn deposit in southern Mongolia is a polymetallic site rich in zinc, lead, and iron. It lies at the boundary between the Neoproterozoic Oortsog Formation limestone and Late Triassic to Early Jurassic granitic intrusions of the Bor-Under Complex.

In this study, we examine the geological framework, mineralogy, alteration features, and geochronology of the skarn system, with a focus on the evolution of ore-forming processes.

Petrographic and mineralogical analyses of skarn samples reveal a complex mineral assemblage that includes garnets (andradite and grossular), wollastonite, hedenbergite, diopside, epidote, carbonates, and quartz. These minerals exhibit hypidioblastic to microgranoblastic textures, indicating thermal metamorphism associated with magmatic intrusion. The skarn bodies display a massive structure with variable ore mineral content (ranging from 5% to 60%), suggesting heterogeneous ore distribution. The mineralization in the skarn evolved through both prograde (with minerals such as pyroxene and wollastonite) and retrograde (including chlorite, epidote, and calcite) stages.

X-ray diffraction (XRD) and scanning electron microscopy (SEM-EDS) analyses identify three main alteration zones: argillic (featuring kaolinite and illite-smectite), propylitic (containing chlorite, epidote, calcite, and quartz), and skarn (with pyroxene, quartz, and calcite). The primary ore minerals comprise magnetite, pyrite, pyrrhotite, chalcopyrite, sphalerite, galena, and bornite, alongside secondary oxidation products such as hematite, goethite, limonite, zincite, and chalcocite. Zoned textures and intergrowths of chalcopyrite and sphalerite indicate multiphase mineralization.

U-Pb dating of zircon grains from the intrusive rocks reveals ages ranging from 250 to 257 million years ago, aligning with post-collisional magmatic events in the region. The timing and composition of hydrothermal events indicate a magmatic-hydrothermal genesis closely associated with granitoid emplacement.

We present a comprehensive analysis of the skarn mineralization, alteration assemblages, and metallogenic implications of the Tsagaan Tolgoi deposit, South Mongolia.

Exploring Two-Liquid Separation of Fluorosilicate Melts: Advancing Phase Relationships and Mineralization Insights for Granite Hosted REE Deposits

William Munro[1], Neil Bennett[2]
[1]University of Toronto, Comox, Canada, [2]University of Toronto, Toronto, Canada

The accelerating demand for rare earth elements (REE), driven by technological advancements, has heightened interest in unconventional deposits such as A-type granite-hosted REE systems. These granites can exhibit extreme REE enrichment, often attributed to fluorosilicate-silicate immiscible magmas. However, the degree of extreme REE enrichment seen in nature has proven challenging to replicate experimentally, motivating further investigation into fluorosilicate immiscible melt evolution.

This study applies novel experimental techniques to investigate thermal gradient-driven separation of immiscible melts and trace element partitioning. The results unexpectedly redefined critical aspects of the fluorosilicate-silicate immiscible system. Most notably, the fluorosilicate-silicate miscibility gap terminates at a peritectic reaction near ~940°C, significantly higher than the previously interpreted ~600°C. The research also constrains temperature-dependent fluorite saturation within the miscibility gap, providing a framework to reinterpret previously misunderstood fast-cooling metastable immiscibility and explain why thermal gradient separation of fluorine-rich immiscible melts was ineffective.

Overall, the study reveals a widening miscibility gap at temperatures above 1300°C, terminating below ~940°C, and prompts a reevaluation of REE enrichment processes in A-type granites. These results support a genetic model where progressive REE-induced solvus depression may allow initially low-REE systems to evolve into the highly enriched compositions observed in nature, although further experimental work is required to confirm this. By redefining the miscibility gap and constraining trace element partitioning controls, this research offers new insights into the genesis of granite-hosted REE systems and advances the broader understanding of fluorine-rich magmatic evolution.

Genesis of a 3.53 Ga Volcanogenic Massive Sulfide Deposit in the Nondweni Greenstone Belt, Kaapvaal Craton

Thendo Netshidzivhe[1,2], Sebastian Tappe[2,3], Allan Wilson[4], Harald Strauss[5], Katie Smart[3,4], Bradley Guy[6], Fanus Viljoen[2]
[1]North-West University, Potchefstroom, South Africa, [2]University of Johannesburg, Johannesburg, South Africa, [3]Institute for Mineralogy, Technical University Bergakademie Freiberg, Freiberg, Germany, [4]University of the Witwatersrand, Johannesburg, South Africa, [5]Institute for Geology and Paleontology, University of Münster, Münster, Germany, [6]Helmholtz Institute Freiberg for Resource Technology, Freiberg, Germany

Volcanogenic massive sulfide (VMS) deposits represent some of the oldest and most significant base metal ores recorded on Earth, which commonly host exploitable quantities of Cu, Zn, Pb, Au and Ag. The association of such VMS-type mineralization with felsic volcanic rock units (massive rhyolites and deformed quartz-feldspar-mica schists) in the Paleoarchean Nondweni Greenstone Belt presents a rare opportunity to investigate the link between early Earth volcanism and seafloor metallogenesis. The Nondweni Greenstone Belt at the southern margin of the Kaapvaal craton (South Africa) consists of mafic-ultramafic volcanic rock sequences, including komatiitic and tholeiitic basalts associated with minor felsic volcanics. We investigated Zn-Cu-Pb-Ag VMS-type mineralization hosted within felsic schists exposed at an abandoned exploration shaft on the St James deposit property, where the mineralogy is dominated by sphalerite, pyrrhotite, chalcopyrite with trace amounts of pyrite, galena and acanthite. We report the first high-precision CA-ID-TIMS zircon age for the felsic volcanic rocks, suggesting a rhyolite magma eruption and VMS-style mineralization age of 3531.91±0.46 Ma. The St James base metal deposit may thus record world's oldest preserved VMS-type mineralization. Rare-earth element modelling provides evidence for a petrogenetic process whereby rhyolite melt formed through 15-17% partial melting of hydrothermally altered basalts at relatively shallow depths. Sulfur isotopic compositions ($\delta^{34}S$ and $\Delta^{33}S$) reveal mantle signatures for the tholeiitic basalts consistent with a juvenile magmatic sulfur source, whereas the rhyolites are characterized by values typical for seawater-altered basaltic oceanic crust. The base metal sulfide mineralization at St James exhibits negative $\Delta^{33}S$ signatures (-0.53 ‰) indicating a mass independent sulfur isotope (S-MIF) signature. The identified S-MIF signature implies incorporation of surficial sulfur into magmatic-hydrothermal fluids that circulated through oceanic crust in an active back-arc environment ca. 3.53 Ga ago. Our results are consistent with a metallogenetic model that involves the operation of plate tectonics during the Paleoarchean.

Two-Stage Ore-Forming Process for Bonanza-Grade Gold Veins in the Brucejack Au-Ag Deposit, British Columbia, Canada

Kevin Man Hoi Ng[1], James Clark[1], Anthony Williams-Jones[1], Jacob Hanley[2], Duncan McLeish[1]
[1]McGill University, Montreal, Canada, [2]Saint Mary's University, Halifax, Canada

An integrated petrographic and microthermometry study was carried out on the bonanza grade gold-bearing veins at the Brucejack epithermal gold-silver deposit in British Columbia, Canada to investigate fluid chemistry, fluid source, and physiochemical conditions of the ore-forming processes of (1) primary bonanza grade electrum and (2) subsequent structural modification. SEM-CL textures reveal that Phase 1 epithermal quartz crystals comprise (1) CL-homogenous cores; (2) oscillatory-zoned overgrowths; and (3) microcrystalline quartz associated with primary electrum mineralisation. This quartz was followed by Phase 2 quartz, which is the immediate host to secondary, structurally modified electrum. Fluid inclusion studies reveal that the primary epithermal gold veins were deposited from relatively saline (7 to 12 eq. wt. % NaCl) magmatic fluids with homogenisation temperatures (Th) ranged from 170 to 200 °C. A decrease in salinity (3 to 6 eq. wt. % NaCl) and Th (150 to 170 °C) was recorded in later quartz overgrowth phases. Epithermal calcite crystals were subsequently deposited at low temperature of 120 to 150 °C from low salinity to non-saline fluids (0-1.7 eq. wt. % NaCl). A resurgence in temperatures is recorded during recrystallisation of Phase 2 quartz associated with post-mineral deformation with Th between 165 and 190 °C. The fluid inclusion assemblage with an electrum inclusion bearing fluid inclusion has a salinity of ~12.5 eq. wt. % NaCl and a Th of 170 °C. Laser ablation ICP-MS analyses of secondary fluid inclusions hosted in quartz crystals detected Au, Ag, As, Sb, Sr, Cu and Pb, indicating that electrum grains were remobilised and captured in secondary fluid inclusions trapped during deformation.

A Remote-Predictive Geologic Map of the Mid-Atlantic Ridge Between 22.6°N - 19.6°N: Implications for Tectonic Evolution and Massive Sulphide Mineralization

Sam Nichols[1], Melissa Anderson[1]
[1]University Of Toronto, Toronto, Canada

The slow-spreading Mid-Atlantic Ridge between 22.6° N and 19.6° N displays a high degree of tectonic complexity, with evidence for ridge segment propagation, transform fault elimination, and ridge jumps. Remote-predictive geological mapping is used for the first time to investigate the evolution of the Mid-Atlantic Ridge at these latitudes to better understand these processes and identify potential areas for future mineral exploration. Using Global Multiresolution Topography data and ship-track bathymetry gathered on cruise FKt230303, a 1:250000 scale map of the Mid-Atlantic Ridge has been developed. Three previously known massive sulfide deposits occur near the ridge, with potential for further mineralization on mapped structures such as two oceanic core complexes and a rifted volcano, which are indicative of crustal-scale faulting and magmatic focusing. The area contains three spreading centers: the TAMMAR Kane Macrosegment (TKMS), and the Central and South Segments, which are all offset by second degree non-transform offsets. Two abandoned spreading centers (ASCs) are also mapped: ASCs North and South. Assuming a full spreading rate of 21 mm/year, evolutions for two offsets are proposed. At 21.5° N, ridge jumps initiated by a change in the regional stress field occurred at ~ 1.9 Ma. This formed the TKMS and Central Segment and caused the death of a transform fault. After, southward propagation of the TKMS began. At 20.6° N, ASC South was discontinuous from the South Segment about a second degree NTO since ~ 15.1 Ma. This NTO was stable until ~ 7.38 - 6.19 Ma, when the South Segment briefly propagated northward until ~ 4.09 - 3.43 Ma, when ASC South began propagating southward. Southward propagation continues today via the Central Segment as a ridge jump from ASC South occurred around 1.9 Ma. Around 457 Ka, magma focusing within the NTO valley constructed the Puy des Folles volcano.

Characteristics of the Epithermal Veins in the South Teresa Deposit, Lepanto, Mankayan District, Philippines

Shion Nohara[1], Pearlyn Manalo[1], Ryohei Takahashi[1], Andrea Agangi[1], Hinako Sato[1], Yasuhiro Awano[1], Akira Imai[2], Jillian Gabo-Ratio[3], Leo Subang[4], Julius Mahinay[4]
[1]Akita University, Akita, Japan, [2]Kyushu University, Fukuoka, Japan, [3]Universtiy of the Philippines, Quezon, Philippines, [4]Lepanto Consolidated Mining Company, Makati, Philippines

The South Teresa deposit is one of the newly delineated epithermal veins located in Lepanto, Mankayan District, Philippines. The veins are characterized by gray quartz with abundant pyrite (stage 1), white quartz with sphalerite, galena, and pyrite (stage 2), and anhydrite with sulfides (stage 3). The veins are hosted in hornblende diorite and andesite porphyry. Alteration haloes composed of illite and kaolinite typically surround the veins. Bulk geochemical analysis reveals that the veins of the stages 2 and 3 have significant mineralization with 0.08 to 5.7 ppm Au, 1.4 to 7.2 ppm of Ag, 43 to 950 ppm Cu, 5.97 to 7400 ppm Zn and 3.0 to 4600 ppm Pb. The Stage 1 veins contain 0.1 to 0.3 ppm of Au, while Ag content is lower than limit of detection.

δ34SCDT values of pyrite in the stage 1 veins range from 0.0 to +2.2 ‰, while those in the stage 2 veins range from -1.9 to + 1.2 ‰. Pyrite from the stage 3 has a δ34SCDT value of -0.1 ‰. Anhydrite from the stage 1 has δ34SCDT values ranging from +16 to +18 ‰, while those from the stage 2 range from +17.0 to +22.0 ‰. Calculated sulfur isotope equilibrium temperatures using pyrite – anhydrite pairs range from 280 to 389 ℃ for the stage 1, 239 to 362 ℃ for the stage 2, and 340 ℃ for the stage 3. Trace element analysis of sphalerite reveals significant amounts of Fe, Cd and Mn contents. FeS content of sphalerite of the stage 2 ranges from 0.13 to 0.95 mol%. These results are consistent with base metal bearing epithermal Au-Ag mineralization derived from an intermediate sulfidation state fluid.

Rhenium Deportment in Samples from the Spremberg-Graustein-Schleife Kupferschiefer Deposit

Zahra Nourizenouz[1], Max Frenzel[1], Bradley Martin Guy[1], Aratz Beranoaguirre[2], Jens Gutzmer[1]
[1]Helmholtz-zentrum Dresden Rossendorf, Helmholtz Institute Freiberg for Resource Technology, Freiberg, Germany, [2]Karlsruhe Institute for Technology, Chair of Economic Geology and Geochemistry, Karlsruhe, Germany

Rhenium is a valuable rare element and is of strategic importance due to its indispensable role in high-temperature superalloys such as those used in jet engines and turbine blades. Global mine production of rhenium is almost entirely a byproduct of copper production from porphyry and sediment-hosted copper deposits, mainly through the recovery of molybdenite concentrates. In the Kupferschiefer sediment-hosted copper deposits of Central Europe, rhenium is recovered during the smelting process of copper concentrates in the flue gas. While the ore is known to contain elevated Re concentrations, the mineralogical controls on its distribution remain poorly constrained. In this study, we present a comprehensive assessment of Re deportment within the Spremberg-Graustein-Schleife Kupferschiefer deposit in Lusatia, Germany.

Representative samples from the mineralization interval of three drill-cores were analyzed for bulk geochemistry, quantitative modal mineralogy by SEM-based automated mineralogy (MLA and TIMA), and sulfide mineral composition by LA-ICP-MS. The results reveal that Cu-S minerals, particularly chalcocite group minerals, in the Kupferschiefer and overlying hanging wall carbonate rocks, are locally enriched in Re. However, in the majority of samples, up to 90% of the total Re content remains unaccounted for within the analyzed sulfide phases. Since molybdenum-bearing minerals were not identified using high resolution SEM-based mineralogy, it is likely that Re is hosted by other phases. In this regard, a strong positive correlation between Re and total organic carbon (TOC) was observed and indicates that Re is likely associated with the organic matter. The hydrophobic nature of this phase and its flotation behavior alongside sulfides during ore processing may contribute to the apparent recovery of Re during smelting.

Tracing Cu-Au Ore-Forming Processes in the Proterozoic Elaine Dorothy Skarn Deposit, Mount Isa (Queensland, Australia), Using In-Situ Sulphide Mineral Compositions

Catherine Nyakecho[1], Avish Kumar[1], Ioan Sanislav[1], Daniel Wiemer[1], Melanie Finch[2]
[1]James Cook University, Townsville, Australia, [2]The University of Melbourne, Melbourne, Australia

Copper-gold-bearing sulfide minerals form under diverse hydrothermal conditions across various geological settings, contributing to economically significant ore deposits. Trace elements incorporated within their crystal lattices provide critical insights into ore-fluid composition, source, and fluid evolution during protracted hydrothermal activity or subsequent remobilization. Skarn deposits form at temperatures between 300°C and 800°C, with sulfides precipitating during the later stages of hydrothermal activity (~250°C–400°C). These sulfide minerals are critical for interpreting the primary ore-fluid composition, offering insights into the chemical evolution, temperature, and metal content during the cooling phase of skarn formation. However, in ancient deposits, tracing ore processes based on sulphide geochemistry remain challenging since multiple complex deformations and hydrothermal activity affect the system.

This study focuses on the Elaine Dorothy Cu-Au-REE skarn, hosted within calcareous metasediments (garnet-diopside skarn) ascribed to the Paleoproterozoic Corella Formation (1750–1742 Ma). Early deformation and associated hydrothermal activities are linked to the calc-alkaline Burstall Granite, the Lunch Creek Gabbro and coeval felsic dikes, all emplaced at ca. 1740 Ma. The timing of this event contradicts previous constraints on the timing of mineralisation in the Mt Isa inlier (~1530 Ma Isan Orogeny). We combine field observations, petrographic studies, and in-situ trace element geochemistry (LA-ICP-MS) of pyrrhotite, pyrite, and chalcopyrite to investigate the paragenetic sequence, metal distribution, and hydrothermal alteration history of Eliane Dorothy. We show that sulphide minerals formed during both prograde to peak (garnet-diopside-wollastonite) and subsequent retrograde metamorphism (amphibole-phlogopite-chlorite). Based on textural evidence, distinct paragenetic sequences are distinguished. Importantly, we find that sulphide minerals are enriched in Ni and Co, suggesting a mafic magmatic fluid origin or contribution.

Future integration of the timing of deformation events with geochronological analysis of mafic rock generations in the vicinity of Elaine Dorothy deposit will potentially refine the understanding of the Mt Isa region's tectonic evolution.

Multi-Method Comparison of Mineral Abundance Estimates for the Sandsloot PGE-Ni-Cu Deposit, Platreef, Northern Bushveld Complex

Giovana Oliveira Pimentel[1], Matthew J. Cracknell[1], Michael Baker[1], Peter Hollings[2], Scott Halley[3], Iain Dalrymple[4]
[1]Centre for Ore Deposit and Earth Sciences, University of Tasmania, Hobart, Australia, [2]Geology Department, Lakehead University, Thunder Bay, Canada, [3]Mineral Mapping Pty Ltd., Hawley Beach, Australia, [4]Anglo American Group Discovery and Geosciences, Perth, Australia

The Platreef is the main host of PGE-Ni-Cu mineralisation in the Northern Limb of the Bushveld Complex, South Africa. The Sandsloot deposit, part of Anglo Platinum's Mogalakwena operation, is typified by pyroxenites separated by metasediment intervals from footwall dolomites of the Transvaal Supergroup. To develop the recently explored underground resources, mineralogical characterisation is essential but is challenging to accomplish in complex deposits like Sandsloot.

We conducted a comparative study across 19 drillholes of quantified mineralogy generated from fusion X-ray fluorescence (XRF), assay whole-rock geochemistry, quantitative X-ray diffraction (QXRD), bulk mineralogy analysis (BMA), Fourier Transform Infrared (FTIR) spectroscopy, and shortwave infrared-visible near infrared (SWIR-VNIR) hyperspectral data. Initially, data were integrated across different spatial scales, via a weighted mean value over 1-meter segments of core for the geochemical and vibrational spectroscopy datasets that corresponded with the 5-meter sampling intervals used for QXRD and BMA analyses. A preliminary calculated mineralogy model was produced from the whole rock data using a linear programming method. In total, 132 samples were compared against QXRD results and regression analysis was applied to assess model performance.

Each mineralogy estimation method identified a distinct range of minerals reflecting the limitations of each technique, particularly in vibrational spectroscopy, where minerals exhibit different responses across various wavelengths. Based on correlations between estimates of the main mineral phases: FTIR demonstrates good correlation (R2 = 0.71-0.94) with traditional mineralogical analysis but lacks a model for olivine; calculate mineralogy performs well for abundant phases (e.g., orthopyroxene, clinopyroxene, plagioclase and serpentine) but poorly for olivine, chlorite and other alteration minerals; hyperspectral imaging lacks reliable matching for specific minerals, particularly clinopyroxene and olivine. Therefore, no single technique can accurately predict all mineral phases. A combination of methods may yield better results by incorporating mineral chemistry into calculations and refining spectral features.

Geochemical Signatures of Apatite in IOCG Systems in the Cloncurry District, Queensland

Ernest Boamah Opoku[1,2], Caroline Tiddy[1,2], Adrienne Brotodewo[1,2], Jennifer Porter[1,2,3], Anthony Budd[1,4]
[1]MinEx CRC, Australian Resources Research Centre (ARRC), Perth, Australia, [2]Future Industries Institute, University of South Australia, Mawson Lakes, Australia, [3]Geological Survey of Western Australia, Perth, Australia, [4]Geoscience Australia, Canberra, Australia

As the world transitions towards a sustainable, green energy future, an increase in deposit discovery is necessary to meet growing global metal demands. To aid in mineral exploration, techniques are being developed to increase the footprint of a system and therefore the size of the exploration target. This can be done through looking at mineral chemistry (e.g. apatite, monazite) to understand proximity to the system and gain insight into how the system formed. Apatite is a common accessory mineral that is being recognised as a valuable indicator mineral as its rare earth element (REE) patterns and trace element ratios have been shown to preserve unique signatures that can be related to different mineralising events (Mao et al., 2016: Econ-Geol). Previous studies have demonstrated this for porphyry Cu-Au (Bouzari et al., 2016: Econ-Geol) and orogenic gold (Zheng et al., 2022: Mineralogy and Petrology) systems however minimal work has been done across iron oxide-copper-gold (IOCG) and iron sulfide-copper-gold (ISCG) deposits (Krneta et al., 2017: Minerals).

In this study, chemistry of apatite from IOCG and ISCG systems in Australia are compared. Kulthor ISCG and Osborne IOCG deposits from Cloncurry District, Queensland are used as case study areas. Scanning electron microscopy (SEM), Electron Probe Micro-Analyzer (EPMA) and laser ablation inductively coupled plasma mass spectrometry (LA-ICP-MS) are used to quantify apatite chemistry and textural features in both background and mineralised samples. This facilitates understanding of physical and chemical changes that have occurred within mineralisation-associated apatite and identification of multiple generations of apatite within the system. Results are compared to apatite geochemical data from other systems in the Cloncurry district and Olympic Province in South Australia. Comparative analysis will be undertaken to identify similarities and differences across the study areas to generate a broad apatite exploration criterion for IOCG and ISCG systems in eastern Australia.

Basin Reconstruction and Dissecting Vein Population: Implications for Exploration of Sedimentary-Hosted Copper Deposits

Chirantan Parui[1], Susanne Schmid[1], Helen McFarlane[1], Giovannni Spampinato[2], Valentin Zuchuat[1]
[1]CSIRO Mineral Resources, Perth, Australia, [2]CSIRO Mineral Resources, Sydney, Australia

In Western Australia, sedimentary-hosted copper (Cu) deposits such as Nifty can be found within the Neoproterozoic Yeneena Basin. The Yeneena Basin formed as an extensional basin that was subsequently inverted during two shortening events: the Miles Orogeny (ca. 680–650 Ma) and the Paterson Orogeny (ca. 580–530 Ma). The mineralisation at the Nifty Cu-deposit is stratabound and vein-hosted and interpreted to have formed as a result of multiple pre-shortening and syn-shortening mineralisation events, corroborated by geochronological data.

This study aims to determine (1) the basin's three-dimensional structural architecture during the different mineralising events, (2) the different vein sets, and (3) the genetic link between vein sets and basin evolution. We have compiled pre-competitive geological and geophysical datasets and collected new structural, mineralogical, geochemical, and isotopic data of veins and host sequences from 17 regional drill holes. Integrating these datasets, we constructed four regional balanced cross-sections to decipher the basin's structural evolution.

Basin reconstruction suggests that the basin was segmented into smaller sub-basins separated by basement highs during the extensional phase. During the basin inversion, the shortening was accommodated primarily by folding, followed by reverse faulting. We have identified four dominant vein sets: Set I (high-angle to bedding), Set II (bed-parallel), Set III (vein strike parallel to hinge lines), and Set IV (vein strike parallel to the shortening direction). There are variable vein compositions, including quartz-vein assemblages and a set of veins associated with different carbonate mineralogy. The dominant geochemical signature of carbonate veins indicates a seawater-brine origin in diagenetic- and epigenetic-related (folding) vein sets, suggesting fluids remobilisation from within the basin under variable temperature regimes.

Linking vein structural, mineralogical, geochemical, and isotopic data with basin evolution suggests that Set I formed during basin development, while Sets II (pre-folding), III and IV (syn- to post-folding) formed during basin inversion.

Using Apatite Chemistry to Identify Proximity to Copper at the Nifty Sediment-Hosted Cu Deposit, Western Australia

Jennifer Porter[1,2], David E Kelsey[2], Fawna J Korhonen[2]
[1]University of South Australia, Adelaide, Australia, [2]Geological Survey of Western Australia, Perth, Australia

The Neoproterozoic Paterson Orogen of Western Australia is the setting of several known copper deposits and recent discoveries (e.g. Winu) have highlighted the importance of this area as a future source of copper. To improve targeting of difficult to detect undercover mineralisation, there has been a focus on re-evaluating the formation histories of known copper deposits, including the sediment-hosted Nifty Cu deposit. Apatite, which forms together with copper ore at Nifty, provides a useful geochemical fingerprint for exploring undercover as it retains chemistry related to the processes of mineralisation. Understanding apatite chemistry in a deposit and how it changes across the footprint of a mineralising system is important to evaluating how this chemical fingerprint can be used to vector to new mineralisation when exploring undercover.

In this study, 25 thick section samples were taken from drillcore across the Nifty mine area from several lithologies including carbonaceous and calcareous shales, pyritic shales, carbonate and silica alteration, and the main Cu mineralisation. Apatite trace element chemistry was obtained using laser ablation inductively coupled plasma mass spectrometry (LA-ICP-MS), and principal component analysis with unsupervised clustering was performed on 37 elements to identify chemically distinct apatite groups within each of the lithologies.

Eight distinct apatite chemistries, including two directly associated with the main Cu mineralisation were identified. These distinct chemistries distinguish apatite with different morphologies and mineral associations, including those associated with oxidised, uranium-bearing fluids from apatite associated with reduced Sn-associated Cu-bearing fluids. The apatite chemistries were also able to predict which apatite is associated with carbonate and silica alteration and which is not directly associated with mineralising fluids. These distinct apatite chemistries support concurrent research suggesting Nifty Cu was formed from more than one mineralising fluid and highlights the utility of apatite when vectoring to ore during sediment-hosted Cu deposit exploration.

Magmatic-Hydrothermal History at the Antakori Cu-Au-Ag Project, Northern Peru: New Ages from U-Pb and Re-Os Geochronology

David Portocarrero[1], Kevin Heather[2], David Cooke[3], Lejun Zhang[3]
[1]Regulus Resources Inc., Lima, Peru, [2]Regulus Resources Inc., La Serena, Chile, [3]Centre for Ore Deposits and Earth Sciences (CODES), University of Tasmania, Hobart, Australia

The Antakori project is located 50 km northwest of the city of Cajamarca in the Hualgayoc district and is part of the Miocene cluster of Cu-Au-Ag deposits of northern Peru, which includes Cerro Corona, Tantahuatay and Yanacocha. The geology consists of a Cretaceous sedimentary basement overlain by mid-Miocene volcanic rocks of the Calipuy Group. The stratigraphy has been intruded by numerous andesite to rhyolite intrusions and breccias that have generated different styles of mineralization including skarn, porphyry and epithermal high and intermediate sulfidation styles. New U-Pb and Re-Os geochronology ages from 12 intrusive phases and 3 molybdenite occurrences helped to define more accurate relationships between magmatism and hydrothermal activity. The results show that there were five main stages of magmatic-hydrothermal activity at Antakori. The first stage was associated with a series of dioritic intrusions emplaced between 15 and 14 Ma that caused thermal metamorphism, porphyry-skarn style alteration and Cu-Au-Ag-(Zn) mineralization. A second stage occurred between 13.5 and 12.5 Ma, characterized by the emplacement of andesitic to dacitic intrusions that generated porphyry-style alteration and mineralization at depth, and an extensive advanced argillic alteration zone with associated high sulfidation epithermal Cu-Au-Ag mineralization at shallower levels in the near-surface volcanic environment. A third stage of magmatic-hydrothermal activity was associated with another group of andesite to rhyodacite intrusions and breccias that formed between 12 and 11.2 Ma, with associated porphyry-style Cu-Au-Ag mineralization. The fourth stage occurred between 11 and 10.4 Ma, marked by the emplacement of rhyodacite flow-dome complexes and phreatic breccias, with associated intermediate sulfidation epithermal Cu-Au-Ag-Pb-Zn mineralization. A final stage occurred between 10 and 9.5 Ma, when series of barren, hypabyssal rhyolite dykes were emplaced, representing the latest magmatic-hydrothermal activity at Antakori. Thus, mineralization at Antakori is the result of multiple overprinting events with magmatic-hydrothermal activity that lasted at least 5.5 m.y.

Multi-Scale Structural and Geochemical Analysis of Gold Mineralization at the REN Deposit, Carlin Trend, Nevada

Amber Prevallet[1], Simon Jowitt[1]
[1]University Of Nevada, Reno, Reno, United States

The REN deposit, located along the northern Carlin Trend in Nevada, hosts 1.5 Moz of inferred and 0.06 Moz of indicated gold resources at an average grade of 6.9 g/t Au (Fiddes et al., 2025). Despite its resource potential, the geologic controls on mineralization are not well constrained. This study aims to develop a comprehensive geological framework for REN by investigating the influence of faults, breccia bodies, and dikes on hydrothermal fluid flow and gold deposition. The deposit is hosted in reactive Lower Plate carbonate rocks, particularly the Devonian Popovich and Silurian-Devonian Roberts Mountains formations, which are favorable for decarbonatization, pyritization, and subsequent gold deposition (Rhys et al., 2015; Cline et al., 2005). High-angle faults such as the Corona and JB3 faults are interpreted to have facilitated hydrothermal fluid flow during Carlin-type mineralization (~42–34 Ma) (Spalding et al., 2005). Intrusive bodies, including Jurassic and Eocene dikes, show compositional and structural alignment with these faults, suggesting a potential genetic link (Spalding et al., 2005). Proposed methods include petrography, geochemistry, and U-Pb zircon and Ar-Ar geochronology to constrain the timing of dike emplacement and faulting. Additional calcite veins and breccia cement analyses will be used to interpret fluid flow pathways and chemical conditions associated with gold precipitation. Comparing mineralized and unmineralized breccia textures will further elucidate their roles in gold localization. This study aims to enhance the understanding of structural and mineralization controls at the REN deposit. The findings aim to clarify the timing and relationship between faulting, intrusion, and gold mineralization, potentially improving exploration strategies in the northern Carlin Trend and contributing to more effective resource evaluation.

Influence of Serpentinization and Mass Balance Constraints on Ni-Co Enrichment in Laterite Profiles in Pomalaa, Sulawesi, Indonesia

Muhammad Fatih Qodri[1,2], Akane Ito[1], Kotaro Yonezu[1], Akira Imai[1], Saefudin Juhri[1], Sufriadin[3], Elwin Elbur[4], Abdul Bari[4]
[1]Department of Earth Resources Engineering, Faculty of Engineering, Kyushu University, Fukuoka, Japan, [2]Institut Teknologi Nasional Yogyakarta, Yogyakarta, Indonesia, [3]Universitas Hasanuddin, Gowa, Indonesia, [4]PT ANTAM Tbk, Jakarta, Indonesia

Indonesia hosts the world's largest lateritic nickel reserves, with Pomalaa, Southeast Sulawesi, being a major mining area developed on ultramafic rocks of the East Sulawesi Ophiolite Belt. These ultramafic rocks have undergone extensive serpentinization that provide a favorable protolith for lateritization. This study examines three representative laterite profiles (Everest, Hummer, and Fortuner) integrating geochemical, mineralogical, and physical property analyses to assess Ni-Co enrichment. The methods employed include X-Ray Diffraction (XRD) for mineralogical identification, X-Ray Fluorescence (XRF) for major and trace element chemical composition and SEM-EDX for chemical composition. Physical property measurements of density, porosity, and magnetic susceptibility also were conducted to analyze the degree of serpentinization along with Loss on Ignition (LOI) data in the bedrocks. Mineralogical results reveal a vertical transition from primary mineral (olivine, orthopyroxene) and serpentine in bedrock to goethite and hematite in limonite zone. physical property analysis indicates that the initial serpentinization process significantly altered the ultramafic bedrock. Analysis of physical properties and LOI indicates that the initial serpentinization significantly altered the ultramafic bedrock. The increasing trend in Loss on Ignition (LOI) from the Everest to Hummer to Fortuner sites suggests a higher degree of hydration associated with serpentinization. Geochemical profiles show pronounced Ni-Co enrichment in the saprolite zones. with Ni concentrations reaching 2.33 wt.% in Everest, 2.61 wt.% in Hummer, and 1.83 wt.% in Fortuner. Cobalt (Co) concentrations exhibit similar enrichment patterns. Furthermore, Mass balance calculations demonstrate that during lateritization, Ni and Co were immobilized and concentrated in the saprolite zone, while other elements like Mg and Si were mobilized and leached. This process leads to a decrease in density and an increase in magnetic susceptibility due to magnetite formation. These results highlight the mineralogical transformations, physical property changes and elemental mobilities that control the economic potential of lateritic Ni-Co deposits in Pomalaa.

Structural Controls and Tectonic Setting of Gold Mineralisation in the Broken River Orocline, Northeastern Australia

Raiza Rodrigues[1], Gideon Rosenbaum[1], Rob Heaslop[2]
[1]The University of Queensland, Brisbane, Australia, [2]Liberation Resources Pty Ltd, Brisbane, Australia

The Broken River Orocline, an orogenic curvature in northeastern Australia, hosts significant gold deposits, with mineralisation episodes linked to the complex structural and tectonic evolution of the orocline during the Paleozoic. Despite the region's recognised mineral potential, the structural controls and timing of mineralisation remain poorly understood. To address this, we conducted an integrated investigation using structural field data, aeromagnetic interpretation, paleostress reconstruction, and geochemical analyses of soil and rock chips. Our results reveal two distinct episodes of gold mineralisation, each associated with different quartz vein systems and tectonic regimes. The first episode, characterised by Au-As ± Sb ± Ag enrichment, occurs in the orocline's northern limb. These veins are controlled by pre-oroclinal structures, including isoclinal folds, axial plane foliations, and reverse faults. Their structural and geochemical signatures resemble those of epizonal orogenic gold systems, possibly formed during contractional deformation within a Middle–Late Devonian accretionary prism and later deformed by oroclinal bending. The second episode is confined to the orocline's hinge zone and features a >148 km² network of Au-Ag-Sb-As-Hg-bearing veins with chalcedony-rich textures, interpreted as a low-sulphidation epithermal system formed at shallow crustal levels. These veins were controlled by NW–SE-trending normal faults developed under NE–SW extensional, and are likely associated with Late Carboniferous–Early Permian felsic magmatism of the Kennedy Igneous Association (KIA). Hydrothermal fluids derived from this magmatism concentrated within the orocline's hinge zone, where pre-existing fabrics were optimally oriented for reactivation during NE–SW extension. An intrusion-related system associated with KIA may have also overprinted older orogenic gold deposits within the northern limb, as indicated by quartz vein exposures associated with magnetic and geochemical anomalies. These findings demonstrate the critical role of structural analysis in understanding the spatial–temporal distribution of gold deposits in orogenic curvatures, with implications for exploration in northeastern Australia and other comparable structurally complex settings.

Structural Analysis of Polyphase Deformation in the South Peruvian Andes and Its Basement: Implications for Cu-Au Mineral Deposit Localization

Maryi Rodriguez-Cuevas[1], Daniel Wiemer[1], Steffen G. Hagemann[2], Nicholas Hayward[3], Ioan Sanislav[1]
[1]Economic Geology Research Centre, James Cook University, Townsville, Australia, [2]Centre for Exploration Targeting, University of Western Australia, Crawley, Australia, [3]Predict Ore Pty Ltd, East Perth, Australia

Trans-lithospheric faults (TLF) acting as primary magma and fluid conduits play a fundamental role in the spatial distribution of mineral deposits. However, why large Cu-Au deposits display a punctuated spatial distribution along their respective principal arc-parallel TLF remains debated: both self-organization and/or high-permeability intersections with inherited basement structures are considered.

The southern Peruvian Andes host world-class Phanerozoic Cu-Au deposits spatially associated with sub-parallel magmatic belts that formed during different accretionary episodes along the western South American margin. Major deposits belonging to individual belts show spatial alignment with deposits of adjacent belts along transverse-lineaments that cross the Andes normal, or oblique to the main direction. If the transverse-lineaments represent long-lived TLFs that controlled the spatial distribution of deposits this would imply that they: (i)were either laterally propagated into newly accreted terrane, or originate from a coherent basement tectonic entity of at least Paleozoic age that underlies the entire region, (ii)have been reactivated repeatedly during main accretionary episodes (iii)have been vertically propagated into younger cover, and (iv)experienced no or only limited lateral displacements or rotations during subsequent reactivation events.

To test these hypotheses, we present field-based structural analyses from selected areas in South Peru to shed new light on the nature, kinematics and relative timing of polyphase deformation affecting Andean basement and overlying lithologies. Correlation of deformational intensities (brittle and/or ductile), metamorphic grade and type and geometry of structural features (e.g., foliations, folds, faults, fractures) helps to improve understanding of the tectonic history and lithospheric architecture. Importantly, we show that basement structures played an important role in controlling successive deformation and repeated magma and magmatic-hydrothermal fluid flow. Furthermore, observed distinct generations of mafic dike swarms, within or nearby TLFs, attest to structural longevity and connectivity of TLFs. Overall, our results allow for advanced predictiveness in Cu-Au mineral system exploration in South Peru.

Newly Discovered Cu-Ni-Sulfide Mineralisation in the 1.8 Ga Stendalen Mafic Intrusion, Ketilidian Orogen, South Greenland

Matthew Roper[1], David Holwell[1], Tiffany Barry[1], Mike Hamilton[2], James Gilbertson[3], Anna Rasch-Christensen[3]
[1]University Of Leicester, United Kingdom, [2]University of Toronto, Canada, [3]Amaroq Minerals, Toronto, Canada

The Stendalen layered mafic intrusion, Greenland, is a 17.5 km², ca. 1300m thick, saucer-shaped meta-gabbroic intrusion hosting prospective magmatic sulfide mineralisation. Stendalen lies within forearc sediments of the accreted Ketilidian orogenic belt on the south margin of the North Atlantic craton. Zircon U-Pb dating of the intrusion has yielded a precise primary crystallisation age of 1804 ± 1.5 Ma; placing Stendalen late within calc-alkaline Julianehåb igneous complex magmatism and indicating emplacement occurred immediately prior to the Ketilidian deformation recorded in the forearc. The mineralisation is considered an atypical magmatic sulfide occurrence, having formed in an arc setting. It appears to share characteristics with similar deposits in the Svecofennian orogenic belt (e.g. Kleva, Älgliden, Espedalen), including low PGE tenors. This is the first study assessing the petrography and mineralisation of the intrusion, utilising drillhole data, field-mapping, thin section and SEM-EDS analysis.

Petrographic analysis shows the intrusion is dominated by plagioclase and hornblende which was largely recrystallised during peak HT-LP Ketilidian metamorphism. The intrusion is underlain by S- and graphite-bearing metasediments which are potentially integral to the generation of mineralisation. The intrusion is extensively crosscut by peraluminous, syn- to late-tectonic granites and granodiorites. Early drilling results reveal that the upper section of the intrusion contains minor disseminated sulfides. Towards the base, the gabbro is taxitic and displays a variety of sulfide textures. The sulfide assemblage comprises pyrrhotite-chalcopyrite-pentlandite with Cu>Ni. The timing of emplacement relative to metamorphism and deformation raises the possibility of sulfide re-mobilisation, currently under investigation.

Understanding Hazelton Group Metallogeny Using Detrital Zircon Geochronology Across Multiple Mineral Districts, Golden Triangle, NW British Columbia

Kate Rubingh[1], Nikola Denisova[1], Nichole Moerhuis[1], Brian McNulty[1], Shaun Barker[1]
[1]Mineral Deposit Research Unit at the The University of British Columbia, Vancouver, Canada

The Hazelton and Stuhini Groups are a volcano-sedimentary succession developed as part of the Stikine Terrane of the Canadian Cordillera, recording the evolution of volcanic arcs during the late Triassic to the middle Jurassic. The co-occurrence of multiple mineral systems including Treaty Creek Cu-Au porphyry, Brucejack mine Au-Ag epithermal and the Ag rich volcanogenic massive sulphide (VMS) deposit at Dolly Varden - proximal to the unconformity between the Stuhini and the Hazelton Groups records an unusual metallogenic diversity around 195 – 185 Ma that existing tectonic models do not adequately explain. Subsequent normal faulting and thrusting related to the Skeena fold and thrust belt have further fragmented the region into distinct structural blocks, adding complexity. Current models emphasizing linear arc parallel structural domains or microplate accretion and rotation fail to fully account for the observed lithological heterogeneity, paleodepth variations, and metal endowment.

A multidisciplinary approach combining new detrital zircon geochronology and trace element data, whole-rock lithogeochemistry, and detailed stratigraphic mapping from multiple mineral districts over a 170 km x 50 km area reveals a complex tectono-stratigraphic framework. More than 50 detrital zircon samples from marine sedimentary and volcaniclastic rock units of the Stuhini and Hazelton Groups were analyzed using LA-ICP-MS (>6000 zircon analyses). The widespread detrital zircon ages, spanning 240 to 165 Ma, record several erosional events and the formation of sub-basins across the Golden Triangle, prior to the development of the Bowser Basin. This complex tectonic and stratigraphic architecture may explain the diversity in the metallogenic environments present in the Golden Triangle. Comparative analysis with modern active arc systems, such as those in Papua New Guinea, suggests that this complexity may reflect dynamic crustal growth processes and structurally partitioned domains, which in turn affect the magmatic and hydrothermal activity observed around these microplates.

Epidote as a Discriminator of Skarn Deposits: Insights from Dolphin W Skarn, King Island, Tasmania

Angela Isaura Santos Costa[1], David R. Cooke[1], Lejun Zhang[1], Yamila Cajal[1], Henry Schaumburg[1]
[1]CODES - University of Tasmania, Hobart, Australia

Epidote chemistry provides valuable insights into the evolution of magmatic-hydrothermal systems. We analyzed epidote samples from Dolphin W skarn, King Island, Tasmania and regional Skipworth Subgroup Volcanics (up to 6.5 km northeast of Dolphin) to assess the use of epidote chemistry as a discriminator of skarn deposits. In the studied samples, endoskarn and volcanic-related epidote replace feldspar or occur as veins, whereas exoskarn epidote replaces prograde-stage garnet.

Our results from LA-ICP-MS analyses show that skarn-related epidote has higher Pb and Mn than volcanic-related epidote, with exoskarn epidote displaying lower Fe/Al ratios (<0.5). Skarn epidote is enriched in W and U, whereas volcanic epidote has higher Cu. Exoskarn epidote is enriched in Sn (>100 ppm) and Bi (>10 ppm), while endoskarn epidote has higher B, Li, As, Sb (>10 ppm). LA-ICP-ToF-MS mapping reveals chemical zoning in exoskarn epidote, indicating multistage precipitation during retrograde skarn alteration. Zones enriched in Fe (>10%) and Mn (>0.3%) correlate with elevated Bi and Sn, forming a distinct geochemical cluster as defined by k-means cluster analysis. U-Pb dating of exoskarn epidote yields an age of 353.4 ± 9.1 Ma, which is slightly older but within analytical uncertainty of the Sandblow Granodiorite (350.8 ± 1.7 Ma).

A comparative LA-ICP-MS analysis of epidote from metamorphic and porphyry-skarn systems (~7,000 compiled spot-analyses) reveals distinct geochemical signatures. Arsenic and Sb are higher for porphyry-related epidote, particularly from Chilean giant deposits (e.g., El Teniente and Collahuasi). Bismuth is a distinctive tracer of skarn-related epidote, as exemplified by Las Bambas Cu skarn, whereas anomalous Sn is characteristic of W-skarn systems (e.g., Dolphin and Kara). In contrast, metamorphic epidote is distinguished by its enrichment in HREE (>10 ppm). Our findings demonstrate that epidote chemistry is a powerful tool for evaluating ore potential in magmatic-hydrothermal systems and for differentiating skarn from other deposit types.

Paragenesis and Trace Element Signatures of Pyrite from Paramanahalli Gold Deposit, Chitradurga Greenstone Belt, Karnataka, India: Implications for Ore Genesis

Manju Sati[1], **Rajagopal Krishnamurthi**[1], Sakthi Sarvanan Chinnasamy[2], Ajit Kumar Sahoo[3]
[1]Department of Earth Sciences, Indian Institute of Technology, Roorkee, India, [2]Department of Earth Sciences, Indian Institute of Technology, Bombay, Mumbai, India, [3]Department of Geology, Banaras Hindu University, Varanasi, India

Gold mineralization at the Paramanahalli, within the Chitradurga Greenstone Belt (CGB), of Western Dharwar Craton (WDC) India, is confined to altered metabasalt and Banded Iron Formation (BIF). Chlorite (Fe-rich)-quartz-biotite-ankerite-rutile-magnetite-pyrite chalcopyrite-pyrrhotite ± gold is the mineral assemblage found in mineralized zones. Pyrite is the dominant sulfide mineral and occurs in both altered rocks and in quartz veins, classified as altered zone hosted and vein-hosted pyrites, respectively. Three distinct generations of pyrite were identified based on textural and geochemical analyses including X-ray element imaging, electron probe microanalysis (EPMA), and LA-ICP-MS: (1) early stage pyrite-I, euhedral grains with silicate and oxide inclusions, (2) main ore stage pyrite-II, subhedral-anhedral grains, closely associated with visible gold and subdivided into Ni-rich (pyrite-IIa) and Co-rich (pyrite-IIb) types, and (3) late stage pyrite-III as anhedral grains and free from mineral inclusions. Trace element variations across pyrite generations reveal decreasing concentrations of As, Ni, Au, Se, Mo, and Te from pyrite-I to pyrite-III. Native gold grain is found to be associated only with altered zone-hosted pyrites (Pyrite-II) from the main ore stage. The inferred paragenetic sequence indicates formation of early pyrite-I, followed by main-stage pyrite-II, and culminates with late-stage pyrite-III. Gold probably occurs as lattice-bound within pyrite-I, and visible gold in pyrite-II formed later during the fluid induced dissolution-reprecipitation processes. Sulfidation, chloritization, and ankeritization are the major wall rock alterations responsible for the formation of sulfides and gold during the main ore stage. These findings emphasize the role of fluids to form visible gold within Pyrite-II and their significance to understand the ore-forming processes in greenstone-hosted gold deposits in CGB, of WDC, India.

Ore-Forming Fluid Characteristics and P–T Condition of Gold Mineralization at Paramanahalli, Chitradurga Greenstone Belt, Karnataka, India

Manju Sati[1]
[1]Department of Earth Sciences, Indian Institute of Technology, Roorkee, Roorkee, India

Paramanahalli gold deposit, Karnataka, India, is hosted by volcano-sedimentary rocks of the Hiriyur Formation within splays of the Chitradurga Shear Zone (CSZ). Gold/sulfide mineralization is predominantly confined to N-S-oriented veins and adjacent altered rocks. Mineralized veins are characterized by a hydrothermal assemblage of chlorite (Fe-rich), ankerite, quartz, pyrite, chalcopyrite, pyrrhotite, gold, and silver, whereas non-mineralized veins consist mainly of quartz and carbonate minerals (ankerite and calcite). Fluid inclusion studies on mineralized quartz veins indicate three primary inclusion types: aqueous biphase (Type Ia), carbonic monophase (Type Ib), and aqueous-carbonic (Type Ic). The homogenization temperatures (Th°C) range from 216°C to 332°C, and salinity from 1.07 to 16.05 wt% NaCl eq. Fluid inclusion studies reveal low to medium salinity (12 wt% NaCl eq.) aqueous-carbonic fluids (H_2O-NaCl-CO_2±CH_4) in mineralized veins, and lesser saline inclusions (1.56–6.44 wt% NaCl eq.) in non-mineralized veins. Microthermometric analyses indicate boiling, fluid mixing and fluid-rock interaction played an important role to the destabilization of gold-bearing sulfide complexes and subsequent gold precipitation with sulfides. The mineralizing fluids, near-neutral to slightly alkaline, transported gold as $Au(HS)_2^-$ complexes. Estimated pressure-temperature conditions for gold mineralization are ~2.2 kbar and 310 °C, while non-mineralized veins formed at ~1.2 kbar and 260 °C. Quartz and calcite from mineralized veins display microstructures indicative of dynamic recrystallization and intracrystalline deformation, suggesting ductile deformation followed by brittle brecciation. Litho-structural controls strongly influenced gold precipitation, extending mineralization into adjacent altered rocks. This study highlights the significance of fluid evolution, deformation processes, and structural controls in the Paramanahalli gold mineralization system, providing insights applicable to orogenic lode gold deposits worldwide.

Porphyry Cu-Au-Ag Mineralization of the Haman Alkaline-Carbonatite Intrusive Complex, South Korea

Chris Sennitt[1], Wan Joong Kim[1]
[1]Korean Metals Exploration Limited, Melbourne, Australia

The Haman Cu-Au-Ag project is located in the southeast of the Korean peninsula and was the main copper-producing region of South Korea up until 1945.

Situated within the Cretaceous Gyeongsang basin, host rocks comprise siltstone, mudstone with evaporites (Haman Formation), and "red bed" sandstones (Jindong Formation) deposited into a shallow lacustrine-playa environment under prevailing semi-arid climatic conditions.

Tonalite was emplaced at 118-110 Ma, followed by monzonite and feldspar porphyry intrusions between 105-95 Ma. The tonalite has cooled and crystallized inwards, with successive reactions involving uralitization of pyroxene to amphibole and albitization of plagioclase, generating residual magmatic fluids enriched in Au, Ag, Cu, Co, W, Bi, As, Mo, and Zn into late monzonite and feldspar porphyry phases. Central cores of the tonalite coincide with intense "bullseye" aeromagnetic anomalies and pyrite-magnetite-biotite-albite alteration. The intrusions are geochemically classified as high-silica adakites and lie within the fertile field for alkaline porphyry Cu-Au intrusions.

Gabbro has intruded the northern margin of the tonalite and was accompanied by NNW striking carbonatite and serpentinite dykes, emplaced between 97-85 Ma. Microgranular magmatic enclaves in the tonalite, monzonite, and gabbro suggest granitic and mafic enclaves crystallized contemporaneously as a result of the injection of a hot dense mafic gabbro melt into the adjacent monzonite magma chamber and still crystallizing warm tonalite. Mineral components of the carbonatite dykes include kaersutite, garnets, chamosite, traces of xenotime and apatite, while serpentinite dykes contain olivine and are suggestive of a mantle hot-spot plume.

Cu-Au-Ag mineralized breccias were deposited along dyke margins and consist of bands of quartz, magnetite, specularite, hematite, carbonate, actinolite, chlorite, and sulphides with a fluidized, brecciated igneous texture. Mineralization was dated at 84.0 Ma and probably deposited from late-stage magmatic hydrothermal fluids associated with the dykes.
The intrusive rock suite at Haman is interpreted as an alkaline-carbonatite porphyry Cu-Au complex.

Intermediate Sulphidation Polymetallic Epithermal Mineralization at Dongil, South Korea

Chris Sennitt[1], Wan Joong Kim[1]
[1]Korean Metals Exploration Limited, Melbourne, Australia

The Dongil Au-Cu-Ag-Pb-Zn deposit is situated near the town of Uiseong, ~220 km southeast of Seoul.

Dongil lies in the Uiseong sub-basin of the Cretaceous Gyeongsang Basin, an area dominated by the Geumseongsan maar-diatreme-ring fracture volcanic complex.
Dongil is hosted in black shale and sandstone (Jeomgog Formation), overlain by shale, mudstone, and sandstone (Sagok Formation) of the Early Cretaceous Hayang Group, deposited into a shallow lacustrine-playa environment under prevailing semi-arid climatic conditions. Coeval volcanic activity is evident, with a rhyodacitic tuff breccia (Gusandong Tuff; 103-97 Ma) interbedded within the Jeomgog Formation and related rhyodacite dome emplaced at Dongil North. Rhyodacite dykes occur with mineralization at Dongil South.
Mineralization consists of 2–11 vein breccias arranged in a sheeted zone, striking north-northwest, 1,800 m long and 20–100 m wide. A mineralized breccia pipe is localized on the margin of the rhyodacite dome at Dongil North. The breccias show a paragenetic sequence of comb quartz (Stage I), hydrofracturing, white quartz (Stage IIa) with pyrite, arsenopyrite, chalcopyrite, sphalerite (Fe-rich), galena under low pH acid conditions, then brecciation with several Pb, Cu, Bi Ag sulphosalts (Stage IIb), electrum, sphalerite (Fe-poor) and late (Stage III) fluorite and calcite under alkaline conditions.

Fluid inclusions show early high-temperature (250-350 °C), moderate salinity (1-7 wt.% NaCl) hydrothermal fluids initially deposited quartz and Cu-Zn-Pb-Fe-As sulphides, followed by boiling, an overpressuring breccia event, then mixing, dilution, and cooling with meteoric waters, which deposited Au-Ag and Ag-Sb-Bi sulphosalts at lower temperatures (200-300 °C).

Sericite (dated at 71-65 Ma) and pyrite form a phyllic assemblage around mineralization, with occasional high-temperature zones implied by garnet, diopside, and magnetite. An outer propylitic assemblage forms a halo around phyllic alteration and is characterized by chlorite, pyrite, and calcite veinlets.

The mineralization is interpreted as intermediate sulphidation Au-Cu-Ag-Pb-Zn epithermal style associated with the rhyodacite dome and dykes.

The Role of Structural Inheritance in Gold Mineralisation Along the Gondwana Supercontinent Margin: Insights from Ravenswood, Queensland

Hugo Serra[1], Ioan Sanislav[1], Steve Harper[2], Brett Davis[3], Gregg Morrison[4], Daniel Wiemer[1]
[1]Economic Geology Research Centre, James Cook University, Townsville, Australia, [2]Ravenswood Gold Pty Ltd, Ravenswood, Australia, [3]Olinda Gold Pty Ltd, The Vines, Australia, [4]Klondike Exploration Services, Townsville, Australia

Supercontinent margins undergo repeated convergence-extension dynamics and global plate kinematic reorganisation resulting in structural inheritance and reactivation. Globally, the (re-)activity of major structures and their intersections is thought to control the spatial distribution of major gold deposits. Furthermore, successive structural events may modify and displace economic gold deposits. To improve predictive exploration targeting, it is crucial to decipher the kinematics and relative timing of first- and lower-order structural architectures relative to gold mineralisation. Here, we focus on the structural evolution of the Jessop Creek Fault (JCF) as a major structure that may have influenced the formation of the Ravenswood Gold deposit – the largest gold mining operation in Queensland, Australia. The region lies within the Paleozoic history of the northeastern Thomson Orogen, part of the Gondwana margin. The deposit occurs at an intersection of major EW, NW-SE, and N-S structures, with gold hosted in NW-striking conjugate quartz-sulphide veins, which record polyphase activation. The latest stage of movement along the N-trending JCF truncates pre-existing structures, mineralisation, and hydrothermal alteration. However, integration of structural, geophysical, and regional deformation data suggests a more complex and multiphase evolution of the JCF, with important implications for fluid flow and gold deposition. We propose that the JCF originated during NE-SW compression, contemporaneous with emplacement of the 420 Ma Jessop Creek Tonalite Complex. The early structural architecture comprised E-W, N-S, and NW-SE faults. Subsequent NW-SE-directed shortening during Devonian-Carboniferous reconfiguration resulted in sinistral transpressional movement along N-trending structures, offsetting pre-existing E-W faults. NW-trending faults likely developed as sinistral strike-slip zones. Late Carboniferous E-W extension reactivated the structural architecture and enabled fluid flow and gold mineralisation. In summary, reactivation of early structures controlled high-permeability pathways for gold precipitation during hydrothermal activity. Our study improves understanding of Gondwana margin-related lithospheric architecture development, critical for ongoing and future exploration strategies.

Spatial Variation of Flake Graphite Grade and Crystallinity in the Namapwia Prospect, Ruangwa District, Lindi, Tanzania

Mikidadi Shaha[1], Andrea Agangi[1], Ryohei Takahashi[1], Pearlyn Manalo[1]
[1]Akita University, Faculty of International Resource Sciences, Akita City, Japan

The Namapwia prospect, located in the Ruangwa district of southern Tanzania, forms part of the Eastern Granulite Nappe Complexes within the Neoproterozoic Mozambique Belt. In this study we examined the metamorphic evolution, crystallinity, and grade variation of flake graphite along the NE–SW strike in the Namapwia prospect. Field observations, petrography, Raman spectroscopy, X-ray diffraction (XRD), and total graphite content (TGC) data were used to evaluate crystallinity of graphite-bearing lithologies.

The prospect is divided into two zones based on mineral assemblages: a biotite-rich zone and a garnet-bearing zone. Graphite occurs in graphite metasandstone, graphite schist, garnet graphite schist, and quartzofeldspathic gneiss. Graphite is commonly disseminated or aligned along foliation planes, associated with quartz, plagioclase, garnet, biotite, and rutile.

Raman spectroscopy was used to estimate both graphitization degree and metamorphic temperatures. Estimated temperatures range from 420 °C to 500 °C, consistent with amphibolite facies metamorphism. R2 ratio decreases from 0.45 in graphite metasandstone to 0.25 in garnet graphite schist, indicating progressive ordering of graphite structure along the strike from NE to SW. XRD crystallinity parameters support these trends, with crystallite sizes (Lc) ranging from 600 Å to 1100 Å, and interlayer spacing (d002) values between 3.350 Å and 3.370 Å. Graphite schist and garnet graphite schist exhibit the highest crystallinity and graphitization.

TGC values also vary throughout the strike length. Graphite metasandstone in the northeast contains 1.4–1.8 wt.% TGC, graphite schist and garnet graphite schist in the central zone yield 2.2–2.8 wt.% TGC, while quartzofeldspathic gneiss in the southwest exhibits lower values (0.3–0.6 wt.% TGC). This distribution highlights a central enrichment zone controlled by lithology and metamorphic grade. Although graphite at Namapwia exhibits excellent crystallinity, its TGC values are relatively low. Further exploration is necessary to delineate additional zones of higher-grade graphite mineralization.

Apatite Geochemistry and Geochronology to Distinguish Between the Fluids Responsible for the Formation of Eloise ISCG Deposit, Australia

Swastik Shinde[1,2], Alex McCoy-West[1,2], Avish Kumar[2], Alkis Kontonikas-Charos[3], Ioan Sanislav[2]
[1]Isotropics geochemistry laboratory, James Cook University, Townsville, Australia, [2]Economic Geology Research Centre, James Cook University, Townsville, Australia, [3]Geological Survey of Queensland, Brisbane, Australia

The processes and fluids involved in the formation of IOCG and ISCG deposits are highly debated. Compositional analyses and isotopic studies of accessory phases are becoming an important tool to decipher these processes and distinguish between different fluid types. We plan to use apatite as a tracer for these purposes. Given that not all IOCG and ISCG deposits are enriched in REE, this is necessary to understand the processes responsible for the selective enrichment of REE in some deposits.

Eloise is an ISCG deposit in Mount Isa Inlier, northwest Queensland, Australia. Previous studies of the deposit have reported a distinct LREE depleted pattern at Eloise. Using a combination of EPMA for major element analysis, and LA-ICP-MS for the trace element analyses and U-Pb dating of apatite and titanite. The major element analysis will focus mainly on F-, Cl-, and OH- given their importance as complexing ligands transporting the REE. On the other hand, the trace element analysis will focus on REE to understand their abundance, along with Arsenic given its concentration in apatite can be used as a proxy for the redox conditions.

Our preliminary results show at least two stages of apatite mineralization. First, an early F-rich stage, represented as core of coarse apatite with relatively constant, LREE depleted patterns. Whereas, the second stage, represented by the rims of coarse apatite, and disseminated fine apatite grains, show a comparatively OH- rich composition with significant variation in LREE compared to the cores. The variation in the As concentration suggest an increase in oxidation conditions in the later event. Preliminary U-Pb geochronology of apatite provides an age of 1558 ± 18 Ma, although the results require careful analyses as Pb-mobility is common in the apatite grains.

Lithostructural Study of Mineralized Formations in the Kéniéba District

Nouhoum Noumouké Sidibé[1]
[1]University Gubkin, Moscou, Russia

The Kéniéba district, located in the Kayes region of western Mali, is part of the Birimian (or West African Paleoproterozoic) period. With major gold deposits as the main source, and other useful metals for which exploration is not currently focused (iron, bauxite, copper, etc.).This region attracts numerous international mining companies, and their activities and discoveries remain insufficient compared to the area's surface area. For this reason, a comprehensive analysis of the lithological and structural aspects of the district's formations is essential to understanding the geological and structural controls of mineralization. This relates to my thesis topic, "Lithostructural studies of the mineralized formations of the Kenieba district." The objective of this research is to identify all the mineralized formations in the area, their relationships with their host rocks, and their structural states, with the aim of highlighting potential areas such as gold and other metals. Methodology Methods Used: Use of previous geophysical data (magnetometry, resistivity). Hyperspectral remote sensing to map mineral alterations. Detailed systematic geological, structural, and geochemical mapping (including soil and termite mound collection) followed by well drilling (for structural and geological boundary accuracy).3D modeling of structural controls. Priority Targets: Intersections of shear zones and lithological contacts. Sectors with high density of sulfide quartz veins. Geochemical anomalies (As, Sb, W) along the structures. Correlation between the large world-class deposits in the area and the collected data. Expectations This study of the Kéniéba district could significantly deepen geological, structural, and metallogenic knowledge. The results will update the area in terms of its geological and structural state and will be the subject of future world-class mining projects.

Mineralogy and Geochemistry of the Yellowknife Lithium Pegmatite Field

Philipp Siebold[1], David Smithson[2], Zhaoshan Chang[1], Shiqiang Huang[1], Mathias Burisch[1]
[1]Colorado School of Mines, Golden, United States, [2]Li-FT Power Ltd., Vancouver, Canada

In this study, the lithium pegmatite dikes and local intrusive suites are investigated to better understand spatial geochemical gradients that may be useful for exploration targeting on the regional-, district- and deposit-scale. We obtained multi-element whole rock geochemical data derived from bore holes, surface composite rock chip samples and channel samples collected from multiple field campaigns. The geochemical data is screened for spatial trends using x-y element plots, GIS and 3D modelling software. Magmatic and hydrothermal muscovite related to Li pegmatites are currently being analyzed for Rb-Sr muscovite dating using LA-ICP-MS. Spodumene-bearing pegmatites at Yellowknife seem to be spatially related to locally enriched intrusive stocks within the large voluminous body of the parental batholiths. These intrusions are characterized by a high abundance of apatite and are strongly enriched in Li, Cs, Rb, Be and other incompatible elements. However, the background Li concentrations of intrusives in the area is generally high, which disqualifies Li for use as a pathfinder element. Apparently, fertile intrusions are characterized by exceptionally high 1/(K/Rb), P and Cs/(Cs+Li+Be). Preliminary field and microscopic observations suggest that the Yellowknife pegmatite field shows a concentric zoning around locally highly fractionated intrusions. A spatial trend from proximal barren pegmatites to more distal spodumene-bearing dikes is recognized. Across the examined dike field, systematic spatial trends of decreasing Nb/Ta and high Sn/W toward the spodumene-bearing dikes are observed. The dikes do not exhibit strong intra-dike zoning but typically are composed of an aplitic border zone and a medium to coarse-grained pegmatitic internal zone. The border zone is composed of primarily quartz, muscovite and accessory phases such as tourmaline, apatite, garnet and columbite-tantalite minerals, whereas the interior zone is primarily composed of quartz, albite, k-spar and spodumene (up to 20 vol.%). Various degrees of greisen and sodic-potassic alteration overprinting the pegmatites can be recognized.

Using In Situ Mineral Chemistry to Extend the Detectable Lithogeochemical Footprint of Magmatic Sulfide Orebodies

William Smith[1], Louise Schoneveld[1], Louisa Stokes[2], Margaux Le Vaillant[1]
[1]Discovery Program, CSIRO Mineral Resources, 26 Dick Perry Avenue, Kensington, Australia, [2]Earth Dynamics Research Group, School of Earth and Planetary Sciences, Curtin University, GPO Box U1985, Perth, Australia

In magmatic sulfide systems, mineral chemistry helps infer ore-forming processes, such as sulfide melt segregation in silicate magma. However, no chemical traits have been established that empirically link mineral chemistries with proximity to sulfide orebodies. Consequently, mineral chemistries are yet to offer compelling guidance towards sulfide accumulations at the prospect scale. Using the Nova-Bollinger Ni-Co-Cu deposit, this research interrogates ferromagnesian silicate chemistry to identify inter- and intra-mineral traits that plausibly reflect interaction with sulfide melt. Mineral chemistries are contextualized with automated mineralogy, synchrotron-based element mapping, and bulk-rock geochemistry.

In the studied ferromagnesian silicates, ratios of first row and post-transition metals differ in sulfide-bearing rocks and weakly correlate with orebody proximity (as computed using three-dimensional geological models). For example, the Co/Mn values of olivine and poikilitic orthopyroxene are consistently lower in sulfide-bearing samples and their Ni/Co values broadly increase with orebody proximity. The Co/Pb values of poikilitic clinopyroxene and symplectic hornblende are also distinctly lower in sulfide-bearing samples. These ratios in minerals comprising sulfide-free samples display strong positive inter-mineral correlations; however, they deviate from linearity in sulfide-bearing samples. For example, Ni/Co values of peritectic orthopyroxene are lower for given Ni/Co values of reactant olivine in sulfide-bearing samples.

Thermodynamic modelling is being utilized to distinguish traits that cannot be explained by magmatic differentiation or contamination, highlighting patterns that plausibly reflect direct interaction with sulfide melt and which may be translated into predictive exploration tools. Comparable datasets from regional magmatic sulfide occurrences—together with published data and an in-house global database—are being interrogated to assess whether these signatures can robustly indicate interaction with sulfide melt. By first identifying prospective traits in situ and then validating their relevance at other sites, the approach aims to enable both confident recognition of indicative signatures in ex situ minerals and apply them in situ as prospect-scale orebody vectors.

Nb-Ta Mineralisation in Pegmatite from the Amapá State, Brazil: First Constraints from U-Pb LA-ICP-MS Geochronology and Mineral Composition on Columbite-Tantalite

Carlos Spier[1,2], Avish Kumar[3], Ivan Belousov[4]
[1]The University of Queensland, Brisbane, Australia, [2]WSP, Brisbane, Australia, [3]James Cook University, Townsville, Australia, [4]University of Tasmania, Hobart, Australia

Columbite-tantalite (coltan) concentrates have been produced by artisanal workers (garimpeiros) from pegmatite and granitic aplite in the Amapá State, Southern Guyana Shield for decades. However, limited information is available about the geological context in which the niobium-tantalum mineralisation was formed. Here, we report U-Pb ages and mineral composition for coltan obtained from a pegmatite that crosscuts mafic-ultramafic rocks of the Paleoarchean Bacuri layered complex, in the Vila Nova region. The pegmatite is highly weathered and consists of a mass of clay minerals containing centimetre-sized quartz and muscovite crystals.

Coltan grains are black and euhedral, with a few clusters of blocky crystals also observed. Their size varies between 200µm and 1mm in length. BSE images reveal that most coltan crystals are homogeneous or show subtle oscillatory zoning, marked by alternations between brighter zones, rich in Ta, and darker zones, rich in Nb. The EPMA data show that coltan grains are rich in Nb (avg. 42.4wt.%) compared to Ta (avg. 13.6wt.%) and exhibit notable variation in concentration between these two cations. However, Fe (avg. 12.3wt.%), Mn (avg. 2.1wt.%) and Ti avg. 1.1wt.%) show relatively uniform concentrations. Micro XRF analyses confirmed that coltan grains are generally homogenous, with some sub-grains varying in Nd and Ta values, and in a few lighter elements, including Ca, P and Al. The chondrite-normalised REE fractionation pattern is characterised by a strong enrichment in heavy rare earth elements compared to the light rare earth elements and a strongly pronounced negative Eu anomaly.

Forty-two U-Pb spot analyses of coltan yielded a weighted mean of 207Pb/206Pb for concordant analyses of 2026 ± 27 Ma (MSWD=1.3), which is close to the 2061 ± 2 Ma (n=84, MSWD=31) concordant age of monazite obtained from the same pegmatite sample. Therefore, the niobium-tantalum mineralisation in the Amapá State, Brazil, formed during the Transamazonian Orogeny.

Fourier transform infrared spectroscopy for geometallurgy modelling - GRE46 gold deposit (NSW, Australia)

Markus Staubmann[1], David Cooke[1], Thomas Rodemann[1], Julie Hunt[1], Matthew Cracknell[1], Ned Howard[2], Karyn Gardner[2]
[1]University Of Tasmania (CODES), Hobart, Australia, [2]Evolution Mining Limited, Sydney, Australia

The mineralogy of a given rock mass is a fundamental geological property that can directly influence our ability to adequately consider, plan and predict the full down-stream environmental and social impacts that a project may encounter. Understanding the mineralogy at the deposit scale provides orebody knowledge that can help inform key project decisions (i.e. extraction sequencing, comminution planning, processing optimization, waste or by product management, and rehabilitation strategies).

The GRE 2.74 Moz gold deposit (34.8 Mt @ 2.45 g/t Au) is one of several low- to intermediate-sulfidation epithermal deposits located in the Cowal Igneous Complex, forming part of the Ordovician to early Silurian Macquarie Arc in central New South Wales (Australia). Gold mineralization at GRE46 is associated with low-density high-grade quartz-carbonate-pyrite veins, hosted within a sequence of calc-alkalic subaqueous volcanic and volcaniclastic rocks. Underground production from the GRE46 deposit began in 2023, introducing a new variable ore type into the process circuit. Understanding the spatial distribution and variability of both primary and alteration mineral assemblages at early stages of deposit characterization has significant implications for mining, processing, metal recovery and waste-rock management.

Fourier transform infrared (FT-IR), and near-infrared (FT-NIR), spectroscopy generates high-resolution data over a wide spectral range with a low signal-to-noise ratio and allows the full spectrum to be utilized in modelling. FT-IR and FT-NIR is applied to conventional drill pulp (p80 <75 µm) samples of the major rock types at GRE46 that have measured quantitative modal mineralogy and comminution parameters. Partial Least Squares (PLS) regression models use the processed spectral data to produce validated predictive mineralogy and comminution models for the deposit. This study highlights the potential for FT-IR and FT-NIR as a rapid, non-consumptive, and cost-effective options for mineralogy modelling and as an upscaling enabler to improve ore body knowledge.

Andradite U-Pb Dating at the Ernest Henry IOCG deposit, NW Queensland: Constraints on the Timing of Cu-Au Mineralization

Jeffrey Steadman[1], Carlos Carrasco Godoy[1,2]
[1]Centre For Ore Deposit and Earth Sciences, University of Tasmania, Hobart, Australia,
[2]CODES Analytical Laboratories, Hobart, Australia

The Ernest Henry deposit, 40 km NE of Cloncurry, Queensland, is the largest iron oxide copper-gold (IOCG) system in the Eastern Succession of the Mount Isa Inlier and the fourth largest in Australia. Despite numerous geochronological investigations of multiple hydrothermal alteration and ore minerals at the deposit (e.g., apatite, titanite, rutile, molybdenite, etc.), a robust age of Cu-Au mineralization remains elusive. A major hindrance is the complex U-Pb behavior in many of the studied minerals, which translates into high uncertainties that cannot resolve independent hydrothermal events. Recently, rare andradite-bearing samples of the host volcanic succession – taken at depth in the deposit – were dated via laser ablation-inductively coupled plasma mass spectrometry (LA-ICPMS). Andradite (and grossular) are robust chronometers in the U-Pb system, with closure temperatures exceeding 800°C, unlike other, more commonly dated minerals at Ernest Henry (e.g., apatite). At Ernest Henry, andradite is intergrown with pyrite, magnetite, quartz, calcite, and chalcopyrite, the latter two of which display textures that suggest either contemporaneous formation with andradite or shortly following. Importantly, unlike the more common Fe-Mn-rich garnet species at Ernest Henry, andradite exhibits minimal retrogression to lower-temperature assemblages (e.g., chlorite). Uranium-Pb dating of andradite yielded a weighted mean age of 1552 ± 6 Ma (2σ, MSWD = 1.57, n = 25), which is older than previous age estimates for Cu-Au mineralization at Ernest Henry (~1530 Ma), younger than peak metamorphism in the region (~1580 Ma), and overlaps recent age estimates for Ernest Henry K-Fe-Mn-rich 'protore' (1555 Ma). Moreover, the low uncertainty of this age indicates that andradite's U-Pb systematics are relatively undisturbed, a feature which can potentially help resolve the ages of geologic events in the Eastern Succession, leading to more robust exploration models in this region.

Development of an Empirical Nickel Targeting Model for Exploration in a Structurally Complex, Poly-Deformed Terrane

Ciara Doherty[1], **Benjamin Stoch**[1], Erin Western[1], Clayton Wright[1]
[1]BHP, Perth, Australia

The Agnew-Wiluna Belt (AWB) is one of the premier regions in the world for potential komatiitic Ni sulphide mineralisation. With extensive Ni-sulphide exploration initiated in 1975, most traditional exploration techniques have been exhausted in the AWB and more surficial targets have already been identified within this jurisdiction. An empirical qualitative and quantitative targeting approach has been developed, whereby numerous datasets are combined and weighted according to their inferred mineral systems control influence, referred to as the Targeting Index (TI). These datasets include the regional-to-local structural analysis, 3D geomodels, an extensive drillhole database, geophysical data suites and geochemical analysis. TI development ensures effective use of all available resources, at various modelling scales, to act as inputs for the generation of new target volumes and allows for future stochastic/simulation and system sensitivity modelling analysis. The TI has been developed and applied to the Leinster Camp scale area, and has been used to identify potential nickel mineralisation and subtle regional trends.

Structural Analysis of the Hyde-Macraes Shear Zone Hosted Gold Deposit, Aotearoa-New Zealand

Madi Styles[1], Melanie Finch[1], Jay Black[1], Matthew Grant[2], Kate McKercher[2], Hamish Blakemore[2]
[1]University Of Melbourne, Melbourne, Australia, [2]OceanaGold, Macraes, New Zealand

Shear zones play an important role in the formation of hydrothermal ore deposits by acting as major conduits for metal-bearing fluids. Deformation enhances porosity and permeability within shear zones, allowing fluids to migrate and accumulate in the Earth's crust. Despite the significance of shear zones in the formation of ore deposits, the processes controlling fluid flow through actively deforming rocks remain poorly understood, largely due to the complex feedback between deformation and fluid migration.

This research focuses on the Hyde-Macraes Shear Zone (HMSZ) in New Zealand's South Island, which hosts the economically significant Macraes gold deposit. Although the lithology and mineralisation of the deposit are well-documented, few have considered how evolving structures within the shear zone control fluid flow and ore deposition. In deforming shear zones, structures that localise mineralisation evolve progressively with increasing strain, influencing where and how fluids are concentrated. Rocks within HMSZ have undergone a complex structural evolution, with quartz-rich and quartz-poor lithologies deforming differently during shearing, creating variations in stress distribution. Since fluids preferentially migrate from high to low pressure zones, this structural evolution likely influenced the formation of fluid pathways. Understanding the relationship between different structures and mineralisation is crucial for understanding fluid flow and ore formation processes.

To understand this relationship, we present 3D reconstructions of Computed Tomography (CT) scans showing the distribution of ore minerals relative to key structural features in the shear zone such as folds, foliation, and shear bands. These results show where auriferous and sulfide-bearing fluids accumulated throughout the HMSZ, highlighting the structural settings most favourable for mineralisation. Our findings contribute to a better understanding of the feedback between deformation and fluid flow in shear zones, with implications for the exploration of structurally controlled ore deposits.

Geological Characterization of an Archean Oxidized Intrusion-Related Au-Cu(-Mo) Deposit: The Upper Beaver Deposit, Kirkland Lake, Canada

Georges Beaudoin[4], Olivier Côté-Mantha[3], Daniel Kontak[1], **Jonathan Sutton**[1,2]
[1]Laurentian University, Sudbury, Canada, [2]Agnico Eagle Mines, Kirkland Lake, Canada, [3]Agnico Eagle Mines, Val-D'Or, Canada, [4]Université Laval, Québec, Canada

The nature and origin of intrusion-related Au systems in the Archean is disputed, partly due to ambiguous features and preservation issues. In contrast, the ca. 2680 Ma Upper Beaver Au-Cu (Mo) deposit (reserves 23.2 Mt grading 3.71 g/t and 0.24% Cu; Agnico Eagle Mines, 2025) of the Abitibi greenstone belt (Ontario, Canada) preserves unequivocal evidence of an Archean magmatic-hydrothermal setting. The host ca. 2703 Ma bimodal volcanic-plutonic rocks of the Blake River Group are cut by a ca. 2680 Ma polyphase hypabyssal intrusive complex with pre-, syn-, and post-ore phases of mixed mafic and felsic compositions. The main phase of regional deformation variably overprints the system. Mineralization occurs in two types: 1) an early concordant replacement-style skarn-like assemblage localized along faults, stratigraphic contacts, and hosted in premineral intrusions; and 2) discordant polymetallic veins forming >90% ore reserve. The early skarn-like and premineral amphibole and plagioclase-amphibole diorites are cut by synmineral amphibole diorite and plagioclase ± quartz-amphibole monzodiorites. The polymetallic (Fe oxide-Cu-Mo-Au) veins are spatially associated with synmineral plagioclase ± quartz-amphibole monzodiorite dikes, which splay from the intrusive stock and are themselves cut by polymetallic veins. Hydrothermal alteration hosting the polymetallic veins includes K-Na feldspar, phyllic and iron-oxide alteration, with a distal barren propylitic facies. Ore petrology and lithogeochemistry define a vein paragenesis: early magnetite-hematite-scheelite veins are cut by quartz-molybdenite-pyrite-gold veins centered on the intrusive complex, with later chalcopyrite-pyrite-anhydrite-sulfosalt-gold-telluride veins extending outside the proximal molybdenite zones. Thus, a spatial zonation of vein types is defined extending from an intrusive center. These veins are cut by a postmineralization plagioclase-amphibole diorite dike, which is subsequently overprinted by quartz-carbonate-tourmaline veins. These features suggest an intrusion-related origin for the deposit, related to an oxidized magmatic complex. This indicates potential for similar ore settings spatially associated with hydrous mafic magmatism in Archean terranes.

Trace Element and Sulfur Isotope Characteristics of the Star of Mangaroon Gold Deposit, Western Australia

Halik Rifki Daeng Taha[1], John Mavrogenes[1]
[1]The Australian National University, Canberra, Australia

The Star of Mangaroon (SOM) gold deposit, located in the Gascoyne Province within the Capricorn Orogen of Western Australia, is a historically significant yet underexplored gold system. This study presents the first detailed petrological and geochemical investigation aimed at characterizing mineralization processes at the SOM deposit. In situ textural, trace element, and sulfur isotope analyses were performed on representative pyrite using Scanning Electron Microscopy (SEM), Laser Ablation Inductively Coupled Plasma–Mass Spectrometry (LA–ICP–MS), and Sensitive High Resolution Ion MicroProbe (SHRIMP).

Pyrite was classified into five textural types: Py-MQZ, Py-Eu, Py-V, Py-TIG, and Py-VRM. Trace element data reveal variation across pyrite types, particularly in Co/Ni ratios. Py-MQZ and Py-Eu exhibit high median Co/Ni ratios (42.13 and 24.43), followed by Py-V (3.83), while Py-VRM and Py-TIG display much lower values (1.13 and 1.20). A consistent trend is observed from Py-Mqz to Py-VRM, marked by decreasing magmatic-related elements (Au, Ag, Bi, Te) and increasing metamorphic-related elements (Ni, W).

Sulfur isotope analyses yielded median $\delta^{34}S$ values of +4.66 ±0.3‰ (Py-MQZ), +5.04 ±0.8‰ (Py-Eu), +5.43 ±0.8‰ (Py-V), +5.57 ±1‰ (Py-TIG), and +6.28 ±0.5‰ (Py-VRM). These values indicate minimal isotopic variation and suggest stable physicochemical conditions and a common sulfur source. The progressive increase from $\delta^{34}S_{py-}$VRM > $\delta^{34}S_{py-}$TIG > $\delta^{34}S_{py-}$V > $\delta^{34}S_{py-}$Eu > $\delta^{34}S_{py-}$Mqz is attributed to isotopic fractionation, transitioning from magmatic to metamorphic influence.

The $\delta^{34}S$ values closely resemble regional granitic sources, particularly the Duchlacher Supersuite (+6.69 ±2‰), implying a granitic reservoir with fractionated sulfur. These isotopic signatures fall within the typical range for orogenic- and magmatic-related gold deposits (+5.4‰ to +11.5‰), such as in the Granites-Tanami Gold Province, supporting a granitic reservoir source for the SOM deposit.

Geochemical Characteristics and Behavior of Hydrothermal Fluids Related to Porphyry Cu-Au Mineralization of the Tujuh Bukit Deposit, Banyuwangi, Indonesia

Ran Takeda[1], Ryohei Takahashi[1], Rob Taube[2], Mochammad Hafid Rahadi[2], Omar Syarief[2], Riansyah Widisaputra[2], Arifudin Idrus[3], Nigel Blamey[4], Tavis Enno[4], Andrea Agangi[1], Pearlyn Manalo[1], Hinako Sato[1]
[1]Akita University, Akita, Japan, [2]PT Bumi Suksesindo, Banyuwangi, Indonesia, [3]Universitas Gadjah Mada, Sleman, Indonesia, [4]Western University, London, Canada

The Tujuh Bukit deposit in East Java, Indonesia, is a telescoping ore system where high-sulfidation epithermal gold-silver mineralization overprinted porphyry copper-gold mineralization. The mineralization in the Tujuh Bukit porphyry Cu-Au deposit occurs as veins, veinlets, stockwork veins, and dissemination mainly in the Old Tonalite. The porphyry-related veins and veinlets were classified into several types based on mineral assemblages and textures, i.e., M veins composed of quartz, magnetite, and hematite, stockwork veins of quartz, B veins of quartz with sulfide centerlines, B veins overprinted by epithermal mineralization, and D veins of pyrite cutting the other veins, in order of formation.

Based on fluid inclusion study, formation temperatures of the M veins, the stockwork veins, the B veins, and the B veins overprinted are 310-320 ℃, 310-460 ℃, 300-470 ℃, and 340-480 ℃, respectively. Bulk gas compositions of fluid inclusions in the M veins and the stockwork veins indicate magmatic signatures, while those in the B veins with or without overprinting indicate both magmatic and meteoric signatures.

Pyrite from the M vein has a δ^{34}SCDT value of +1.6 ‰, while those from the stockwork veins are -0.5 and +0.5 ‰. δ^{34}SCDT values of pyrite and chalcopyrite from the B veins range from -2.3 to -1.6 ‰, while those of pyrite from the B veins overprinted and the D veins range from -2.2 to -1.3 ‰ and -4.8 to -4.4 ‰, respectively. δ^{34}SCDT values of pyrite and prismatic enargite which fill fractures range from -4.0 to -2.0 ‰ and -7.7 to -1.9 ‰, respectively. Euhedral covellite which fills fractures has δ^{34}SCDT values of -8.4 ‰, but covellite disseminated in the host rocks has δ^{34}SCDT ranging from -4.0 to -1.9 ‰. The overall S-isotopic data indicates a gradual decrease from early to late stages of mineralization.

A Structural Ore-Control Model for Carlin-Type Gold Deposits, Youjiang Basin, China

Qinping Tan[1], Jianzhong Liu[2], Daiwei Tan[2], Zhuojun Xie[1], Yong Xia[1]
[1]Institute Of Geochemistry, Chinese Academy Of Sciences, Guiyang, China, [2]Bureau of Geology and Mineral Exploration and Development Guizhou Province, Guizhou, China, Guiyang, China

Carlin-type Au deposits are hydrothermal replacement bodies hosted primarily within carbonate rocks, in which Au occurs mainly as a solid solution within disseminated pyrite. The Youjiang Basin in Southwest China represents the second largest known concentration of these deposits globally, second only to Nevada, USA. The orebodies of the Youjiang Carlin-type Au deposits can be classified into two types: "fault-controlled" and "strata-bound". Fault-controlled orebodies are localized along high-angle reverse faults. In contrast, stratabound orebodies derive their name from their layered or layer-like morphology.

This study utilizes fault-related fold theory to propose a structural ore-control model explaining these two types of orebodies. Central to this model is the development of bedding-parallel detachment faults, typically localized along interfaces such as unconformities or major lithological contacts. These interfaces exhibit a marked competency contrast, separating an underlying competent footwall unit (e.g., limestone or dolostone) from an overlying incompetent hanging wall sequence (e.g., an argillaceous-calcareous clastic succession).

During regional deformation, the incompetent hanging wall shortens and thickens significantly, while the competent footwall remains relatively undeformed. This differential strain concentrates slip along the detachment surface. Continued shortening within the hanging wall creates detachment folds above this primary slip surface. Where folding intensity exceeds the rock mass rupture strength, bedding-parallel slip transitions to strata-cutting thrusting initiated from the detachment horizon. This process generates a fault-propagation fold geometry, characterized by thrust displacement diminishing towards the fault tip as slip is progressively converted into folding of the hanging wall.

In this model, detachment folds are linked to the development of “strata-bound” orebodies, while fault-propagation folds control the formation of “fault-controlled” types. This model stresses the importance of these specific fault-related folds, originating from competency contrasts between footwall and hanging wall units, in localizing mineralization. This understanding provides a basis for potential breakthroughs in exploration for Youjiang Carlin-type gold deposits.

What Controls High-Grade Carlin-Type Gold Deposits?: Structural and Geochemical Factors from the Leeville Underground Complex, North Carlin Trend, Nevada, USA

Koray Tasbicen[1], Elizabeth Holley[1], Kendle Fraley[2], Elizabeth D. Stock[3], Joseph Becker[2]
[1]Colorado School Of Mines, Golden, United States, [2]Nevada Gold Mines, Elko, United States, [3]Barrick Gold Inc., Toronto, Canada

The Leeville Underground Complex (LUC) hosts high-grade Carlin-type gold deposits (CTGDs), each reflecting different physical and chemical controls over time and space. The primary host rock, the Wispy unit (WS) of the Devonian Popovich Formation (Dp), is highly porous, permeable, and enriched in reactive iron. Other Siluro-Devonian sedimentary rocks either lack geochemical reactivity or porosity; therefore, they require additional ground preparation factors for Eocene Carlin-type gold mineralization.

The West Leeville (WL) and turf deposits are hosted in the Dp-WS, with stratabound mineralization gently deepening northward. At WL, a NE-plunging anticline thickens the Dp-WS, cut by NW-trending strike-slip faults that accommodate dike swarms of various ages (Jurassic–Eocene). In southern WL, gold mineralization follows the thickened WS unit along a NE trend, while in northern WL, it transitions to a NNW trend, obliquely cutting the anticline. At the inflection zone, mineralization is vertical and deep, following a NW-trending dike swarm.

At western turf, the Silurian Roberts Mountain Formation (Srm) hosts vertical and locally brecciated gold mineralization. The ore body geometry is controlled by an NNE-plunging anticline and a NW-trending vertical dike swarm cutting its core. Additionally, a major pre-Eocene metasomatic fluid-flow event reduces the volume of rocks that could potentially be mineralized. Hence, the intersection of structures and the metasomatic event creates high-grade zones that are narrow and vertical.

This study documents structural and geochemical controls on the Eocene Carlin-type gold mineralization at the LUC. These controls can be used as vectors supporting the brownfield and greenfield exploration of CTGDs. Deposit- to district-scale structures and geochemical proxies can also guide operational mine planning and production decisions. Understanding the cumulative factors in the CTGDs would improve the genetic model, iteratively comparing the factors with other deposit types with similar geological controls.

Sulfur Mineralizations of the Tichibanine Deposit (Bou-Azzer Inlier, Central Anti-Atlas, Morocco): Geology, Petrographic Investigations, and Sulfide Ore Chemistry

Marieme Jabbour[1], Abdeslam Toummite[1], Moha ikenne[2], Said Ilmen[3], Yassine Lmahfoudi[4], Laurent Sapor[4], Lhou Maacha[4], Mohamed Zouhair[4], Marieme Jabbour[2], Ilya Prokopyev[5]
[1]Moulay Ismail University, Errachdia, Morocco, [2]Laboratory of Applied Geology and Geo-Environment (LAGAGE) Department of Geology, Faculty of Sciences. B.P 8106, Dakhla city, 80 000 Agadir, Morocco, Agadir, Morocco, [3]Polydisciplinary Faculty of Ouarzazatate, Ibn Zohr University, Ouarzazate, Morocco, Ouarzazate, Morocco, [4]MANAGEM Group, Twin Center, Tour A, BP 5199, Casablanca. Morocco, Casablanca, Morocco, [5]Sobolev Institute of Geology and Mineralogy, Siberian Branch of the Russian Academy of Sciences, Russia, Siberian, Russia

The origins of numerous copper deposits within the Neoproterozoic to Cambrian formations of the Bou-Azzer inlier, remain a subject of ongoing debated. To contribute to this discussion, structural, petrographic, and geochemical investigations were conducted on the Tichibanine copper deposit. This deposit, situated in the extreme ESE of the inlier, is located within the northern domain, specifically the Tichibanine Ben-Lagrad Group. This domain is bounded to the north by the Ediacaran volcano-sedimentary formations of Ouarzazate Group and to the south by the sedimentary formations of Tiddiline. At the group scale, several mineralized zones, including Tichibanine, Tanaldat, and Ben-Lagrad, have been identified. Copper mineralization occurs both within altered volcanic lithologies (e.g., epidotized basalts and chloritized dacites) and in association with quartz and carbonate veins, which exhibit surficial malachite occurrences. Mineralized structures at Tichibanine display a variety of orientations, predominantly E-W, and define lithological contacts between chloritized dacites and spilitized basalts. These structures are sub-parallel to the major N120 fault and exhibit folding proximal to it. Field observations, drill core logging, and scanning electron microscopy petrography revealed diverse mineralization modes, including sub-massive veins, veinlets and disseminated forms within the rock mass. The mineral assemblage comprises chalcopyrite, pyrite, sphalerite, sulfosalts (+/- arsenopyrite), and iron oxides (magnetite, ilmenite, and hematite). Electron probe microanalysis (EPMA) point analyses were performed on ore samples to determine ore mineral chemistry. Additionally, laser ablation inductively coupled plasma mass spectrometry (LA-ICP-MS) analyses were conducted on the pyrite and chalcopyrite from Tichibanine. Preliminary analysis of the LA-ICP-MS data indicates significant trace elements variability within the pyrites. Two pyrite generations were identified, with concentrations of As, Se, Te, Cu, Zn, and Pb decreasing from early-stage pyrite (pyrite 1) to late-stage pyrite (pyrite 2).

Reinterpretation of Field Data from Ag-Pb-Zn (F-Ba) Mineralization in the Sierra de Santa Rosa, Melchor Muzquiz, Coahuila, Mexico

Jose Tavizon[1], Antonio Arribas[1]
[1]University of Texas at El Paso, El Paso, United States

The following work is a reinterpretation of data obtained in an exploration project in Melchor Muzquiz Coahuila, in the Sierra de Santa Rosa, from April 2019 to July 2019. To date, the type of deposit occurring in the area has not been fully defined due to limited alteration and mineralization evidence. Polymetallic mineralization is manifested as barite and fluorite replacements in mantos, in addition to calcite veins with galena nodules and disseminated silver in calcareous horizons; anomalous concentrations of critical elements such as Ga, Cd, and W can be detected in some key zones. Three fluid migration events are proposed for different evolutionary stages of the Sabinas basin. Terraspec infrared spectrometric data identified illite-smectite, montmorillonite and kaolinite-smectite clays present in the mineralized structures; this association is typical of low temperature (<200°C) mineralizing fluids. Kaolinite is commonly formed through the reaction of sulfuric acid derived from the oxidation of sulfides by the action of meteoric waters. Old mine workings with greater mineral evidence exhibit anomalous surface Ag-Pb-Zn values; the flow of metallic ions from the surface to the subsoil has left in its path halos of meteoric clays between the mineralized structures and potentially enriched zones in favorable horizons within the boundary between the water table and the oxidized, vadose zone. The margin of the El Cedral fault is interpreted as a replica of the regional La Babia fault, the northeastern limit of the basin, and was the potential conduit for fluids confined to the basin floor. A Mississippi-Valley-type mineralization model is proposed for Sierra de Santa Rosa Ag-Pb-Zn deposits; further fluid inclusion analysis in the calcite veins mineralization may provide additional genetic evidence associated with these deposits.

Unravelling Rubidium Enrichment in Highly Fractionated Pegmatites: Insights from the High-Grade Mt Edon Deposit, Western Australia

Yoram Teitler[1], Behnam Sadeghi[1], Heta Lampinen[1], Louise Schoneveld[1], Michael Verrall[1], Bahman Rashidi[2]
[1]CSIRO Mineral Resources, Kensington 6151, Australia, [2]Everest Metals Corporation Ltd, Perth 6000, Australia

The Mt Edon rubidium (Rb) deposit is located in the southern part of the Paynes Find Greenstone Belt within the Murchison Domain of the Youanmi Terrane (Yilgarn Craton, Western Australia). The area hosts one of the highest-grade rubidium pegmatite deposit reported to date, with an inferred mineral resource of 3.6 Mt grading 0.22% Rb_2O and 0.07% Li_2O (at a 0.10% Rb_2O cut-off), including a high-grade zone of 1.3 Mt grading 0.33% Rb2O and 0.07% Li2O (0.25% Rb2O cut-off). Mt Edon provides a valuable opportunity to investigate Rb enrichment and mineral deportment within highly fractionated Lithium-Cesium-Tantalum (LCT) systems.

The deposit is characterised by a 1.2 km-long pegmatite corridor comprising NE-trending, folded pegmatite sills. The pegmatites, which display mineralogical zoning, are mainly characterised by a medium-grained albite–quartz–muscovite assemblage, with local segregations of microcline–quartz–muscovite and small pods of lepidolite. Spodumene, petalite, zinnwaldite and cookeite occur as accessory phases. Muscovite exhibits the highest Rb concentrations, up to ~7,600 ppm, followed by microcline (~5,000 ppm) and cookeite (~2,000 ppm).

This study aims to further assess the nature and distribution of rubidium and lithium mineralisation at Mt Edon, providing a comprehensive characterisation of Rb and Li mineral deportment to support the optimisation of potential recovery pathways. Mineralogical and chemical characterisation has been undertaken from drill core to micro scale. By combining systematic hyperspectral characterisation with quantitative XRD, micro-XRF mapping, electron probe microanalysis (EPMA), and LA-ICP-MS mineral analysis, this project delivers new insights into the geochemical signatures and mineral deportment of rubidium and its associated pathfinder elements. These findings will contribute to a better understanding of rubidium enrichment in highly fractionated pegmatites and inform exploration strategies for Rb-rich systems.

Molybdenum Isotope Compositions and Formation Mechanisms in Metallic Sulfides in Porphyry Deposits

Yamei Tian[1,2], Jifeng Xu[1,3], Kang Cao[4], Feng Huang[1], Jianming Zhu[3], Alex McCoy-West[2,5]
[1]State Key Laboratory of Geological Processes and Mineral Resources and School of Earth Science and Resources, China University Of Geosciences, Beijing, China, [2]IsoTropics Geochemistry Laboratory, Earth and Environmental Science, James Cook University, Townsville, Australia, [3]Frontiers Science Center for Deep-time Digital Earth, China University of Geosciences, Beijing, China, [4]State Key Laboratory of Geological Processes and Mineral Resources and School of Earth Science and Resources, China University of Geosciences, Wuhan, China, [5]Economic Geology Research Centre, James Cook University, Townsville, Australia

Molybdenite-bearing porphyry deposits are the primary source of molybdenum (Mo) for industrialized society and a major host of Mo in the upper continental crust. Numerous studies have reported the Mo isotope compositions (δ98/95Mo, normalized to NIST 3134) of molybdenite in magmatic-hydrothermal systems, revealing considerable variations (Greber et al., 2015; Li et al., 2019). In contrast, Mo isotope compositions of metal sulfides associated with molybdenite in magmatic-hydrothermal systems, such as pyrite (Py), chalcopyrite (Ccp), and pyrrhotite (Po), remain poorly constrained. The Pulang porphyry Cu-Au (-Mo) deposit in Yunnan, Southwest China, one of the largest porphyry Cu deposits in China, offers an excellent opportunity to investigate Mo isotope behavior due to its well-developed hydrothermal alteration zones and distinct paragenetic sequences. The metallogenic stages, from older to younger, are classified as K-silicate, epidote-chlorite, chlorite-illite, and quartz-illite stages (Cao et al., 2019). Py, Ccp and Po from Pulang deposit exhibit wide ranges of Mo concentrations and isotope compositions. Compared to the earlier epidote-chlorite stage (δ98/95Mo from -0.57‰ to 3.76‰ for Py; 0.12‰ to 2.6‰ for Ccp), Py (0.68‰ to 1.75‰) and Ccp (-0.31‰ to 0.94‰) from the later chlorite-illite stage display lighter Mo isotopic signatures. For Po, Mo isotope compositions from the lower-temperature quartz-illite stage are heavier than those from the chlorite-illite stage. These results suggest that the Mo isotope compositions of hydrothermal sulfides are partly controlled by crystal lattice conditions, precipitation sequence, and temperature-dependent alteration stages within magmatic-hydrothermal systems. Therefore, Mo isotopes in metallic sulfides have the potential to provide critical constraints on metal-rich fluid precipitation, the delineation of favorable mineralization zones, and the evolutionary pathways of complex magmatic-hydrothermal systems.

Reference

[1] Greber et al. (2015) Lithos 190-191, 104-110. [2] Li et al. (2019). EG 114, 981-992. [3] Cao et al. (2019) EG 114, 275-301.

Geochemical Signatures from Porphyry Cu Systems in Sonora, Mexico

Oscar Daniel Torres-Alarcon[1], David Hernández-Uribe[1], Martín Valencia-Moreno[2], Cecilia Santillana-Villa[2]
[1]University Of Illinois - Chicago, Chicago, United States, [2]Universidad Nacional Autonoma de Mexico - Estacion Regional del Noroeste, Hermosillo, Mexico

Porphyry Cu deposits (PCDs) stand as the primary global source of copper. While PCDs are relatively well described, less attention has been given to the magmatic conditions required to form this type of ore deposit. In this study, we constrain the geochemical signatures of the most representative PCDs and barren rocks in Sonora, Mexico, including the Buenavista del Cobre PCD (~7140 Mt of Cu) in the Cananea mining district. Barren rocks yield pressures-temperatures of ~0.3-2.7 kbar and ~613-831 °C, consistent with emplacement depths at 1 to 9 km. PCDs formed between ~621 to ~837 °C. Zircon chondrite-normalized rare earth elements (REE) patterns show heavy-REE enrichment in barren rocks and PCDs. However, barren rocks display a stronger negative Eu anomaly, indicating either greater plagioclase fractionation or different redox conditions. U-Pb geochronology yield weighted mean ages of 73–57 Ma in barren rocks and 80–57 Ma in PCDs, consistent with the magmatism in the Cretaceous-Eocene Mexican Magmatic Arc. PCDs exhibit higher oxygen fugacity (fO_2; ~1.2 ΔFMQ), H_2O content (~7.4 wt%), and Eu/EuN* (0.5) than barren rocks (~0.5 ΔFMQ, ~6.3 wt% and ~0.4, respectively). This indicates an evident higher magmatic fertility in PCD-related intrusions compared with barren rocks. The zircon copper fertility index (ZCFI) coefficient varies widely but is higher in PCDs (~3.7) compared to barren rocks (~2.4). Notably, the PCDs's ZCFI yield values below the proposed fertility limit of 0.45. Our results provide the first state-of-the-art quantitative constraints of the geochemical signatures of PCDs and barren rocks from different mining districts in Sonora, Mexico. We find that higher H_2O content, fO_2, and Eu/EuN* are key proxies for magma fertility.

Quantitative Mineralogical Characterisation of Complex Ores to Interpret Behavior in Processing

Djoan Kate Tungpalan[1], Terrence Rosal[1], Paolo Aiman Ramirez[1], Jillian Gabo-Ratio[2]
[1]Department of Mining, Metallurgical and Materials Engineering, University of the Philippines Diliman, Quezon City, Philippines, [2]National Institute of Geological Sciences, University of the Philippines Diliman, Quezon City, Philippines

This research demonstrates the crucial role of quantitative mineralogical measurement for a more efficient resource development. Three different types of ore were quantitatively characterised using the MLA. The characteristics were then used to interpret the potential leaching behavior of the ores. One ore is a refractory gold ore. The gold occurs primarily as tellurides, which implies that cyanide extraction is not practicable. The grain size distribution showed that the gold grains are finer than 30 microns. It was also found that majority of the gold is locked in pyrite. These suggest that the gold is poorly liberated which could slow down the kinetics of gold dissolution. The second ore is an arsenic-copper ore. The primary arsenic minerals are enargite and tennantite, while the primary copper minerals are enargite, tennantite and chalcopyrite. These minerals occur in varying particle sizes, but the concentration and degree of mineral liberation increase as the particle size becomes finer. The MLA measurement showed that the non-arsenic, copper-bearing minerals are closely associated with the arsenic-bearing minerals. This suggests that flotation would not be applicable in separating the arsenic-bearing minerals from the non-arsenic, copper-bearing minerals. A third ore is also an arsenic-copper ore obtained from the oxidized area of the same deposit. The primary arsenic minerals are enargite and scorodite. MLA images showed the occurrence of enargite in varying particle size fractions, and most enargite grains are well liberated. These characteristics, favor the dissolution of arsenic. However, separating the non-arsenic, copper-bearing minerals from the arsenic-bearing minerals is a challenge as the main copper mineral is enargite. The characterisation provided valuable knowledge to better understand the ores and their leaching behavior. This crucial knowledge estimates the recovery efficiency of the target metals which is essential in designing appropriate processing routes for future development of the ore resources.

Mineralogy and Geochemistry of REE-Enriched Cryolite-Molybdenite Melt in a Porphyry Molybdenum Deposit, Trans-Pecos Texas, USA

Mert Ugurhan[1], J. Richard Kyle[1], Brent A. Elliott[1]
[1]Bureau of Economic Geology, The University of Texas at Austin, Austin, United States

The Cave Peak deposit is a Paleogene diatreme-hosted, fluorine-rich molybdenum porphyry system enriched in Nb, REE, and other critical minerals. The Cave Peak intrusive system was emplaced into Paleozoic strata near the western boundary of the north-trending Salt Basin graben during incipient regional extension that culminated in the Rio Grande Rift in central New Mexico. Cave Peak is genetically related to the nearby but unmineralized Marble Canyon Stock and is a highly evolved endmember of this alkalic magmatic system. The Cave Peak intrusive system consists of an outer rhyolite breccia mass with a complex intrusive core, which has been overprinted by multiple hydrothermal events that formed the molybdenum and associated mineralization. Molybdenite is the primary Mo-bearing mineral and typically occurs in stockwork veinlets, commonly with quartz, biotite, and fluorite. Anomalous Cu, Pb, Zn, Sn, and W concentrations are also present.

Unusual molybdenite-bearing aluminofluoride veins are locally present within the Cave Peak deposit, where disseminated molybdenite spherules occur within a matrix of cryolite, ralstonite, elpasolite, pachnolite, chiolite, fluorite, biotite, and amphibole. Geochemical trends and spatial relationships suggest aluminofluoride melt formed during the late-stage of alkali feldspar granite evolution, when the residual melt became enriched in F, Na and Al, depleted in Si and K, and saturated with molybdenite. Whole-rock geochemistry of a representative cryolite-molybdenite sample revealed concentrations of fluorine, Na_2O, Al_2O_3, and molybdenum exceeding 10% (analytical detection limit), 28.9%, 22.5%, and 0.6%, respectively. REE+Y concentrations in the sample are elevated, reaching 3295 ppm. The sample is also enriched in the trace elements of lithium and bismuth, with concentrations reaching 221 ppm and 68 ppm, respectively. Energy-dispersive spectra obtained from scanning electron microscopy revealed the presence of monazite in another cryolite-molybdenite sample.

Interrogating the Plutonic-Volcanic Connections for Economic Copper Deposit Potential Across the Sunda-Banda Arc

Sri Budhi Utami[1], Esti Handini[2], Sarah Wood[1], Heather Handley[3], Louise Goode[4], Gideon Rosenbaum[1], Teresa Ubide[1]
[1]University Of Queensland, Brisbane, Australia, [2]Universitas Gadjah Mada, Yogyakarta, Indonesia, [3]University of Twente, Enschede, The Netherlands, [4]Independent Researcher, Bristol, United Kingdom

Copper is a critical metal that is vital for the transition towards clean and renewable energy. Given the rapidly worsening effects of climate change, there is strong motivation to meet global demands for copper by exploring for new deposits. Volcanic arcs have the potential to host significant copper porphyries, which is demonstrated by the recent discovery of the Onto porphyry in Sumbawa, Indonesia. Sumbawa is part of the Sunda-Banda Volcanic Arc, and the region hosts several porphyry copper deposits as well as numerous highly active volcanoes, including Tambora and Merapi. How and why the same volcanic arc can produce porphyry deposits and active volcanoes is still poorly understood. Hence, we will evaluate the copper accumulation potential of magmas along this arc by interrogating magmatic mineral records of several volcanoes and the Batu Hijau copper porphyry deposit for evidence of processes and geochemical fingerprints. Recent advances in micro-analytical techniques e.g. trace element mapping of magmatic phases using LA-ICP-MS, and thermodynamic modelling of storage conditions and volatile systematics, means we can track micro-scale magmatic histories relevant to volcanism vs. pluton/porphyry formation, and their roles in copper accumulation. We will also evaluate the geodynamic factors that can lead to the formation of porphyry deposits. Implications of our findings can advance understanding of the role of magmatism and crustal geodynamics in ore deposit formation.

Defining the Syn-Kinematic, Amphibolite-Facies Sequeirinho-Style IOCGs, Carajás Mineral Province

Charlys V. S. Neves[1], Roberto F. Weinberg, Leandro D. Campos, Andrew G. Tomkins, Carlos E. Ganade, Ivan Belousov, Jeffrey Oalmann
[1]Monash University, Melbourne, Australia

Most of the classification schemes for Iron Oxide Copper-Gold (IOCG) deposits are too broad, encompassing multiple deposit styles, obscuring possible genetic differences. The Carajás Mineral Province hosts two distinct IOCG types: the Salobo and Sequeirinho styles. We investigate the Sequeirinho-style and its Ni-Co-rich analogs, to define its style and to propose a model for its formation.

The ore bodies are sigmoidal, brecciated lenses enveloped by steeply dipping mylonitic rocks where mineralization occurs either following the mylonitic foliation or as massive breccias, also often deformed. Microstructural and mineralogic evidence indicates mineralization was syn-kinematic and syn-peak metamorphism at mid-amphibolite-facies.

The development of deposit-scale alteration haloes depends on the ore body size, with the transition from preserved protoliths (e.g., granitoids, basalts, metapelites) to mineralization occurring from mm-scale on smaller bodies to m-scale on world-class deposits. The mineralized rocks define a paragenetic sequence comprising an early stage that can be of two kinds: both kinds display F-Cl-apatite, magnetite, and allanite, but the dominant silicate can be either actinolite (±K-Cl-hastingsite) or biotite/chlorite. This is followed by a transitional sulfide stage with pyrite, (Ni-Co)pyrrhotite, and REE-epidote, more marked in the Ni-Co-rich deposits; and by the proper Cu-Au mineralization, with chalcopyrite, epidote, chlorite, F-apatite, Fe-actinolite, and gold. The mineral chemistry evidences an evolution from reduced and chlorine-rich fluid to oxidized and Cl-poor and likely F-rich fluid.

The chalcopyrite-rich Sequeirinho-style IOCGs constrast with many IOCGs worldwide, including the Salobo-style counterpart in Carajas, which is represented by 'blue breccias' (bornite-chalcocite-covellite-magnetite) and garnet-biotite-grunerite alteration haloes. Sequeirinho-style deposits are likely associated with saline fluids released during amphibolite-facies metamorphism. Reverse shear zones acted as fluid pathways, promoting interaction with several lithologies, including mafic-ultramafic rocks (resulting in Ni-Co-rich deposits). In this scenario, fluid overpressure pulses triggered brecciation and mineralization. This model reconciles structural control with alteration paragenesis, offering new exploration vectors.

Nb and Zr Mineralisation Driven by Carbonatite Metasomatism at the Mt Weld Carbonatite, Western Australia

Arthur Vicentini de Oliveira[1], Denis Fougerouse[1], Axel K. Schmitt[2], Peter D. Kinny[1], Anderson M. Santos[3], Artur P. Deditius[3], Ganesh Bhat[4]
[1]Curtin University, Perth, Australia, [2]John de Laeter Research Centre, Curtin University, Perth, Australia, [3]Murdoch University, Perth, Australia, [4]Lynas Rare Earths Ltd, Perth, Australia

The growing demand for Nb in green technologies, coupled with supply chain risks, solidifies it as a critical metal, currently sourced exclusively from carbonatites. Zirconium is another critical metal commonly associated with Nb, but with restricted economic in carbonatites. Instead, Zr-bearing minerals (e.g., baddeleyite, zircon) are key targets for petrochronological studies in such systems. However, many carbonatites are cumulates overprinted by carbo-hydrothermal fluids, hindering the understanding of the petrogenetic relationships of pyrochlore (main Nb mineral) and Zr-minerals during magmatic and post-magmatic evolution.

Here, we examine the Mt Weld carbonatite (Western Australia), a world-class REE deposit with significant Nb-Ta and Zr resources. The intrusion consists of an outer calcite carbonatite containing phoscorite clasts (olivine/biotite-apatite-magnetite cumulates) and inner ferroan-dolomite carbonatite. Early magmatic phases (calcite carbonatite, olivine phoscorite) were overprinted by carbo-hydrothermal fluids linked to REE-mineralisation. Alkali-rich fluids altered olivine to biotite in phoscorites and transformed magmatic baddeleyite ($\delta^{18}O$=0.15‰) into metasomatic zircon ($\delta^{18}O$=4.61‰) and, with prolonged fluid activity, into hydrothermal zircon ($\delta^{18}O$=3.97‰). The formation of pyrochlore and Nb-ilmenite (<2.5wt.% Nb_2O_5) is also related to metasomatised biotite phoscorite. Incomplete replacement of Nb-bearing baddeleyite (<2.2wt% Nb_2O_5) by zircon with associated pyrochlore±REE-rich carbonates demonstrates that baddeleyite may be an important Nb carrier during magmatic stages subsequently released to form pyrochlore.

These observations confirm that carbo-hydrothermal fluids at Mt Weld are Si-rich, as seen by common phlogopite-amphibole veinlets (often carrying fine-grained pyrochlore, zircon, and Nb-ilmenite) and uncommon quartz veins in the ferroan dolomite carbonatite, favouring formation of zircon. Similar processes in large Nb carbonatite deposits (e.g., Araxá, Catalão I, and St. Honoré) suggest SiO_2 may enhance Nb mineralisation during phoscorite metasomatism. These findings highlight the importance of carbo-hydrothermal fluids in carbonatites, not only for REE deposits, but also as a significant factor for Nb remobilisation and concentration to form higher grade deposits.

Beneath the Surface: The Geology of the Magdalena VHMS Deposit, Iberian Pyrite Belt, Spain

Poliana Vidal Salgado[1], Michael Baker[1], Yamila Cajal[1], Jeffrey Steadman[1], David Le Madec[2], James Royall[2]
[1]CODES/University of Tasmania, Hobart, Australia, [2]Sandfire Resources - MATSA, Almonaster La Real, Spain

The Magdalena deposit is a volcanic-hosted massive sulfide (VHMS) deposit located in Almonaster la Real, Spain, owned and operated by Sandfire Resources-MATSA. It lies within the northern section of the Iberian Pyrite Belt (IPB), one of the world's most significant metallogenic provinces, known for its numerous massive sulfide ore bodies. Although a relatively recent discovery, Magdalena is part of a broad cluster of significant world-class VHMS deposits in the northern IPB, which includes Aguas Teñidas and Rio Tinto.

Magdalena is a blind deposit, consisting of ore bodies found 160 – 1,000m beneath the surface, and contains 25.4 Mt @ 2.2% Cu, 2.5% Zn, 0.8% Pb, and 37.5 g/t Ag (Sandfire Resources Report, 2024). The deposit is hosted within the Volcanic-Sedimentary Complex (VSC) of the IPB. However, the geology of the deposit remains undocumented and poorly understood.

Detailed core-logging reveals that the stratigraphy of Magdalena includes rhyolites to rhyodacites, and andesites to basalts, interbedded with green to purple shales and slates, characteristic of the middle and upper VSC. The ore-envelope lithology is dominated by volcaniclastic facies, including quartzose clasts, and alteration assemblages containing quartz, feldspar, and layers or aggregates of white mica, with minor chlorite reflecting the hydrothermal alteration associated with the mineralization process.

Mineralization is polymetallic and highly variable, displaying a range of textural and compositional ore types. These include: i) pyrite-rich aggregate layers alternating with quartz–white mica-rich volcaniclastic rocks; ii) massive, laminated sulfide ore with compositional banding between pyrite, sphalerite, galena and chalcopyrite; iii) massive pyrite-dominant ore, with coarse to fine grained pyrite; and iv) brecciated massive sulfide zones. Subordinate veins of chalcopyrite–quartz–carbonate, quartz, and quartz–chlorite–feldspar–pyrite are also observed either within the ore body or distally. This variability suggests a complex depositional and post-depositional history, likely influenced by syn-volcanic and diagenetic processes typical of VHMS systems in the IPB.

Paragenesis and Hydrothermal Alteration Characterization Using Short-Wave Infrared (SWIR) in the Intermediate Sulfidation Epithermal Deposit of Ajaruni, Huacullani, Puno, Peru

Rosa Luz Cornejo[1], Milene Aracely Escarsena[1], Fernando Ruiz Guevara[1], Raul Montesinos[2], **Brayan Paul Vilcanqui**[1]
[1]Universida Nacional Del Altiplano Puno (UNAP), Peru, [2]Colorado School of Mines, USA

The Ajaruni deposit is an intermediate sulfidation epithermal Ag-Pb-Zn deposit located in southern Peru, with estimated reserves of approximately 60 Moz of Ag. The main mineralization is hosted within Miocene andesitic volcanic rocks and primarily occurs along veins and hydrothermal breccias. This study integrates geological mapping, petro-mineralogical analysis, geochemistry, and short-wave infrared (SWIR) spectroscopy to characterize lithological units, mineral paragenesis, and hydrothermal alteration patterns. The mineral paragenesis is divided into five stages: (I) an early low-temperature assemblage dominated by chlorite, carbonates, and quartz; (II) barite and hematite deposition; (III) the main sulfide stage with pyrite, chalcopyrite, sphalerite, galena, argentite, and Ag sulfosalts, accompanied by white micas; (IV) late-stage amethyst quartz and carbonates; and (V) a supergene alteration stage marked by manganese oxides and jarosite, reflecting surface oxidation and weathering processes. SWIR analysis of white micas in the mineralized zones reveals Al-OH absorption features between 2200 and 2220 nm, indicating a composition ranging from muscovite to phengite, with a crystallinity index (CI) between 0.28 and 4.7, higher near veins and breccia zones, reflecting intense high-temperature hydrothermal fluid activity. Multi-element geochemical analysis shows elevated silver concentrations (>100 ppm), Pb and Zn contents >1%, and anomalous Cu values averaging 1000 ppm, confirming a well-developed polymetallic mineralization system. Additionally, the presence of a phreatomagmatic breccia with clasts containing truncated quartz veinlets and pyrite sutures suggests a potential connection to deeper porphyry-style mineralization. This multidisciplinary approach enhances the understanding of the Ajaruni deposit and provides valuable insights for the exploration of similar epithermal systems in southern Peru.

Exploration of Lithium Minerals in Lead-Zinc-Barites Mineralization at Ubaru Area, Ebonyi State, Southeastern Nigeria

Paul Visaa[1], Egesi Ndukauba[2]
[1]University of Port Harcourt, Port Harcourt, Nigeria, [2]University of Port Harcourt, Port Harcourt, Nigeria

This study assesses the lithium potential of structurally controlled lead-zinc-barites vein systems at Ubaru, located in the Lower Benue Trough, southeastern Nigeria. The area is underlain by Albian-aged shales of the Asu River Group and hosts NW–SE and NE–SW trending quartz-barite veins enriched in galena, sphalerite, and fluorite. While traditionally exploited for base metals, the site remains unexplored for critical metals such as lithium. Thirty-two samples were collected from hydrothermal veins, altered wall rocks, and pegmatitic zones. Petrographic analysis revealed the presence of sericitized feldspar, muscovite, tourmaline, and quartz within veins and veinlets, with minor accessory fluorite and chlorite. X-ray fluorescence (XRF) analysis confirmed significant concentrations of Pb (up to 2.13 wt.%), Zn (up to 3.42 wt.%), Ba (4.87–6.54 wt.%), and Mn (0.84 wt.%). Though lithium-specific ICP-MS data is pending, pathfinder elements such as Rb (210-430 ppm), Cs (58–144 ppm), and F (530–1,100 ppm) indicate evolved, volatile-rich hydrothermal systems—typical of lithium-cesium-tantalum (LCT)-type mineralization. Environmental water samples from nearby streams and mine pits revealed lead concentrations between 0.016–0.048 mg/L and zinc between 1.23–3.52 mg/L—below WHO permissible limits but indicative of metal mobility. Total dissolved solids ranged from 321 to 528 mg/L, while calcium hardness exceeded 260 mg/L in some locations, suggesting active water-rock interaction. Soils from vein-adjacent sites showed depletion in potassium (0.81 wt.%) and sodium (0.58 wt.%) relative to farmlands, consistent with base-metal leaching and cation exchange processes. This study proposes that the Ubaru vein systems may host cryptic lithium enrichment associated with pegmatite-altered hydrothermal zones. Follow-up studies using ICP-MS, XRD, and fluid inclusion geothermometry are underway to establish lithium occurrence, temperature of formation (estimated 152–240°C), and ore genesis. The results contribute novel data to Nigeria's orebody knowledge and align with global strategies for critical mineral exploration in polymetallic terrains.

Apatite Petrochronology and Geochemistry – Tracking Mineral System Evolution at the Ernest Henry Iron Oxide-Copper-Gold (IOCG) Deposit, Northwest Queensland, Australia

Xin Wang[1], Alkis Kontonikas-Charos[2,3], Elena Belousova[2], Lisa Kearney[2,3], Courteney Dhnaram[2], Vladimir Lisitsin[2], Renjie Zhou[1]
[1]School of the Environment, The University of Queensland, Brisbane, Australia, [2]Geological Survey of Queensland, Department of Natural Resources and Mines, Manufacturing and Regional and Rural Development, Queensland Government, Brisbane, Australia, [3]School of Earth and Atmospheric Sciences, Queensland University of Technology, Brisbane, Australia

Understanding multistage alteration and mineralization in iron oxide-copper-gold (IOCG) systems remains an attractive topic in economic geology. Geochronology without sufficient mineralogical context can limit petrologic interpretations. Our study employs integrated apatite petrochronology with detailed geochemistry at the Ernest Henry deposit (NW Queensland) to provide both temporal and chemical constraints on fluid evolution. As one of the most economically significant IOCG deposits in Australia with exceptional preservation and exposure, Ernest Henry serves as a global type locality offering critical insights into IOCG formation processes worldwide.

Petrographic observations, in situ U-Pb geochronology, and compositional analyses reveal distinct apatite populations correlating with multiple hydrothermal events. Early regional Na ± Ca alteration (dated at ~1570-1560 Ma) produced Apatite-1 with high Na and Cl content, reflecting element remobilization/leaching by acidic fluids. Apatite-2 from the Inter-Lens (a pre-mineralization structure located at depth) is petrogenetically and geochemically similar to Apatite-1 but with elevated F, S, and As. Apatite associated with pre-ore K-Fe-(Mn) alteration is influenced by regional post-peak metamorphism, while pervasive K-Fe alteration within the ore breccia produced apatite displaying magmatic-like REY signatures and enrichment in F and S. Ore stage apatites (dated at ~1529-1521 Ma) have the highest As and F contents, documenting continued fluid evolution and mineralization.

Apatite groups also exhibit compositional variations corresponding to their spatial distribution, notably increasing F, S, As, Mn, Cu, Fe, and Ba concentrations coupled with decreasing Na and Cl from distal areas (>2 km from the orebody) toward mineralized zones. Such spatially and temporally resolved geochemical trends can serve as effective proxies for identifying and distinguishing between regional and ore-forming processes within a long-lived mineral system. The integration of apatite petrochronology and mineral chemistry not only illuminates the formation framework of this IOCG deposit, but also contributes to a holistic methodological approach for investigating complex, multistage mineral systems globally.

Fertile or Futile? Using Plagioclase Composition and Texture to Assess Copper Fertility in Magmas

Catharine Wang[1], Teresa Ubide[1], Daniela Parra-Encalada[1], Ian Campbell[2], Steve Barnes[3], Brenainn Simpson[4], Joel Fitzherbert[4], Jonathon Hoye[5]
[1]The University Of Queensland School of Environment, Brisbane, Australia, [2]Australian National University Research School of Earth Sciences, Canberra, Australia, [3]CSIRO Perth, Perth, Australia, [4]Geological Survey of New South Wales, Australia, [5]Evolution Mining, Australia

Despite the growing demand for copper in the green energy transition, new discoveries of copper deposits have plummeted in the past decade. As a result, research is needed to identify new mineral indicators that can support the exploration of poorly exposed deposits. Plagioclase has become a subject of interest due to its abundance in arc magmas and sensitivity to crystallisation conditions. Plagioclase form and preserve distinct growth zones due to its broad stability and slow diffusion, making it a valuable tracer of magma evolution, and potentially copper fertility. Hydrothermal alteration has hindered the development of plagioclase as a porphyry indicator mineral, but high-resolution techniques now allow precise in situ analysis. This study explores plagioclase textural features and chemical indices that may preserve magmatic processes conducive to mineralisation.

First, we build a global geochemical compilation of plagioclase from porphyry deposits and barren magmatic systems. Then, we compare the compilation with new data from two porphyry copper systems – the Yerington Cu-Mo deposit (USA) and the Northparkes Cu-Au deposit (Australia). Both deposits have well-characterised pre-, syn- and post-mineralisation stages that can be used to track variations in plagioclase with the evolution of copper fertility. We analyse crystal texture through polarised light imaging and analyse major and trace elements with electron microprobe and laser ablation mass spectrometry.

We find that the anorthite content (Ca-rich plagioclase end-member), Fe, Ba, and Sr vary between the mineralised and unmineralised intrusions. Barren plagioclase typically has patchy simple zoning and resorption textures indicative of disequilibrium. In contrast, pre- and syn-mineralisation plagioclase shows fine oscillatory zoning with normally zoned cores, indicating crystallisation from relatively evolved melts and multiple episodes of magma ascent and recharge with similar magmas. These differences in composition and texture provide potential fertility indicators that could improve copper exploration outcomes.

In-Situ Geochemical Signatures of Apatite in Carbonatite-Related REE Deposits: Insights into Magmatic Sources and Fluid Overprints

Chunwan Wei[1,2], Ross Chandler[1], John Mavrogenes[1], Cheng Xu[2]
[1]Australian National University, Canberra, Australia, [2]Peking University, Beijing, China

Carbonatite-associated rare earth element (REE) deposits commonly form in tectonically active settings, with both primary magmatic processes and post-magmatic hydrothermal events contributing to ore formation. Apatite, a key accessory mineral in most carbonatites, serves as a sensitive record of these events. This study integrates cathodoluminescence (CL) imaging with in-situ trace element and Sr-Nd-O isotopic analyses of apatite from three large carbonatites that host significant REE mineralization: Bayan Obo (China), the Gifford Creek Carbonatite Complex (GCCC, Australia), and Miaoya (China).

Apatites from each carbonatite show diverse CL textures, from oscillatory zoning to mottled or patchy overgrowths, which correlate with differences in REE and Sr contents and help to distinguish magmatic from hydrothermal generations. Magmatic apatites show near-CHUR to slightly enriched isotopic signatures. When integrated with isotopic studies from other global ore-hosting carbonatites, these results suggest that crustal material may have contributed to ore formation in these systems, but only to a limited extent.

Hydrothermal apatite generally shows elevated $\delta^{18}O$_SMOW values, consistent with the involvement of ^{18}O-enriched metamorphic fluids. Sr and Nd isotope signatures, however, vary with fluid chemistry. The hydrothermal apatite that displays minimal isotopic change from the associated magmatic apatite suggests interaction with inert CO_2-rich fluids, while the hydrothermal apatites that display a wide range of $^{87}Sr/^{86}Sr$ ratios indicates interaction with F- or Cl-bearing fluids capable of mobilizing Sr. Bayan Obo and GCCC, both of Mesoproterozoic age, display a broader range of εNd(t) values than the younger Miaoya deposit, likely due to Sm/Nd fractionation during fluid-mediated REE redistribution. Although $^{143}Nd/^{144}Nd$ ratios remain robust, variation in Sm/Nd can bias age and source interpretations. Collectively, these results offer insights into the role of multiple fluid events in shaping REE mineralization within carbonatite deposits.

Episodic Magmatism Contributes to Sub-Seafloor Copper Mineralization: Insights from Textures and Geochemistry of Zoned Pyrite in the Ashele VMS Deposit

Bing Xiao[1,2,3], Huayong Chen[1,2,3], Yuzhou Feng[4], Yuanming Pan[4]
[1]Guangzhou Institute of Geochemistry, Chinese Academy of Sciences, Guangzhou , China, [2]University of Chinese Academy of Sciences, Beijing, China, [3]Guangdong Provincial Key Laboratory of Mineral Physics and Materials, Guangzhou, China, [4]University of Saskatchewan, Saskatoon, Canada

We combine textural evidence, in situ analysis of sulfur isotopic composition with trace elements of pyrite from the well-preserved, large-tonnage Ashele volcanogenic massive sulfide (VMS) deposit (Central Asian Orogenic Belt, NW China) to elucidate the origin of the ore-forming materials and evaluate controlling factors of VMS deposits. Three texturally and compositionally distinct types of pyrite have been recognized. The first type of growth zones with high Cu, As, and volatile elements (e.g., Hg and Tl) concentrations and negative δ34S values (-7.83‰–-0.35‰) most likely formed from the input of magmatic volatiles from degassing of the underlying magmatic systems. The second type of Cu-As-rich growth zones is not enriched in volatile elements and has positive δ34S values (2.59‰–6.56‰), which is interpreted to have precipitated during fluid boiling. The third type of growth zones without Cu-As enrichment but having positive δ34S values (0.30‰–9.76‰) was probably formed without boiling or magmatic volatile input. Our results suggest that multi-stage magmatic degassing or fluid boiling accompanied by Cu precipitation recorded in individual pyrite grains can reveal episodic magmatic-hydrothermal activities, which are essential factors in forming large-tonnage VMS deposits.

Structural and Hydrothermal Controls on Mineralization of the Bronson Slope Porphyry Cu-Au-Mo Deposit, British Columbia: Implications for Exploration Targeting.

Marcus Adam[2], Antonio Arribas[1], William Therkeld[2], **Luis Yagual**[1]
[1]University of Texas at El Paso, El Paso, United States, [2]Seabridge Gold, Vancouver, Canada

Bronson Slope is a Cu-Au-Mo porphyry deposit located in the Stikine Terrane of the Golden Triangle District, northwestern British Columbia. This district hosts world-class porphyry and epithermal deposits, including KSM, Brucejack, and Red Chris. The deposit has an inferred resource of 517.3 Mt grading 0.33 g/t Au and 0.09% Cu. Mineralization is hosted within multiphase monzonite and quartz monzonite porphyritic intrusions emplaced into volcano-sedimentary rocks of the Triassic Stuhini Group. The mineralized body trends NW-SE, dips 60°–70° SW, and extends ~1 km in length, 0.5 km in width, and over 900 m in depth. This study integrates lithological, structural, spectral, and ICP-MS geochemical data from 27 drill holes (25,246 m), complemented by geochronology, remote sensing, airborne magnetics, and magnetotellurics. A 3D model developed in Leapfrog Geo defines the deposit's geometry, stratigraphy, ore distribution, and geophysical features.

Hypogene alteration, veining, and mineralization exhibit a well-developed zoning with a central Cu-Au-rich core, surrounded by a Mo envelope and an outer Zn-Pb shell, crosscut by a NW-SE gold anomaly. Two distinct prograde hypogene alteration phases are identified: (1) pervasive biotite alteration in diorite and metasediments with biotite veining swarms and (2) magnetite-biotite alteration in early monzonite, characterized by intense quartz–magnetite stockwork veining. Intermineral monzonite is moderately altered by biotite–magnetite and cut by sheeted quartz–magnetite ± molybdenite veins. Late andesite dikes, weakly pervasively altered to chlorite, crosscut lithological units. Biotite alteration is overprinted by green mica-chlorite in the core and white mica-chlorite alteration at peripheries, which are structurally controlled along NW-SE structures.

Economic mineralization occurs as chalcopyrite-pyrite-gold and pyrite-chalcopyrite-gold quartz-carbonate veins. A weakly developed ~250 m thick supergene profile is dominated by goethite and minor chalcocite along fractures. High magnetic responses correspond to barren quartz–magnetite veins, while low magnetic and resistive signatures are linked to chalcopyrite–pyrite veining, indicating overprinting by a younger mineralizing event.

Structural Model of the La Plata VHMS, in the Western Cordillera of Ecuador

Antonio Arribas[1], Marcelo Llerena[2], **Luis Yagual**[1]
[1]University of Texas at El Paso, El Paso, United States, [2]Universidad Central del Ecuador, Quito, Ecuador

La Plata is a polymetallic, volcanic-hosted massive sulfide deposit (VHMS) dominated by masses of sphalerite, chalcopyrite, pyrite, and galena, with associated gold and silver mineralization, and bounded by quartz, sericite, and chlorite-pyrite alteration halos. The deposit is hosted by volcanic and volcano-sedimentary units of the Paleocene-Eocene Macuchi Unit in the Western Cordillera of Ecuador. The La Plata deposit is located within a 5-km-long and 250-m-wide north-south trending mineralized corridor controlled by dextral-reverse faults. It is affected by at least one tectonic deformation event, divided into two phases. The first phase developed in a dextral strike-slip regime, associated with asymmetric sigmoidal boudin structures generated by stretching in a ductile regime, accompanied by NE-SW dolerite intrusions, while the second phase exhibits a progressive deformation linked to the first phase with the generation of transverse transfer faults, compressional structures associated with duplex, fault-fold geometries, as well as andesitic and rhyodacite intrusions parallel to the fault zones. East-west sinistral and northeast-southwest dextral transfer faults separate the corridor into five structural blocks, from south to north: La Mina Sur, La Mina Norte, Quebrada Tajo, Guatuza, and San José. This study presents a structural model of the La Plata deposit using the implicit modeling method in Leapfrog Geo. The model integrates lithological, geochemical, and structural information collected on terrain and subsurface. The project comprises 1,079 mapping points, 77 trenches, and 127 drill holes drilled from 1996 to 2017 and defines the geometry, distribution, and spatial orientation of the host units, ore zone, and primary and secondary structural surfaces. The resulting model will be useful for the estimation of the mineral resource at La Plata.

Magma Fertility Assessment in the Atacama Region: Implications for Cu and Au Prospectivity and Geodynamics During the Cretaceous

Colomba Trinidad Yañez Labarca[1], Xavier Coddou Henriquez[1], Gonzalo Henriquez Ascencio[1], Julietta Zamora Tapia[1], Javier Fuentes Sepúlveda[1]
[1]Universidad Del Desarrollo, Santiago, Chile

The Coastal Cordillera in the Atacama region of Northern Chile presents an important level of mining activity; however, it does not present the same level and number of studies for its characterization as the Eocene-Oligocene deposits (e.g. Chuquicamata-Escondida-Collahuasi). In a context of need to find more deposits, as a response to the increasing demand on critical elements, different fertility parameters have been proposed to evaluate the potential of magmas for Cu and Au mineralization. In this work, we utilize new parameters proposed by Loucks et al., 2024 for zircon and whole-rock chemical analyses and use the opportunity of integrating both zircon and whole-rock analyses by compiling geochemical data in the interval 26°-29,5°S in order to identify fertile sites as well as to understand known occurrences of Cu and Au mineralization in the region. Our data compilation consists of 105 whole-rock analyses and 1701 zircon spot analyses spanning the 140-90 Ma age range. The most fertile sites were identified in the 100-90 Ma interval for Cu and Au assessment, using Sr/Y, Al2O3/TiO2 and newly proposed (Eu/Eu*)/YbN x 104 + 5(Ce/√U x Ti)) copper fertility indicators, as well as Ba/Zr, Th/Yb and Nb/Y indicators of gold fertility. Although Cu and Au fertility can develop under different conditions,we propose a geodynamic model of evolution along the northern Chile Cretaceous arc wherein magmatic fertility for both elements arose during compressional events that took place during 120-95 Ma linked to the development of a "flat-slab" subduction, generated as south Atlantic seafloor spreading caused the paleo-Andean continent margin to over-ride the western subduction zone,inducing low-angle subduction and compressive stress conditions conducive to Cu-Au mineralization at 100-90 Ma.

Late Jurassic Epithermal Ag-Au Mineralization in Southeast Coastal China: Evidence from U-Pb Dating of Hydrothermal Monazite at the Dongji Deposit

Lina Yi[1], Xianwu Bi[1], Ming-Liang Huang[1], Wei Gao[1], Yanwen Tang[1], Xingchun Zhang[1]
[1]Institute of Geochemistry Chinese Academy of Science, Guiyang, China

The Southeast Coastal Metallogenic Belt of China is extensively covered by Yanshanian volcanic rocks and hosts numerous epithermal Ag-Au deposits, the ages of which remain highly controversy. This severely hinders understanding of ore genesis and the relationship between regional mineralization events and geodynamic settings. The Dongji Ag-Au deposit (Au: 12.5 t, 4.27 g/t) in the Zhenghe ore district of Fujian province is a typical epithermal Ag-Au deposit hosted within quartz porphyry. In this study, hydrothermal monazite intergrowing intimately with Au-bearing pyrites, which is distinctly different from sparsely distributed magmatic monazite, was identified for the first time in typical ore samples from the 170-m level of the Dongji deposit. In-situ U-Pb isotopic dating of this hydrothermal monazite was performed using Laser Ablation Inductively Coupled Plasma Mass Spectrometry (LA-ICP-MS) with a beam diameter of 10 μm. The analyses yielded a high-precision, low-dispersion weighted mean U-Pb age of 150.5 ± 1.7 Ma (n = 28, MSWD = 0.30). This age is indistinguishable, within analytical error, from the formation age of the host quartz porphyry (154 ± 2 Ma, indicating that the mineralization is genetically linked to the emplacement of the quartz porphyry intrusion. This result constrains the main mineralization age of the Dongji Ag-Au deposit to the Late Jurassic (~150 Ma), which is distinct from the known Early Cretaceous epithermal mineralization event (e.g., Zijinshan, ~100 Ma) and probably defines a new Late Jurassic Ag-Au mineralization event in the Southeast Coastal Metallogenic Belt of China. This discovery provides crucial new chronological constraints on regional epithermal mineralization and their genetic linkage to concealed porphyry Cu systems in Southeast China.

Syn- and Late-Orogenic Hydrothermal Si-Al-Ca Alteration of Ultramafic Rocks: Evidence for Ni-Co Downgrading in Outokumpu, Eastern Finland

Christoforos Zamparas[1,2], Clifford Patten[3], Agnes Kontny[4], Alexandre Peillod[1,2], Jochen Kolb[1,2]
[1]Institute of Applied Geosciences, Karlsruhe Institute of Technology (KIT), Adenauerring 20b, 76131 Karlsruhe, Germany, [2]Laboratory of Environmental and Raw Materials Analysis (LERA), Karlsruhe Institute of Technology (KIT), Adenauerring 20b, 76131 Karlsruhe, Germany, [3]Institute of Mineralogy and Petrography, University of Innsbruck, Innrain 52, 6020 Innsbruck, Austria, [4]Institute for Applied Geosciences, Karlsruhe Institute of Technology (KIT), Adenauerring 20a, 76131 Karlsruhe, Germany

The Outokumpu Cu-Zn-Co-Ni deposit is a Paleoproterozoic ultramafic-hosted VMS deposit in Eastern Finland. It is polygenetic and characterized by metal remobilization through deformation and regional amphibolite-facies metamorphism. Samples were retrieved from the Outokumpu Deep Drill Hole, which crosscuts Archean pegmatitic granitoids, the Outokumpu ophiolite, and overlying turbiditic schist and gneiss. The samples investigated belong to rocks of the Outokumpu assemblage, i.e., ultramafic rock (serpentinite, metaperidotite), diopside and/or tremolite calc-silicate rock (skarn) and black schist. This study utilizes petrography, pyrrhotite EPMA and LA-ICP-MS data. The focus is to distinguish different remobilization steps of Fe-sulfides during retrogression. During prograde metamorphism (M_1-M_2), pyrrhotite in skarn and black schist breaks down. This manifests Ni-Co-depletion in residual pyrrhotite, which lacks significant pentlandite exsolutions. Granoblastic and coarse (0.5-5.0 mm), Ni-Co-rich pyrrhotite in serpentinite defines peak (M_3, >550-670°C, 2-4 kbar) to early post-peak (static-M_3) metamorphic stages with antigorite-olivine-anthophyllite assemblages. Contemporaneously and following the peak conditions, tremolite and, subsequently, talc and chlorite (M_{4}-$_{5}$) indicate a Si-Al-Ca-metasomatism in both ultramafic rocks during retrograde evolution. Granular pentlandite exsolutions in pyrrhotite formed below 550°C. A second generation of Ni-Co-poor µm-scale pyrrhotite is hosted in the tremolite-chlorite-talc-carbonate-graphite alteration zone and formed in reaction zones between the peak-metamorphic serpentine and the hydrothermal alteration assemblage by retrograde mineral reactions. Below 480-500°C, the increased SiO_2-activity (ca. 0.001-0.01 m) suppresses magnetite growth from pyrrhotite. Fluid-saturated greenschist-facies conditions, testified from thermodynamic calculations, and high sulfur fugacity (fS_2~1E-8.8 Pa) drive the replacement of pyrrhotite to serpentine and tochilinite, a rare hydrous Mg-Fe-sulfide. Petrographic and geochemical evidence indicate Ni-Co-ore mineral pyrrhotite has undergone downgrading during the retrograde metasomatic evolution. This sets a precedent for ore deposits in similar geotectonic environments.

Devolatilization of Carbonate-Rich and Carbonaceous Rocks as Fluid Source for the Outokumpu Cu-Zn-Co-Ni-Ag-Au Deposit, Eastern Finland

Christoforos Zamparas[1,2], Clifford Patten[3], Agnes Kontny[4], Alexandre Peillod[1,2], Jochen Kolb[1,2]
[1]Institute of Applied Geosciences, Karlsruhe Institute of Technology (KIT), Adenauerring 20b, 76131 Karlsruhe, Germany, [2]Laboratory of Environmental and Raw Materials Analysis (LERA), Karlsruhe Institute of Technology (KIT), Adenauerring 20b, 76131 Karlsruhe, Germany, [3]Institute of Mineralogy and Petrography, University of Innsbruck, Innrain 52, 6020 Innsbruck, Austria, [4]Institute of Applied Geosciences, Karlsruhe Institute of Technology (KIT), Adenauerring 20a, 76131 Karlsruhe, Germany

Ultramafic-hosted VMS deposits have been systematically overlooked. Reasonably, textural overprint by deformation and high-grade metamorphism conceal the origin of the ore. Such case is the polygenetic Outokumpu Cu-Zn-Co-Ni-Ag-Au deposit. Samples retrieved from the Outokumpu Deep Drill Hole, which crosscuts the Outokumpu ophiolitic complex, and former mines (Vuonos, Kotalahti) are studied. The ophiolite contains the homonymous assemblage, i.e., ultramafic rocks, calc-silicate rock (diopside-tremolite skarn), and quartz-rock, enclosed within a thin black schist envelope as m- to km-scale lenses. Petrographic relationships, µ-XRF, sulfide and oxide EPMA and LA-ICP-MS, and thermodynamic modeling are utilized. Devolatilization of the diopside-tremolite skarn and the black schist during prograde metamorphism mobilized S and the critical elements Co, Ni, Bi, Te, and Sn that subsequently precipitated in the form of sulfides in the ultramafic rocks. Devolatilization is additionally indicated by molybdenite flakes in graphite-rich portions of altered rock and elevated Ga-contents of Cr-V-Ti-Al-Zn-poor metamorphic magnetite. Initial thermodynamic calculations show >20 and >25 mol% H_2O and CO_2, respectively, at 300°C, with dissolved H_2S as the mobilizing fluid.

Nickel contents in pyrrhotite are different from pyrrhotite in orthomagmatic deposits (10-100 times less Ni) but have similar Ni-Co concentration ranges (~0.01-0.50 wt.%) and Co/Ni-ratios (~0.02-3) to ultramafic-hosted deposits (Troodos, Cyprus). Additionally, Bi, Pb, Sn, and Te have high concentrations in the whole ore mineral range (up to ~0.1-0.2 wt.% for each element), dissimilar from ultramafic-hosted VMS deposits. These features support the hypothesis that the proto-ore originated from listvenite-birbirite seafloor alteration and leaching of ultramafic rocks. Additionally, they address for the first time the contribution of critical elements such as Bi, Pb, Sn, Mo, and Ga by crustal metamorphic fluids during orogeny in the well-studied Outokumpu deposit.

Fluid-Rock Interaction Recorded in Apatite: Implications for Source and Evolution of Ore-Forming Fluids in the World-Class Jiaodong Gold Deposits, China

Liang Zhang[1], Shu-Qi Yang[1], Gui-Ying Li[1], Li-Qiang Yang[1], Jing Zhang[1]
[1]State Key Laboratory of Geological Processes and Mineral Resources & School of Gemmology, China University of Geosciences, China

The giant Jiaodong gold province, with >5,700 tons of proven gold resources, is the largest in China. However, the source of auriferous fluids and the gold mineralization process remains debated. Traditional C-H-O-S-Pb isotope data have failed to resolve this issue. Apatite, commonly found in gold-hosting granites, records fluid–rock interactions during alteration, offering a potential tracer for ore-forming fluid sources and evolution. In this study, we analyzed textural features, in-situ trace elements, and oxygen isotope compositions of unaltered, altered, and hydrothermal apatite from host granites and syn-gold diorite beneath major orebodies using CL, LA-ICP-MS, and SHRIMP.

Magmatic apatite in granites is typically subhedral to euhedral with yellow-green CL. Altered apatite shows fractures with purple-gray CL in altered zones. Rare hydrothermal apatite displays oscillatory zoning and commonly overgrows or crosscuts earlier apatite. Chondrite-normalized REE patterns of magmatic apatite are relatively flat with strong negative Eu anomalies. During alteration, fluids preferentially leached LREE and HREE, while MREE remained. Breakdown of feldspars released Eu, producing hydrothermal apatite with positive Eu anomalies and depletion in LREE and HREE. These changes, alongside a hydrothermal assemblage of muscovite, quartz, pyrite, polymetallic sulfides, and gold, suggest a transition from oxidizing to reducing conditions.

In syn-gold diorite, magmatic apatite shows brown CL and a right-leaning REE pattern with a slight negative Eu anomaly, whereas its altered zones show green CL, reduced Cl and S contents, and inclusions of monazite and celestite, suggesting an oxidizing fluid. Although mixing with the dominant reducing auriferous H2O–CO2–NaCl±CH4 fluid may have aided gold deposition, the limited oxidizing dolerite fluid cannot be the main gold source.

Oxygen isotope data from syn-gold hydrothermal apatite from granite overlap with values from oceanic crust altered by high-temperature melts/fluids, pointing to a Paleo-Pacific Plate contribution to the Jiaodong gold mineralization.

Supported by National-Key-Research-Program (2023YFC2906902) and National-Natural-Science-Foundation (42272071).

Genesis of Pegmatite-Type High-Purity Quartz Deposits in Tysfjord, N Norway

Haoyang Zhou[1,2], Axel Müller[2]
[1]Guangzhou Institute of Geochemistry, Chinese Academy of Sciences, Guangzhou, China,
[2]Natural History Museum, University of Oslo, Oslo, Norway

High-purity quartz (total trace elements < 50 µg/g) is a raw material for the high-tech and low-carbon industries. So far, high-purity quartz is mainly sourced from granites and granitic pegmatites. While granitic pegmatites are common in the crust, those becoming high-purity quartz deposits are very rare. Thus, the genesis of pegmatite-type high-purity quartz deposits remains poorly understood. Tysfjord in N Norway is a tectonic window in Caledonides, exposing granitic gneiss that belong to the Baltica basement. There are more than 30 pegmatites with a Nb-Y-F (NYF) affinity in Tysfjord, which are the globally largest known intra-plutonic NYF pegmatites and produce the world's second-best high-purity quartz. This work summarizes the geology and quartz chemistry of the Tysfjord pegmatites, aiming at improving our understanding on the genesis of pegmatite-type high-purity quartz deposits. The pegmatites formed during two events: Paleoproterozoic (1772–1755 Ma) granitic magmatism and Caledonian orogenic metamorphism (400–379 Ma). The Paleoproterozoic pegmatites, which show strong shearing of the pegmatite bodies and breakdown of pegmatite minerals, formed from highly fractionated melts of the ~1.8 Ga granitic magmatism and were deformed at 410–400 Ma under peak conditions of up to ~12 kbar and ~730 °C during the Caledonian orogeny. In contrast, the well-zoned Caledonian pegmatites are not deformed and formed from anatectic melts during post-collisional extension of the Caledonian belt. We determined the trace element contents of quartz from both groups of pegmatites. Both pegmatite groups have high-purity quartz. The chemistry of quartz from the Paleoproterozoic pegmatites has been modified by the Caledonian metamorphism and became homogeneous via dynamic recrystallization, whereas the chemistry of quartz from the Caledonian pegmatites shows typical primary magmatic signatures and is largely heterogeneous. Our work reveals that both Paleoproterozoic magmatism and Caledonian metamorphism are important for the formation of high-purity quartz deposits within the Tysfjord pegmatites.

Zr-Nb-REE Enriched Alkaline Silicate Magmatism in Victoria

Tom Andrews[1], Zsanett Pinter[2]
[1]Geological Survey of Victoria, Melbourne, Australia, [2]Commonwealth Scientific and Industrial Research Organisation (CSIRO), Clayton, Australia

Zr-Nb-REE-rich alkaline silicate magmatism is widespread in the Lachlan Fold Belt, Victoria, southeast Australia.

Samples were collected from five alkaline igneous fields, elevated in radiometric Th and U. Whole rock geochemical characterisation was completed on all samples. Geochronology was completed on representative samples. Mineralogy and paragenesis of mineralised samples were investigated using XRF mapping, TIMA and EPMA-CL.

Fractionation, alteration, silica saturation, alkalinity and mineralogy vary across and within the fields. Mineralization is focused on altered portions of the most fractionated units, and with late dykes.

The timing of mineralization varies between each field from primary igneous, to detueric, to magmatic-hydrothermal. Mineralization is heterogeneous and is unlikely to be effectively tested by current sampling.

A new mineralised unit was discovered following up on a regional radiometric anomaly. Similar untested anomalies exist in the public statewide airborne radiometric dataset.

New geochronology links alkaline magmatism with two periods of regional extension.

1. Triassic extension due to rollback in the New England Fold Belt, represented by the Mt Leinster Igneous Complex, Gonzaga Monzonite, Gallows Hill Phonolite and the Toongi deposit in New South Wales.

2. Jurassic initiation of Gondwana and Australia-Antarctica rifting, with the Den Hills Formation representing the distal, low-degree melt portion of the proximal, high-degree melts of the Karoo and Farrar large igneous provinces and Tasmanian dolerites.

Separation of alkaline fields from the rift axis implicates distance in producing enriched, low-degree partial melts over proximal, diluted, high-degree melts.

The mantle beneath Victorian has been fertilised by repeated Phanerozoic subduction, with evidence of enrichment recorded in mantle xenoliths. The enriched mantle is primed to produce volatile + Zr-Nb-REE rich low-degree partial melts upon decompression.

Victorian alkaline igneous fields likely have unrealised correlatives across the Tasmanides, linked by repeated Phanerozoic subduction and extension. Mineral exploration should be focused on the hangingwall of translithospheric faults.

Building Recognition, Trust and Confidence through Good Practice: An Example from the Research Ireland Centre for Applied Geosciences (iCRAG)

Shonny Lehane[1], Aoife Brady[1], Elspeth Sinclair[1]
[1]iCRAG, Ireland

In a time where geoscience is actively being defunded, building recognition, trust and confidence through the good practice of geoscience research centres has never been more important.

The Research Ireland Centre for Applied Geoscience (iCRAG), is now widely recognised worldwide as a Centre for geoscience research by both academic and industry partners. It is a successful and impactful centre, hosted by University College Dublin, and comprises a multidisciplinary group of over 150 researchers across eleven institutions. As a co-funded Centre it is successfully delivering cutting-edge applied and fundamental research, and navigated multiple stakeholders and often competing or conflicting priorities.

iCRAG places strong value on the increased level of impact and engagement it's researchers enjoy as a result of their association with the Centre. This is largely due to the unique Centre Operations who ensure compliance to national and international standards and policies, and create an inclusive and supportive environment for researchers to foster interdisciplinary project ideation and collaborative approaches.

Following the AGI Guidelines for Ethical Professional Conduct guidelines, to which SEG along with over 30 professional organisations are signatories, we will demonstrate how the Centre's practice adheres to these recommendations. We will demonstrate how iCRAG provides supporting frameworks and pathways that benefit the individuals, partners and the organisation. We will further elucidate how building culture reinforces trust and confidence; internally, with research partners, and the broader public. Viewing iCRAG as a segment of the broader geoscience ecosystem in Ireland, Europe and globally, we will demonstrate how research centres, and iCRAG specifically, are an integral part of this ecosystem bridging the gap between traditional academia and industry.

Quantifying Economic Value of Air Quality Improvement Schemes in Quarry Mining: A Social and Environmental Perspective

Tochukwu Ngwu[1], Komsoon Somprasong[2], Chinwe Oramah[3]
[1]Chiang Mai University, Chiang Phueak, Mueang Chiang Mai, Thailand, [2]Chiang Mai University, Chiang Phueak, Mueang Chiang Mai, Thailand, [3]Arctic University of Norway, Harstad, Norway

The quarrying industry has played a crucial role in global economic advancement for several decades, especially in the construction and energy sectors. However, quarrying operations contribute to significant environmental degradation, predominantly air pollution, which has extensive implications for ecosystem health. Consequently, various regulatory frameworks and environmental protection schemes are stipulated by policymakers and regulatory bodies to ensure that quarrying operations are environmentally sustainable. While these schemes ensure compliance with environmental policies, there has been limited emphasis on the quantification of its resulting economic value, this leads to an inability to ascertain its efficiency and capacity to deliver measurable value for stakeholders. To address this, the study evaluates the efficiency of air quality improvement schemes in two quarry mines, quantifies its environmental and social benefits in monetary terms and determines the economic value generated. Additionally, we conduct a comparative analysis of the selected quarries to explore the key factors influencing the variations in economic value derived from the schemes. To account for uncertainty in estimating the derived economic benefits, we apply Monte Carlo Simulation to analyze the sensitivity of results to changes in key assumptions—specifically, pollutant reduction levels and the size of the affected population. This approach facilitates an understanding of the impact of each key assumption on the overall economic value generated by the schemes. Ultimately, this study will provide new insights into the broader economic contributions of air quality improvement schemes, highlighting their inputs beyond regulatory compliance. The integration of Monte Carlo Simulation in the economic evaluation will offer a data-driven approach for evaluating environmental and social returns under uncertainties, facilitating more informed decision-making and strategic resource allocation in quarry operations.

Systems-Thinking to Tackle the Impending Economic Geologist Crisis

Elizabeth Sharman[1], Nicole Januszczak[2]
[1]BHP, London, United Kingdom, [2]BHP, Toronto, Canada

Globally there are decreasing numbers of geoscientists studying in post-secondary institutions and working professionally, particularly those specializing in economic geology. In the context of decreased discovery success, and predicted increased demand for critical metals, this represents a workforce crisis.

Mineral Systems consider what critical geological processes need to be active and effective for a deposit to form and supports opportunities proceeding through the exploration pipeline. We can apply the same process-thinking to the people pipeline for geoscientists.

In mineral systems we start with the original source of metal, the human resource equivalent is supporting the interest and understanding of geoscience in school-age children and their parents. Generation of a medium to move metal has its counterpart in supporting university training of geoscientists, and concentration of metal into medium is the provision of quality economic geology education as part of this training. This optimizes the potential for successful migration of a metal-rich medium, and of well-trained geoscientists into the profession. This is important across all disciplines, but to find the critical metals required globally we need to recruit some of those geoscientists into the exploration industry, just as we separate metals once they have migrated through the crust. The enrichment of metals, physically and chemically, to form the highest quality systems is, in workforces, represented by on-going training and development. The final critical process in mineral systems this is the preservation of mineralization at a mineable depth – for people it is creating the processes to ensure geoscientists stay within the minerals industry.

We need success and collaboration across all critical processes, across all pillars of geoscience – industry, academia, government, and associations - for Skilled Geoscientists to form, develop, transition to the workforce and thrive; and ensure we can provide the people to power the next world class discovery.

Resourcing Western Australia - Mineral, Geochemical and Isotopic Datasets to Accelerate the Discovery of Critical Mineral Resources in Western Australia

Rose Turnbull[1], Erin Gray[1], Imogen Fielding[1], Jack Lowrey[1], Sandra Romano[1], Tim Ivanic[1], Matt de Paoli[1], Fawna Korhonen[1], Klaus Gessner[1]
[1]Geological Survey Of Western Australia, Perth, Australia

With global demand increasing for the critical mineral resources essential for the clean energy transition, the Western Australian government is investing significantly in data acquisition initiatives that support exploration activities to accelerate the discovery and extraction of critical mineral resources across the State. The collection and integration of multiple precompetitive geoscience datasets is a critical component of exploration success, by enabling us to better understand mineral systems, where these are likely to be located across the state, and improving our ability to detect them undercover.

The Geological Survey of Western Australia (GSWA) is the custodian of several significant datasets, (e.g., geological maps, geophysics, geochronology, geochemistry) and is internationally recognised for our ability to collect and freely disseminate the large and specialised datasets required to advance exploration. In addition to increasing the data density and spatial coverage of existing datasets, we are actively investing in the acquisition of complementary statewide datasets, notably isotopic tracers, that will provide an increased understanding of the 4D geological evolution of WA, with an emphasis on imaging major crustal boundaries through time. An additional investment is the acquisition of a State-funded Tescan Integrated Mineral Analyser (TIMA), which will provide mineral characterisation to support several regional projects, including the Heavy minerals map of WA, and characterization of mine waste materials. This presentation will present preliminary results from these two nascent projects.

An inhouse TIMA will also increase GSWA's ability to identify minerals in-situ within ore deposits to target direct dating of mineralization. This presentation will also provide an overview of the current mineralogical, geochemical and isotopic datasets that exist for WA, including our newly compiled geochronology database of mineralization ages, and updated isotope maps and datasets (Hf, Nd, O, Pb) that inform on lithospheric architecture and crustal and evolution.

Secondary Prospectivity of Mine Wastes from the Thalanga, Highway-Reward and Balcooma Volcanic-Hosted Massive Sulphide Deposits

Rosie Blannin[1], Allan Silva Gomes[1], Steve Chingwaru[1], Kam Bhowany[1], Anita Parbhakar-Fox[1]
[1]WH Bryan Mining Geology Research Centre, Sustainable Minerals Institute, The University of Queensland, Brisbane, Australia

Volcanic-hosted massive sulphide (VHMS) deposits are associated with high sulphide contents. Mine wastes resulting from mining and processing of VHMS deposits may present a high risk for acid and metalliferous drainage. VHMS deposits are associated with critical metals such as Co, In, Se, Bi, Te and Sb, which are not typically recovered and report to mine wastes. Reprocessing of VHMS-derived mine wastes to recover critical metals would likely provide the benefit of reducing environmental risks. The Thalanga Cu-Pb-Zn-Au-Ag VHMS deposit was mined intermittently from 1989-2022 for the recovery of Zn, Pb and Cu. Ores from the nearby Highway-Reward and Balcooma Cu-Au VHMS deposits were processed at the site from 1998-2008 to recover Cu and Au. Mine wastes from the Thalanga site were sampled and characterised, including tailings (n=210), waste rock (n=24), run-of-mine (n=14) and secondary efflorescent salts (n=8), to investigate critical metal endowment. The Thalanga mine wastes contain elevated contents of Au, Zn, Pb, Cu and the critical metals Te, Bi, Sb, Se, As, Mo, In, Sn and Co, relative to crustal abundance. The tailings from the different ore sources have distinct chemical signatures; Thalanga tailings are more enriched in Pb, Zn, Ag and Au, while Balcooma and Highway-Reward tailings have higher Te, Se, As, Bi and Co contents. Mine waste mineralogy is dominated by quartz, pyrite and mica, secondary phases such as jarosite, Fe-hydroxides and Fe-sulphates, and sulphides including sphalerite, chalcopyrite and galena. The critical metals are predominantly hosted by sulphide minerals and secondary phases. The efflorescent salt samples are enriched in Co, Cu and Zn relative to the tailings and predominantly comprise sulphate minerals such as jarosite, tamarugite and alunogen. This research will allow targeted processing tests to investigate metal recovery and reduction of potential environmental impacts from the high pyrite contents in the Thalanga mine wastes.

Cobaltite-Pyrite-Pyrrhotite Association: Triassic Metasedimentary Rocks in the South Kitakami Belt, Japan

June Born Born[1], Takuya Echigo[1], Yasushi Watanabe[1]
[1]Akita University, Akita, Japan

Pyrite framboids coated with cobaltite in sulfide-rich nodules are confirmed from the pelitic and psammitic slate of the Upper Triassic Isatomae Formation in the Ishinomaki district, southernmost part of the South Kitakami Belt, Japan. The nodules are a few cm (~3 cm) in size and included in low-grade metamorphic products of prehnite-pumpellyite facies. The rocks contain grains of quartz, albite, K-feldspar, apatite, epidote, pumpellyite/prehnite, magnetite and titanite with minor rutile in matrix minerals of chlorite, illite and biotite. The main sulfide minerals included in the nodules and disseminated in the rocks are pyrite, cobaltite, and pyrrhotite, with trace pentlandite, galena, and electrum. The framboidal pyrites (~5 μm) are commonly coated with cobaltite. Minute electrum grains (~2 μm) appear next to the compact anhedral pyrite and cobaltite grains in the nodules, surrounded by albite, K-feldspar, titanite, and pumpellyite. Carbonaceous materials are identified as seams and/or nodules, associated with chlorite and biotite in the samples. A bulk rock composition of the nodules-bearing sample shows significant enrichment of As, Co, Ni, Cr, Cu, and Zn. The chemical compositions of anhedral compact pyrites determined by EMPA show a strong negative correlation (R2=0.992) between Fe+Co+Ni+Mo versus As+S. Also, a strong negative correlation between As and S (R2=0.996) in compact cobaltite shows the substitution of As for S. These results suggest that diagenetic framboidal and compact pyrite is the main metal source for Isatomae Formation. Cobaltite, coating framboidal pyrite, may have formed by adsorption of cobalt and arsenic in a reducing environment during the diagenesis, and compact cobaltite as well as electrum due to the recrystallization of pyrite and transformation from pyrite to pyrrhotite during the metamorphism.

Authigenic iron oxyhydroxide rims attenuate deleterious element fluxes during sulphide oxidation in historical gold mine tailings

Steve Chingwaru[1,2], Bjorn Von der Heyden[1], Margreth Tadie[1]
[1]Stellenbosch University, Stellenbosch, South Africa, [2]BRC, Sustainable Minerals Institute, Unversity of Queensland, Brisbane, Australia

Historical gold tailings pose significant environmental risks worldwide, primarily due to acid mine drainage and the release of toxic elements triggered by the oxidation of residual sulphides. The formation of metastable secondary iron oxyhydroxides during this process affects the mobility of these elements, yet the mass transfer between sulphide cores, oxyhydroxide rims, and the surrounding environment remains poorly understood. This study uses a gold tailings storage facility in South Africa's Klerksdorp goldfields as a natural laboratory to explore how these iron oxyhydroxide rims influence the mobility of Co, Ni, As, Pb, Zn, Au, and Cu following prolonged surface exposure. Employing a multi-analytical approach, LA ICP-MS, automated mineralogy, and wet chemistry, the study quantifies elemental release versus retention. Despite comprising less than 1% of the mineralogy, iron oxyhydroxide rims significantly retain As, Ni, Cu, and Zn (up to 15 %), while Co, Au, and Pb show minimal retention (<3%), indicating higher mobility. Results suggest that current natural immobilisation is inadequate for environmental compliance, underscoring the need for targeted remediation, including reprocessing and containment strategies to mitigate risk and recover valuable metals.

Process Mineralogy of Historical Sn Mine Waste: Recovery of W, In, and Sn from the Herberton Mineral Field, Australia

Steve Chingwaru[1], Olivia Mejías González[1], Anita Parbhakar-Fox[1]
[1]1. W. H. Bryan Mining & Geology Research Centre, Sustainable Minerals Institute, The University of Queensland, Brisbane, Australia

Critical metals such as tungsten (W), indium (In), antimony (Sb), and tin (Sn) are essential to modern technologies, including renewable energy, electronics, defence, and advanced manufacturing. With W, In, and Sb listed on global critical minerals lists due to high supply risk and economic importance, secondary sources like historical mine waste from granitoid-related Sn deposits offer a promising, sustainable alternative. However, these materials are often mineralogically complex, necessitating an integrated process mineralogy approach to accurately assess their recovery potential and economic value. This study examines legacy mine waste from the Herberton Mineral Field (HMF), Queensland, Australia, a historic Sn district by combining geochemical (ICP-MS), mineralogical (XRD, MLA, SEM), and trace-element (LA-ICP-MS) analyses from eight spatially distributed sites. Geochemical results reveal significant enrichment: W (1.6–11,650 ppm), Sn (6–81,600 ppm), In (0.004–1,155 ppm), and Sb (0.41–5,200 ppm), with concentrations up to 2,500 times crustal averages. Deportment analysis shows In is mainly hosted in chalcopyrite (4–65%), sphalerite (3–92%), cassiterite (2–26%), tetrahedrite (17%), scorodite (16–59%), and Fe-oxyhydroxides (0.05–56%). Sb occurs in arsenopyrite (3–64%) and secondary phases (1.3–83%). Sn is dominantly hosted in cassiterite (>60%), while W occurs in cassiterite (<28%), pyrite (0.04–4%), schwertmannite (21%), Fe-oxyhydroxides (~4%), and scorodite (0.33–9%). Given the dominance of dense mineral phases (>4.1 g/cm^3), gravity separation was simulated using particle density and size data, with metal recovery modelled via mass balance and multi-mineral equations. Simulated gravity concentration achieved recoveries of ~80% Sn, ~59% In, ~45% W, and ~12% Sb, with overall mineral recovery between 24% and 83%. This study highlights the underutilised reprocessing potential of HMF waste and presents a transferable framework for assessing similar polymetallic mine wastes globally, offering a scalable, data-driven strategy for critical metal recovery and environmental remediation from legacy tailings.

Understanding Bismuth as Critical Mineral: From Source to Sink

Cendi Dana[1,2], Evo Dika[5], Paul Duuring[3,4], Ignacio Gonzalez-Alvarez[2,4], Steven Hollis[1], Yuan Mei[2]

[1]School of GeoSciences, The University of Edinburgh, Edinburgh EH9 3FE, United Kingdom, [2]CSIRO, Mineral Resources, Discovery Program, Kensington 6151, Australia, [3]Geological Survey of Western Australia, Department of Mines, Industry Regulation and Safety, East Perth 6004, Australia , [4]Centre for Exploration Targeting, University of Western Australia, Crawley 6009, Australia, [5]Ministry of Villages and Development of Disadvantaged Regions of Republic Indonesia, Depok 16451, Indonesia

Bismuth has been identified as a critical metal in several countries due to its significant role in the transition to green energy and the economy. Despite its importance, the behavior of bismuth within the geological cycle remains poorly understood. This study offers a comprehensive review of bismuth, covering its origin, mineralogical characteristics, evolution within the Earth system, and enrichment during ore-forming processes. Bismuth is primarily concentrated in the continental crust and associated with felsic rocks, occurring in various mineral deposits. Currently, there are 240 recognized bismuth-containing mineral species, predominantly sulfides and sulfosalts. Besides sulfide ores, bismuth can be sourced from residual materials and industrial and mining waste. The primary method of bismuth production is pyrometallurgy, though efforts are being made to develop more environmentally friendly separation technologies. The complex mineralogy of bismuth presents significant challenges in its processing and refining. Bismuth has substantial potential in mineral exploration, as demonstrated by its historical and ongoing use as a pathfinder element for other commodities. A significant risk to bismuth supply is the concentration of 70% of global production in a single country. Despite bismuth enrichment being reported in various deposits worldwide, many exploration programs and resource calculations do not account for bismuth grades. Reporting bismuth grades for all known bismuth-bearing deposits could alleviate future supply risks. Recent technological advancements are important for extracting bismuth from secondary sources, meeting the criteria of being economically viable and scalable, with minimal environmental impact.

Mineralogical Controls on Critical Elements (Bi, In, Te) and Ag Deportment in the Ruwai Zn-Pb-Ag Skarn Deposit, Central Kalimantan, Indonesia

Shelly Mardhia Faizy[1], Bradley M. Guy[1], Ebert Doreen[1], Joachim Krause[1], Valby van Schijndel[2], Arifudin Idrus[3], Ernowo Ernowo[4], Agata V. Kindangen[5], Anggara Widyastanto[6], Max Frenzel[1]
[1]Helmholtz-Zentrum Dresden-Rossendorf, Helmholtz Institute Freiberg for Resource Technology, Germany, [2]GFZ Helmholtz Centre for Geosciences, Germany, [3]Department of Geological Engineering, Universitas Gadjah Mada, Indonesia, [4]National Research and Innovation Agency, Indonesia, [5]Center for Mineral, Coal and Geothermal Resources, Geological Agency, Indonesia, [6]PT. Kapuas Prima Coal, Indonesia

Skarn deposits make significant contributions to the global supply of Zn and Pb and are recognized for their potential to yield critical elements as by-products. However, only few studies have focussed on the mineralogical controls on the distributions of these elements within the ores, hindering efforts to optimize their recovery.

To address this research gap, this study focuses on the Zn-Pb-Ag Ruwai skarn deposit, the largest polymetallic skarn deposit in Indonesia, with a total resource of 14 Mt. of ore at ~5 wt.% Zn, ~3% wt.% Pb, ~108 g/t Ag. Ore samples were selected based on their mineralogical and compositional variability, with a particular focus on Bi, In, Te and Ag concentrations. By integrating bulk geochemical, mineralogical, and microanalytical data, the study investigated the geological/mineralogical controls on the deportment and mineralogical partitioning of these elements across different ore types.

The main ore minerals are galena and sphalerite, with minor amounts of pyrite, chalcopyrite and pyrrhotite, and trace occurrences of Ag-Bi sulfosalts and tellurides. Integration of modal mineralogy with EPMA and LA ICP-MS results, validated by bulk geochemical data, reveals highly variable deportments for the critical elements across the sample suite. This variability mostly reflects differences in both mineral abundances and the concentrations of the critical elements within the samples. In general, galena is the most important host for Ag and Bi; sphalerite carries In; tellurides along with small contributions from galena host Te. Minor contributions also come from Ag-Bi bearing minerals. Moreover, the observed correlation between modal mineralogy and elemental deportment suggests a potential regularity in the partitioning of these element across the different mineral phases that can be used for deportment predictions.

Natural Leachability of Critical Metals - Investigating AMD and Metal Recovery at Normanby Mine

Afreen Iqbal[1]
[1]University Of Queensland, Brisbane, Australia

Mining is the largest export industry in Australia, and concurrently the most polluting. There are over 50,000 mine waste sites around the country, many of which are poorly managed or discarded, and present the risk of generating acid and metalliferous drainage (AMD). The environmental hazards of these waste piles further include leaching of heavy metals, acidifying surface and groundwater, pollution of soil and sediment and potential toxicity to plant or aquatic life dependent on the area. Reprocessing this waste can reduce AMD production and provide a source of critical minerals. This study aims to determine the AMD potential of the abandoned Normanby mine in southeast Queensland and evaluate the recovery potential of critical metals from the site. A geochemical analysis of the site found multiple enriched metalloids including Ag, As, Bi, Cu and Sb at concentrations 10 times the average crustal abundance, all of which are listed as critical minerals for Australia. This investigation will utilise novel methods such as a 100-day accelerated kinetic leach column test for six waste rock samples to determine acid generating potential and metal enrichment and contribute to a protocol for the characterisation and management of legacy mine waste across Australia.

From Source to Sink, Geochemical and Mineralogical Dynamics of Sediment-Associated Metal(loid)s in an Acid Mine Drainage-Affected River

Rich Crane[2], Karen A Hudson-Edwards[2], Adam J Jarvis[5], **Elin Jennings**[1], William M Mayes[4], Patrizia Onnis[3]
[1]W.H. Bryan Mining & Geology Research Centre, Sustainable Minerals Institute, University Of Queensland, Brisbane, Australia, [2]Camborne School of Mines, Department of Earth and Environmental Sciences, University of Exeter, Penryn, UK, [3]Department of Chemical and Geological Sciences, University of Cagliari, Monserrato, Italy, [4]School of Environmental Sciences, University of Hull, Hull, UK, [5]School of Engineering, Newcastle University, Newcastle Upon Tyne, UK

Acid mine drainage (AMD) remains a major environmental concern, with long-term contamination of river systems requiring costly remediation efforts. While there is significant research on aqueous contamination from AMD, there is limited information on the spatial variation of AMD-generated metal(loid)s in the particulate phase in the affected rivers and their associated mineralogical phases. Specifically, the significance of Fe oxides to the sediment load of AMD-affected rivers and their role in particle size and mineralogy in contaminant source, transport and deposition in these rivers. This study addresses these gaps by examining the distribution of arsenic (As), copper (Cu), zinc (Zn), and iron (Fe) in sediments from a river catchment affected by AMD, into the estuary. Through a comprehensive analysis of particle size fractions, high concentrations of As, Cu, Zn, and Fe were recorded downstream of the AMD discharge, with fine sediments (<0.063 mm) hosting these contaminants. These metal(loid)s were predominantly associated with secondary Fe oxides. The study revealed that detrital and secondary mineral phases play a crucial role in contaminant transport and remobilisation, with sulfide and sulfate minerals hosting Zn, and Fe oxides being the primary hosts for As and Cu. This work fills a key knowledge gap by providing new insights into the complex interactions between metal(loid)s, particle size, and secondary mineralogy in AMD-affected riverine environments. The findings underscore the importance of considering these factors in future remediation efforts, particularly in the face of challenges such as climate change, which may further influence sediment stability and the mobilisation of contaminants.

Extracting Critical Metals from Mine Waste Using Molten Alkali Salts

Kathryn Keane[1], Michael Anenburg[1]
[1]Australian National University, Canberra, Australia

The mining industry globally has produced large quantities of tailings, often containing valuable resources. With the progression of technology, these previously untapped resources are becoming increasingly desired, and due to a difference in form, extraction techniques need to be adjusted for these secondary sources. This is particularly relevant to rare earth elements (REE). Uranium mine tailings have been found to contain potentially economically viable concentrations of REE, such as tailings from the Mary Kathleen and Olympic Dam mines. However, investigations into REE-tailings extraction methods have yielded mixed results, and are not currently sufficient economically for large scale application. Another issue in REE extraction from tailings is the availability of processing infrastructure, as creating new facilities would be extremely costly—any feasible methods should consider the use of existing equipment. We are therefore devising a novel extraction method to separate REE from fine-milled tailings, and do so using readily available equipment. We use high temperature roasting of tailings with alkali salts to induce flux melting and recrystallisation of REE into predictable and consistent mineral phases. Preliminary results indicate methods have been successful, with REE mineral grains increasing in size relative to starting tailings.

DERT (Diagnostic Environmental Reporting Tool): A Unified Geospatial Platform Transforming Mine Remediation and Environmental Compliance

Samuel Boone[2], Keith Dimech[1], **Fabian Kohlmann**[1], Wayne Noble[1]
[1]Lithodat Pty Ltd, Melbourne, Australia, [2]The University of Sydney, Sydney, Australia

Effective remediation and secondary metal recovery from mine wastes rely heavily on timely, accessible, and standardised environmental contamination data. Traditional static reporting formats and fragmented datasets frequently hinder efficient decision making, environmental compliance, and risk management, leading to prolonged site closures, delayed approvals, and increased remediation costs. With global mining activity intensifying alongside stricter environmental regulations and heightened community expectations, a novel, integrated solution for environmental reporting is critical.

The Diagnostic Environmental Reporting Tool (DERT) provides a powerful, unified digital platform designed specifically to address these challenges within the minerals industry. Built on robust, four-dimensional (3D plus temporal) geospatial technology, DERT systematically standardises, validates, and archives diverse environmental geochemical datasets, facilitating rapid interrogation and actionable insights via intuitive, browser-based mapping interfaces.

DERT's immediate benefits for mine remediation include automated ingestion of laboratory data, dynamic classification of contamination against relevant regional environmental standards, rapid anomaly detection, and interactive dashboards for streamlined decision-making. By securely integrating historical and contemporary geochemical data into a cloud-based relational model, DERT enables precise environmental assessments, significantly reduces uncertainty, and promotes economically viable recovery of secondary metals from mining residues.

The platform's flexible relational architecture also supports global interoperability, enabling seamless collaboration across industry, consultancy, regulatory bodies, and researchers. Ultimately, DERT accelerates regulatory approvals, reduces operational expenses, enhances transparency, and fosters responsible environmental stewardship across mining projects.

Through its transformative approach to contamination reporting and data management, DERT not only optimises remediation outcomes but also reshapes the perception of mining - from environmentally burdensome to proactive, innovative, and sustainably profitable.

Using Vegetation as a Tool for Environmental Monitoring in Legacy Mines

Maria Mavroudi[1], Rushaniia Gubaidullina[1], Michael Tost[1]
[1]Montanuniversitaet Leoben, Austria

Remote sensing has proven invaluable to geosciences, supporting applications from mineral exploration to environmental monitoring. In densely vegetated terrain, canopy reflectance often masks the spectral signal of underlying soils and rocks. The vegetation's spectral response can act as an indirect proxy for soil quality, allowing plants to indicate information and thus serve as a preliminary tool for environmental monitoring. Legacy mine sites that were never adequately rehabilitated often pose serious environmental problems due to the weathering of the waste dumps, discharging toxic elements like heavy metals into surrounding soils and waters.

This study investigates how contaminated soils alter the growing plant's health and how those stress-induced changes are reflected in vegetation spectral signatures relative to uncontaminated, healthy plants. These changes (anomalies) are detected using vegetation indices through a handheld hyperspectral sensor. The pilot study focuses on a legacy tailings site where geochemical analyses reveal elevated arsenic concentrations in both soil and water. In the study, laboratory pot tests are conducted with four soil treatments: naturally contaminated tailings-derived and uncontaminated soil from the pilot site, clean and artificially contaminated soil matching the pilot site's heavy-metal load. Seedlings are grown in each substrate until their leaves reach a consistent, analyzable size, at which point hyperspectral measurements begin. Throughout the experiment, macroscopic indicators such as leaf length and visible color changes are recorded to complement the results. Field- and lab-derived vegetation spectra are analysed to expose differences between contaminated and reference sites. Important conclusions regarding vegetation indices are derived from the laboratory tests. The next step is to apply these indices to satellite imagery, generating arsenic contamination maps for AMD-affected pilot sites, which will be validated with in-situ geochemical data. The results should deliver a non-invasive, transferable framework providing insights for generating maps of arsenic-contaminated zones in vegetated mining-affected landscapes.

Mine Waste as a Resource: Critical Minerals in Mine Wastes in New Mexico U.S.A.

Bonnie Frey[1], Zohreh Kazemi Motlagh[1], Richard Kelley[1], **Virginia McLemore**[1], Evan Owen[1], Amy Trivitt[1]
[1]New Mexico Tech/NMBGMR, Socorro, United States

There are tens of thousands of inactive mine features throughout the United States (including coal, uranium, metals, potash, and industrial minerals). Many of these mines have existing mine wastes, generated during mineral production, which could have potential for critical minerals, especially since the actual mineral production was generally for precious and base metals and not critical minerals. The first step is to inventory and prioritize the mine wastes. Second step is to sample, characterize and estimate the quantity of critical minerals in these mine wastes to determine if there is enough material that could be reprocessed. Characterization includes determining the type of deposit, mineralogy, geochemistry, and size of the mine waste. The mine history of the area also needs to be summarized to better understand the deposits. In New Mexico (southwestern United States), mine wastes including waters are being characterized for critical minerals. Tonnage estimates of the important critical minerals are calculated. For example, the Copper Flat tailings contains an estimated 402,900 kg of Cu, 1,280 kg of Bi, 6,880 kg of Co, 15,600 kg of Ga, and 53,400 kg of Zn. The Black Hawk district contains unusual arsenide five-element veins (Ag, Ni, Co, As, U, Bi and other elements) and mine waste samples are elevated in Ag (>200 ppm), Co (290 ppm), Ni (2,130 ppm), and Zn (8,463 ppm). In New Mexico, there are an estimated 94 million tons of uranium tailings that contain vanadium, cobalt, arsenic, and possibly rare earth elements. Even mining-influenced waters could have critical minerals. Future mining of mine wastes that potentially contain critical minerals will directly benefit the economy of the local areas. Possible re-mining of mine wastes could clean up these sites and pay for reclamation. These characterization projects included training of younger, professional geologists and students in economic and reclamation geology.

Cobalt Trace Element Mineralogy and Geometallurgy in the Mount Lyell Copper Mine Tailings, Tasmania, Australia

David Cooke[1], Mohammadbagher Fathi[1], **Owen Missen**[1], Emrecan Yurdakul[1]
[1]Centre for Ore Deposit and Earth Sciences (CODES), University of Tasmania, Hobart, Australia

Cobalt (Co) is a critical metal due to its key role in battery technologies. It is rare for cobalt to be the primary commodity at a mine site (e.g. cobaltite deposit at Mount Cobalt, Queensland, Australia). Instead, most of the world's Co is produced as a by-product, and significant concentrations of copper (Cu) are contained in secondary resources, which may be a key component of ensuring Co supplies. Pyrite is the most common secondary source of Co (may contain >5 wt.%) and is the most abundant sulfide in metalliferous tailings dams generated from Cu mining operations. In this study, we analysed fresh tailings samples collected from the Princess Creek tailings dam (storing 20 years' worth of tailings from the Mount Lyell Copper Mine, Tasmania, Australia) and weathered tailings samples from sediment banks of the King River, which mine tailings were flushed down from 1914-1994. Mineralogy and trace element systematics were analysed using X-ray fluorescence (XRF), scanning electron microscopy (SEM) including automated mineralogy and laser ablation inductively coupled plasma mass spectrometry (LA-ICPMS) for trace element mineralogy analysis. Gravity and flotation experiments, followed by the same analytical techniques, were used for the geometallurgical testwork. LA-ICPMS spot analysis shows that Co concentrations in pyrite (>40% of the tailings modal mineralogy) regularly exceeded 0.1 wt.% with a maximum of 0.66 wt.% Co). The applied processing route has proven successful in producing a concentrate of cobaltiferous pyrite. Comparison of fresh (Princess Creek) and weathered (sediment banks and delta of the King River) tailings was undertaken to contrast the deportment of cobalt and other critical metals in fresh and weathered pyritic tailings. Fresh samples from the tailings impoundment contain a lower portion of iron (oxyhydr)oxides than weathered samples, meaning that different Co processing and extraction pathways may be required depending on the Co distribution.

Recycling Lithium-Rich Wastes: Perspectives from Mine Waste Rock and Li-Rich Enamels and Lithium Aluminosilicate Glass (LAS) Industrial Non-Compliant Materials

Giovanni Baldi[5], Pilario Costagliola[2], Giuseppe Finto[2], Pierfranco Lattanzi[1], Marta Morana[2], **Guia Morelli**[1], Laura Niccolai[5], Giovanna Poggi[3], Simona Raneri[2], Massimiliano Reginelli[4], Valentina Rimondi[2]
[1]National Research Council - CNR, Geoscience and Earth Resources Institute - IGG, Florence, Italy, [2]Department of Earth Science, University of Florence, Florence, Italy, [3]Department of Chemistry, University of Florence, Florence, Italy, [4]Eurit s.r.l., Italy, [5]Colorobbia Consulting s.r.l., Vinci, Florence, Italy

Lithium is classified as a critical raw material under the European Critical Materials Act. Demand of lithium minerals necessary for the green energy transition and for electric cars battery production is expected to increase about eight times in the next future due to the EU plan to dismiss fossil-fuel vehicles by 2035.

Reserves of Li are limited with prediction of Li shortages. To meet the increased demand alternative sources are needed. Recycling is focused on batteries and cell manufacturing scraps, while other Li rich materials as mine or industrial wastes have received limited attention. For example, glass and ceramics are the second-largest sector for global lithium use (18%) after batteries.

From this perspective, the ARTU and LiCylce projects provide an opportunity to reflect on the possibility of defining a laboratory-scale lithium extraction process from mining and industrial wastes for industrial scale-up. Preliminary characterization of non-compliant industrial enamels and LAS (lithium aluminosilicate glass) from a Tuscany-based company (Colorobbia) showed Li tenors (up to 14400 mg/kg for enamels, 48690 mg/kg for LAS) comparable to the main mined mineral spodumene; additionally, mine wastes from the extraction of rhyolite and eurite, and kaolin from mines based in Tuscany contained up to 182 mg/kg and 3041 mg/kg, respectively.

The feasibility of extracting lithium from waste is part of the projects. In addition, the projects will encourage reflection on the storage of a critical mass of lithium to be usable at industrial scale, including the development of specific protocols for the possible use of mine excavated volumes to store Li-rich wastes.

This contribute wants to open a discussion for the future development of innovative methods for the recovery of lithium from industrial and mine wastes, to find new advanced recycling strategies, to promote circular wastes management and to promote alternative environmentally safe and sustainable extraction processes.

Process Mineralogical Constraints on the Recovery of Platinum Group Metals via Direct Leaching of UG2 Ore in the Bushveld Complex

Nonhlanhla Ndlovu[1], Nonkuselo Madlakana[1], Glen Nwaila[1]
[1]University Of The Witwatersrand, Johannesburg, South Africa

As global efforts to reduce carbon emissions intensify, the demand for platinum group metals (PGMs) continues to rise. This is driven by their unique physicochemical properties such as good corrosion and oxidation resistance, high melting temperatures, electronic and catalytic properties, and their important role in platinum-catalysed hydrogen-powered fuel cells for electric vehicles. Traditionally, PGMs have been recovered via a crush–mill–float–smelt process followed by multi-stage leaching process to recover base metals and PGMs. However, the complex mineralogy and texture coupled with declining ore grades, often result in suboptimal metal recoveries during flotation and create mounting uncertainties in long-term PGM supply. Direct leaching has attracted significant attention as a potentially cost-effective and environmentally benign alternative to the conventional processing route, particularly for low-grade, mineralogically complex ores. The crux of successful direct leaching hinges on a rigorous mineralogical investigation that explain the distribution, association, and textural attributes of PGM-bearing minerals. In this study, we employed a process mineralogy framework to evaluate the feasibility of direct leaching over froth flotation. Representative samples from the UG2 Reef were characterised using the TESCAN Integrated Mineral Analyser (TIMA) for mineralogical properties and inductively coupled plasma mass spectrometry (ICP-MS) for chemical composition. TIMA analysis revealed that PGM minerals are mostly fine-grained and occur within chromite and base-metal sulphides, resulting in poor liberation and limited lixiviant access. ICP-MS confirmed high concentrations of Ni, Cu, and Co, which may compete with PGMs during leaching, increasing reagent consumption and lowering overall recovery. Understanding these textural complexities and PGM associations is important for modulating key parameters such as particle size distribution, leaching reagent type(s), conducive solution pH and temperature and overall leach kinetics in order to improve recovery of PGMs. Insights gained from this mineralogical investigation offer guidance for developing more efficient direct leaching process route of complex and low-grade PGM ores.

Antimony Fate During Schwertmannite Transformation: Implications for Passive Remediation in Acid Mine Drainage–Impacted Environments

Mohammad Rastegari[1], Niloofar Karimian[2,3], Scott Johnston[1,4], Edward Burton[1]
[1]Faculty of Science and Engineering, Southern Cross University, Lismore, Australia, [2]School of Earth, Atmosphere & Environment, Monash University, Melbourne, Australia, [3]CSIRO Mineral Resources, Clayton South, Melbourne, Australia, [4]Catchments, Coasts and Communities Cluster, Southern Cross University, Lismore, Australia

Antimony (Sb) contamination frequently arises from the oxidative weathering of Sb-bearing sulfide minerals during mining and ore processing. In acid mine drainage (AMD) environments, schwertmannite is a commonly occurring metastable Fe(III) oxyhydroxysulfate mineral. Over time, it transforms into more stable phases such as goethite and jarosite. Schwertmannite plays a key role in regulating Sb mobility; however, its potential use—and that of its transformation products—as part of passive remediation strategies remains poorly characterized. This study investigated the geochemical behaviour of Sb(V)-bearing schwertmannite and its transformation under AMD conditions. X-ray diffraction (XRD) and Fe K-edge EXAFS spectroscopy showed that schwertmannite transformed to goethite at a higher pH ($\approx$ 3) and to jarosite at a lower pH (< 2.5) over time. Antimony K-edge EXAFS spectroscopy revealed that Sb(V) was sequestered within the schwertmannite structure through partial substitution for Fe(III). During transformation, Sb(V) mobility initially increased, but it subsequently remained structurally incorporated in the transformation products via the same substitution mechanism. Dissolution experiments indicated that Sb(V) incorporation significantly reduced schwertmannite solubility, thereby decreasing Sb mobility. Furthermore, the retention of Sb within the newly formed, more stable mineral phases suggests that schwertmannite transformation creates a long-term sink for Sb in AMD environments. These findings offer new insights into the capacity of schwertmannite and its transformation products to immobilize Sb through natural geochemical processes. This work contributes to evaluating the effectiveness of in-situ iron mineral passive remediation approaches and informs strategies for managing secondary metal contaminants across the mining value chain.

Evaluation of the Geochemistry and Mineralogy of Mine Waste in Nevada, USA

Joseph Rosal[1], Simon Jowitt[1], Jesse Scholpp[1]
[1]University Of Nevada, Reno, Reno, United States

The increased demand for critical metals has led to a coincident increase in assessing the potential of alternative and secondary sources of these critical commodities. One of the most prospective secondary sources of critical metals is mine waste. Although humans have been mining for thousands of years, only twelve metals were mined between the Bronze Age and the end of the Medieval period. In comparison, modern standards of living require nearly all of the elements on the periodic table, although many metals that are classified as critical today were not mined at significant levels before the turn of the century. Waste rock from historic mine sites therefore, could be enriched in critical co- and by-product metals such as tellurium, indium, and gallium, which are typically produced as byproducts of Cu and Zn production, respectively. Modern technology also allows for the mining of large tonnage, low-grade deposits, and the cutoff grade for many commodities has historically been higher than it is today. This means that mineralization that is economic today may have been considered waste at the time of operation. Poor metal recovery during processing can also lead to waste material being enriched with critical metals.

This study focuses on the characterization of mine waste samples from a number of current and legacy mine sites in Nevada, USA, a state with a significant history of mining that contains likely more than 300,000 individual mine sites and occurrences. This characterization will allow the identification of the critical metal potential of individual sites as well as the distribution of critical metals among different mineral phases within the waste to be analyzed. These results will feed into both state- and nation-wide assessments of the critical, base, and precious metal potential of mine wastes across the USA.

Geoenvironmental Indicators Using Line Scan Hyperspectral and XRF Core Analyses for Early AMD Prediction in Northern Queensland Ore Deposits, Australia

Enrique Saez Salgado[1], Anita Parbhakar-Fox[1], Nathan Fox[1], Laura Jackson[1]
[1]W.H. Bryan Mining & Geology Research Centre, Sustainable Minerals Institute, The University of Queensland, Brisbane, Australia

Acid and Metalliferous Drainage (AMD) constitutes a major environmental risk in the mining industry, with the potential to cause long-term degradation of soil and water systems. Early identification of AMD risk is essential for implementing effective mitigation strategies during mine planning. This study evaluates the application of line-scan hyperspectral mineralogy, in combination with continuous-scanning X-ray fluorescence (XRF), to predict AMD potential using drill core samples from four mineral deposits in Northern Queensland, Australia. Samples were analysed using the HyLogger™ and Minalyzer core scanner (CS) systems to assess mineralogical indicators associated with AMD, focusing on the presence of potentially acid-forming (PAF) minerals and those with acid-neutralising capacity (ANC). The HyLogger™ Geoenvironmental Index (Hy-GI) was applied to delineate AMD risk domains by integrating continuous downhole thermal infrared (TIR) hyperspectral mineral data with sulphur concentrations obtained from geochemical assays. Of the samples analysed, 46 from the Ernest Henry, Wolfram Camp, and Baal Gammon deposits—characterised by high sulphide content and limited carbonate buffering—exhibited high AMD potential. In contrast, 24 samples from the Mount Isa deposit demonstrated high neutralisation potential. Comparison with traditional static AMD assessment methods, such as Acid-Base Accounting (ABA) and Net Acid Generation (NAG), demonstrated that the combination of hyperspectral carbonate intensity data and sulphur content serves as an effective proxy for identifying both acid-generating and neutralising materials. The Hy-GI indicator successfully differentiated between acid-generating and neutralising zones, serving as a robust early predictor of AMD potential and confirming the value of integrating hyperspectral mineralogy with geochemical data for early-stage AMD assessment. This approach offers a rapid, cost-effective, and non-destructive method for mine waste characterisation and supports more sustainable resource development through improved environmental risk forecasting and mine planning.

Predictive Geometallurgy and Mine Waste Valorisation: Circular and Eco-friendly Recovery of Gold in the Witwatersrand Goldfields (South Africa)

Xolile Carol Simelane[1], Sehliselo Ndlovu[2], Glen Tsundukani Nwaila[1]
[1]University Of The Witwatersrand, Johannesburg, South Africa, [2]University of Arizona, Tuscon, United State of America

Gold has held a position of immense value throughout human history, and the Witwatersrand Basin in South Africa stands as a testament to this, hosting the largest known gold anomaly on Earth. Despite the extraction of significant portion of gold from this region, mining of primary high-grade ores has plateaued, leaving behind a legacy of large volumes of mine waste. However, historical mine waste dumps, present a unique opportunity for the implementation of circular economy initiatives and sustainable gold recovery practices. In recent years, various studies have explored methods for optimising gold recovery from the abundant waste materials in the Witwatersrand goldfields. However, the study outcomes have not been adopted by the mining industry due to their high energy requirements and challenges in findings integration thus hindering industrial process change. This study introduces a novel hydrometallurgical approach centred on the agglomeration of low-grade ores and tailings. The study areas are located ~ 60 km east and west from Johannesburg which is at the heart of Gauteng and where mine waste reprocessing operations are situated. Our study approach offers a promising avenue for recovering gold from secondary sources in an environmentally friendly and economically viable manner. Preliminary results reveal that the tailings consist mainly of major elements silica, aluminium and iron with trace concentrations of sulphur and uranium which correlates with mineralogical data obtained from the TIMA and QXRD, showing high content of silicates (quartz, muscovite), iron oxides (hematite) and low concentration of sulphides (pyrite) and uranium rich minerals (uraninite), typically associated with gold. With this data and further gold characterization, we intend to design a predictive geometallurgical flowsheet that improves gold recovery efficiency. Efficacy will stem from the use of eco-friendly reagents (i.e. glycine), in the dissolution of gold during the agglomeration process and compared with conventional cyanide-based reagents.

Metal Isotopic Monitoring of Environmental, Extraction, Distribution of Metal Associated with Mine Practice

Adam Simon[1], Ryan Mathur[2], Karr McCurdy, Lee Ann Munk
[1]University Of Michigan, Ann Ar, United States, [2]Juniata College, Huntingdon, United States, [3]VectOres Science, Inc., Huntingdon, United States, [4]University of Alaska Fairbanks, Fairbanks, United States

Quantitatively determining the source(s) of metals in ground and surface water is important for monitoring the environment during mining, smelting, refining, and reclamation operations. We developed a hydrogeochemical technique that uses the isotopic composition of metals in water at parts per trillion concentrations to trace the source(s) of a given metal or metals and the minerals from which they were derived. The technique is based on the comparison of metal isotope ratios of, for example, 65Cu/63Cu, 66Zn/64Zn, 56Fe/54Fe, 109Ag/107Ag and 124Sn/122Sn, in environmental samples and from samples tied directly to mining activity, mine waste, ore, and tailings. The technique can be used by mining and other firms such as industrial, governmental, and agricultural companies to distinguish among multiple possible sources of metals when conducting environmental impact assessments. The technique has been used to demonstrate that 1) copper and zinc in stream water draining the Red Dog lead-zinc mine in Alaska were not sourced from tailings, 2) copper in stream water draining a copper deposit in China was most likely sourced from pyrite in tailings, and 3) silver in forest soils downwind of a smelter in the Czech Republic can be traced to smelter emissions. The technique has also been used to fingerprint the source(s) of metals in the supply chain. For this application, the isotopic composition of multiple metals including copper, lead, silver, tin and zinc, and metal concentrations, identify the geographic source of metals such as gold and tin in the supply chain. Application of this technology distinguishes gold sources within mining districts and possibly from individual deposits helping to detect illicit gold sources and sales.

Indian Iron Ore Waste Mine tailings for Critical (Li) and REE minerals -Case Studies

Perumala Venkata Sunder Raju[1], Basant Rath[2]
[1]National Geophysical Research Institute, Hyderabad, India, [2]National Mineral Development Corporation, Hyderabad, India

Banded Iron Formations (BIFs) are among the most significant rock types of the Precambrian era, widely distributed across global shield regions. India ranks fourth in global iron ore production, contributing approximately 246 million tons, or 8% of the world's total. Enhancing the grade of iron ore is critical for efficient mineral beneficiation, where comminution plays a key role. Advanced in-situ mineral characterization techniques such as Light Microscopy (LM), Scanning Electron Microscopy (SEM), Powder X-ray Diffraction (PXRD), and automated tools like the Mineral Liberation Analyzer (MLA) provide precise data for developing effective beneficiation strategies. These methods have been successfully implemented at both laboratory and pilot scales. Furthermore, the exploration of Rare Earth Elements (REEs) and other critical minerals in weathered iron ores, mine tailings, and waste dumps has revealed significant potential. In the Ubrani area of Karnataka, lithium-bearing petalite has been identified within titaniferous magnetite seams. The lithium appears to be associated with hematitized V-Ti-Mg-rich zones and likely formed during late-stage hydrothermal processes, coinciding with a major orogenic gold mineralization event (2.45–2.60 Ga) in the Dharwar Craton. These findings open new avenues for critical mineral exploration in waste tailings as its cost effective

Unlocking Australia's Re-Mining Potential

Jane Thorne[1], Zhehan Weng[1,2], Callum Kucka[1], Apsara Jayasekara[1], Anita Parbhakar-Fox[3], Dane Lamb[2], Gavin M. Mudd[2,4], Sarlae R.B. McAlpine[1], Allison Britt[1]
[1]Geoscience Australia, Canberra, Australia, [2]School of Engineering, RMIT University, Melbourne, Australia, [3]WH Bryan Mining Geology Research Centre, Sustainable Minerals Institute, The University of Queensland, Brisbane, Australia, [4]Critical Minerals Intelligence Centre, British Geological Survey, United Kingdom

Critical minerals and strategic materials, essential for modern technologies, are often found at low concentrations within Australian ore deposits. Conventional ore processing historically had limited recovery for these materials, allowing them to end up in mine waste. Assessment of re-mining potential supports sustainable recovery of critical minerals from mine waste, to realise more value from Australia's already discovered mineral deposits.

Under the Australian Government's 35-year Resourcing Australia's Prosperity initiative, Geoscience Australia is assessing Australia's re-mining potential. In collaboration with Australia's state and territory governments, RMIT University, and the University of Queensland, this project represents a pioneering effort to sustainably recover critical minerals from secondary sources and aims to enhance resilience, sustainability and efficiency in Australia's mineral resources sector. The project involves a national-scale assessment of mine waste, identifying new opportunities for critical minerals supply from previous mining and processing activities. It provides precise geospatial and technical information delivered via the digital Atlas of Australian Re-mining Potential (Geoscience Australia, 2024) and includes geochemical analyses of selected case study sites nationwide.

Future work will focus on understanding Australia's secondary resource potential through rapid estimations of critical mineral potential at selected sites, extrapolated from existing geochemical understanding and publicly reported resource information. Continued sampling, supported by geophysical surveys and geochemical analysis, will characterise additional sites and develop new methodologies and tools to support both the extraction and modern management of reprocessed mine waste.

Outcomes of assessing Australia's re-mining potential will include a comprehensive view of the nation's minerals inventory, presenting commercial, strategic, and environmental benefits. Re-processing mining and mineral-processing wastes is a significant step towards realising a circular economy, affirming Australia as a reliable and responsible supplier of critical minerals to global partners.

Reference:

Geoscience Australia. (2024). Atlas of Australian Re-mining Potential, Geoscience Australia, Canberra. Retrieved 29 April 2025 from https://portal.ga.gov.au/persona/australian-remining-potential

Drone-based VNIR–SWIR Hyperspectral Imaging for Mapping and Monitoring Environmental Conditions in Legacy Mine Sites

Justus Constantin Hildebrandt[3], Friederike Koerting[3], Steven Micklethwaite[1], Andres Ortega Lucero[1,2], Ekaterina Savinova[1], **Victor Tolentino**[1,2]
[1]W.H. Bryan Mining Geology Research Centre, Sustainable Minerals Institute, The University of Queensland, Brisbane, Australia, [2]Camborne School of Mines, University of Exeter, Penryn, United Kingdom, [3]HySpex, Norsk Elektro Optikk AS, Oslo, Norway

While mining is essential for sourcing the materials needed for the energy transition, there are widespread negative environmental impacts associated with its activities. Legacy mine sites, in particular, are a major environmental concern which have attracted international attention. The scope and complexity of this challenge requires effective solutions that can reliably map and monitor changes in the environment. Drone-mounted hyperspectral imaging (HSI) is an emergent technique that senses, at centimetre-spatial resolution, infrared radiation across many narrow, contiguous spectral bands for each pixel in an image. We explore the capabilities of drone-mounted HSI for mapping surface patterns related to contamination dispersal in the former Mary Kathleen U-REE mine site, in Queensland, Australia. Hyperspectral data across the visible to near-infrared (VNIR) and short-wave infrared (SWIR) wavelength ranges (400–2500 nm) were collected over selected areas of the site. Analyses were performed using data-driven (Spectral Angle Mapper; SAM) and knowledge-based (Band Ratios; BRs) spectral processing techniques. The SAM technique identifies contamination dispersion patterns linked to the precipitation of sulphate-rich evaporative sediments, and differentiates mineral compositions within visually similar areas. The BR technique highlights reactive surfaces and clay mixtures, reinforcing key patterns identified by SAM. The results indicate that drone-based HSI can capture and distinguish complex surface trends, demonstrating its potential to enhance the monitoring and rehabilitation efforts at legacy sites. It also shows that by mounting HSI cameras on drones, the spatial resolution of the data is significantly improved compared to other platforms (e.g., centimetre-scale versus decimetre-scale for satellite and manned aircraft), enhancing the ability to detect subtle mineral and surface variations.

Repurposing Historical Iron Slag as a Sustainable Source of Rare Earth Elements: A Case Study from Centre Furnace in Pennsylvania

Jacqueline Silva[1], Maureen Feineman
[1]University Of Texas At El Paso, El Paso, United States

Rare earth elements (REEs; including the lanthanides, scandium, and yttrium) are critical to the global transition to renewable energy and advanced technologies, as they are essential components of wind turbine magnets, solar panel glass, display screens, and electric vehicle batteries. However, primary REE deposits are often of low grade and environmentally costly to develop. This study investigates the potential for historical iron smelting slag to serve as a secondary, low-impact source of REEs.

We focus on slag from the Centre Furnace in Centre County, Pennsylvania—a major 18th to early 20th century iron production site. Slag is a silicate-rich, glassy waste product generated during smelting, formed by the reaction of iron ore with limestone flux. Typically discarded or used for aggregate, slag has rarely been evaluated as a potential source of critical minerals. Our geochemical analysis reveals total REE concentrations ranging from 600 to 1,000 ppm in the Centre Furnace slag, exceeding the generally accepted threshold (~300 ppm) for REE resource interest by a factor of 2–3. Importantly, the amorphous and pre-processed nature of slag suggests that REE extraction could be more efficient and cost-effective compared to conventional hard-rock sources.

This case study highlights the potential of historical smelting waste as a viable REE resource and underscores the broader opportunity for re-evaluating legacy mine wastes in the context of critical mineral supply and environmental remediation. Ongoing work aims to determine whether this enrichment is a localized anomaly or characteristic of broader slag deposits across Pennsylvania's historic iron districts

A New Discovery Frontier for Copper/BHT Deposits in the Thomson Orogen Revealed by Merging Public Research and Precompetitive Data

John Anderson[1]
[1]Austrike Resources Pty Ltd, Glenalta, Australia

With increasing demand for mineral resources, there are two principal avenues to improving Australia's discovery rate: 1) Revitalising brownfields provinces with disruptive geological and targeting ideas; and 2) Applying innovative approaches to the network of research outcomes and precompetitive data to reveal new greenfields frontiers. Experienced mineral explorers will provide the questions and handrails for modern technologies including AI.

The Darling Headwaters offer such new frontiers. A 6Ma drainage reactivation provides a showcase for the 2011 National Geochemical Survey of Australia and the 2023 Heavy Mineral Map of Australia. The pathfinder matrix of mineral suites (including sulphides), geochemistry and mineral hardnesses provides far-seeing target vectoring and proximity measures ranging down to direct targeting. Vectoring with target minerals and metals is a step-change improvement to flagging empirical tactics like potential field targeting.
The overlapping pathfinder dispersion fields for multiple catchments in the Darling Headwaters are untangled by careful analysis of different pathfinder suites, geomorphic vergence/history and scant drill geology. This approach delineates a nickel-copper target area with pentlandite chalcopyrite sphalerite lead-ins and a new tin-tungsten belt with cassiterite scheelite galena Sn W Pb lead-ins.

Potential for greenfields copper and Broken Hill Type provinces is also indicated in the Thomson Orogen under the margins of the Eromanga Basin between Thargominda and Cunnamulla. Gahnite chalcopyrite pyrrhotite tongxinite (Cu_2Zn) rhodonite Pb S Hg pathfinders vector into a 300 x 150km source region with potential for Proterozoic basement highs under shallower covered parts of the Eromanga. This region includes metasediment domains, largely interpreted from geophysics by previous researchers, and co-incident gravity and magnetic features consistent with BHT signatures.

The prospective region warrants a collaborative program by government and industry of data acquisition, research and strategic drilling aimed at producing the next generation of discoveries for copper, zinc and other metals as needs evolve.

REE Pattern Shape Analysis Using Lambda Coefficients

Michael Anenburg[1], Lucas Tatnell[2]
[1]Australian National University, Australia, [2]University of Western Australia, Australia

Rare earth element (REE: La–Lu) in rocks, minerals, and other terrestrial materials are often normalised to reference baselines such as chondrite asteroids or primitive mantle compositions. Normalised REE concentrations typically appear smooth when plotted with REE along the x axis (either as atomic number or ionic radius). These smooth lines are termed REE patterns and have varying slopes, curvatures (e.g., convex or concave) and other fine scale structures that can provide information about the material the REE pattern is from. The two elements cerium and europium can anomalously deviate from the smooth lines owing to redox effects. Numerical quantification of these shapes, curvatures, and anomalies is traditionally done with element ratios. For example, the La/Lu ratio gives information on whether a pattern is sloping up or down. The Sm/Lu ratio, on the other hand, provides information on curvature concavity. The europium anomaly is typically quantified as Eu/Eu*, with Eu* being the interpolated geometric mean of its neighbouring elements using √(Sm×Gd). In addition, some REE patterns contain a tetrad effect signature, where four groups of consecutive elements show internal fractionation, for which similar quantification methods, based on element ratios, have been developed. Here, we present an alternative method to quantify these shapes. We use the lambda coefficient method developed by O'Neill (2016) to represent patterns as independent numbers representing REE abundance, slope, quadratic curvature, sinusoidality, and higher order curvature ("W-type"). We also show an accurate method to extract the complete set of tetrad coefficients. Cerium and europium anomalies are calculated relative to the curvature, avoiding potentially erroneous linear interpolations. Lambda coefficients are useful parameters for tracing geochemical evolution, petrological modelling, discrimination efforts, dimension reduction, and many other applications in mineralogy, petrology, and geochemistry.

O'Neill, H. S. C. (2016). Journal of Petrology, 57(8), 1463–1508.

Spectral and Structural Analysis of Satellite Imagery for Lithium Pegmatite Exploration in Northwestern Somalia

Hooshang Asadi Haroni[1], Maryam Veiskarami[2]
[1]Tethyan Gold Exploration, Brisbane, Australia, [2]Zamin Pars Gohar (GeoPersia), Isfahan, Iran

Lithium (Li) is a critical element in the production of rechargeable Li-ion batteries, with growing demand driven by the rapid expansion of the electric vehicle industry. This demand has intensified global exploration for economically viable Li deposits. Key lithium-bearing minerals of economic importance include spodumene ($LiAlSi_2O_6$), lepidolite ($K(Li,Al)_3(Si,Al)_4O_{10}(F,OH)_2$), and petalite ($LiAlSi_4O_{10}$), typically hosted in pegmatites. This study employed spectral and structural analysis of moderate- to high-resolution satellite imagery data to identify potential pegmatite-hosted Li mineralization in northwestern Somalia, a region with partial vegetation cover. The remote sensing approach focused on detecting diagnostic Li-indicator minerals, including spodumene, white micas (e.g., lepidolite, muscovite), clay alteration minerals (e.g., smectite, illite, kaolinite), ferric iron oxides, formed as weathering products of Li-bearing minerals, and silica, which helps distinguish silica-rich granites and pegmatites from surrounding rocks. Image processing targeted distinct spectral features across key electromagnetic spectrum regions of Shortwave Infrared (SWIR, 1.4 to 1.6 μm and 2.1–2.3 μm), indicative of OH^-, Al^{3+}, Li^+, and Mg^{2+} components, Visible and Near Infrared (VNIR, 0.490–0.665 μm) associated with ferric iron oxides, and Thermal Infrared (TIR, 8.65–9.10 μm) representing Si–O stretching vibrations. The Spectral Angle Mapper classification algorithm was applied to multispectral ASTER and hyperspectral PRISMA imagery using reference spectra from the USGS spectral library to identify Li-bearing minerals and associated clays. Ferric iron oxides were mapped using red/blue band ratios from high-resolution WorldView-2 and Sentinel-2A VNIR data. Silica-rich lithologies were delineated using the Band 11/Band 12 ratio from ASTER TIR data. Additionally, structural analysis identified favorable open-space settings such as fold hinges and axial planes within an E–W trending metamorphic unit that likely facilitated the intrusion of Li-rich pegmatites/granites. These features, extracted from WorldView-2 and Sentinel-2A data, were integrated with the mapped mineral indicators and delineated first- and second-priority Li potential targets for field validation.

Rapid Fingerprinting of Cassiterite Paragenesis via µXRF for Exploration of Li Pegmatites

Jason Bennett[1,2], Melvin Hartley[1], James Waldron[1], Evie Burton[1]
[1]Portable Spectral Services, Perth, Australia, [2]University of Western Australia, Perth, Australia

Cassiterite, nominally SnO_2, is often detected in the heavy mineral fraction of stream sediment surveys. Its presence could be used as an indicator for nearby Li-Ta-Nb-Cs pegmatite systems. However, cassiterite is also commonly found in other mineral systems unrelated to Li mineralisation, such as Sn-W greisen and vein-hosted deposits or polymetallic Cu-Pb-Zn-Sn skarn deposits. Recently, Bennett et al (2024) demonstrated that the minor element contents of cassiterite can be used to fingerprint the paragenetic system from which the crystals originated. Pegmatites in particular can be identified by their high Nb and Ta contents concomitant with low Nb/Ta ratios, which were acquired via costly and time consuming EPMA and LA-ICP-MS analyses.

Here we demonstrate via applied case studies that these data can be acquired rapidly via the Bruker M4+ Tornado µXRF. Following routing sample mapping and phase identification, the geochemistry of each cassiterite crystal is extracted via a custom grain segmentation algorithm. This allows for fast assessment of LCT pegmatite fertility from the presence of cassiterite in stream sediment surveys. We also demonstrate how this process can be used on loose grains without mounting in epoxy resin. This allows for select crystals to be collated with other samples along with reference materials into the same mount for subsequent trace element analysis or geochronology.

References:

Bennett, J., Kemp, A.I.S., Hagemann, S.G., Fiorentini, M.L., Roberts, M.P., 2024. Systematic trends in the substitution mechanisms of minor elements in cassiterite. Australian Journal of Earth Sciences 71, 1098–1124. doi: https://doi.org/10.1080/08120099.2024.2325396

Fast-Tracking Brownfields Exploration at Geita: Unleashing the Power of 3D Seismic and Cutting-Edge Seismic Innovations

Sian Bright[1], Shimba Kwelwa[2], Ockert Terblanche[2], Sasha Ziramov[1], Andrew Pethick[3]
[1]Fleet Space Technologies, Perth, Australia, [2]AngloGold Ashanti, Tanzania, [3]Digital Earth Consultants, Perth, Australia

Greenstone belts, historically recognised and extensively explored for their gold resources, remain prime candidates for brownfields exploration. However, revitalising an area that has already undergone extensive mining and exploration activity can be challenging. The key lies in harnessing the wealth of geological data gathered over time to uncover fresh insights and provide confidence in exploration concepts.

This case study focuses on the Geita Gold Mine, Tanzania. Where reprocessing of legacy 3D seismic data using Kirchhoff Pre-stack Depth Migration coupled with the use of Fleet's patented SEISOMICS workflow, has yielded valuable information about the local geology and mineral system. The SEISOMICS workflow uses variations in the texture of the seismic data to automatically extract geological features and produces 3D geobodies that can be related back to the known geology.

The Geita deposits are hosted in a package of iron-rich supracrustal rocks intruded by the Nyankanga intrusive complex, the area has been affected by seven deformation events resulting in complex patterns of folding, ductile shearing and brittle faulting (Kwelwa et al, 2018). Gold mineralisation is controlled by the intersection of iron-rich stratigraphy and the moderately north-dipping Geita structural trend. The SEISOMICS shows two distinct seismic domains that correspond with iron-rich stratigraphic packages. Viewing the domains as 3D geobodies highlights the distribution of iron-rich stratigraphy in a volume of around 20km^3 and allows visualisation of major folding and structural offsets of these units.

The intersection of predicted iron-rich stratigraphy with the north-dipping Geita structural trend is used as a guide for targeting down-plunge extensions of known mineralisation, significantly reducing the exploration search space and highlighting the most prospective areas to target beneath current mining. The highest priority target is a shallow west plunging zone beneath the Geita Hill Mine area that has potential to extend mineralisation to over 1km below surface.

Machine Learning Approach to Automated Mineral Identification from RGB Reflected Light Microscopy

Kevin Brown[1], Shaun Barker[1], Brian McNulty[1], Lindsey Heagy[1], Cassady Harraden[1]
[1]Mineral Deposit Research Unit, Department of Earth, Ocean and Atmospheric Sciences, The University of British Columbia, Vancouver, Canada

This study assesses the performance of common machine learning (ML) methods in classifying mineral phases in reflected light microscopy imagery. Mineralogy, modal composition, and classified maps are generated from pixel-based analysis of RGB inputs from petrographic image data generated using a Zeiss Axioscan 7 Geo. Methods are divided into unsupervised clustering and supervised classification approaches to quantify efficacy and workflow requirements. Methods tested include k-means clustering, k-nearest-neighbor, and random forest supervised classification. Unsupervised k-means optimization is based on within-cluster sum-of-squared distance. Supervised methods are trained on 107 single-mineral phase cropped image patches. Target mineral phases include pyrite, chalcopyrite, bornite, chalcocite/digenite, magnetite, hematite, titanite, and a silicate background class to isolate opaque phases. Samples are sourced from the Red Chris and Lorraine porphyry deposits. Training data is generated using conventional reflected light petrography to determine mineral phase and validated with overlaid mineralogy maps from energy dispersive spectroscopy (EDS) generated from automated scanning electron microscopy (SEM). The training data is then generated as single-phase, homogeneous patches cropped from whole samples. Method success is evaluated using pixel-level assessment of reserved patches for testing (80-20 split) to quantify accuracy. K-Means demonstrates a viable means for quickly assessing the number of mineral phases present, while random forest provides the addition of target mineral detection, modal abundances, and labelled mineralogical maps.

By leveraging the convenience of standard petrographic analysis with the scalability of ML algorithms, reflected light imaging offers a rapid and cost-effective means of quantifying opaque mineral assemblages and modal abundances through pixel-based classification. This approach utilizes petrographic thin sections and digital microscopy to quantitatively classify opaque minerals, avoiding the expense and specialty equipment needed for SEM automated mineralogy. In contrast to labour intensive techniques such as point counting and manual petrographic review, ML analysis is a rapid alternative for high-throughput and field-adjacent mineralogical assessments.

Exploring Regolith-Hosted Rare Earth Element Deposits: A Practical Methodology for Mineral Exploration

Paulina Vergara[1], **Nicolás Bustos**[1], Carlos Marquardt[1]
[1]Pontificia Universidad Católica De Chile, Santiago, Chile

The global demand for rare earth elements (REEs)—especially the heavy REEs (HREEs)—continues to grow each year, driven by the increasing production of electronic devices and the accelerating green transition. In 1985, global production was just 50,000 metric tons of REE oxides; today, it stands at around 300,000 metric tons, and demand is expected to exceed that figure by 2030. Nowadays, China dominates the REE supply chain, accounting for approximately 65% of global production and 85% of global processing, a situation with geopolitical implications. This highlights the urgent need for new stakeholders to enter the market, which requires effective mineral exploration strategies. This study presents a practical methodology for exploring regolith-hosted REE deposits (RH-REED) in tectonically active zones. By integrating multiple data layers—including climate conditions, lithology, structural geology, whole-rock geochemistry, and spectral imagery—it is possible to identify promising target areas for field-based exploration. Applying these techniques can support the discovery of new prospects for the future extraction of REEs from RH-REED.

Hydrothermal Alteration Zones Wstimated by ASTER Satellite Image and ASD TerraSpec on the Eastern Zone of Cerro de Pasco, Peru

Evelyn Caiza[1]
[1]Central University of Ecuador, Quito, Ecuador

Remote sensing and field spectroscopy are increasingly vital for mineral exploration in rugged, poorly accessible terrains. In central Peru's Cerro de Pasco region—host to porphyry, polymetallic, and orogenic gold systems across multiple metallogenic epochs—we apply a combined satellite and field-based spectral approach to identify hydrothermal alteration zones as vectors for new exploration.

This study integrates ASTER satellite imagery with spectral data collected using an ASD TerraSpec from representative outcrops to delineate alteration zones associated with mineralized systems. Four complementary techniques were employed: RGB composite mapping of iron oxides and clays, band ratio analysis to detect alteration assemblages, SWIR-based spectral indices, and Spectral Angle Mapper (SAM) classification. The SAM method, combining ASTER data with field-derived spectra, enabled refined discrimination of hydrothermal halos.

Validation of results across all methods highlighted consistent mineralogical zoning, allowing the identification of eleven new exploration targets. These include seven high-sulfidation systems with advanced and intermediate argillic alteration; three porphyry-related systems displaying phyllic, propylitic, and advanced argillic zones; and one stratabound polymetallic system characterized by carbonate-rich phyllic-propylitic assemblages.

Our results demonstrate the effectiveness of integrating satellite spectral data with ground-truthing to delineate mineral alteration in geologically and structurally complex terrains. This workflow enhances early-stage targeting where traditional mapping is constrained and underscores the potential of remote sensing and spectroscopy as strategic tools in exploration programs across the Central Andes.

LoopConverter: A Tool to Standardise Geological Data for 3D Geological Modelling

Rabii Chaarani[1,2]
[1]Northern Territory Geological Survey, Darwin, Australia, [2]Monash University, Melbourne, Australia

The Loop platform, like any modelling system, requires input datasets to be correctly formatted to process the data and perform the necessary computations to deconstruct geological data and generate 3d geological models. Geological data integrated in GIS databases are distributed in non-standardised datasets. Manually formatting and correcting GIS datasets from different sources becomes time-consuming, especially when dealing with regional geological maps and data. The LoopConverter Python package is designed to solve this problem by allowing the implementation of data converters that would enable automated formatting of any GIS database. We will discuss how to implement a converter to prepare a dataset for ingestion by Map2Loop, using the Henbury geological map from the Northern Territory as an example. Our example showcases the complete Loop workflow that allows increased automation in 3d modelling, subsequently rapid hypothesis testing by generating 3d models while avoiding time-consuming manual data formatting.

New Insights of Alteration Mineral Mapping in the Skarn-Porphyry Systems

Huayong Chen[1]
[1]Guangzhou Institute of Geochemistry, Chinese Academy of Sciences, China

The Edong ore district in southeastern Hubei Province is a famous skarn-porphyry ore deposits region in China. In recent years, mineral exploration in the deep area has been paid attention, which requires new exploration methods and approaches. In this study, the latest comprehensive research results for the alteration minerals in the Tonglvshan Cu-Fe-Au skarn deposit, Jiguanzui Cu-Au skarn deposit and Tongshankou Cu-Mo-W skarn-porphyry deposit were presented, especially based on the application of the short wave infrared spectrum (SWIR). On the basis of the assemblages of alteration minerals, paragenesis and alteration zoning, trends of the main alteration mineral SWIR features were identified, and criteria for exploration were further extracted for each skarn (-porphyry) deposit. For instance, high Pos2250 values (Pos2250>2253 nm) of Fe-rich chlorite, high crystallinity (Pos2170>2170 nm, Dep2170>0.18) of kaolinite group, anomalous high and low Pos2200 (<2202 nm, or >2212 nm) values of white mica-montmorillonite, emerging of kaolinite, dickite and saponite, can be used as vectors towards mineralization center in the Tonglvshan Cu-Fe-Au deposit; the high Pos2200 (>2209 nm) values of white mica-montmorillonite can be used as vectors for the Jiguanzui Cu-Au deposit; and the high frequency of chlorite Pos2250 (>2249 nm) and Pos2335 (> 2333nm) can be used as vectors for the Tongshankou Cu-Mo-W deposit. These research findings show that the alteration minerals can provide effective exploration indicators for the skarn-porphyry deposits in Edong district, and these results have been preliminarily applied and verified in the Tonglvshan Cu-Fe-Au deposit.

Microinclusion Geochemistry in White Micas from Porphyry Deposits and Transitional Environments: A Case Study from Globe-Miami Copper District, USA

Axel Cima[1], Paula Montoya Lopera[1], Ivan Belousov[1,2], David R. Cooke[1], Matthew Cracknell[1], Adam Gorecki[3], Samantha Scher[3,4]
[1]Centre for Ore Deposit and Earth Sciences (CODES), University of Tasmania, Hobart, Australia, [2]CODES Analytical Laboratories (CAL), University of Tasmania, Hobart, Australia, [3]BHP, Tucson, United States, [4]LKI Consulting, Washington D.C, United States

White micas are key alteration minerals commonly associated with phyllic alteration zones and also play an important role in the sericite-chlorite zone of porphyry systems. In economic geology, particularly in porphyry deposit exploration, the term "white mica" typically refers to minerals within the phengite solid-solution series, with compositions ranging between the end members muscovite (K[Al2][AlSi3]O10[OH]2), Mg-celadonite (K[Mg2][Si4]O10[OH]2), and Fe-celadonite (K[Fe2][Si4]O10[OH]2).

Efforts have been conducted to understand white mica as a vectoring tool for mineralization in porphyry environments, focusing on metal zonation patterns associated with pH gradients of mineralizing fluids and distinguishing different phyllic alteration events. However, the mineralization footprint recorded by white mica, along with the potential influence of mineral microinclusions on its chemical composition and implications for exploration, remains poorly understood.

In this work, we present new LA-ICP-MS data from white mica related to the Ocelot porphyry copper deposit, Globe-Miami district, Arizona, USA, we aim to understand the influence of mineralizing hydrothermal fluids with the occurrence of diverse microinclusion phases locked within white mica. For this purpose, white mica from different host-rocks, alteration styles and distance to mineralization were sampled and analyzed.

Preliminary results revealed variations in white mica chemistry with a contrast between primary magmatic muscovite from host-rocks and hydrothermal phengitic and illitic white micas related to phyllic and intermediate argillic alteration. A high abundance of mineral microinclusions (zircon, rutile, Cu-Fe sulphides, and REE bearing phosphates including apatite and monazites) was detected in white micas proximal to areas with high Cu and Au contents demonstrating a potential link between these microinclusions and white mica associated mineralization in the Ocelot deposit.

Breaking Through the Basalt Barrier: Unveiling the Mt Read Volcanics with Ambient Noise Tomography

Tjaart De Wit[1], George Taylor[1], Ashley Morris[1]
[1]Ambient Resources, Hobart, Australia

Ambient noise tomography (ANT) was employed across Special Exploration Licence 11/2021 near Guildford, western Tasmania, to generate a S-wave velocity model of the upper 2 km of the crust. This study, conducted in 2023, aimed to delineate the thickness of post-Paleozoic basalt cover and map the sub-basalt geology relevant to exploring the prospective Mount Read Volcanics (MRV) beneath. Our S-wave velocity model reveals significant basalt thicknesses (potentially >500 m) northeast of Mount Pearse and southeast of Waratah, posing a considerable challenge to conventional exploration techniques targeting MRV-hosted mineralisation. Furthermore, the ANT data effectively maps the contact between Proterozoic basement in the west and the Paleozoic MRV to the east, providing crucial structural context. Complementary analysis using Rayleigh wave ellipticity and horizontal-to-vertical spectral ratio (HVSR) methods supports the spatial trends in basalt thickness identified by ANT, although HVSR generally suggests a thinner cover (<380 m). The insights gained from this passive seismic survey provide critical information for designing future exploration programmes within the tenement. The enhanced understanding of cover thickness and sub-basalt architecture will guide targeted exploration strategies to effectively evaluate the potential of the MRV beneath this challenging cover sequence.

Regolith Landform Mapping Using Support Vector Machine and Artificial Neural Network: Study Case of Sissingue Gold Project in Côte d'Ivoire

Hamed Diaby[1], Prof. Kouamé Fernand Koffi[1]
[1]Université Félix Houphouet Boigny, Abidjan, Côte d'Ivoire

The use of regolith remains poorly understood and Côte d'Ivoire has no Regolith Landform Map (RLM). Since 1999, at the Sissingué Gold project (SGP), the geochemical campaigns have come up against the problem of understanding the dispersion of the gold anomaly. For this reason, our objective is to develop an RLM of the SGP, located in northern Côte d'Ivoire, to understand the distribution of the anomaly. The approach is based on remote sensing and Geographic Information Systems (GIS). During this research, 13 data sets based on Landsat 8, Aster, Sentinel 1A images, and the ALOS PALSAR DTM were tested using two machine learning algorithms, the Support Vector Machine (SVM) and the Artificial Neural Network (ANN). The best predictions were made using the Landsat 8 image merged with the Sentinel-1A VH image. Geometric and radiometric corrections were made to achieve a level of global accuracy of 87% for ANN and 86% for SVM. However, the field validations are higher for the SVM than for the ANN. Spectral analysis was also used to identify minerals and clay such as goethite, haematite, kaolinite, sulphides and pyrite using the Matched Filtering (MF) and Mixture-Tune Matched Filtering (MTMF) algorithms. These minerals were correlated with the RLM and geological map. The RLM was developed from the SVM best prediction map by overlaying the non-genetic field map and based on interpretation. The regolith scheme used is Relict Erosional and Depositional. The various relict, erosional and depositional units were identified by assigning each subclass to a regolith unit. The geochemical results from soil and drilling were included in the GIS and overlaid with the regolith map for the creation of a geological model to decrypt the gold mineral dispersion. The prospect anomalies are residual but reveal transported units in places, which explains the lack of surface geochemical results.

Planning, Calibrating and Running a Drone Magnetic Survey on a Gold Project Côte d'Ivoire Central Part

Hamed Aly Diaby[1]
[1]Université Félix Houphouet Boigny, Abidjan, Côte d'Ivoire

The Yaouré Gold project in the central part of Côte d'Ivoire has planned and designed a drone magnetic survey to see the potential extent of gold mineralisation from the mine pit. The goal of this project is to accurately configure and record good-quality data with the drone towing magnetometer in high-vegetated and hilly areas. We conducted the first Airbird Magnetic (GEM advanced towed magnetometer U25A) survey in Côte d'Ivoire using a 10 m tow cable with success.

The survey was carried out with a DJI Matrice 300 RTK drone, flying at 75 m above ground level with 50 m line spacing. Flight stability was a key factor as the wind can affect the magnetometer rear wing thus altering the quality of data.

An average speed of 14 m/s was selected based on tests to conduct the mission for flat ground and 12m/s for hilly places. Sensor orientation was set to a roll angle of 30° counterclockwise and at a vertical angle to get a better amplitude signal toward the East-West line. For the tie line, the sensor was oriented at 30° clockwise with an angle of 45° backwards toward the nose of the magnetometer. These settings are important to get a good quality signal and a magnetic field close to the regional value. In total 945 km line was completed mostly with magnetic field amplitude above 170. Few gaps were noted and reflow later before at the end of the project. Diurnal corrections were performed using the GEM 19T base station magnetometer with 3 s recording time. The GEM magnetometer U25A model cycling time was set at 1Hz, which means a measurement was taken each 1.4m at an average speed of 14.2 m/s.

Good quality data was accomplished, and Total Magnetic field image and derivatives maps were generated.

Mineral Mapping Through a Ni-Cu Ore-Bearing Intrusion

Lily Dickson[1,2], Louise Schoneveld[2], Tobias Schlegel[2]
[1]University Of Leeds, Leeds, United Kingdom, [2]CSIRO (Mineral Resources), Kensington, Perth, Australia

Layered intrusions can host several ore deposit types, including magmatic Ni-Cu sulphide deposits. Exploration of these deposits can be difficult as they represent small targets with ore bodies that are usually confined to the base or within a small layer of the host intrusion. Subtle changes in mineral chemistries of the major silicate minerals, minor and trace minerals have the potential to track interactions with the sulphide melt. Therefore, the variability in mineral chemistries and in modal mineral abundances could potentially be used as a vectoring tool towards Ni-Cu sulphide ore horizons in layered intrusions.

The Mirabela layered intrusion, located in NE Brazil [1, 2], is the pilot case study for this research. The intrusion is host to the stratiform disseminated Ni-sulphide bearing Santa Rita orebody which is up to 180 m in thickness and lies just below the ultramafic-gabbro contact [2]. Six drillholes from Mirabela have been sampled across, and where possible, above and below the orebody and these samples were analysed petrographically and using a Tescan integrated mineralogy analyser (TIMA). The TIMA data was processed to accurately identify and quantify the modal mineralogy and individual mineral chemistries throughout 43 samples. This data was evaluated using CSIRO's XTMineral app to test for drillhole-scale mineral abundance gradients. Together with trace element data collected from pyroxene, amphibole and biotite obtained via LA-ICP-MS analysis, modal mineralogy data was used to assess the mineral zonation across the Mirabela intrusion and to investigate potential links between sulphide mineralisation and a hydrous component.

[1] da Silva et al. (1996). Congresso Brasileiro De Geolόgia, 39, 217-220.
[2] Barnes et al. (2011). Economic Geology, 106, 1083-1110.

Why a Foundational Data Layer is Critical to Unleash AI for Mineral Exploration and Mining

Mason Dykstra[1]
[1]MinersAI, Lafayette, United States

The mineral exploration and mining industries are at a pivotal moment where integrating machine learning (ML), data analytics, and artificial intelligence (AI) is becoming essential to driving new discoveries and operational efficiencies. However, the full potential of these advanced technologies remains untapped due to a fundamental barrier: the lack of a unified, standardized foundational data layer. Today, geological data is typically fragmented, heterogeneous, and scattered across numerous databases, companies, and institutions. The data frequently varies in quality, format, accessibility, and comprehensiveness, making it challenging for ML and AI algorithms to effectively identify meaningful patterns and insights. Without a robust, standardized data foundation, mining and exploration companies face significant hurdles in leveraging advanced computational tools, ultimately limiting their discovery potential and operational efficiency.

A Foundational Data Layer will:
1. Enhance Data Quality and Accessibility
By establishing global standards and protocols, ensures consistent quality, format, and interoperability of diverse datasets, significantly accelerating analytical processes.
2. Unlock Advanced Analytics and ML Capabilities
Machine learning algorithms thrive on large, high-quality datasets. With a standardized foundational layer, companies can effectively train predictive models, identify previously hidden patterns, and rapidly pinpoint prospective exploration targets.
3. Reduce Cost and Increase Efficiency
By removing repetitive data cleaning and transformation tasks, reduces the cost and time required for analyses, allowing teams to focus on higher-value tasks like interpretation and strategic decision-making.
4. Enable Collaboration and Innovation
A standardized data environment fosters collaboration across organizations and institutions, facilitating shared innovation and collective intelligence.
5. Regulatory and Sustainability Compliance
With increasing global focus on sustainability and ESG metrics, a foundational data layer simplifies compliance and transparency, helping companies proactively manage risks and adhere to regulatory requirements.
This presentation will present the reasons for, and roadmap to a foundational data layer for the industry.

Rapid Cu-Porphyry Indicator Mineral Characterization by µXRF: A Case Study Investigating Benchtop-µXRFs as a Prospective Automated Indicator Mineral Analytical Tool

Ben Eaton[1], Shaun Barker[1], Lindsey Heagy[1]
[1]University Of British Columbia, Vancouver, Canada

Developing rapid and cost-effective methods for characterizing indicator minerals is beneficial to the minerals exploration industry facing the challenge of exploring for deposits buried under sedimentary cover. Porphyry indicator minerals from stream sediment samples proximal to the Lorraine alkalic copper deposit in northern British Columbia, Canada, were characterized by scanning electron microscopy and benchtop micro-X-ray-fluorescence (µXRF) at the Mineral Deposit Research Unit, University of British Columbia. The aim of this analysis was to identify and characterize indicator minerals common to copper porphyry deposits and develop improvements to existing methods. Several different mineral sample preparation, analytical instrumentation, and mineral identification techniques were tested and compared. Scanning electron microscope characterization of heavy mineral concentrates fines (50-250 µm) proved to be an effective, but timely and costly, quantitative indicator mineral method. Indicator minerals' modal mineralogy, grain counts, grain size, grain shape, inclusions, and association were determined. Benchtop µXRF characterization of porphyry indicator minerals was found to be a promising tool to aid in exploring for porphyry copper deposits under sedimentary cover. Several key porphyry indicator minerals were readily visually identified by µXRF chemical maps (apatite, zircon, rutile, magnetite) in the prepared heavy mineral concentrates fines. Mineral identification was also rapidly derived from unquantified µXRF counts data of compositionally distinct indicator minerals using simple and interpretable unsupervised machine learning methods. Finally, random forest supervised machine learning methods were found to have the potential to assist in quantitative porphyry indicator mineral identification by µXRF, addressing difficulties in identifying minerals by µXRF due to multi-mineral mixed pixels.

Fluorite Deposits in Mongolia: Geochemical Characterization and Exploration Implications

Jamsran Erdenebayar[1,3], Mayuko Fukuyama[2], Ogata Takeyuki[3], Amarjargal Byambajav[3], Baatar Tumenbayar[4]
[1]MUST, Ulaanbaatar, Mongolia, [2]Akita University, Akita, Japan, [3]MiReSLab, Akita, Japan, [4]Researcher-consultant, Paleontological Institute of Mongolia Academy of Sicence, Mongolia, Ulaanbaatar, Mongolia

Over 600 fluorite deposits and occurrences have been identified in Mongolia, primarily in the eastern and central regions. These deposits are mostly hydrothermal in origin and can be divided into two genetic groups: (1) those related to volcanic activity, such as epithermal and carbonatite-hosted types, and (2) those associated with deeper magmatic systems, including greisen, pegmatite, and alkaline intrusive.

This study examines yttrium (Y), ytterbium (Yb), barium (Ba), and strontium (Sr) concentrations in fluorite to differentiate deposit types. We analyzed fluorite samples for these trace elements using LA-ICP-MS.

The results show significant differences in REE patterns and Sr content across the various types of deposits.

Fluorite from Dornod Sand, hosted in interlayered sedimentary rocks, shows low total REE and Sr concentrations with subdued europium (Eu) anomalies, typical of low-temperature epithermal hydrothermal systems.

Khalzan Buregtei fluorite is an accessory mineral in a syenitic alkaline intrusive complex. Trace element data indicate a magmatic to late-magmatic origin. Its low REE and Sr content suggests crystallization from evolved F-rich alkaline melts rather than shallow hydrothermal fluids.

The Mushugai deposit presents two generations of fluorite. One shows strong LREE enrichment and a weak positive Eu anomaly with a positive Sr-REE correlation, reflecting precipitation from carbonatitic melts. The second displays HREE enrichment, a pronounced negative Eu anomaly, and a negative Sr-REE trend, suggesting a secondary hydrothermal overprint from depleted fluids.

Fluorite from the Tumen Tsogt greisen system contains low total REE concentrations (<25 ppm), with weak or absent Eu anomalies. Its formation is attributed to pneumatolytic-hydrothermal fluids derived from rare-metal granitic magmas.

These results provide a robust geochemical framework for classifying fluorite deposits in Mongolia. The integration of REE patterns, Sr–REE trends, and fluid chemistry enhances genetic interpretations and guides exploration targeting both industrial-grade fluorite and REE-enriched mineralization.

Selective Calibrating Phosphatic Ore Resource Models to Mill Feedstock Head Grade Accurately Using Machine Learning

Farzi Yusufali[2], David M. First[1], Danial Hasan[1]
[1]Stratum AI, Toronto, Canada, [2]Stratum AI, Brisbane, Australia

Machine learning can create value in the mining industry by allowing multiple data types, of varying dataset size and granularity, interact with each other to calibrate a resource model prediction to mill processing measurements. The authors will introduce attention-learning neural network modelling, chosen features considered and included, and results of mill calibrating phosphate grade predictions to measured sorted feed via intermediary steps like cleaning and sorting. The authors present the results of this mill calibration technique in the Al Khabra area, an Arqah Phosporite member in northern region of the Kingdom of Saudi Arabia. The authors demonstrate how statistically significant non-linear correlated features, like segregation rate history, sand concentration, lithologies for ore and waste, seam number, etc. are used as inputs for models. Limited feature selection is crucial to ensure no overfitting. The predictions calibrated for this asset are phosphate (P2O5), silicon oxide (SiO2), and calcium (CaO). The ground truth was 10 months of monthly head grade measurements. The benchmark to improve against was the Kriging model's grade predictions for the aforementioned minerals. The results are the following: 1) for P2O5, there was 39% less monthly deviation, 2) for SiO2, there was 64% less monthly deviation, and 3) for CaO, there was 36% less monthly reduction. The results demonstrate that machine learning-based calibration techniques, especially non-linear methods, are more accurate in predicting mill outputs over the existing methods (where multiple intermediary steps introduce error). Suggested improvements for mill calibration include visual logging and improving sampling techniques at the crusher level.

Thermodynamic Modelling and Trace Element Partitioning as Predictive Tools for REE-Enrichment in Alkaline-Silicate Igneous Systems

Corinne Frigo[1], Owen Weller[1], Caroline Soderman[1], Charles Beard[2], Teresa Ubide[3]
[1]Department of Earth Sciences, University Of Cambridge, United Kingdom, [2]Department of Earth Sciences, Utrecht University, Netherlands, [3]School of the Environment, University of Queensland, Australia

The increasing demand for rare earth elements (REEs) in green technologies calls for more effective strategies to identify environments where they may have been concentrated. Alkaline-silicate igneous systems (and associated carbonatites) are key hosts of REE deposits, yet many contain no economic mineralisation. This study evaluates the potential of coupling thermodynamic phase equilibria modelling with empirical trace element partitioning equations to predict optimal conditions for residual enrichment of the REE during the crystallisation of alkaline-silicate systems.

We test this approach using a case study of the Erongo Complex in Namibia, focusing on a suite of late intrusive plugs of basanite to foidite composition [1]. Using the bulk composition of the most primitive basanite as a starting melt, we simulate batch and fractional crystallisation at varying pressure-temperature conditions with the software MAGEMin [2] and a new thermodynamic dataset for alkaline-silicate systems [3]. These simulations predict melt and mineral equilibria, modal abundances, and compositions, which we test against observations from natural samples. Empirical partitioning equations [e.g., 4–6] are used to estimate REE concentrations in melt and coexisting mineral phases. These results are compared with whole-rock REE data and LA-ICP-MS maps of zoned clinopyroxene grains, known as effective recorders of melt evolution [4]. By confirming that major and trace element trends observed in nature can be reproduced, the thermodynamic models can be used to forward model the 'sweet spot' for optimal REE enrichment [7].

References:

[1] Trumbull RB et al. (2003) J. Petrology 44
[2] Riel N et al. (2022) Geochem. Geophys. Geosyst. 23
[3] Weller O et al. (2024) J. Petrology 65
[4] Beard C et al. (2019) J. Petrology 60
[5] Baudouin C et al. (2020) Contrib. Min. & Pet. 175
[6] Salazar-Naranjo AF & Vlach SRF (2025) Chem. Geo. 677
[7] Soderman et al. (2025) Nature Geoscience, doi:10.1038/s41561-025-01695-3

Smarter Use of Geochemistry and Machine Learning to Support Geological Domaining and Geometallurgical Modelling of Porphyry Cu Deposits

Melissa Gregory[1], Tracie Burrows[1], Paul Greenhill[1], Luis Torres[1], Kathy Zunica[1], Ian Lipton[1]
[1]AMC Consultants, Melbourne, Australia

Consistent definition of geological domains is essential for geological modelling. The traditional methods for geological model construction are typically reliant upon subjective visual geological logging or a limited geochemical assay suite. This results in models that poorly represent the true characteristics of the rocks.

Whole rock multi-element geochemistry provides a fingerprint of the elemental composition of drillhole samples that is objective and amenable to powerful data science techniques. Applying dimension reduction and grouping analysis to process high-dimensional data can generate practical solutions for domaining and guide robust sample selection for orebody knowledge studies, for example, metallurgical, geotechnical or environmental.

Grouping materials with similar characteristics using these techniques results in better geological interpretations and improved understanding of geological processes and controls on mineralization. Groups of samples with like geochemical characteristics are defined using a combination of grouping algorithms and geological information and these groups are then interpreted spatially in two and three dimensions. Since they are based on high-quality assay data, these groups provide a higher level of consistency and repeatability than visual logs alone.

To convert a geological model into a geometallurgical model requires the integration of the drillhole database with metallurgical test work. Selecting representative metallurgical samples is critical for establishing relationships between the composition of the rocks and mineral processing response, such as metal recovery and concentrate grade. Machine learning techniques facilitate modelling of predictive relationships between metallurgical test results and the geochemical characteristics of the ore. This enables the results of a limited number of metallurgical tests to be leveraged against the larger drillhole database to provide predictions of ore processing response that can be deployed into every block in a resource block model.

We will present examples of this approach to improve interpretation, support sample selection and build geometallurgical models in porphyry Cu deposits.

Ore Vectoring Using Machine Learning: Case Study of Copper Occurrences in Northern Ireland

Felix Sihombing[1,2], Laurence Robb[1,4], Hannah Hughes[3], Richard Palin[1]
[1]University of Oxford, Oxford, United Kingdom, [2]Universitas Indonesia, Depok, Indonesia, [3]Camborne School of Mines, University of Exeter, Penryn, United Kingdom, [4]University of the Witwatersrand, Johannesburg, South Africa

Machine learning can be implemented in mineral exploration mapping activities, such as for mineral prospectivity mapping (MPM) and geochemical anomaly detection. The approach can also be integrated into other targeting frameworks such as mineral exploration targeting (MET) and exploration information system (EIS). In this study, we present a new approach for applying machine learning in mineral exploration by using it for ore vectoring. In this approach, machine learning is tasked with predicting numerical distance from proximal to distal zones, which could help indicate the center of mineralization. We demonstrate this method using case study of copper occurrence in Northern Ireland. The new and emerging techniques of graph deep learning was employed for this approach. Additionally, we employ random forest (RF) which has been more commonly used in MPM study. Our study demonstrates that graph deep learning techniques can properly generate vectoring maps aligned well with the geological setting of the deployment region. In contrast RF produces map that poorly reflect geological condition and more difficult to interpret for guiding the exploration. Additionally, one prospective zone mapped by graph deep learning can be confirmed by a newer data from the deployment region. This 'virtual' discovery highlights the potential for combining ore vectoring with graph deep learning can be useful for future exploration efforts.

Anomaly Detection Using Variogram and Machine Learning Can Reveal Potential Orefields: Case Studies from the United Kingdom and Australia

Felix Sihombing[1,2], Laurence Robb[1,4], Hannah Hughes[3], Richard Palin[1]
[1]University Of Oxford, Oxford, United Kingdom, [2]Universitas Indonesia, Depok, Indonesia,
[3]Camborne School of Mines, University of Exeter, Penryn, Cornwall, United Kingdom,
[4]University of the Witwatersrand, Johannesburg, South Africa

The use of machine learning for mineral exploration has increased in recent years, with most applications focusing on supervised learning approaches for mineral prospectivity mapping; however, unsupervised learning approaches for anomaly detection also have significant potential and deserves further investigation. In this study we integrate variogram modelling as a preprocessing tool for machine learning-based anomaly detection. The method is demonstrated through a case study on detecting large orefields in the using national-scale geochemistry survey from United Kingdom (UK) and Australia. Multicomponent geochemical stream sediment data comprising major and minor oxides, and a variety of base and strategic metals were used to produce variograms. These variogram data was then input into an unsupervised learning method to produce data clusters which were compared with known orefields. Our results from UK case study demonstrate the potential to detect geochemical anomalies that may be spatially related to orefields, whether exposed or hidden. Additionally, the approach can identify similarities between such anomalies, which can aid mineral exploration by enabling comparisons between orefields and previously underexplored areas. One particular observation is that the method highlights several anomalous regions in Scotland that are not currently considered orefields, indicating a larger untapped potential in the northern part of UK, compared to the south However, the method could not be replicated successfully using the Australian dataset, likely due to lower data density and other contributing factors. This highlights the need either for a new geochemical mapping campaign with denser sampling comparable to that of the UK, or for the compilation of levelized, publicly available geochemical datasets. Nevertheless, we envisage that this method can be applied successfully in other countries that have the required data density, and that it may also be suitable for project-scale exploration, where even higher data density is available.

Spectral and Compositional Characterization of Lithium-Bearing Zinnwaldite Mineralization in Eastern Tasmania

Wei Hong[1,2], David R Cooke[1,2], Lejun Zhang[1,2], Ivan Belousov[1,2], Russell Fulton[3], Jeffrey Oalmann[1]
[1]Centre for Ore Deposit and Earth Sciences (CODES), University of Tasmania, Hobart, Australia, [2]Regional Research Collaboration (RRC) Program for Environmentally Sustainable Production of Critical Metals, University of Tasmania, Hobart, Australia, [3]TinOne Resources Inc., Hobart, Australia

Lithium has many industrial applications, but as approximately 75% of its demand is for rechargeable batteries, it has become a major exploration target over the past decade. Large Paleozoic granite batholiths in eastern Tasmania intruded into a north-trending magmatic arc parallel to the active continental margin of eastern Australia. Some of the highly fractionated post-collisional granitoids potentially host critical mineral resources, including tin, tungsten, lead, zinc, silver, indium, and/or bismuth. Recently, high grade Li ores (up to 1.2 wt.% Li2O) have been discovered to associate with greisen alteration in eastern Tasmania. These mica-rich greisens are related to leucogranitic intrusions (~375 Ma) emplaced in a post-collisional extensional setting and the mineral potential of such greisens is yet to be constrained. We have characterized the spectral and chemical compositions of Li-rich minerals from altered granites at the Dead Pig prospect, to determine the genesis of greisen-hosted Li mineralization and support the exploration for this new type of Li resource. Petrographic, SEM and microprobe analyses reveal that Li-bearing mineral in the greisens is mainly zinnwaldite, intergrown with an alteration assemblage of topaz + albite + K-feldspar + quartz overprinted by fluorite ± kaolinite ± dickite. LA-ICP-MS analyses show that the zinnwaldite has an average of 1.32 wt.% Li, 12.1 wt.% Fe, 7,400 ppm Mn, 6,900 ppm Rb, 280 ppm Cs, 100 ppm Nb, and 80 ppm Sn. The zinnwaldite is characterized by extremely high Rb/Sr and low Nb/Ta ratios, with F averaged at 7.6 wt.%. Although easily misrecognized as biotite due to its black color and other similar properties, zinnwaldite shows distinctive spectral (short-wave and thermal infrared) reflectance positions that distinguish it from other micas. Our results provide insights into the origin of greisen-hosted Li mineralization and contribute to the development of a new spectral tool to aid Li exploration.

Utilising In-situ Lu-Hf Dating to Constrain Timing of Mineralisation in the Mount Isa Inlier: A Case Study from Maronan.

Helena Hordern[1], David Murphy[1], John Caulfield[1], Elena Belousova[2]
[1]Queensland University of Technology, Brisbane, Australia, [2]Geological Survey of Queensland, Brisbane, Australia

The Mount Isa Inlier in northwest Queensland hosts a diverse range of mineral deposit types including Pb-Zn Broken Hill-type deposits (e.g. Maronan), Iron Oxide-Copper-Gold (IOCG) deposits (e.g. Mount Dore and Mount Colin), and sediment-hosted copper deposits (e.g. Mount Oxide). Conventional geochronological techniques such as Re-Os on molybdenite, U-Pb on apatite and Ar-Ar on mica are commonly used and well-established. However, these systems can be reset by later thermal overprinting events, complicating interpretations of the timing of mineralisation.

This project seeks to further develop the Lu-Hf dating techniques as a key method of dating apatite and garnet associated with mineralisation across the Mount Isa region. Crucially, the Lu-Hf system is less susceptible to thermal resetting. Using this method, we aim to refine age constraints of ore-forming processes, including in overprinted terranes.

The initial phase of the project focuses on the Maronan deposit, a Broken Hill-type deposit in the Inlier. Core samples sourced from the Geological Survey of Queensland were used to produce thin sections for detailed petrographic imaging and geochemical analysis. Techniques including optical microscopy, Raman spectroscopy, EPMA, cold-CL, TIMA phase, BSE and CL imaging and trace element analysis through LA-ICP-MS are employed to characterise apatite and garnet and assess their relationship to mineralisation. Samples showing the strongest textural and geochemical association with mineralisation events will be selected for Lu-Hf geochronology.

Future work will apply this workflow to remaining deposits. Our aim is to assess the Lu-Hf method as a versatile tool for dating mineralisation across a range of deposit types. Ultimately, this research will enhance our understanding of mineralisation processes in the Mount Isa Inlier, with the potential to contribute to global exploration. If garnet and apatite show to possess unique geochemical fingerprints related to ore formation, they may serve as valuable mineralisation vectors in the search for critical minerals.

Revealing High Geological Spatial Variability at Bench Scale – Implications Upstream and Downstream in the Mining Value Chain

John Jackson[1], Fred Blaine[2], Emmanuel Schnetzler[3]
[1]Imdex Ltd, Brisbane, Australia, [2]Imdex Ltd, Perth, Australia, [3]Imdex Ltd, San Francisco, USA

The development and deployment of BLASTDOG—a semi-autonomous system for acquiring high-resolution, real-time subsurface data from blast holes or grade control holes in open-cut mining—has revealed significantly greater geological variability than previously expected across a range of “bulk” metalliferous deposits, including coal, iron ore, and porphyry copper deposits.

The BLASTDOG probe integrates three primary geophysical sensors—total count natural gamma, magnetic susceptibility, and inductive conductivity—to infer rock type, including lithology and alteration. An eight-arm caliper is also used to identify rock mass defects such as bedding planes, fractures, and joints. Using examples from these deposit types, the spatial variability is revealed through geostatistical variography and with 3D modelling and visualisation. In nearly all deployments to date, this approach has revealed a higher variability of the geology than expected by the site geologists.

We explore the implications of this variability across the mining value chain. Upstream, it affects data acquisition, the transformation of data into actionable geological information, and its integration into mine and processing design during project development, including feedback loops into grade control. Downstream, the observed variability influences operational performance in drilling and blasting, geotechnical assessments, and mineral processing.

Innovative Remote Sensing and GIS-Based Approach for the Identification and Mineralogical Characterization of Mine Tailings in South Africa

Gaone Joseph[1], Linda Iaccheri[2], Stephanie Enslin[3]
[1]University Of Witwatersrand, Johannesburg, South Africa, [2]University of Witwatersrand, Johannesburg, South Africa, [3]University of Witwatersrand, Johannesburg, South Africa

Mine tailings are widely distributed across Southern Africa, reflecting the region's rich mining history. Recently, a comprehensive geospatial inventory detailing the precise location and extent of tailings in South Africa was developed by the University of Cape Town. However, this database excluded some abandoned tailings with no prior history. These older and abandoned tailings are viewed as potential sources of critical minerals and environmental liabilities. Tailings, therefore, demand renewed attention through modern, scalable technologies.

This study presents a technology-driven, GIS- and remote sensing-based workflow for the identification and mineralogical characterization of mine tailings across South Africa. Using Google Earth Pro for initial visual interpretation based on morphological features such as shape and surface discoloration, over 50 abandoned tailings sites were mapped, with particular emphasis on the Witwatersrand Basin and the Barberton Greenstone Belt.

To characterize the tailings, multispectral satellite data from ASTER and Sentinel-2 were processed in ArcMap using band ratios and other spectral enhancement techniques to extract spectral signatures linked to mineralogical variation. A subset of tailings from the West Rand Goldfields and Fairview Gold Mine underwent further validation through laboratory techniques, including X-ray Diffraction (XRD), the Tescan Integrated Mineral Analyser (TIMA), and petrography. This bridged remote characterization with micro-scale mineralogical analysis.

By integrating open-source satellite data, advanced GIS tools, and analytical mineralogy, this research demonstrates how innovative geoscience technologies can be applied at multiple spatial scales to address both environmental risks and resource recovery potential. Preliminary findings suggest that spectral patterns from known tailings, grounded in local geology, can be used to classify unknown or abandoned tailings—especially those with limited or no historical documentation. This adds the older and abandoned tailings to the already developed national tailings inventory and promotes a data-informed approach to reprocessing, environmental management, and future exploration in legacy mining landscapes

Capabilities of Advanced Mineral Identification and Characterisation System (AMICS)

Andrew Kostryzhev[1]
[1]University Of Queensland, Australia

Advanced Mineral Identification and Characterisation System (AMICS) is based on energy dispersive X-ray spectroscopy utilised in a scanning electron microscope (SEM-EDS). A specially designed software coordinates the sample positioning inside the microscope chamber, EDS spectra acquisition, database management, phase identification, and results visualisation in the form of maps, graphs and tables ready for reporting. Characterisation can be conducted automatically for up to 28 mounted samples with or without coating. Although, many options exist for an experience researcher to influence the analysis process: (i) extract a spectrum for each acquisition point and determine chemical composition at this point, (ii) visualise mineral maps for each phase or a group of phases to characterise their interactions, (iii) determine the particle geometry-chemistry relationship, (iv) conduct statistical analysis of EDS spectra, define presence of solute elements, and visualise distribution of particles with a particular chemical composition, (v) manually develop the spectra data-base for unknown phases and solid solutions. All these approaches can be used to fully analyse the sample chemistry or facilitate the following study of the sample with other analytical techniques, such as XRF, EPMA, ICPMS, or SIMS. The combination of modern instrumentation with dedicated software allows analysis of large area (tens of mm2) at high resolution (particle sizes down to 1 micron). Characterisation of fine mineral bodies is critical in studies of pre-processed materials such as mine tailings, fine grained waste, or clays. Therefore, application of AMICS is growing in geology, mining, and mineral processing. This presentation will outline the system capabilities and showcase recent results from a number of projects, in particular: (i) identification of commercially valuable metallic elements in mine tailings, clays, and electronic waste; (ii) mineral phase balance in rocks, soils and dust, (iii) characterisation of laboratory produced oxide nano-particles and pyrometallurgical slags.

Novel Applications of Time of Flight (TOF) Mass Spectrometry to Compositional Mapping

Alice Macdonald[1], Jeffrey Oalmann[1], Jack Ward[1], Carlos Carrasco Godoy[1], Ivan Belousov[1], Rebecca Carey[1]
[1]Centre for Ore Deposites and Earth Sciences (CODES), University Of Tasmania, Hobart, Australia

Laser ablation inductively coupled mass spectrometry (LA-ICP-MS) is increasingly becoming an integral tool for mineral exploration. Popular applications include using mineral chemistry to assess fertility and as a tool for vectoring to mineral deposits; along with compositional maps, where the spatial distributions of elements can provide constraints on mineral growth histories and metal deportment. Although widespread, quadrupole instruments have many limitations, particularly for compositional mapping where analytical times can be lengthy and measured isotope choice is limited. Coupling a laser ablation system with ultra-fast washout to a time of flight (TOF) mass spectrometer makes it possible to measure most of the periodic table (Na – U) quasi-simultaneously from a single laser pulse allowing for high resolution maps of large areas to be produced substantially faster than with a quadrupole mass spectrometer. Here, we showcase some applications of compositional maps obtained with LA-ICP-TOF-MS, including a new method to map and quantify groundmass compositions, applicable to a range of sample types. These groundmass compositions can be used as proxies for melt composition, providing constraints on magma sources and processes. Results from preliminary testing of this method, comparison groundmass compositions obtained by LA-ICP-TOF-MS mapping to bulk rock compositions obtained by XRF of unaltered volcanic rocks from New Zealand and Iceland will be presented. Our approach highlights the strength of using LA-ICP-TOF-MS to rapidly collect high-resolution compositional data and indicates where improvements to quantifying these data can be made.

Getting the Most out of Drill Hole Data: Meaningful and Robust Boundary Detection Tools

Ryan Manton[1], Gerhard Visser[1], Lequn Zhang[1], Rebecca Montsion[1]
[1]Csiro, Perth, Australia

A diverse range of geoscientific data is collected to represent geological features and processes relevant to deposit formation. However, tools for identifying features in drill hole/core data are limited due to complex data relationships and geologist confidence in machine learning methods. Data Mosaic, a multiscale boundary detection tool from the CSIRO has been re-designed to identify tricky geologic features amidst multidisciplinary data noise. Various use cases are being tested to assess the suitability for different geological questions and data types. Boundaries like lithologic contacts, alteration zones, ore grade domains, and metallurgical domains have unique statistical requirements based on their context. Data types such as hyperspectral, geochemical, petrophysical, and electrical scanning technologies have inherent biases and appropriateness based on analytical methods and correlations. Using complex systems mathematics, Data Mosaic integrates information across scales, providing quantitative support for decision-making and higher confidence constraints for modelling workflows.

Systematics and Mineral Systems: How Controlled Vocabularies Enable System Mapping for Sedimentary-Hosted Copper Systems

Helen McFarlane[1,2], Mark Lindsay[1,3,4]
[1]CSIRO, Kensington, Australia, [2]Mineral Exploration Cooperative Research Centre (MinEx CRC), Bentley, Australia, [3]School of Earth Science, The University of Western Australia, Crawley, Australia, [4]ARC ITTC for Data Analytics for Resources and Environment, Crawley, Australia

The mineral systems approach is widely applied in mineral exploration as a conceptual framework capturing the geological processes controlling mineral formation. Such models have been developed based on both empirical data and geological observations to define the essential components and processes during metallogenesis, including sources, pathways and architecture, drivers, ore deposition and preservation. The components are described in natural language and are variably translated into, or integrated with, geoscience datasets (maps, experimental data, geophysical datasets) and mappable proxies to generate predictions of mineral prospectivity. The challenge is to find internal consistency between concepts and data to test plausibility and identify uncertainties. This study applies a controlled vocabulary of consistently defined mineral systems concepts. Ore-forming processes and their relationships are described using standardised terms that enhance interoperability, analysis and systems understanding. Such an approach takes us toward the translation of mineral systems models into machine-readable formats, allowing conceptual validity tests of these models.

This study investigates the application of the mineral systems critical components controlled vocabulary in sedimentary-hosted copper systems, establishing knowledge graphs and system dynamics modelling to understand concepts, entities and relations. In the case of classic sedimentary-hosted copper systems, the understanding of these critical components is predominantly derived from world-renowned deposits, including the Kupferschiefer (Europe) and Katangan (Central African Copperbelt) deposits. Using this holistic approach to visualise and map the components relative to system dynamics and associated empirical data, we interrogate the plausibility of such conceptual models in Australian sedimentary-hosted copper systems and assess the intersection and overprinting of mineralising events in terranes that have often witnessed multiple episodes and styles of mineralisation during their geological evolution.

Identifying Mineralisation Within the Sunrise Dam Gold Mine Using Geomechanical Targeting Technology

John Mclellan[1,2], Christine Shore[3], Perkins David[3]
[1]Director, GMEX (Geological Modelling for Exploration), Townsville, Australia, [2]Senior Research Fellow, Adjunct, James Cook University, Townville, Australia, [3]AngloGold Ashanti Australia, Perth, Australia

Drill targeting within mine sequences can be difficult, particularly when mineralisation is located in narrow structural corridors, shear-zones, fracture systems and/or veins. Structural analysis and interpretations can be very useful in aiding the understanding of the geology, building a three-dimensional model and its deformational history. However, even with this knowledge targeting is always difficult. Understanding how deformation events have affected the rock units and consequent structure development, is key to highlighting structural targets for investigation.

Here, we use the currently accepted geological model for the Sunrise Dam Gold Mine (SDGM) in Western Australia, to develop a three-dimensional mesh representing the main architecture of the deposit. A 3D hexahedral model was developed and subjected to boundary conditions to simulate the main D4 deformation event (NE compression), most likely to have been responsible for the high grade steeply orientated mineralisation domains. Initial models were run for sensitivity analysis, and the known positions of ore lodes were used as validation for several input parameters. The parameter validation strongly suggested that mineralisation was a) closely related to areas of increased shear strain in the adjacent shear zones, and b) failure state (both shear and tensile) within the volcaniclastic packages and more competent intrusives, which often displayed volumes of increased pore fluid pressure.

The geomechanical modelling of the system has highlighted several key target areas that localise shear strain, increased pore fluid pressures and tensile failure. Shear strain localisation in and around the more competent intrusive rocks appear to be important, even from early stages of the modelling. The structural architecture accommodates most of the movement within the system, however, this has accommodated partitioning of strain into lithological units such as the volcaniclastic units and intrusive bodies, where there is a good correlation with areas of tensile failure, high pore pressures and known ore bodies.

Meso-Scale Imaging and Whole-Rock Quantification of Indium in Mine Waste from a Granite-Related Deposit Using Laser-Induced Breakdown Spectroscopy (LIBS)

Olivia Mejías, Alejandro Fayad[2,4], Melissa Narbey[2,4], Martin Valenzuela[3], Anita Parbhakar-Fox[1,4]
[1]SMI-BRC, The University of Queensland, Brisbane, Australia, [2]AXT Pty Ltd, Perth, Australia, [3]Huelva University, Huelva, Spain, [4]ARC Training Centre in Critical Resources for the Future, Brisbane, Australia

Laser-Induced Breakdown Spectroscopy (LIBS) is a rapid, in-situ, semi-quantitative analytical technique based on the spectroscopic analysis of atomic emissions from laser-induced plasmas. In this study, LIBS was applied for meso-scale geochemical mapping using the ECORE benchtop system on centimetre-scale slabs of indium (In)-rich waste rock from the Baal Gammon mine, a granite-related deposit in Australia. Selected samples were further used to develop LIBS-based calibration models for quantifying indium in whole-rock pressed powder discs, using both benchtop and handheld LIBS instruments.

A range of elements were detected alongside indium, including those relevant to prospectivity (e.g., copper, tin, silver) and environmental monitoring (e.g., arsenic). Elemental imaging of the ~451 nm indium emission line revealed detailed spatial patterns of indium enrichment, including high intensities along chalcopyrite grain boundaries, within lamellar textures, at mineral interfaces—particularly between sulphide–sulphide and magnetite–sulphide contacts—and as micro-inclusions. In terms of relative abundance, LIBS signal intensities were highest in sphalerite, followed by stannite and chalcopyrite, with pyrrhotite showing the lowest response. These trends were consistent with mineral-specific indium concentrations obtained through correlative LA-ICP-MS analysis.

LIBS-based indium calibration models demonstrated high predictive accuracy for whole-rock sulphidic matrices. For the benchtop system, the models achieved an R^2 of 0.99, a root mean squared error of prediction (RMSE-P) of 73 ppm In, and a limit of detection (LOD) of 4 ppm In. The handheld prototype yielded an R^2 of 0.98, an RMSE-P of 102 ppm In, and a LOD of 28 ppm In. Optimal performance of the handheld instrument was achieved by restricting data processing to high-stability regions, and avoiding pre-ablated areas. Baseline correction was essential, whereas smoothing algorithms proved ineffective for trace-level elements such as indium.

This study highlights LIBS as a promising, high-throughput, and spatially resolved technique for rapid geochemical imaging and quantification of critical elements in complex mine waste matrices.

Enhancing Usability of Public Data: 3D for Free

Gem Midgley[1], Matthew Greenwood[2], Craig Annison[3], James Reid[1], Stephanie Howe[1]
[1]Mira Geoscience Asia Pacific, Perth, Australia, [2]Geological Survey of Queensland, Brisbane, Australia, [3]Xcalibur Smart Mapping, Perth, Australia

Pre-competitive data are an immensely valuable and often under-utilised resource in Australia. A variety of initiatives have made wireline petrophysics, drillhole assays, hyperspectral data and a range of surface and airborne geophysical data available, coupled with value-added products including geological interpretations, geophysical inversions, and multidisciplinary data compilations. All are downloadable from online portal systems, without cost, however some improvements could be made to simplify accessibility, visualisation and utilisation.

This work takes the recently-released AusAEM data for NE Queensland from the GSQ data portal and compiles it into a Geoscience ANALYST workspace; making it more accessible by removing the need for expensive proprietary software whilst increasing the usability of data and interpretations by accurately representing them in 3D space.

The workflow involves 'drag and drop' importation of the AusAEM TEMPEST airborne electromagnetic data with direct import of grids (including coordinate system metadata attribution). The Geoscience ANALYST import functionality allows creation of both line objects and draped meshes from the database deliverables. Geoscience Australia layered earth inversion conductivity model outputs require minor editing before import; then they can be visualised as 3D draped sections conforming to flight paths and the data displayed with different colour stretches. Acquisition and processing reports are attached to data objects to ensure metadata is retained.

In only a few steps, free data is made freely accessible in free 3D software, and the GEOH5 open format ensures it will continue to be available to future users. Multidisciplinary applications include:

- Geological Mapping of basement lithology and structure, and regolith thickness and character
- Mineral Exploration via mapping and quantification of discrete conductors, alteration zones, and host and stratigraphic lithologies
- Hydrogeological Investigations of groundwater resources

Additional value can be gained from further data integration through simple import of other GEOH5 datasets available from the GSQ data portal into Geoscience ANALYST.

Automated Quantitative Mineralogy Applied to the Hydrothermal Footprint of the Paulo Afonso Cu Deposit, Carajás Mineral Province

Victor Nogueira[1], Adalene Silva[1], Catarina Toledo[1], Guilherme Teixeira De Siqueira e Silva[2,1], Victor Mota e Nogueira[3], Érico Zacchi[1]
[1]University Of Brasilia, Brasília, Brazil, [2]Vale S.A., Brazil, [3]AXT PTY Ldta, Australia

In the context of modern mineral exploration, automated quantitative mineralogy is emerging as a powerful tool to enhance data quality and mineral system understanding—an increasingly crucial factor as exploration strategies shift towards data-rich, model-driven approaches. To characterize the hydrothermal footprint of the Paulo Afonso Cu-Au deposit, a total of 46 samples were investigated through petrographic analysis. A selected subset of 12 samples was further analyzed using the Elemission ECORE (a high-speed LIBS scanner), with results validated using the TESCAN TIMA (SEM-EDS). Automated quantitative mineralogy was applied to process the data obtained from ECORE and TIMA, using supervised machine learning algorithms for mineral classification, resulting in detailed mineralogical data for each sample. The Carajás Mineral Province, located in the southeastern portion of the Amazon Craton, Brazil, hosts several IOCG-type mineralizations. Among them, the Paulo Afonso deposit stands out as a Cu-Au deposit located approximately 16 km from Salobo—a world-class copper deposit—and is associated with the Cinzento Shear Zone. At Paulo Afonso, the host rocks are metamorphosed granites that have been intensely altered by hydrothermal processes. A distal sodic alteration zone was identified, mainly characterized by albitization, overlain by a potassic zone dominated by biotite formation. In proximal zones, a chloritic halos were observed. Calcic alteration becomes more prominent in the mineralized zone, with amphiboles forming a Ca-Fe (actinolite + magnetite) envelope around the mineralization, which occurs as disseminated chalcopyrite, in veins and hydrothermal breccias. The results obtained using the ECORE scanner demonstrate the effectiveness of the method as a rapid, preparation-free tool with high efficiency in the chemical and textural characterization of hydrothermal alteration associated with Cu-Au mineralization in the Paulo Afonso deposit. These findings highlight the promising application of automated quantitative mineralogy for defining mineralogical footprints, offering significant potential for mineral exploration strategies.

Refining the understanding of the hydrothermal systems footprint using automated quantitative mineralogy in the Carajás Mineral Province.

Victor Mota e Nogueira[1], Adalene Moreira Silva[2], Catarina L. B. Toledo[2], Moriá Caroline de Araújo[2], Luana Machado Campos[2,3], Juliana de Araújo Carvalho[2,4], Guilherme Teixeira[2,4], Juliana Natal Berbert[2], Tiago Pedrosa Lyra[2], Érico Natal Pedro Zacchi[2], Daniel Shkromada de Oliveira[2], Lucas Nunes Moraes[2], Ana Luiza Chaves[2], Miguel Oliveira da Costa[2]
[1]Automated Mineralogy Incubator (AMI), Perth, Australia, [2]Institute of Geosciences - University of Brasilia, Brasilia, Brazil, [3]Anglo American Ltda., [4]VALE S/A, Parauapebas, Brazil

To better utilise petrophysical data for defining the hydrothermal footprint, mineralogy should be used as a common language to explain the distinct physicochemical properties involved in targeting mineralised systems. Methods that employ new technologies for large-scale geological material characterisation are essential for defining the mineral footprint. The Carajás Mineral Province, located in the southeastern sector of the Amazonian Craton, represents the largest and best-preserved Archean sector of the craton and known for hosting several IOCG type deposits. Despite the geographical proximity, each deposit has distinctive petrophysical, geochemical and geophysical characteristics.

We propose a new workflow that integrates automated quantitative mineralogy on 100 samples, analysed using the Elemission ECORE high speed LIBS scanner and the TESCAN TIMA SEM-EDS system, together with petrophysical and geochemical data from the Salobo, Paulo Afonso and Santa Inês Gabbro targets. Samples representing different lithotypes and distinct petrophysical characteristics were primarily selected for laser-induced breakdown spectroscopy (LIBS) scanning using the ECORE equipment. For validating the LIBS classification, key samples were selected for a high-resolution analysis using TIMA. Mineral abundance data, mineral and elemental maps, mineral association and grain size were generated for each sample.

For each target, representative samples of the host rock, distal and proximal parts of the hydrothermal alteration envelope, and the mineralised alteration zone were analysed to investigate variations in rock physical properties based on contrasts observed in the petrophysical data. The mineral footprint emerged as a critical factor in characterising hydrothermal system alterations. This highlights the value of combining ECORE and TIMA, an efficient workflow for mineral characterisation that can be scaled to enhance understanding of the hydrothermal footprint and improve confidence in exploration surveys.

Mineralogical and Textural Characterization of the Ag-Pb-Cu-Au Maronan Deposit (NW-QLD) Using Laser-Induced Breakdown Spectroscopy (LIBS)

Daniel Patias[1,2], Nathan Fox[1], Jaime Poblete[1,3], Andrew Barker[3], Zofia Swierczek[2]
[1]Sustainable Mineral Institute - The University of Queensland, Brisbane, Australia, [2]AXT Pty Ltd, Perth, Australia, [3]Maronan Metals Limited, Sydney, Australia

Laser-induced breakdown spectroscopy (LIBS) provides fast (40 seconds per cm^2), high-resolution chemical mapping at a microscopic scale (50 μm pixel resolution). It gives detailed insights into mineral distribution and textures, helping to better understand ore bodies and make informed decisions in the mining industry.

In this study, LIBS was used to analyze drill core specimens from the Proterozoic sediment-hosted Maronan deposit, with a JORC 2012 total resource of about 118 Moz Ag, 2 Mt of Pb, 272 000 t of Cu and 0.76 Moz of Au (ASX: MMA March 12th, 2024). The aim was to improve the understanding of rock type variability, alteration styles, cross-cutting relationships and their association with mineralization, supporting exploration and future metallurgical decisions. A total of 30 drill core specimens were analyzed, with additional validation performed through SEM-TIMA scanning. Four distinct mineralogical and alteration domains were identified: (1) a Mn-rich domain exhibiting skarn-like characteristics, with variable amounts of magnetite-pyrrhotite, carbonate minerals, and quartz, showing notable Ag and Pb mineralization with occasional minor Cu; (2) a carbonate-rich domain containing Mn-rich minerals such as rhodonite and Ca-Mn garnet, frequently associated with Pb (galena) and Ag mineralization (sulfosalts); (3) a K-Mg-Na-rich domain associated with Cu mineralization, containing albite, muscovite, biotite, chlorite, amphibole, and pyrite; and (4) a silica-rich domain dominated by quartz, typically barren.

LIBS identified minerals in fine-grained rocks, revealing Cu and Pb mineralization linked to specific alteration styles and rock types. Some Cu-rich specimens show early sodic alteration cross-cut by late K-Mg-Ca phases. Lead-enriched samples are usually associated with carbonate ± apatite ± fluorite, overprinting Mn-rich skarn minerals. Qualitative chemical and semi-quantitative mineral maps improved geological interpretation, matched assays data, and provided additional insights compared to traditional methods such as geochemical assays, XRF scanning, or SEM-based mineralogy, highlighting LIBS' potential for ore body knowledge studies.

Advanced Hydrogeological Modeling and Sensitivity Analysis to Optimize Dewatering and Characterize Water Flow at the Mingomba Cu Deposit

Gloria Quispe Oruro[1]
[1]Stanford University, Stanford, United States

This research applies advanced 3D geological modeling and hydrogeological simulation techniques to support dewatering strategy design during shaft development at the Mingomba Cu deposit on the Copperbelt (explored and developed by KoBold Metals). Leveraging Petrel, a software widely used in the oil and gas sector, the study demonstrates how cross-disciplinary technologies can be adapted to mineral development, enabling high-resolution characterization of groundwater flow in structurally complex and fracture-dominated environments. Stratigraphic and structural datasets from geotechnical boreholes are integrated into a dynamic simulation framework, where hydraulic properties are assigned using the Kozeny-Carman relationship, and Darcy's Law governs transient flow across heterogeneous stratigraphic units.

Innovative computational workflows underpin the project's methodology: Monte Carlo simulations are employed to quantify uncertainty in key parameters such as permeability, porosity, and pressure response. Additionally, a sensitivity analysis identifies which variables exert the most influence on model outcomes, providing actionable insights to prioritize future data collection and optimize the value of new measurements. This targeted approach enhances the robustness and predictive power of the simulation, especially in scenarios of operational uncertainty.

Model calibration draws upon geostatistical methods and historical analogs from the adjacent Konkola mine, where previous dewatering campaigns led to increased near-surface permeability due to subsidence. Scenario-based planning supported by the model enables just-in-time dewatering localized at active mining fronts, helping to minimize risk and improve development efficiency.

This work exemplifies how technological innovation, including the repurposing of petroleum industry tools and probabilistic analysis, significantly enhances decision-making in mineral development. In particular, this modeling framework contributes to the efficient and responsible extraction of critical minerals like copper to enable the global energy transition. It reflects the broader trend of integrating computational methods across scales to transform how we discover, develop, and manage next-generation mineral systems.

Data-Driven Geometallurgy at Tujuh Bukit: Integrating Hyperspectral Core Imaging, SEM/MLA, Dynamic Rebound Hardness, and Machine Learning for Processing Design Optimisation

Mochammad Hafid Rahadi[1], Riansyah Widisaputra[1], Nada Salsabila Deva[1], Alexsei Robert Taube[1]
[1]Merdeka Copper Gold, Banyuwangi, Indonesia

In the evolving discipline of geometallurgy, the integration of high-resolution mineralogical datasets and data-driven analytics has reshaped the understanding of ore deportment and its impact on mineral processing. This study extends a previous machine learning-based silicate alteration framework to support metallurgical design by incorporating multi-source datasets, including hyperspectral core imaging (Corescan), quantitative mineralogy (SEM/MLA), multi-element geochemistry (Geochem), and dynamic rebound hardness (Equotip) data.

The research focuses on the Tujuh Bukit Cu-Au porphyry deposit, Banyuwangi, East Java, Indonesia, which hosts 1.7 Bt at 0.5% Cu and 0.5g/t Au. The geological setting comprises multi-phase diorite and tonalite intrusions within a volcano-sedimentary host rock, exhibiting potassic, phyllic, and propylitic alteration overprinted by structurally controlled advanced argillic alteration. These overprints influence the mineralogical expression of copper and gold, creating heterogeneous ore domains with variable flotation responses.

Since 2019, 156,909 linear metres of Corescan, along with Equotip data, have been used to build robust alteration facies through unsupervised machine learning. Nine key alteration minerals and hardness features were clustered using the K-means algorithm, creating consistent domains. These domains were then predicted in historical drill data (without Corescan or Equotip) via Neural Network and AdaBoost algorithms using geochemical and/or Equotip inputs, expanding the 3D alteration model coverage.

Geochem-based sulphide calculations and SEM/MLA analyses were integrated to validate sulphide associations across domains. Results show a strong correlation between alteration and metal deportment: in primary porphyry zones, Cu and Au are mainly hosted in chalcopyrite and bornite. As advanced argillic overprinting progresses, Cu and Au become associated with covellite, chalcocite, enargite, digenite, and occasionally occur as fine inclusions within pyrite and gangue minerals such as quartz and muscovite.

By integrating machine learning with multi-source datasets, this study establishes robust geometallurgical domains, supporting process design optimisation and demonstrating machine learning's scalability in geoscience.

Using PHAST Reaction-Transport Modeling to Study the Formation of Zebra Limestone

Andrea Salas Rivera[1], Kathleen Olivas[1], Rebecca Navarrette[1], Amanda L. Labrado[2], Benjamin Brunner[1]
[1]The University of Texas At El Paso, El Paso, United States, [2]Applied Physics Laboratory, University of Washington, Seattle, United States

Zebra rock formation is often associated with carbonate-hosted lead-zinc Mississippi Valley Type (MVT) ore deposits and hydrothermal dolomite-hosted hydrocarbon reservoirs. These features are controlled by complex interactions between fluid chemistry, temperature, and pressure. Theoretically, such conditions can be reconstructed using Reaction-Transport Modeling (RTM), which simultaneously simulates solute transport and geochemical reactions. However, most RTMs are developed to address specific case studies, often requiring custom code or proprietary software. This presents limitations such as low computational efficiency, limited portability across platforms, and poor documentation, ultimately making these models difficult for other researchers to reuse or adapt.

PHAST, a modeling tool that integrates PHREEQC, a geochemical modeling software with HST3D, a solute transport model, is a free and open-source program developed by the USGS. It is designed to simulate three-dimensional groundwater flow, solute transport, and multicomponent geochemical reactions. PHAST offers a user-friendly interface, accessible tutorials, and multithreaded processing, which improves computational performance. These advantages raise the question of whether PHAST can be applied more broadly, for instance, to simulate the geochemical processes involved in ore deposit formation.

This study aims to adapt PHAST to predict the development of zebra limestone, which is characterized by alternating millimeter- to centimeter-scale bands of light and dark calcite crystals. These features have been observed at the Gypsum Valley salt diapir in the Paradox Basin, Colorado. By using PHAST, we explore whether this tool can be utilized to yield insights into ore formation processes.

From Basement Rocks to Mineral Systems: Demonstrating the Value of In-Situ Zircon (U-Th)/He Thermochronology

Paola Salas Rivera[1], Carla Martinez[2], Bernardo Elizondo[2], Maximilian Ehrenfels[3], Daniel Stockli[3]
[1]The University Of Texas At El Paso, El Paso, United States, [2]The University of Texas at Rio Grande Valley, Edinburg, United States, [3]The University of Texas at Austin, Austin, United States

Many economically significant mineral systems, such as porphyry copper, tin granites, alkaline intrusions, and pegmatites, are closely associated with intrusive magmas. Constraining the timing and thermal evolution of these intrusions is critical for understanding ore-forming processes and enhancing exploration models. This study highlights the application of in-situ laser ablation zircon (U-Th)/He (ZHe) thermochronology, a rapid HF-free analytical technique, for resolving the thermal histories of such systems with high spatial resolution.

As part of an undergraduate research initiative through the AGeS-DiG program, we tested the capabilities of in-situ ZHe analysis on zircon grains collected from the Llano Uplift in Central Texas, USA, where Mesoproterozoic granites are unconformably overlain by Paleozoic sedimentary rocks. While the primary objective focused on explaining the timing of exhumation related to the Great Unconformity, the analytical approach demonstrated significant potential for broader applications in economic geology studies and mineral exploration.

Our results reveal a phase of rapid basement exhumation at ~850 Ma, associated with Rodinia rifting, and three subsequent Phanerozoic burial-exhumation cycles. These findings, supported by numerical modeling in HeFTy, demonstrate the effectiveness of in-situ ZHe thermochronology in reconstructing complex thermal histories relevant to intrusion-related mineralization. This investigation highlights the value of deploying advanced thermochronology techniques, which offer faster, high quality data acquisition and refined thermal constraints to better understand intrusion-related mineral systems worldwide.

UAV-borne Hyperspectral Data Across Tailings and Waste at Memi Mine in the Republic of Cyprus

Katerina Savinova[1], Andres Ortega Lucero, Victor Tolentino, Rikke Vestergaard, Friederike Koerting, Constantin Hilderbrand, Ioulia Georgiadou Gavrilovic, Nick Cook
[1]The University of Queensland, Brisbane, Australia

The EU-funded M4Mining project (www.m4mining.eu; Horizon Europe scheme) promotes sustainable mining practices through integrated remote sensing data from uncrewed aerial vehicles (UAVs) and satellites for regular monitoring of mining-related masses across the entire life of mine chain. The overarching goal is to develop comprehensive remote sensing solutions for active and legacy mine sites, including tailings monitoring.

In this case study, we present the results of hyperspectral imagery (HSI) integrated with geochemical and mineralogical data from a field campaign in April 2024 at the Memi Mine, a legacy pyrite mine in the Republic of Cyprus. Here, there is processed waste and tailings from the production of sulphur concentrate, derived from iron pyrite ore mining at Memi and adjacent mines. HSI data were collected with a Mjolnir VS-620 using Visible to Near- and Shortwave Infrared (VNIR-SWIR) sensors with a combined wavelength range of 400-2500 nm. As part of the M4Mining project development, the drone-based hyperspectral data are displayed as 3D surfaces using co-acquired LiDAR data and integrated into a single full-spectrum hyperspectral mesh. Several control points were identified for ground-truth validation and spectral sampling using a portable field spectrometer.

The reflectance-corrected HSI is being assessed and compared by applying different classification methods for minerals known to occur in areas affected by acid mine drainage and secondary iron-bearing mineral precipitation, as well as for mineral sequestration potential. The geochemical and mineralogical characterisation of the samples is integrated and presented, aiding in the interpretation and validation of the collected HSI data and mineral classifications. Surface patterns of hematite, goethite, and jarosite, with only minor carbonate, are interpreted from the hyperspectral data. XRD results further demonstrate the presence of variably altered minerals that are associated with a VHMS deposit.

The Application of Unsupervised Machine Learning to Identify Geochemical Associations with Geological Domains of Nickel Laterite Deposit in Sorowako, Indonesia

Alya Sevy Setiawan[1], Fadlin Fadlin[1], Reza Febri Agus Setyawan[2]
[1]Jenderal Soedirman University, Banyumas, Indonesia, [2]PT Vale Indonesia Tbk, East Luwu, Indonesia

Unsupervised machine learning, including Principal Component Analysis (PCA) and K-Means Clustering was employed to identify complex multi-chemistry relationships and trends within geological domains by reducing data dimensionality and performing multivariate grouping. The study utilized drill hole geochemical data from two prospect areas: "AS" and "AL". This study used 1120 samples from Prospect AS and 1053 samples from Prospect AL with different geological settings, dunite, and lherzolite bedrock, resulting in more diverse geochemical characteristics. Elements used include Ni, Co, Fe, SiO_2, MgO, Mn, Cr, Al, and Ca. The results of PCA and clustering were interpreted based on their spatial distribution and geochemical characteristics of nickel laterite deposits, with geological domains serving as validation references to assess the objectivity and applicability of the method for exploration purposes. Robust PCA with Box-Cox (No Trim) transformation revealed that one principal component captured 96.7% of the total data variance in Hill AS, while two principal components accounted for 85.82% in Hill AL. K-Means clustering, performed using an unsupervised method, grouped data based on laterite geochemical characteristics. PCA and clustering results showed that element distributions varied by geological domain. Spatial analysis of PCA and clustering identified multi-chemistry relationships and geochemical trends in laterite deposits, with optimal cluster numbers at k = 3 and k = 4. The three-group model identified Limonite, Saprolite, and Bedrock domains, while the four-group model revealed an additional transitional domain between saprolite and bedrock, referred to as "saprock" formed through lateritization processes. The method also identified high MgO and SiO_2 contents in dunite bedrock, attributed to olivine ($(Mg,Fe)_2SiO_4$) dominance forming Mg-rich clay minerals, whereas higher Al content in lherzolite bedrock dominated by pyroxene, generated Al-rich clay minerals. Cross-matrix validation showed average accuracies of 87.9% in Hill AS and 80.13% in Hill AL, indicating strong model performance.

Determination of Iron Oxidation State in Biotite Using Electron Probe Microanalysis

Shahed Shahrestani[1], Ioan Sanislav[1], Yang Liu[1], Kevin Blake[1], Avish Kumar[1], Shane Askew[1]
[1]Economic Geology Research Centre (EGRU), College of Science and Engineering, James Cook University, Townsville, Qld 4811, Australia, Townsville, Australia

The oxidation state of iron in minerals and silicate glasses is an influential indicator of redox conditions during magma generation and crystallisation. This results from oxygen fugacity, temperature, pressure, and melt composition. The Fe^{3+}/Fe^{2+} ratio, in particular, affects the local structure of silicate melts, thus influencing their physical and chemical properties, including viscosity, degree of polymerisation, heat capacity, density, and phase equilibrium. Several techniques have been proposed to accurately determine the iron oxidation state in minerals. In this study, a JXA-iHP200F electron probe microanalysis instrument was calibrated using garnet samples including almandine (Fe^{2+}-rich) and andradite (Fe^{3+}-rich), under conditions of 20nA beam current, 15 Kv accelerating voltage, 10 μm bean size, TAP as crystal, and 30 ms dwell time on peaks. The flank method was applied which encompasses searching for maximum and minimum positions of difference spectra between almandine and andradite and calculating the ratio of Lβ/Lα at those positions. Lβ/Lα values are sensitive to iron oxidation state, and thus to the Fe^{2+} contribution in specimen. A regression line between Lβ/Lα and Fe^{2+} has been successfully established for biotite samples collected from various global locations. This calibration may be applied in geological investigations.

Automated Coal Picking in MWD Logs

Katie Silversides[1], Rian Dutch[1], Nurassyl Battalgazy[2], John Jackson[2], Tom Carmichael[1]
[1]Datarock, Melbourne, Australia, [2]IMDEX, Brisbane, Australia

Measure while drilling (MWD) data can be routinely collected from drill rigs without requiring additional scanning equipment or time. MWD can provide useful information about the rock drilled, particularly about the relative hardness. However, MWD data is noisy and can be difficult to interpret. Additionally, when collected during blast hole drilling the dataset can quickly become very large.

This study uses data MWD collected during blast hole drilling in a coal mine. The aim is to identify holes where the drill stops at the top of the coal seam and use these to model the coal seam. Accurately knowing the depth of the coal helps prevent damaging the coal when blasting.

A manually labelled dataset was used to create a MWD library containing two classes, touch coals and not coal. A 1D convolutional neural network (CNN) was trained on this dataset. The trained model was then applied to holes from different benches, processing the last 3m of each hole to determine if a touch coal was present. Additional rules were then applied to identify the exact boundary depth and eliminate some of the false picks.

The coal pick model had an accuracy of 86% when compared to the manual labels, with most of the errors being missed touch coals (13%). These missed coals most often occurred where the MWD did not continue far enough into the coal. Another cause of errors was artefacts in the MWD. Where the touch coals were correctly labelled, 89.9% were within 30cm of the manually labelled coal, and 99.2% were within 1m.

The manual and CNN coal picks were then used to create surfaces representing the top of the coal seam. The two surfaces were highly similar, indicating that the CNN method of identifying coal picks is suitable for use in this case.

Breakthrough Hydrogeochemical Techniques for Exploration and Heap Leaching

Adam Simon[1], Ryan Mathur[2], Karr McCurdy[3], Lee Ann Munk[4]
[1]University Of Michigan, Ann Arbor, United States, [2]Juniata College, Huntingdon, United States, [3]VectOres Science, Inc., Huntingdon, United States, [4]University of Alaska Fairbanks, Fairbanks, United States

We developed a novel hydrogeochemical technique that de-risks the discovery of epithermal, porphyry, iron oxide-copper-gold (IOCG), volcanogenic massive sulfide (VMS), magmatic sulfide, and sediment-hosted deposits. The technique uses the ratios of metal isotopes at parts per trillion concentrations (e.g., 65Cu/63Cu; 66Zn/64Zn; 56Fe/54Fe; 109Ag/107Ag; 124Sn/122Sn) in ground and surface water as a probe for the presence of specific sulfide minerals beneath cover. The technique accurately distinguishes among, for example, chalcopyrite, chalcocite, and bornite, and hypogene vs. supergene chalcocite. The primary data are effective in the identification of concealed prospects and in the locating of drilling targets. The technique requires 250 mL of water, and results and interpretation are typically available two to three weeks after sample collection. We have successfully used the hydrogeochemical technique to help mining companies conduct brownfield expansion projects and greenfield exploration worldwide. The technique is also used to monitor mineralogy during copper heap leaching operations. This is done by using the copper isotope ratio of leach solutions and head mineral material, which accurately distinguishes among enargite, chalcopyrite, covellite and chalcocite. These data constrain the relative proportions of copper sulfide minerals that provide copper during hydrometallurgical dissolution and allow operators to tailor lixiviant chemistry to increase metal recovery. Ores that have experienced complex, multiple enrichment cycles contain various ratios of each mineral phase because, for example, covellite and chalcocite can yield different amounts of copper during leaching and quantifying the minerals responsible for providing copper directly impacts the method of extraction chosen for different ore types because not all ores of similar grade leach at the same rate. This information quantitatively assesses the leaching effectiveness of microbes and to constrain the process(es) involved in metal release considering there are a variety of mechanisms operating, including diffusion, redox reactions, and ion conduction.

High Velocity = High Prospectivity? A Seismic Window into Southeast Tasmania's Natural Hydrogen Potential

George Taylor[1,2], Tjaart de Wit[1,2]
[1]Institute of Mine Seismology, Hobart, Australia, [2]Ambient Resources, Hobart, Australia

Natural hydrogen has the potential to provide a new, clean energy source that may substitute for fossil fuels in many applications. Initial studies into natural hydrogen formation implied that resource generation required a set of complex and rare geological conditions, but more recent observations have indicated the potential for hydrogen to be a far more common occurrence throughout the Earth's crust, particularly in association with traditional hydrocarbon plays. In this study we present the results of a passive seismic imaging study to assess the potential for a natural hydrogen resource beneath north Bruny Island in south-east Tasmania. We use ambient noise surface wave tomography to create a 3D S-wave velocity model of the top 2 km of the Earth's crust beneath Bruny Island and image the geological structure of the thick Paleozoic sedimentary basin that constitutes the island. Our S-wave velocity model clearly distinguishes between the thick Permian – Triassic sedimentary sequence in the northern part of the island, and the thinner sediments in the south. Most notably, we have discovered an elongate high S-wave velocity structure located ~1800 m beneath the south-eastern portion of the island, that we interpret as a large section of up-lifted metamorphic basement rocks of likely Proterozoic age. In addition, we image several probable fault structures affecting the basement rocks that are likely the result of the extensional tectonic regime that affected Tasmania between late Cretaceous – Paleocene.

Multispectral Survey Based on Remote Sensing for Chromite Prospecting in Oman Ophiolite

Cezar Augusto Teixeira Falcão Filho[1], Kim Braga Poletto[1], Eduardo Antonio Abrahão Filho Abrahão Filho[2], Luan Oliveira Lima[2], Parham Ahrabian[2], Ahmed Mohamed Al Rahbi[2], Mohammad Al Maktoumi[2]
[1]Geoterramar - Remote Sensing, Ilhéus, Brazil, [2]Oman Chromite Company, Sohar, Oman

This abstract presents the results of a collaborative mineral exploration campaign conducted by Geoterramar and Oman Chromite Company (OCC) in northern Oman's ophiolitic belt, a region recognized for its chromite mineralization potential. The 104 km² study area exhibits optimal conditions for remote sensing analysis due to minimal vegetation cover, limited anthropogenic disturbance, and an arid climate that reduces atmospheric interference.

High-resolution WorldView-3 multispectral imagery was utilized to develop a predictive model for chromite exploration targets. The dataset, consisting of the OR2A product dated 05/23/2021, was atmospherically compensated (AComp) and geometrically corrected/orthorectified by DigitalGlobe Inc. In addition to panchromatic (0.31 m) and VNIR multispectral (1.24 m) capabilities, the imagery includes SWIR bands positioned at diagnostic wavelengths for detecting Al-OH, Mg-OH, Fe-OH-bearing alteration minerals, and carbonate minerals.

Advanced digital image processing techniques were employed, including pan-sharpening, downscaling, band stacking, principal component analysis (PCA), and band ratioing, to enhance mineralogical signatures. Predictive spatial models for chromite exploration were developed using mean shift segmentation, Isodata clustering, Maximum Likelihood classification, and machine learning algorithms such as Support Vector Machine (SVM), Random Forest, and Artificial Neural Networks (ANN). The final predictive model was generated through map algebra integration.

Principal components PC3 and PC5 proved particularly effective in delineating the spatial relationship between dunite conduits and gabbro dikes, with all mapped chromite occurrences located along these conduit structures—confirming both lithological and structural controls on ore emplacement. The proximity of gabbro dikes to mineralized zones further suggests a metallogenic association among dunite conduits, gabbro intrusions, and chromite mineralization, establishing a robust framework for future prospecting efforts.

The integration of multiple machine learning models with spectral and spatial data significantly improved the identification of areas with high chromite mineralization potential, contributing to more efficient resource allocation and exploration planning.

Mapping Hydrothermal Alteration for Gold Prospecting in British Columbia, Canada: Integrating Multispectral Remote Sensing and Geophysical Data

Cezar Augusto Teixeira Falcão Filho[1], Daniel Mariano[2], Harrison Cookenboo[2], Henrique da Fonseca Bylaardt[2]
[1]Geoterramar - Remote Sensing and Environmental Analysis, Ilhéus, Brazil, [2]Avant Exploration, Vancouver, Canada

Advanced remote sensing techniques and geophysical data were integrated to identify prioritized targets and their associated hydrothermal alteration zones as potential indicators of gold and copper mineralized rocks. The study area is located in northern British Columbia, Canada, near the border of the Stikine and Peace River Regional Districts. This region lies within British Columbia's Toodoggone district, part of the "Golden Horseshoe" area in the world-renowned Golden Triangle of the Omineca/Liard mining division.

The Feature-Oriented Principal Component Selection (Crosta technique) was applied to Landsat 8 multispectral data. A critical alteration zone was identified where intensely argillized rocks are heavily iron-stained. This zone was used to train machine learning algorithms to detect other regions with similar multispectral signatures. Four specific pixels were selected for training to ensure the models remained focused on these spectral characteristics. This approach successfully identified 47 points across 16 potential alteration zones rich in clays and iron oxides.

The original airborne magnetic survey data, acquired by Precision GeoSurveys Inc., were reprocessed using modern techniques to enhance interpretative value. Using Oasis Geosoft software, the reprocessing refined the magnetic data, with the 2nd Vertical Derivative Map highlighting shallow magnetic features. Gradient and inclination maps, along with their derivatives, confirmed contacts and defined major magnetic domains based on their "worm" patterns. Key interpretations include:
A prominent fault with an NNW-SSE strike along the western ridge.
A mylonitic zone in the hanging wall of the Kutcho Fault.
Low-magnetic sedimentary rocks and moderate- to high-magnetic marine volcanics, interpreted as interbedded units of the Hazelton Group.
These structural and lithological patterns, combined with alteration mapping, helped confirm prospective geological domains and mineralization targets.

Machine Learning-Driven Geometallurgical Evaluation of the Lowermost Manganese Bed in the Kalahari Manganese Field, South Africa

Louis van Eeden[3]
[1]Stellenbosch University, Stellenbosch, South Africa, [2]African Rainbow Minerals Geometallurgy Research Chair, Stellenbosch, South Africa, [3]TECT Geological Consulting, Somerset West, South Africa

The Kalahari Manganese Field (KMF), located in the Northern Cape, South Africa, boasts a significant portion, approximately 77%, of the world's known land-based manganese reserves. The focus of the study is on the lower most Mn seam at one of the Mn mines, containing ore of lower grade (<36 wt% Mn), situated in the eastern portion of the KMF.

Using a random forest regression (RFR) algorithm, predictive models were developed to quantify the relationships between ore textures, geochemical characteristics, and comminution behaviour. Six drill cores logged at 50 cm intervals were analysed using XRF data, with textural variations serving as key predictive variables. The Mn content was successfully predicted using ore textures, achieving a high correlation ($R^2 = 0.92$), with "Massive" and "Hematite Lutite" textures contributing most to model accuracy.

This predictive model has direct implications for underground mining optimization. By integrating machine learning into geometallurgical workflows, mining operations can refine ore classification and extraction strategies, reducing off-seam mining and minimizing waste. The model enables real-time decision-making, reducing reliance on costly and time-consuming laboratory analyses.

Comminution behaviour, assessed through relative crushability tests using a jaw crusher, highlighted mineralogy as the dominant factor influencing grindability. Manganese oxide-rich textures exhibit higher crushability, guiding efficient processing strategies.
By leveraging textural analysis and predictive algorithms, this study supports sustainable mining practices through improved resource utilization, operational efficiency, and cost reduction. Machine learning-driven ore classification enhances mine planning, ensuring targeted extraction and extended mine life.

A Multi-Scale Integrated Geological Modeling Approach of the Northern Norrbotten Ore Province, with a Focus on Iron Oxide-Apatite Mineral Systems

Ervin Veress[1], Tobias E. Bauer[1], Oskar Rydman[1], Joel BH. Andersson[1], Maxim Smirnov[1], Thorkild M. Rasmussen[1], Irvine R. Annesley[2]
[1]Department of Civil, Environmental and Natural Resources Engineering, Luleå University of Technology, Luleå, Sweden, [2]GéoRessources, Université de Lorraine, CNRS, CREGU, Vandœuvre lès Nancy, France

Critical raw materials, particularly metals, are at the center of the energy transition as the shift towards a renewable energy infrastructure accelerates. The subsurface has long been utilized for energy production and resource extraction, and geoscientists play a key role in providing the geological knowledge required to locate, assess, and manage the mineral resources essential for future energy systems. Explorers need to locate more deposits, in particular giant orebodies, to minimize the environmental footprint of mining and ensure sustainable practices. However, most giant ore deposits with surface expressions have already been discovered, and prospectors must work in new frontiers to explore and understand deeply buried mineral systems. Collaborations across and beyond traditional geoscience disciplines are becoming increasingly important as discoveries require multidisciplinary and integrated exploration tools. These tools should integrate the mineral system paradigm with existing geological and geophysical models and interrogate and characterize the spatial and temporal relationships of its components.

The Northern Norrbotten Ore Province (NNOP) is located in the northernmost part of Sweden and is regarded as a major metallogenic province hosting more than 40 iron oxide-apatite (IOA) deposits, including the giant Kiirunavaara deposit. More than a century of mining, exploration, and geological surveying in the NNOP has led to an outstanding repository of geoscientific data, encompassing geological mapping, deep drill core archives, high-resolution geophysical datasets, and detailed three-dimensional geological models at the local scale. Our study leverages this solid conceptual framework and geoscientific knowledge, proposing an integrated geological modeling workflow to characterize the spatial and temporal characteristics of the IOA mineral system components from the NNOP. The integrated workflow facilitates the interpretation of ore deposit expressions and their components within the broader mineralizing system context, and improves the predictive capability of exploration programs to delineate geologically favourable zones for mineralization at a regional scale.

Application and Optimization of Blasting Parameters in Selected Rock Quarries of Southwest and Northcentral Nigeria

Paul Visaa[1], Joseph Ifeanyi Nwosu[2], Ndukauba Egesi[3], Josephine Onwualu-John[4], K Wordu[5], C.C Asadu[6]

[1]University of Port Harcourt, Port Harcourt, Nigeria, [2]University of Port Harcourt, Port Harcourt, Nigeria, [3]University of Port Harcourt, Port Harcourt, Nigeria, [4]University of Port Harcourt, Port Harcourt, Nigeria, [5]University of Port Harcourt, Port Harcourt, Nigeria, [6]Ministry of Mines and steel development Abuja, Nigeria, Abuja, Nigeria

Blasting operations in the Auchi, Igarra, Okene, and Lokoja mineral districts of southwest and northcentral Nigeria frequently encounter challenges associated with poor rock fragmentation. Boulders ranging from 1 to 2 meters in diameter are common, often requiring costly re-blasting to reduce them to manageable sizes for crushing. Field investigations across eight mining pits revealed that quarries often utilize excessively high benches relative to the blasthole diameters employed, resulting in burdens that cannot be effectively fragmented. Moreover, there is a general absence of properly calculated burden, spacing, and charge quantities for each blasthole, as well as inadequate firing sequences. To address these issues, rock samples were collected from each pit and subjected to compressive strength analysis to determine the mechanical properties required for optimized blasting design. The predominant rock types identified include granite, marble, gneiss, and migmatite. The compressive strength of the granites ranged from 23 MPa to 26 MPa (class factor 5), while that of the marbles varied from 16 MPa to 20 MPa (rock factor 7). Blasthole diameters observed ranged between 40 mm and 80 mm, with bench heights from 10 m to 18 m. Analytical results suggest that the maximum recommended bench height for an 80 mm diameter blasthole is 7.2 m in granite and 11.2 m in marble quarries. Based on these findings, we propose appropriate bench heights, burden, spacing, and charge quantities tailored to each blasthole size and rock type. Additionally, a trapezoidal firing sequence is recommended to improve blast efficiency by promoting better fragmentation, enhancing material heaping for efficient loading, and minimizing the risk of fly-rock. The adoption of these optimized parameters is expected to reduce operational costs, enhance productivity, and ensure safer blasting practices across quarries in these regions.

Pyrite Trace Element Behavior in IOCG Deposits: An Unsupervised Learning Approach for Vectoring and Ore Paragenesis

Shouchun Yu[1], Jeffrey A. Steadman[1], Matthew J. Cracknell[1], Sebastien Meffre[1]
[1]Centre for Ore Deposit and Earth Sciences, University of Tasmania, Hobart, Australia

Iron oxide copper-gold (IOCG) deposits are a class of mineral deposits that can contain significant quantities of strategic and critical minerals, essential for the renewable energy transition. However, the high degree of complexity and internal variability of IOCG deposits has inhibited the development of a robust genetic model. Over the last decade, trace elements in pyrite have been used successfully in ore deposit exploration as geochemical pathfinders to ores, including for some IOCG deposits. Here we present the results of a comparison of unsupervised learning methods applied to pyrite LA-ICP-MS trace element maps from three IOCG and affiliated provinces in Australia: Arthur Metamorphic Complex (Alpine prospect), TAS; Gawler Craton (Hillside deposit, Prominent Hill deposit, and Con Ryan prospect), SA; and Cloncurry District (Ernest Henry deposit, E1 prospect, and Starra prospect). We aim to cluster the pyrite LA-ICP-MS maps in order to assess the degree of compositional complexity among samples from different systems, and to test whether certain common elemental patterns could be defined. The LA-ICP-MS maps show elemental zoning, which represents variability in temperature and chemistry during pyrite crystal growth. For most samples, hierarchical clustering works better than K-means by highlighting elemental zoning in pyrite, while some K-means clustering results tended to emphasize noise. For sites like Ernest Henry, Hillside, and Prominent Hill, clusters from both methods indicate proximity to the orebody. However, some clusters appear only in specific deposits while other clusters are common across multiple deposits. Hillside and Prominent Hill, or Ernest Henry and E1 samples show similar clusters, suggesting geological and fluid composition similarities. This study provides critical insights for improving exploration strategy and resource estimation, as well as the conceptual development of ore genesis processes through a rapidly emerging tool.

An Apatite to Unravel Magma Chamber Dynamics, Mantle Sources and Metallogenesis in the Bushveld Complex

Peace Zowa[1], Ben Hayes[2], Grant Bybee[2]
[1]University of the Witwatersrand | RES, Johannesburg, South Africa, [2]University of the Witwatersrand, Johannesburg, South Africa

The metallogenesis, magmas and mantle sources responsible for the formation of the Rustenburg Layered Suite (RLS) of the Large Igneous Province (BLIP) remain exceptionally controversial despite decades of research. Recent work using zircon and bulk-rock Lu-Hf isotope data show variations in isotopes and suggests that the BLIP intrusions were produced by the melting of several distinct reservoirs. We present systematic in situ Sr-Nd isotopic compositions, major (Cl, F, OH) and trace element abundances in apatite, from the coeval satellite intrusions, which include the Phalaborwa Complex, Marble Hall, Lindeques Drift, Roodekraal intrusions, Molopo Farms Complex and Uitkomst Complex. Understanding the geochemical properties of these intrusions is essential in providing insights into the metallogenesis of the BLIP. Advances in LA-MC-ICP-MS technology now enable detailed analysis of apatite geochemistry, providing new insights into magmatic processes. Apatite is capable of retaining important geochemical information and is a repository of REES. We provide a comprehensive in-situ data set for apatite from the BLIP using LA-MC-ICP-MS, Electron Microprobe, and a Tescan Integrated Mineral Analyser. Major element compositions show that Molopo Farm Complex is Cl-rich, while the Roodekraal intrusions, Marble Hall, Uitkomst Complex, Lindeques Drift and Phalaborwa Complex show variable F-rich apatite compositions. Trace element data, although variable for these intrusions, generally show a negative slope suggestive of LREEs enrichment consistent with the apatite trace element trends that have been previously reported for carbonatites and mafic-ultramafic rocks. Our in-situ Sr-Nd isotopes show that apatite from the different intrusions record variable initial isotopic compositions at 2.06 Ga for 87Sr/86Sr (0.7030 to 0.7130) and εNd (-10.2 to -2.5). The apatite data provide evidence of in situ Sr-Nd isotopic variability for these intrusions. On the basis of this new isotopic data, we provide insights into the magmatic processes involved in forming the BLIP.

Connection and Collaboration – Building a Resilient Geoscience Community and Future Workforce

Kaylene Camuti[1], **Ned Howard**[1], Jenna McGovern[1], Martin Robinson[1], Doug Young[1], Jaime Livesey[1]
[1]Australian Institute of Geoscientists, Education Committee, Sydney, Australia

A sustainable future relies on geoscience. Geoscientists are central to finding critical minerals, supporting the energy transition, managing water resources, and building safe, sustainable communities. But the ability to meet these challenges is increasingly at risk because geoscience education is under threat.

In Australia, as in many other countries, falling enrolments have led to shrinking course content and the closure of geoscience programs. In the past five years, approximately 15% of tertiary geoscience teaching capacity has been lost to program closures alone. Eroded teaching and research capacity compromises not only education, but Australia's long-term capability in critical sectors.

In response, the Education Committee of the Australian Institute of Geoscientists (AIG) initiated a highly collaborative Geoscience Education Action Plan, involving experts and stakeholders from schools, academia, employers, students and early career geoscientists. Two themes emerged from the discussion: connection and collaboration.

Connection is essential to attract and retain future geoscientists. This includes:
- Supporting established, impactful outreach programs in schools (e.g. AusEarthEd, TESEP, CORE, Geoscience Pathways).
- Cultivating student cohorts early in university through immersive field experiences.
- Engaging students through mentoring and exposure to geoscientists and educators conveying the relevance and career opportunities of geoscience.

Collaboration is vital to reverse the downward trend. No single group can solve this alone. Stronger alignment between educators, industry, government, outreach organisations and professional bodies is needed to:
- Share expertise and resources to deliver high-quality learning experiences and achieve shared goals.
- Drive structural change in education policy and administration.
- Raise public awareness of the societal value of geoscience.

The Australian experience offers insight for the global geoscience community. This presentation will outline the collaborative approach being taken, and argue that building deep connections—with students, communities and between institutions—is not just desirable, but essential.